The Automotive Chassis

The Automotive Chassis: Engineering Principles

SECOND EDITION

Chassis and vehicle overall
Wheel suspensions and types of drive
Axle kinematics and elastokinematics
Steering – Springing – Tyres
Construction and calculations advice

Prof. Dipl.-Ing. Jörnsen Reimpell
Dipl.-Ing. Helmut Stoll
Prof. Dr.-Ing. Jürgen W. Betzler

Translated from the German by AGET Limited

BUTTERWORTH
HEINEMANN

OXFORD AUCKLAND BOSTON JOHANNESBURG MELBOURNE NEW DELHI

Butterworth-Heinemann
Linacre House, Jordan Hill, Oxford OX2 8DP
225 Wildwood Avenue, Woburn, MA 01801-2041
A division of Reed Educational and Professional Publishing Ltd

A member of the Reed Elsevier plc group

Original copyright 1986 Vogel-Buchverlag, Würzburg
Fourth German edition published by Vogel-Buchverlag, Würzburg 1999
First English edition published by Arnold 1996
Second edition published by Butterworth-Heinemann 2001

British Library Cataloguing in Publication Data
A catalogue record for this book is available from the British Library

Library of Congress Cataloguing in Publication Data
A catalogue record for this book is available from the Library of Congress

ISBN 0 7506 5054 0

Composition by Cambrian Typesetters, Frimley, Surrey
Printed and bound in Great Britain by Biddles, Guildford & Kings Lynn

Contents

Preface

This translation of the fourth German edition is published by Butterworth-Heinemann as the second English edition of *The Automotive Chassis*.

We are fortunate to have Prof. Dr.-Ing. Jürgen W. Betzler as co-author; he has been an expert in the field of chassis/simulation technology and design studies at the University of Cologne since 1994. Jointly, we revised *The Automotive Chassis: Engineering Principles* to include a large number of technical innovations.

The clear and easy descriptions, many example designs and calculations and the inclusion of 434 illustrations and tables are easily understood and have, over the years, proven to be the best way of imparting information.

The authors' many years of experience in chassis engineering support the practical bias and will help engineers, inspectors, students and technicians in companies operating in the automotive industry and its suppliers to understand the context. The comprehensive index of key words and numerous cross-references make this book an invaluable reference work.

We should like to thank Dipl.-Ing. Achim Clasen for collating the test results in the Automotive Engineering Laboratory at the Technical University in Cologne and Sabine Jansen M.A. for her hard work in converting the symbols.

Cologne/Rösrath

Jörnsen Reimpell
Helmut Stoll
Jürgen W. Betzler

1

Types of suspension and drive

This chapter deals with the principles relating to drives and suspensions.

1.1 General characteristics of wheel suspensions

The suspension of modern vehicles need to satisfy a number of requirements whose aims partly conflict because of different operating conditions (loaded/unloaded, acceleration/braking, level/uneven road, straight running/cornering).

The forces and moments that operate in the wheel contact area must be directed into the body. The kingpin offset and disturbing force lever arm in the case of the longitudinal forces, the castor offset in the case of the lateral forces, and the radial load moment arm in the case of the vertical forces are important elements whose effects interact as a result of, for example, the angle of the steering axis.

Sufficient vertical spring travel, possibly combined with the horizontal movement of the wheel away from an uneven area of the road (kinematic wheel) is required for reasons of *ride comfort*. The recession suspension should also be compliant for the purpose of reducing the rolling stiffness of the tyres and short-stroke movements in a longitudinal direction resulting from the road surface (longitudinal compliance, Fig. 1.1), but without affecting the development of lateral wheel forces and hence steering precision, for which the most rigid wheel suspension is required. This requirement is undermined as a result of the necessary flexibility that results from disturbing wheel movements generated by longitudinal forces arising from driving and braking operations.

For the purpose of ensuring the optimum *handling characteristics* of the vehicle in a steady state as well as a transient state, the wheels must be in a defined position with respect to the road surface for the purpose of generating the necessary lateral forces. The build-up and size of the lateral wheel forces are determined

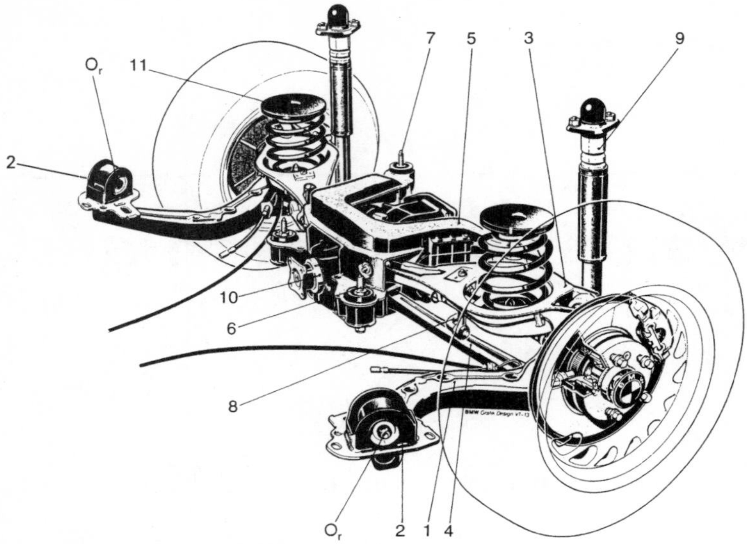

Fig. 1.1 A multi-link rear axle – a type of suspension system which is progressively replacing the semi-trailing arm axle, and consists of at least one trailing arm on each side. This arm is guided by two (or even three) transverse control arms (Figs 1.62 and 1.77). The trailing arm simultaneously serves as a wheel hub carrier and (on four-wheel steering) allows the minor angle movements required to steer the rear wheels. The main advantages are, however, its good kinematic and elastokinematic characteristics.

BMW calls the design shown in the illustration and fitted in the 3-series (1997) a 'central arm axle'. The trailing arms 1 are made from GGG40 cast iron; they absorb all longitudinal forces and braking moments as well as transfering them via the points 2 – the centres of which also form the radius arm axes (Figs 3.158 and 3.159) – on the body. The lateral forces generated at the centre of tyre contact are absorbed at the subframe 5, which is fastened to the body with four rubber bushes (items 6 and 7) via the transverse control arms 3 and 4. The upper arms 3 carry the minibloc springs 11 and the joints of the anti-roll bar 8. Consequently, this is the place where the majority of the vertical forces are transferred between the axle and the body.

The shock absorbers, which carry the additional polyurethane springs 9 at the top (Fig. 5.50), are fastened in a good position behind the axle centre at the ends of the trailing arms. For reasons of noise, the differential 10 is attached elastically to the subframe 5 at three points (with two rubber bearings at the front and one hydro bearing at the back). When viewed from the top and the back, the transverse control arms are positioned at an angle so that, together with the differing rubber hardness of the bearings at points 2, they achieve the desired elastokinematic characteristics. These are:

- toe-in under braking forces (Figs 3.64 and 3.82);
- lateral force compliance understeer during cornering (Figs 3.79 and 3.80);
- prevention of torque steer effects (see Section 2.10.4);
- lane change and straight running stability.

For reasons of space, the front eyes 2 are pressed into parts 1 and bolted to the attachment bracket. Elongated holes are also provided in this part so toe-in can be set. In the case of the E46 model series (from 1998 onwards), the upper transverse arm is made of aluminium for reasons of weight (reduction of unsprung masses).

by specific toe-in and camber changes of the wheels depending on the jounce and movement of the body as a result of the axle kinematics (roll steer) and operative forces (compliance steer). This makes it possible for specific operating conditions such as load and traction to be taken into consideration. By establishing the relevant geometry and kinematics of the axle, it is also possible to prevent the undesirable diving or lifting of the body during braking or accelerating and to ensure that the vehicle does not exhibit any tendency to oversteer and displays predictable transition behaviour for the driver.

Other requirements are:

- independent movement of each of the wheels on an axle (not guaranteed in the case of rigid axles);
- small, unsprung masses of the suspension in order to keep wheel load fluctuation as low as possible (important for driving safety);
- the introduction of wheel forces into the body in a manner favourable to the flow of forces;
- the necessary room and expenditure for construction purposes, bearing in mind the necessary tolerances with regard to geometry and stability;
- ease of use;
- behaviour with regard to the passive safety of passengers and other road users;
- costs.

The requirements with regard to the steerability of an axle and the possible transmission of driving torque essentially determine the design of the axis.

Vehicle suspensions can be divided into rigid axles (with a rigid connection of the wheels to an axle), independent wheel suspensions in which the wheels are suspended independently of each other, and semi-rigid axles, a form of axle that combines the characteristics of rigid axles and independent wheel suspensions.

On all rigid axles (Fig. 1.23), the axle beam casing also moves over the entire spring travel. Consequently, the space that has to be provided above this reduces the boot at the rear and makes it more difficult to house the spare wheel. At the front, the axle casing would be located under the engine, and to achieve sufficient jounce travel the engine would have to be raised or moved further back. For this reason, rigid front axles are found only on commercial vehicles and four-wheel drive, general-purpose passenger cars (Figs 1.3 and 1.4).

With regard to independent wheel suspensions, it should be noted that the design possibilities with regard to the satisfaction of the above requirements and the need to find a design which is suitable for the load paths, increase with the number of wheel control elements (links) with a corresponding increase in their planes of articulation. In particular, independent wheel suspensions include:

- Longitudinal link and semi-trailing arm axles (Figs 1.13 and 1.15), which require hardly any overhead room and consequently permit a wide luggage space with a level floor, but which can have considerable diagonal springing.
- Wheel controlling suspension and shock-absorber struts (Figs 1.8 and 1.57), which certainly occupy much space in terms of height, but which require little space at the side and in the middle of the vehicle (can be used for the engine

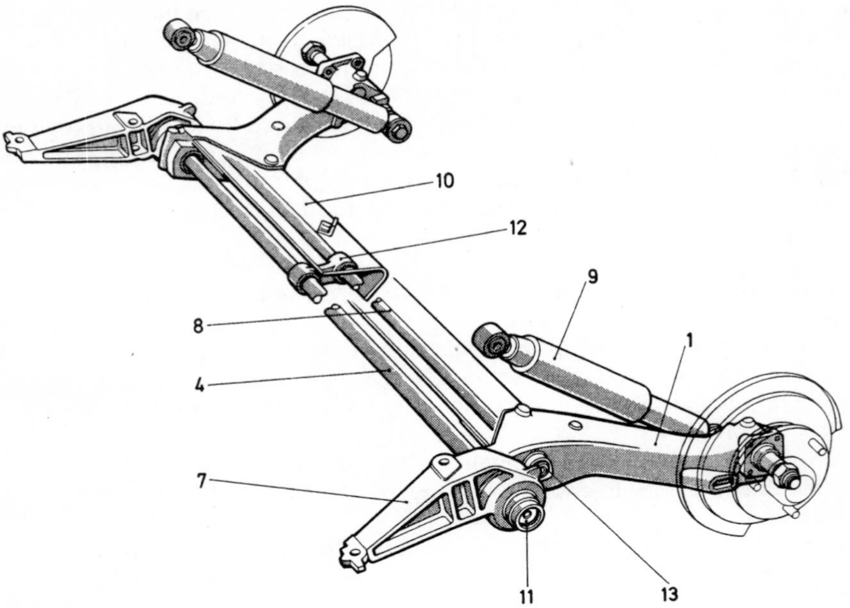

Fig. 1.2 An extremely compact four-bar twist beam axle by Renault, with two torsion bar springs both for the left and right axle sides (items 4 and 8). The V-shape profile of the cross-member 10 has arms of different lengths, is resistant to bending but less torsionally stiff and absorbs all moments generated by vertical, lateral and braking forces. It also partially replaces the anti-roll bar.

At 23.4 mm, the rear bars 8 are thicker than the front ones (∅ 20.8 mm, item 4). On the outside, part 8 grips into the trailing links 1 with the serrated profile 13 and on the inside they grip into the connector 12. When the wheels reach full bump, a pure torque is generated in part 12, which transmits it to the front bars 4, subjecting them to torsion. On the outside (as shown in Fig. 1.63) the bars with the serrated profile 11 grip into the mounting brackets 7 to which the rotating trailing links are attached. The pivots also represent a favourably positioned pitch centre O_r (Fig. 3.159). The mounting brackets (and therefore the whole axle) are fixed to the floor pan with only four screws.

On parallel springing, all four bars work, whereas on reciprocal springing, the connector 12 remains inactive and only the thick rear bars 8 and the cross-member 10 are subject to torsion.

The layout of the bars means soft body springing and high roll stability can be achieved, leading to a reduction of the body roll pitch during cornering.

To create a wide boot without side encroachments, the pressurized monotube shock absorbers 9 are inclined to the front and therefore are able to transmit forces upwards to the side members of the floor pan.

or axle drive) and determine the steering angle (then also called McPherson suspension struts).

- Double wishbone suspensions (Fig. 1.7).
- Multi-link suspensions (Figs 1.1, 1.18 and 1.19), which can have up to five guide links per wheel and which offer the greatest design scope with regard to

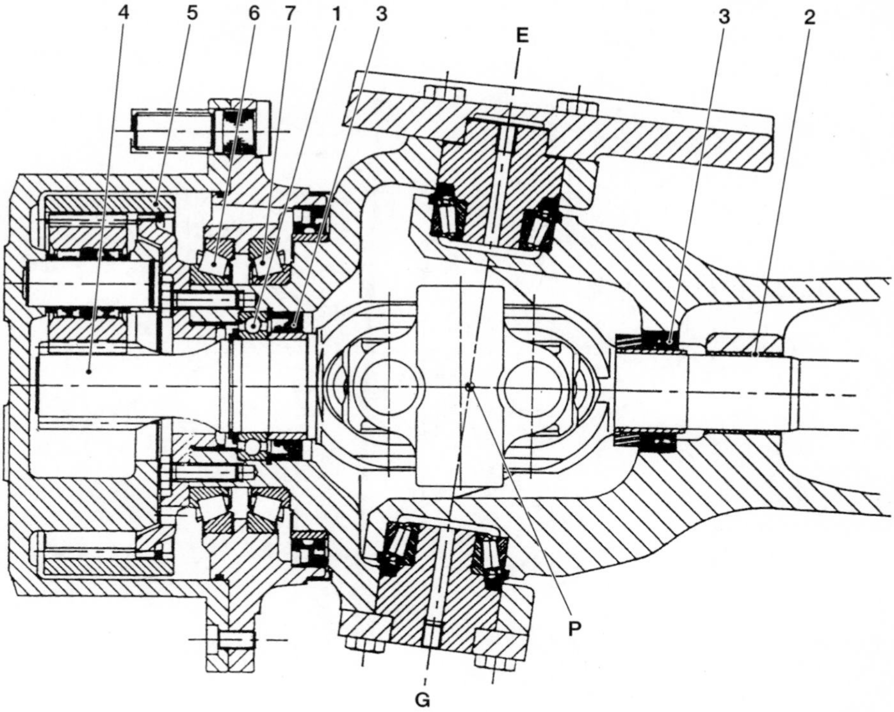

Fig. 1.3 Driven, rigid steering axle with dual joint made by the company GKN – Birfield AG for four-wheel drive special-purpose vehicles, tractors and construction machinery.

The dual joint is centred over the bearings 1 and 2 in the region of the fork carriers; these are protected against fouling by the radial sealing rings 3. Bearing 1 serves as a fixed bearing and bearing 2 as a movable bearing. The drive shaft 4 is also a sun gear for the planetary gear with the internal-geared wheel 5. Vertical, lateral and longitudinal forces are transmitted by both tapered-roller bearings 6 and 7. Steering takes place about the steering axis EG.

the geometric definition of the kingpin offset, pneumatic trail, kinematic behaviour with regard to toe-in, camber and track changes, braking/starting torque behaviour and elastokinematic properties.

In the case of twist-beam axles (Figs 1.2, 1.31 and 1.58), both sides of the wheels are connected by means of a flexurally rigid, but torsionally flexible beam. On the whole, these axles save a great deal of space and are cheap, but offer limited potential for the achievement of kinematic and elastokinematic balance because of the functional duality of the function in the components and require the existence of adequate clearance in the region of the connecting beam. They are mainly used as a form of rear wheel suspension in front-wheel drive

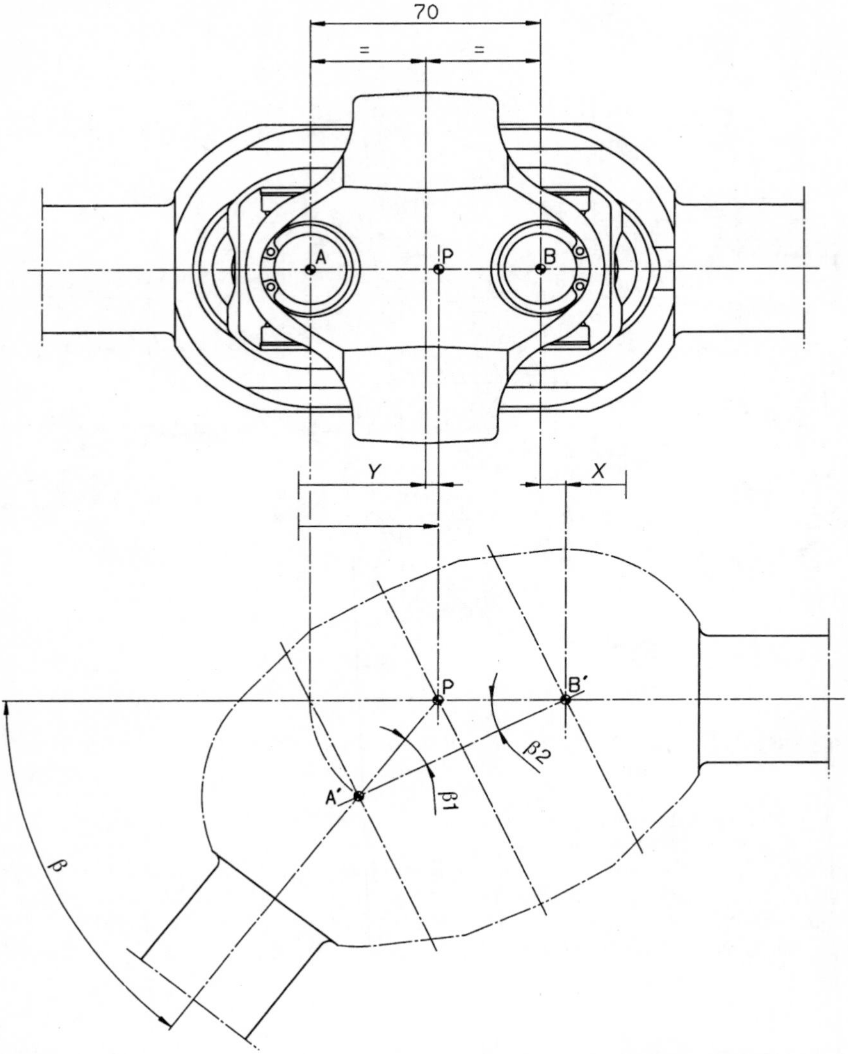

Fig. 1.4 Top view of the dual joint (Fig. 1.3). The wheel end of the axle is turned about point P in the middle of the steering pivot during steering. The individual joints are constrained at points A and B so that point A is displaced to position A′, P is displaced to P′ and B is displaced along the drive axle by the distance *X* to B′. In order to assimilate the variable bending angle β resulting from the longitudinal displacement of point B, the mid-point of the joint P is displaced by the distance *Y*. The adjustment value *Y* depends on the distance between the joints and the steering angle at which constant velocity is to exist. Where large steering angles can be reached (up to 60°), there should be constant velocity at the maximum steering angle.

The adjustment value *Y* and the longitudinal displacement *X* should be taken into consideration in the design of the axle.

vehicles up to the middle class and, occasionally, the upper middle class, for example, the Audi A6, and some high-capacity cars.

1.2 Independent wheel suspensions – general

1.2.1 Requirements

The chassis of a passenger car must be able to handle the engine power installed. Ever-improving acceleration, higher peak and cornering speeds, and deceleration lead to significantly increased requirements for safer chassis. Independent wheel suspensions follow this trend. Their main advantages are:

- little space requirement;
- a kinematic and/or elastokinematic toe-in change, tending towards understeering is possible (see Section 3.6);
- easier steerability with existing drive;
- low weight;
- no mutual wheel influence.

The last two characteristics are important for good road-holding, especially on bends with an uneven road surface.

Transverse arms and trailing arms ensure the desired kinematic behaviour of the rebounding and jouncing wheels and also transfer the wheel loadings to the body (Fig. 1.5). Lateral forces also generate a moment which, with unfavourable link arrangement, has the disadvantage of reinforcing the roll of the body during cornering. The suspension control arms require bushes that yield under load and can also influence the springing. This effect is either reinforced by twisting the rubber parts in the bearing elements, or the friction

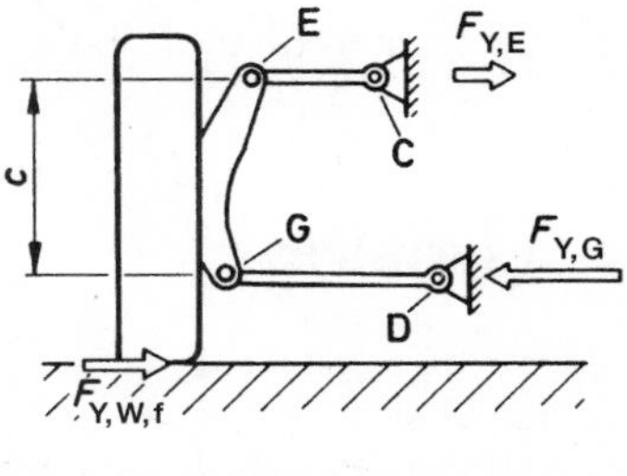

Fig. 1.5 On front independent wheel suspensions, the lateral cornering force $F_{Y,W,f}$ causes the reaction forces $F_{Y,E}$ and $F_{Y,G}$ in the links joining the axle with the body. Moments are generated on both the outside and the inside of the bend and these adversely affect the roll pitch of the body. The effective distance c between points E and G on a double wishbone suspension should be as large as possible to achieve small forces in the body and link bearings and to limit the deformation of the rubber elements fitted.

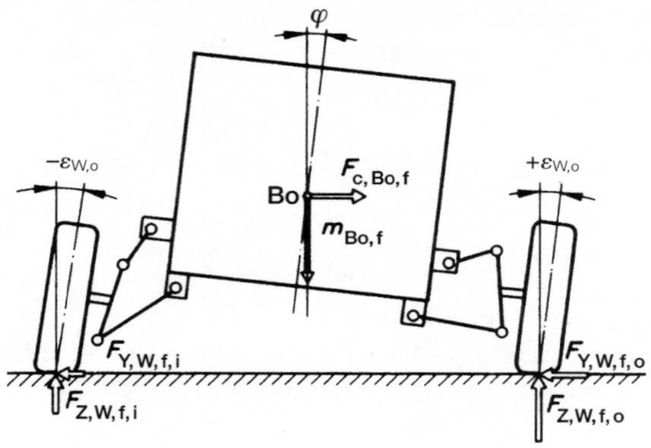

Fig. 1.6 If the body inclines by the angle φ during cornering, the outer independently suspended wheel takes on a positive camber $\varepsilon_{W,o}$ and the inner wheel takes on a negative camber $\varepsilon_{W,i}$. The ability of the tyres to transfer the lateral forces $F_{Y,W,f,o}$ or $F_{Y,W,f,i}$ decreases causing a greater required slip angle (Fig. 3.53 and Equation 2.16), $m_{Bo,f}$ is the proportion of the weight of the body over the front axle and $F_{c,Bo,f}$ the centrifugal force acting at the level of the centre of gravity Bo. One wheel rebounds and the other bumps, i.e. this vehicle has 'reciprocal springing', that is:

$$F_{Z,W,f,o} = F_{Z,W,f} + \Delta F_{Z,W,f}$$
$$F_{Z,W,f,i} = F_{Z,W,f} - \Delta F_{Z,W,f}$$

increases due to the parts rubbing together (Fig. 1.11), and the driving comfort decreases.

The wheels incline with the body (Fig. 1.6). The wheel on the outside of the bend, which has to absorb most of the lateral force, goes into a positive camber and the inner wheel into a negative camber, which reduces the lateral grip of the tyres. To avoid this, the kinematic change of camber needs to be adjusted to take account of this behaviour (see Section 3.5.4) and the body roll in the bend should be kept as small as possible. This can be achieved with harder springs, additional anti-roll bars or a body roll centre located high up in the vehicle (Sections 3.4.3 and 5.4.3).

1.2.2 Double wishbone suspensions

The last two characteristics above are most easily achieved using a double wishbone suspension (Fig. 1.7). This consists of two transverse links (control arms) either side of the vehicle, which are mounted to rotate on the frame, suspension subframe or body and, in the case of the front axle, are connected on the outside to the steering knuckle or swivel heads via ball joints. The greater the effective distance *c* between the transverse links (Fig. 1.5), the smaller the forces in the suspension control arms and their mountings become, i.e. component deformation is smaller and wheel control more precise.

The main advantages of the double wishbone suspension are its kinematic

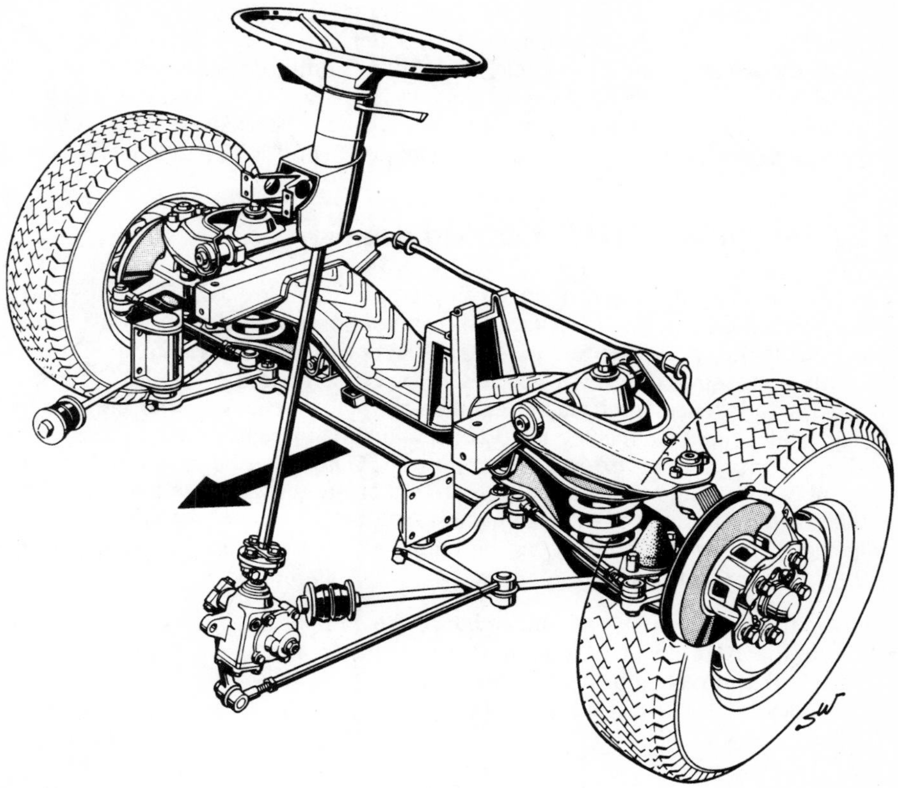

Fig. 1.7 Front axle on the VW light commercial vehicle Lt 28 to 35 with an opposed steering square. A cross-member serves as a subframe and is screwed to the frame from below. Springs, bump/rebound-travel stops, shock absorbers and both pairs of control arms are supported at this force centre. Only the anti-roll bar, steering gear, idler arm and the tie-rods of the lower control arms are fastened to the longitudinal members of the frame. The rods have longitudinally elastic rubber bushings at the front that absorb the dynamic rolling hardness of the radial tyres and reduce lift on uneven road surfaces.

possibilities. The positions of the suspension control arms relative to one another – in other words the size of the angles α and β (Fig. 3.24) – can determine both the height of the body roll centre and the pitch pole (angles α' and β', Fig. 3.155). Moreover, the different wishbone lengths can influence the angle movements of the compressing and rebounding wheels, i.e. the change of camber and, irrespective of this, to a certain extent also the track width change (Figs 3.50 and 3.7). With shorter upper suspension control arms the compressing wheels go into negative camber and the rebounding wheels into positive. This counteracts the change of camber caused by the roll pitch of the body (Fig. 1.6). The vehicle pitch pole O indicated in Fig. 6.16 is located behind the wheels on the front axle and in front of the wheels on the rear axle. If O_r can be located over the wheel centre (Fig. 3.161), it produces not only a better anti-dive mechanism, but also reduces the squat on the driven rear axles (or lift on the front axles). These are also the reasons why the double wishbone suspension is used as the rear axle on more and more passenger cars, irrespective of the type of drive, and why it is progressively replacing the semi-trailing link axle (Figs 1.1, 1.62 and 1.77).

1.2.3 McPherson struts and strut dampers

The McPherson strut is a further development of double wishbone suspension. The upper transverse link is replaced by a pivot point on the wheel house panel, which takes the end of the piston rod and the coil spring. Forces from all directions are concentrated at this point and these cause bending stress in the piston rod. To avoid detrimental elastic camber and caster changes, the normal rod diameter of 11 mm (in the shock absorber) must be increased to at least 18 mm. With a piston diameter of usually 30 mm or 32 mm the damper works on the twin-tube system and can be non-pressurized or pressurized (see Section 5.8).

The main advantage of the McPherson strut is that all the parts providing the suspension and wheel control can be combined into one assembly. As can be seen in Fig. 1.8, this includes:

- the spring seat 3 to take the underside of the coil spring;
- the auxiliary spring 11 or a bump stop (see Fig. 5.49);
- the rebound-travel stop (Fig. 5.54);
- the underslung anti-roll bar (7) via rod 5;
- the steering knuckle.

The steering knuckle can be welded, brazed or bolted (Fig. 5.53) firmly to the outer tube (Fig. 1.56). Further advantages are:

- lower forces in the body-side mounting points E and D due to a large effective distance *c* (Fig. 1.5);
- short distance *b* between points G and N (Fig. 3.30);
- long spring travel;
- three bearing positions no longer needed;
- better design options on the front crumple zone;
- space at the side permitting a wide engine compartment; which
- makes it easy to fit transverse engines (Fig. 1.50).

Nowadays, design measures have ensured that the advantages are not outweighed by the inevitable disadvantages on the front axle. These disadvantages are:

- Less favourable kinematic characteristics (Sections 3.3 and 3.5.2).
- Introduction of forces and vibrations into the inner wheel house panel and therefore into a relatively elastic area of the front end of the vehicle.
- It is more difficult to insulate against road noise – an upper strut mount is necessary (Fig. 1.9), which should be as decoupled as possible (Fig. 1.10, item 10 in Fig. 1.8 and item 6 in Fig. 1.56).

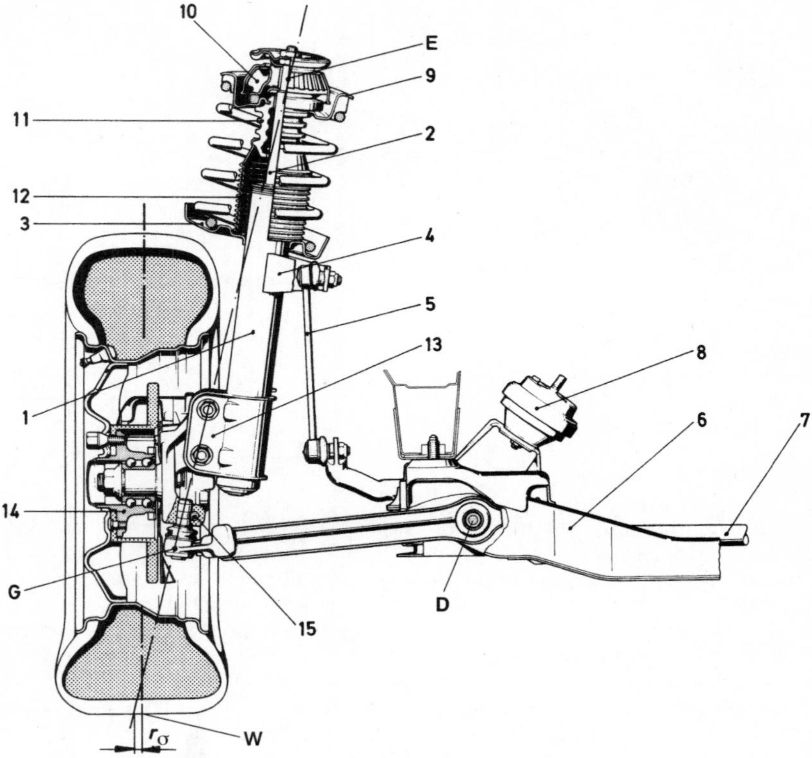

Fig. 1.8 Rear view of the left-hand side of the McPherson front axle on the Opel Omega (1999) with negative kingpin offset at ground (scrub radius) r_σ and pendulum-linked anti-roll bar. The coil spring is offset from the McPherson strut to decrease friction between piston rod 2 and the rod guide. Part 2 and the upper spring seat 9 are fixed to the inner wheel house panel via the decoupled strut mount 10.

The additional elastomer spring 11 is joined to seat 9 from the inside, and on the underside it carries the dust boot 12, which contacts the spring seat 3 and protects the chrome-plated piston rod 2. When the wheel bottoms out, the elastomer spring rests on the cap of the supporting tube 1. Brackets 4 and 13 are welded to part 1, on which the upper ball joint of the anti-roll bar rod 5 is fastened from inside. Bracket 13 takes the steering knuckle in between the U-shaped side arms.

The upper hole of bracket 13 has been designed as an elongated hole so that the camber can be set precisely at the factory (see Fig. 3.102). A second-generation double-row angular (contact) ball bearing (item 14) controls the wheel.

The ball pivot of the guiding joint G is joined to the steering knuckle by means of clamping forces. The transverse screw 15 grips into a ring groove of the joint bolt and prevents it from slipping out in the event of the screw loosening.

The subframe 6 is fixed to the body. In addition to the transverse control arms, details of which are given in Ref. 5, Section 10.4, it also takes the engine mounts 8 and the back of the anti-roll bar 7. The drop centre rim is asymmetrical to allow negative wheel offset (not shown) at ground (scrub radius) (Figs 2.10, 2.11 and 2.23).

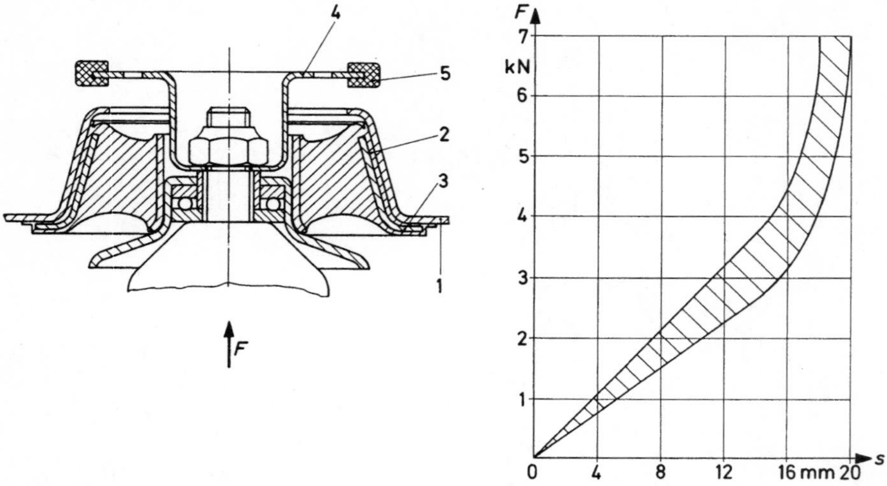

Fig. 1.9 McPherson strut mount on the VW Golf III with a thrust ball bearing, which permits the rotary movement of the McPherson strut whereas the rubber anchorage improves noise insulation. Initially the deflection curve remains linear and then becomes highly progressive in the main work area, which is between 3 kN and 4 kN. The graph shows the scatter. Springing and damping forces are absorbed together so the support bearing is not decoupled (as in Fig. 1.10).

In the car final assembly line the complete strut mount is pressed into a conical sheet metal insert on the wheel house inside panel 1. The rubber layer 2 on the outside of the bearing ensures a firm seat and the edge 3 gives the necessary hold in the vertical direction. The rubber ring 5 clamped on plate 4 operates when the wheel rebounds fully and so provides the necessary security (figure: Lemförder Fahrwerktechnik AG).

- The friction between piston rod and guide impairs the springing effect; it can be reduced by shortening distance *b* (Figs 1.11 and 3.30).
- In the case of high-mounted rack and pinion steering, long tie rods and, consequently, more expensive steering systems are required (Figs 1.57 and 4.1); in addition, there is the unfavourable introduction of tie-rod forces in the middle of the shock-absorbing strut (see Section 4.2.4) plus additional steering elasticity.
- Greater sensitivity of the front axle to tyre imbalance and radial runout (see Section 2.5 and Refs 1 and 4).
- Greater clearance height requirement.
- Sometimes the space between the tyres and the damping element (Fig. 1.41) is very limited.

This final constraint, however, is only important on front-wheel drive vehicles as it may cause problems with fitting snow chains. On non-driven wheels, at most

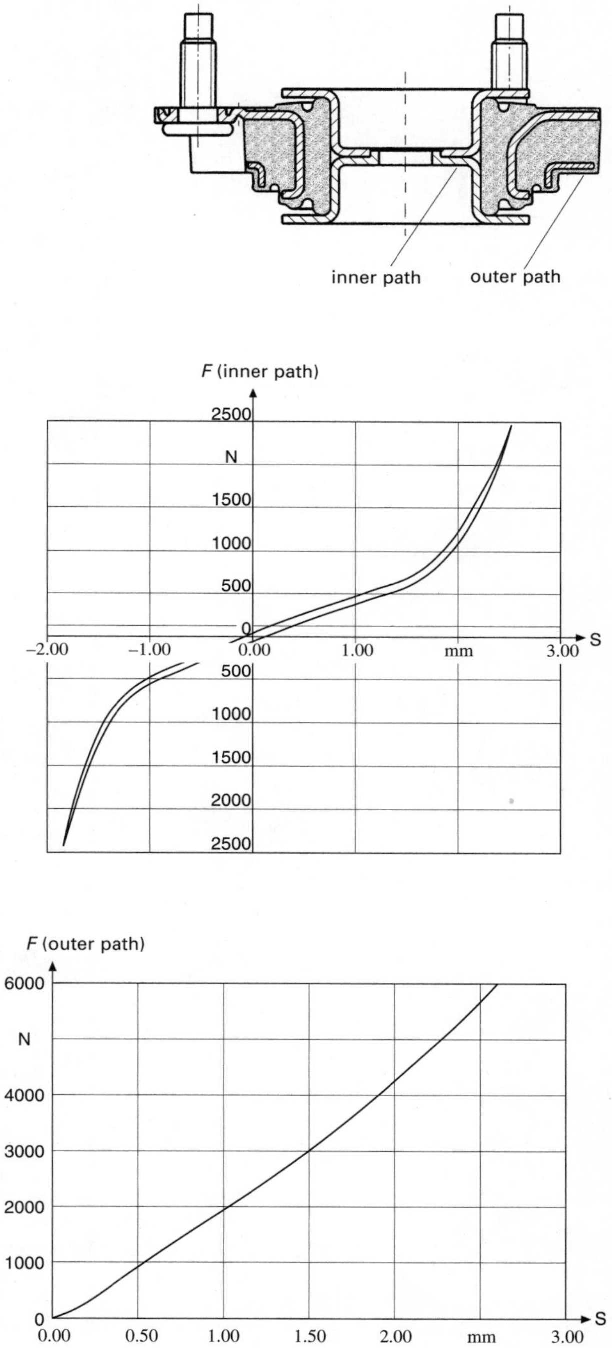

Fig. 1.10 The dual path top mount support of the Ford Focus (1998) manufactured by ContiTech Formteile GmbH. The body spring and shock-absorber forces are introduced into the body along two paths with variable rigidity. In this way, it is possible to design the shock-absorber bearing (inner element) in the region of small amplitudes with little rigidity and thus achieve good insulation from vibration and noise as well as improve the roll behaviour of the body. With larger forces of approximately 700 N and above, progression cams, which increase the rigidity of the bearing, come into play. A continuous transition between the two levels of rigidity is important for reasons of comfort. The bearing must have a high level of rigidity in a transverse direction in order to ensure that unwanted displacements and hence changes in wheel position do not occur. The forces of the body springs are directed along the outer path, which has a considerably higher level of rigidity.

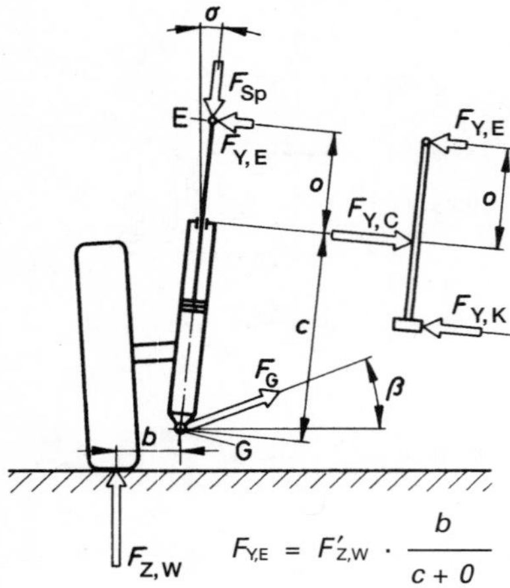

Fig. 1.11 If lateral force $F_{Z,W}$ moves lever arm b round guiding joint G, the lateral force F_{Sp} continually acts in the body-side fixing point E of the McPherson strut as a result of the force $F_{Y,E}$. This generates the reaction forces $F_{Y,C}$ and $F_{Y,K}$ on the piston rod guide and piston. This is $F_{Y,C} + F_{Y,E} = F_{Y,K}$ and the greater this force becomes, the further the frictional force F_{fr} increases in the piston rod guide and the greater the change in vertical force needed for it to rip away.

As the piston has a large diameter and also slides in shock-absorber fluid, lateral force $F_{Y,K}$ plays only a subordinate role (see Fig. 5.54). $F_{Y,K}$ can be reduced by offsetting the springs at an angle and shortening the distance b (see Figs 1.56 and 3.30, and Equation 3.4a).

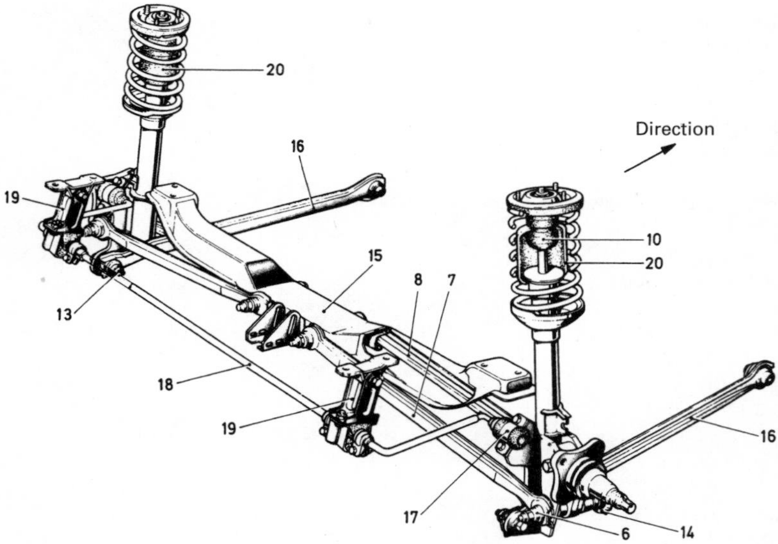

Fig. 1.12 The McPherson strut rear axle on the Lancia Delta with equal length transverse links of profiled steel trunnion-mounted close to the centre the cross-members 7 and 8. As large a distance as possible is needed between points 6 and 14 on the wheel hub carrier to ensure unimpaired straight running. The fixing points 13 of the longitudinal links 16 are behind the wheel centre, exactly like mounting points 17 of the anti-roll bar 18. The back of the anti-roll bar is flexibly joined to the body via tabs 19. The additional springs 10 attached to the top of the McPherson struts are covered by the dust tube 20. The cross-member 15 helps to fix the assembly to the body. An important criterion for dimensioning the control arm 16 is reverse drive against an obstruction.

the lack of space prevents wider tyres being fitted. If such tyres are absolutely necessary, disc-type wheels with a smaller wheel offset *e* are needed and these lead to a detrimentally larger positive or smaller negative kingpin offset at ground r_o (Figs 2.8 and 3.102).

McPherson struts have become widely used as front axles, but they are also fitted as the rear suspension on front-wheel drive vehicles (e.g. Ford Mondeo sedan). The vehicle tail, which has been raised for aerodynamic reasons, allows a larger bearing span between the piston rod guide and piston. On the rear axle (Fig. 1.12):

- The upper strut mount is no longer necessary, as no steering movements occur.
- Longer cross-members, which reach almost to the vehicle centre, can be used, producing better camber and track width change (Figs 3.15 and 3.48) and a body roll centre that sinks less under load (Fig. 3.30).
- The outer points of the braces can be drawn a long way into the wheel to achieve a shorter distance *b*.
- The boot can be dropped and, in the case of damper struts, also widened.
- However, rubber stiffness and the corresponding distance of the braces on the hub carriers (points 6 and 14 in Fig. 1.12) are needed to ensure that there is no unintentional elastic self-steer (Figs 3.79 and 3.80).

1.2.4 Rear axle trailing-arm suspension

This suspension – also known as a crank axle – consists of a control arm lying longitudinally in the driving direction and mounted to rotate on a suspension subframe or on the body on both sides of the vehicle (Figs 1.13 and 1.63). The control arm has to withstand forces in all directions, and is therefore highly subject to bending and torsional stress (Fig. 1.14). Moreover, no camber and toe-in changes are caused by vertical and lateral forces.

The trailing-arm axle is relatively simple and is popular on front-wheel drive vehicles. It offers the advantage that the car body floor pan can be flat and the fuel tank and/or spare wheel can be positioned between the suspension control arms. If the pivot axes lie parallel to the floor, the bump and rebound-travel wheels undergo no track width, camber or toe-in change, and the wheel base simply shortens slightly. If torsion springs are applied, the length of the control arm can be used to influence the progressivity of the springing to achieve better vibration behaviour under load. The control arm pivots also provide the radius-arm axis O; i.e. during braking the tail end is drawn down at this point (Fig. 3.159).

The tendency to oversteer as a result of the deformation of the link (arm) when subject to a lateral force, the roll centre at floor level (Fig. 3.33), the extremely small possibility of a kinematic and elastokinematic effect on the position of the wheels and the inclination of the wheels during cornering consistent with the inclination of the body outwards (unwanted positive camber) are disadvantages.

Fig. 1.13 Trailing-arm rear suspension of the Mercedes-Benz A class (1997). In order to minimize the amount of room required, the coil spring and monotube gas-pressure shock absorber are directly supported by the chassis subframe. The connecting tube is stress optimized oval shaped in order to withstand the high bending moments from longitudinal and lateral wheel forces which occur in the course of driving. The torsion-bar stabilizer proceeds directly from the shock-absorber attachment for reasons of weight and ease of assembly. When establishing the spring/shock-absorber properties, the line along which the forces act and which is altered by the lift of the wheel is to be taken into consideration, as a disadvantageous load-path can occur with jounce. The two front subframes are hydraulically damped in order to achieve a good level of comfort (hydromounts). The chassis subframe can make minor elastokinematic control movements. When designing subframe mounts, it is necessary to ensure that they retain their defined properties with regard to strength and geometry even with unfavourable conditions of use (e.g. low temperatures) and for a sufficiently long period of time, because variations in the configuration have a direct effect on vehicle performance. The longitudinal arms which run on tapered-roller bearings and which are subject to both flexural as well as torsional stress are designed in the form of a parallelogram linkage. In this way, the inherent disadvantage of a trailing arm axle – unwanted toe-in as a result of the deformation of the link when subject to a lateral force – is reduced by 75%, according to works specifications.

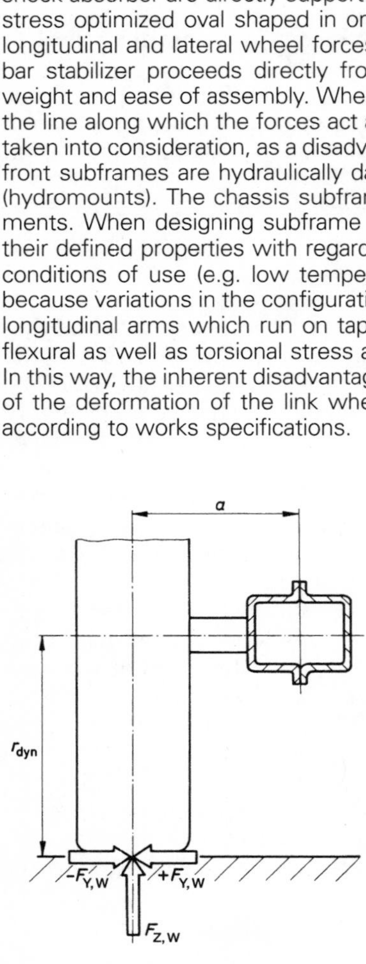

Fig. 1.14 On rear axle trailing-link suspensions, the vertical force $F_{Z,W}$ together with the lateral forces $F_{Y,W}$ cause bending and torsional stress, making a corresponding (hollow) profile, e.g. a closed box profile necessary. A force from inside causes the largest torsional moment (see Chapter 4 in Ref. [3]):

$$T = F_{Z,W} \times a + F_{Y,W} \times r_{dyn}$$

1.2.5 Semi-trailing-arm rear axles

This is a special type of trailing-arm axle, which is fitted mainly in rear-wheel and four-wheel drive passenger cars, but which is also found on front-wheel drive vehicles (Fig. 1.15). Seen from the top (Fig. 1.16), the control arm axis of rotation $\overline{EG}$ is diagonally positioned at an angle $\alpha = 10°$ to $25°$, and from the rear an angle $\beta \leqslant 5°$ can still be achieved (Fig. 3.36). When the wheels bump and

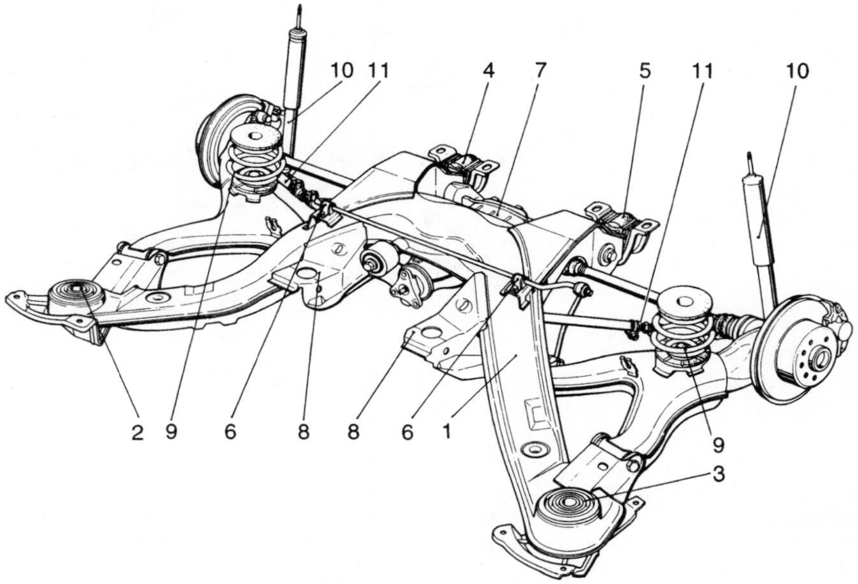

Fig. 1.15 Tilted-(Multiple) Staft Steering Rear Axle of the Opel Omega (1999), a further development of the tilted shaft steering axle. The differential casing of the rear-axle drive is above three elastic bearings, noise-isolated, connected with subframe (1), and this subframe is again, with four specially developed elastomer bearings on the installation (pos. 2 to 5). On top of part seated are the bearings (6) for the back of the stabilizer. Both of the extension arms (8) take up the inner bearings of the tilted shafts, which carry the barrel-shaped helical springs (9). In order to get a flat bottom of the luggage trunk, they were transferred to the front of the axle drive shafts. The transmission i_{Sp} (wheel to spring, see equation 5.14 and paragraph 5.3.2 in (3)), becomes thereby with 1.5 comparatively large. The shock absorbers (10) are seated behind the centre of the axle, the transmission is with $i_D = 0.86$ favourable.

The angle of sweep of the tilted shafts amounts to alpha = 10° (Fig.3.35) and the Dachwinkel, assume roof or top angle beta = 1°35′. Both of these angles change dynamically under the influence of the additional tilted shaft (11). These support the sideforces, coming from the wheel carriers directly against the subframe (1). They raise the lateral stability of the vehicle, and provide an absolute neutral elastic steering under side-forces and also, that in driving mode, favourable toe-in alterations appear during spring deflection, and also under load (Fig. 3.20). The described reaction of load alteration in paragraph 2.12 disappears – in connection with the arrangement and adaptation of bearings 2 to 5 – almost entirely.

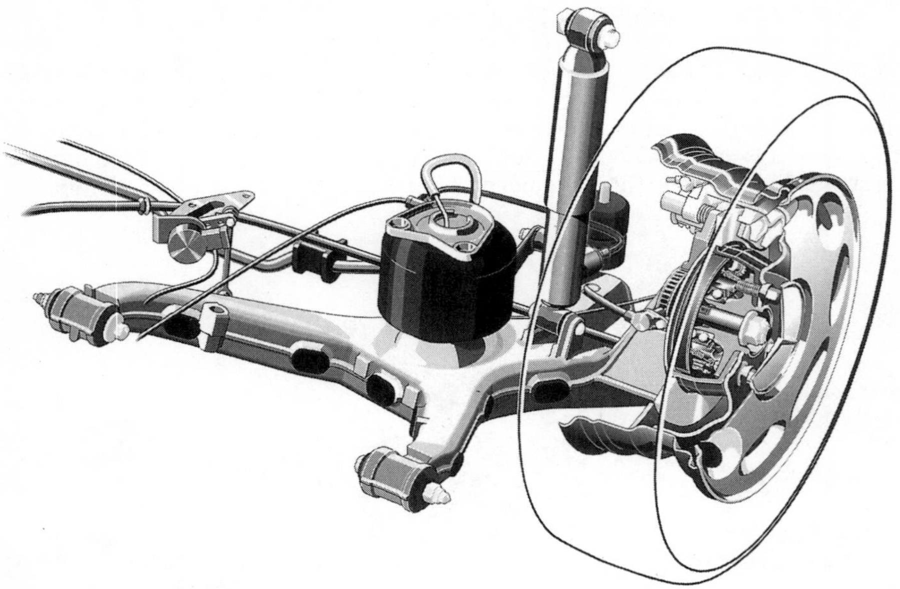

Fig. 1.16 Flat, non-driven air-suspended semi-trailing-arm rear axle of the Mercedes-Benz V class, whose driven front axle with spring-and-shock absorber strut has conventional coil springs. The air-spring bellows are supplied by an electrically powered compressor. The individual wheel adjustment permits the lowering or lifting of the vehicle as well as a constant vehicle height, regardless of – even one-sided – loading. It is also possible to counteract body tilt during cornering. The damping properties of the shock absorbers are affected by spring bellow pressure depending on the load. The short rolling lobe air-spring elements make a low load floor possible; its rolling movement during compression and rebound results in self-cleaning. In the case of semi-trailing arm axles, roll understeer of the rear axle can be achieved (Fig. 3.73) by means of a negative verticle angle of pivot-axis inclination (Fig. 3.36); the kinematic toe-in alteration is also reduced (Fig. 3.49).

rebound-travel they cause spatial movement, so the drive shafts need two joints per side with angular mobility and length compensation (Fig. 1.17). The horizontal and vertical angles determine the roll steer properties.

When the control arm is a certain length, the following kinematic characteristics can be positively affected by angles α and β (Fig. 3.20):

- height of the roll centre;
- position of the radius-arm axis;
- change of camber;
- toe-in change;

Camber and toe-in changes increase the bigger the angles α and β: semi-trailing axles have an elastokinematic tendency to oversteering.

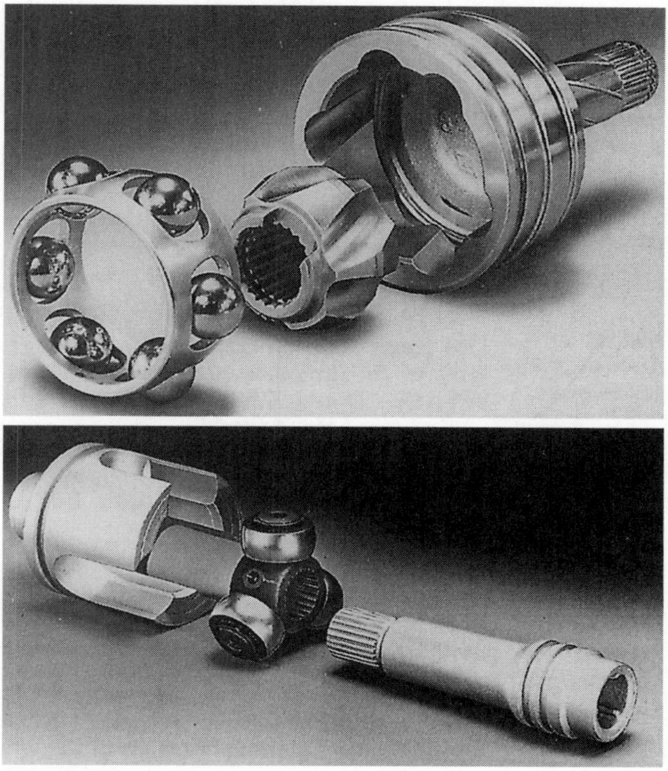

Fig. 1.17 Constant velocity sliding joints by GKN Automotive. In front-drive vehicles, considerable articulation angles of the drive axles occur, sometimes even during straight running, as a result of the installation situation, short propshafts and lifting movements of the body due to torque steer effects. These result in force and moment non-conformities and losses which lead to unwanted vibration. The full-load sliding ball joint (top, also see Fig. 1.53) permits bending angles of up to 22° and displacements of up to 45 mm. Forces are transmitted by means of six balls that run on intersecting tracks. In the rubber–metal tripod sliding joint (bottom), three rollers on needle bearings run in cylindrically machined tracks. With bending angles of up to 25° and displacements of up to 55 mm, these joints run particularly smoothly and hence quietly.

1.2.6 Multi-link suspension

A form of multi-link suspension was first developed by Mercedes-Benz in 1982 for the 190 series. Driven and non-driven multi-link front and rear suspensions have since been used (Figs 1.1, 1.18, 1.19 and 1.44).

Up to five links are used to control wheel forces and torque depending on the geometry, kinematics, elastokinematics and force application of the axle. As the

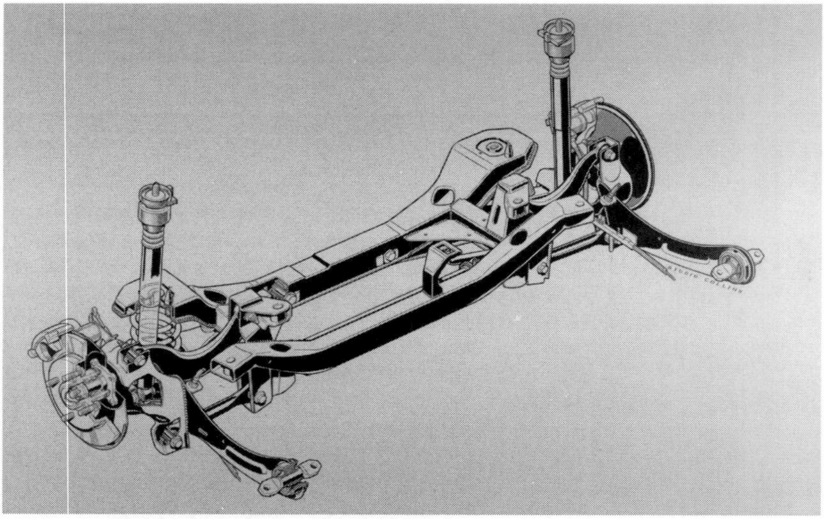

Fig. 1.18 Multi-link suspension of Ford Werke AG. Derived from the Mondeo Turnier model series, multi-link suspension is used by Ford for the first time in the Focus models (1998) in the segment of C class vehicles. This is called the 'control sword axle' after the shape of the longitudinal link. As there are five load paths available here instead of the two that exist in twist-beam axles and trailing arm axles, there is great potential for improvement with regard to the adjustment of riding comfort, driving safety and noise and vibration insulation. As a result of a very elastic front arm bush, the high level of longitudinal flexibility necessary for riding comfort is achieved. At the same time, very rigid and accurate wheel control for increased driving safety is ensured by the transverse link, even at the stability limit. The longitudinal link is subject to torsional stress during wheel lift and to buckling stress when reversing. By using moulded parts, it was possible to reduce the unsprung masses by 3.5 kg per wheel.

arrangement of links is almost a matter of choice depending on the amount of available space, there is extraordinarily wide scope for design. In addition to the known benefits of independent wheel suspensions, with the relevant configuration the front and rear systems also offer the following advantages:

- Free and independent establishment of the kingpin offset, disturbing force and torque developed by the radial load.
- Considerable opportunities for balancing the pitching movements of vehicles during braking and acceleration (up to more than 100% anti-dive, anti-lift and anti-squat possible).
- Advantageous wheel control with regard to toe-in, camber and track width behaviour from the point of view of tyre force build-up, and tyre wear as a function of jounce with almost free definition of the roll centre and hence a very good possibility of balancing the self-steering properties.
- Wide scope for design with regard to elastokinematic compensation from the

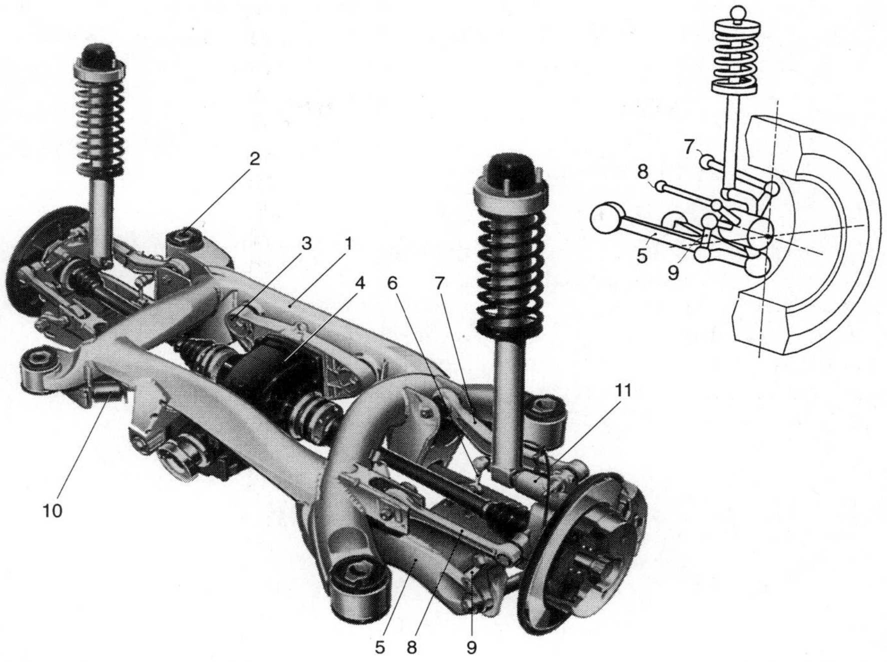

Fig. 1.19 Multi-link rear suspension of the BMW 5 series (E39, 1996). For the first time in large-scale car production, mainly aluminium is used for the suspension system derived from the geometry of the BMW 7 series.

The subframe (rear-axle support) (1), produced from welded aluminium tubes, is attached to the bodywork by means of four large rubber mounts (2). These are soft in a longitudinal direction for the purposes of riding comfort and noise insulation and rigid in a transverse direction to achieve accurate wheel control. The differential gear also has compliant mounts (3). The wheel carrier is mounted on a U-shaped arm (5) at the bottom and on the transverse link (7) and inclined guide link (8) at the top. As a result of this inclined position, an instantaneous centre is produced between the transverse link and guide link outside the vehicle which leads to the desired brake understeer during cornering and the elastokinematic compensation of deformation of the rubber bearings and components. The driving and braking torque of the wheel carrier (11) is borne by the 'integral' link (9) on the swinging arm (5), which is subject to additional torsional stress as a result. This design makes it possible to ensure longitudinally elastic control of the swinging arm on the guide bearing (10) for reasons of comfort, without braking or driving torque twisting the guide bearings as would be the case with torque borne by pairs of longitudinal links. The stabilizer behind presses on the swinging arm (5) by means of the stabilizer link (6), whereas the twin-tube gas-pressure shock absorber, whose outer tube is also made of aluminium, and the suspension springs provide a favourably large spring base attached directly to the wheel carrier (11). For reasons of weight, the wheel discs are also made of aluminium plate. The wheel carrier is made of shell cast aluminium. The rear axle of the station wagon BMW Tourer is largely similar in design. However, the shock absorber extends from the U-shaped swinging arm in order to allow for a wide and low loading area.

point of view of (a) specific elastokinematic toe-in changes under lateral and longitudinal forces and (b) longitudinal elasticity with a view to riding comfort (high running wheel comfort) with accurate wheel control.

As a result of the more open design, the wheel forces can be optimally controlled, i.e. without superposition, and introduced into the bodywork in an advantageous way with wide distances between the supports.

The disadvantages are:

- increased expenditure as a result of the high number of links and bearings;
- higher production and assembly costs;
- the possibility of kinematic overcorrection of the axle resulting in necessary deformation of the bearings during vertical or longitudinal movements;
- greater sensitivity to wear of the link bearings;
- high requirements with regard to the observation of tolerances relating to geometry and rigidity.

1.3 Rigid and semi-rigid crank axles

1.3.1 Rigid axles

Rigid axles (Fig. 1.20) can have a whole series of disadvantages that are a consideration in passenger cars, but which can be accepted in commercial vehicles:

- Mutual wheel influence (Fig. 1.21).
- The space requirement above the beam corresponding to the spring bump travel.
- Limited potential for kinematic and elastokinematic fine-tuning.
- Weight – if the differential is located in the axle casing (Fig. 1.20), it produces a tendency for wheel hop to occur on bumpy roads.
- The wheel load changes during traction (Fig. 1.22) and (particularly on twin tyres) there is a poor support base b_{Sp} for the body, which can only be improved following costly design work (Fig. 1.42).

The effective distance b_{Sp} of the springs is generally less than the tracking width b_r, so the projected spring rate c_φ is lower (Fig. 1.23). As can be seen in Fig. 1.61, the springs, and/or suspension dampers, for this reason should be mounted as far apart as possible (see also Section 5.3 and Chapter 6 in Ref. [3]).

The centrifugal force ($F_{c,Bo}$, Fig. 1.6) acting on the body's centre of gravity during cornering increases the roll pitch where there is a rigid axle (see Section 5.4.3.5).

Thanks to highly developed suspension parts and the appropriate design of the springing and damping, it has been possible to improve the behaviour of rigid drive axles. Nevertheless, they are no longer found in standard-design passenger cars, but only on four-wheel drive and special all-terrain vehicles (Figs 1.43 and 1.68).

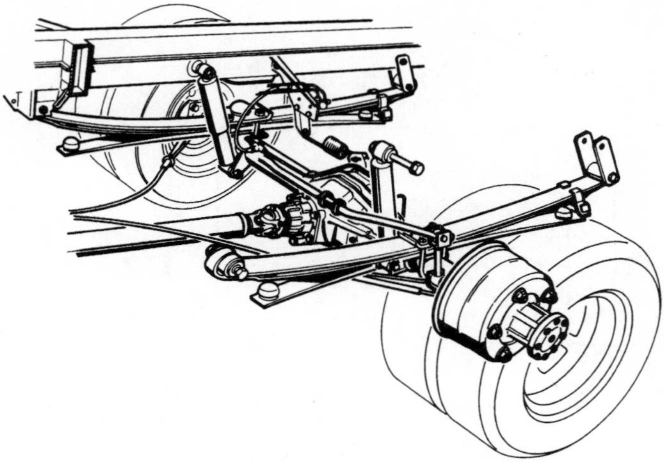

Fig. 1.20 Rear axle on the VW LT light commercial vehicle. The long, parabola-shaped rolled-out, dual leaf springs cushion the frame well and are progressive. The rubber buffers of the support springs come into play when the vehicle is laden. Spring travel is limited by the compression stops located over the spring centres, which are supported on the side-members. The spring leaves are prevented from shifting against one another by the spring clips located behind them, which open downwards (see also Fig. 1.68).

The anti-roll bar is fixed outside the axle casing. The benefits of this can be seen in Fig. 1.23. The shock absorbers, however, are unfortunately located a long way to the inside and are also angled forwards so that they can be fixed to the frame side-members (Fig. 5.23).

Fig. 1.21 Mutual influence of the two wheels of a rigid axle when travelling along a road with pot-holes, shown as 'mutually-opposed springing'. One wheel extends along the path s_2 and the other compresses along the path s_1.

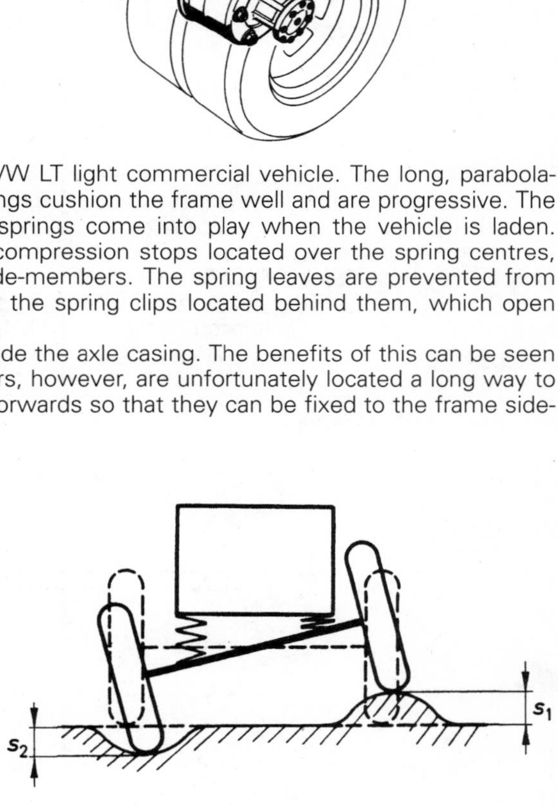

Because of its weight, the driven rigid axle is outperformed on uneven roads (and especially on bends) by independent wheel suspension, although the deficiency in road-holding can be partly overcome with pressurized mono-tube dampers. These are more expensive, but on the compressive stroke, the valve characteristic can be set to be harder without a perceptible loss of comfort. With this, a responsive damping force is already opposing the compressing wheels.

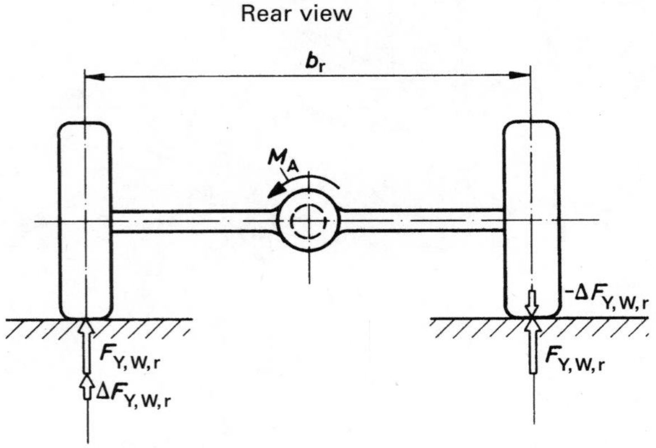

Fig. 1.22 If the differential is located in the body of the rigid axle, the driving torque M_A coming from the engine is absorbed at the centres of tyre contact, resulting in changes to vertical force $\pm\Delta F_{Y,W,r}$.

In the example, M_A would place an additional load on the left rear wheel ($F_{Y,W,r} + \Delta F_{Y,W,r}$) and reduce the vertical force ($F_{Y,W,r} - \Delta F_{Y,W,r}$) on the right one.

On a right-hand bend the right wheel could spin prematurely, leading to a loss in lateral force in the entire axle and the car tail suddenly breaking away (Fig. 2.37; see also Section 6.5 in Ref. [3]).

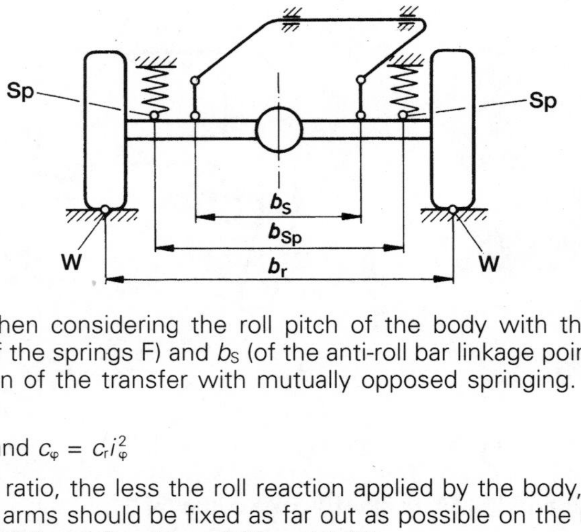

Fig. 1.23 When considering the roll pitch of the body with the rigid axle the distances b_{Sp} (of the springs F) and b_S (of the anti-roll bar linkage points) are included in the calculation of the transfer with mutually opposed springing. i_φ is squared to give the rate c_φ:

$$i_\varphi = b_r/b_{Sp} \text{ and } c_\varphi = c_r i_\varphi^2$$

The greater the ratio, the less the roll reaction applied by the body, i.e. the springs and anti-roll bar arms should be fixed as far out as possible on the rigid axle casing (see Section 5.4.3.5 and Equations 5.20 and 5.21).

This is the simplest and perhaps also the most economic way of overcoming the main disadvantage of rigid axles. Section 5.6.4 contains further details.

In contrast to standard-design vehicles, the use of the rigid rear axle in front-wheel drive vehicles has advantages rather than disadvantages (Fig. 1.24). As Section 6.1.3 explains, the rigid rear axle weighs no more than a comparable independent wheel suspension and also gives the option of raising the body roll centre (which is better for this type of drive, see Fig. 3.42). Further advantages, including those for driven axles, are:

- they are simple and economical to manufacture;
- there are no changes to track width, toe-in and camber on full bump/rebound-travel, thus giving
- low tyre wear and sure-footed road holding;
- there is no change to wheel camber when the body rolls during cornering (Figs 1.6 and 3.54), therefore there is constant lateral force transmission of tyres;
- the absorption of lateral force moment $M_Y = F_{T,X}\ h_{Ro,r}$ by a transverse link, which can be placed at almost any height (e.g. Panhard rod, Fig. 1.25);
- optimal force transfer due to large spring track width bsp
- the lateral force compliance steering can be tuned towards under- or over-steering (Figs 3.81 and 1.29).

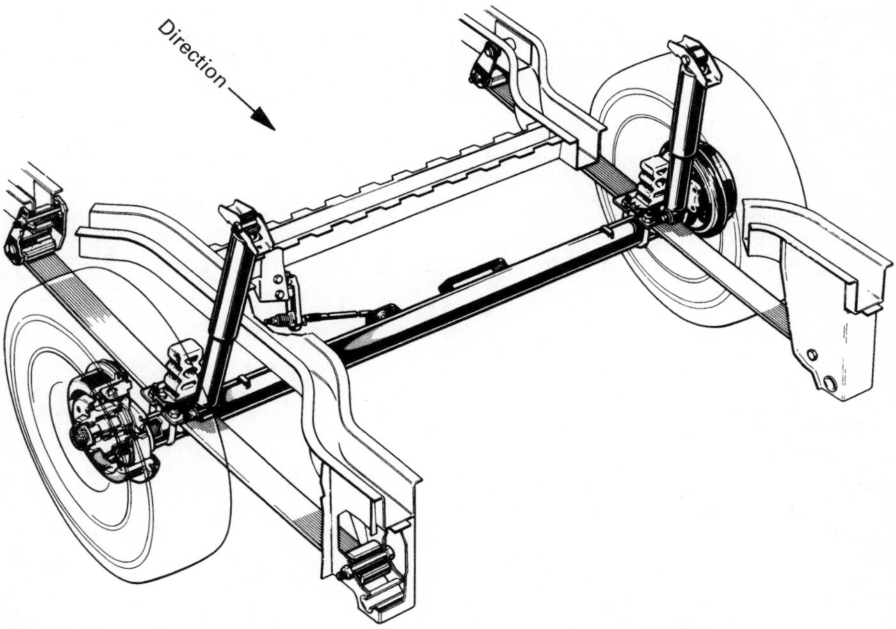

Fig. 1.24 The rear axle on a Ford Escort Express delivery vehicle. Single leaf springs carry the axle and support the body well at four points. The shock absorbers (fitted vertically) are located close to the wheel, made possible by slim wheel-carriers/hub units. The additional elastomer springs sit over the axle tube and act on the side members of the body when at full bump.

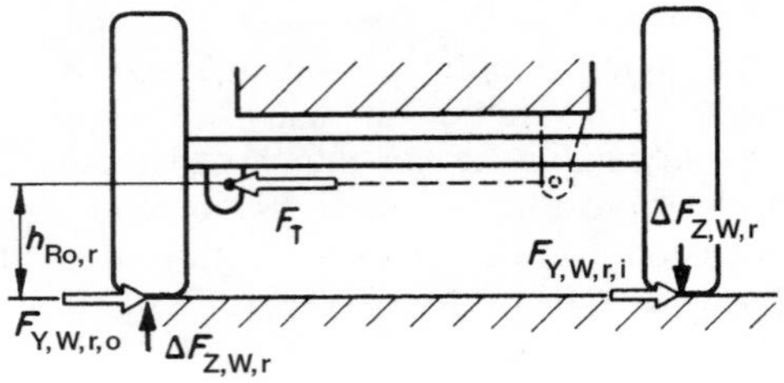

Fig. 1.25 On rigid axles the axle body absorbs the bending moments which arise as a result of lateral forces. Only the force F_T occurs between the suspension and the body, and its size corresponds to the lateral forces $F_{Y,W,r,o}$ and $F_{Y,W,r,i}$. On a horizontal Panhard rod, the distance $h_{Ro,r}$ is also the height of the body roll centre. The higher this is above ground, the greater the wheel force change $\pm\Delta F_{Z,Wr}$.

There are many options for attaching a rigid axle rear suspension beneath the body or chassis frame. Longitudinal leaf springs are often used as a single suspension control arm, which is both supporting and springing at the same time, as these can absorb forces in all three directions as well as drive-off and braking moments (Figs 1.26 and 5.20). This economical type of rear suspension also has the advantage that the load area on lorries and the body of passenger cars can be supported in two places at the back: at the level of the rear seat and under the boot (Fig. 1.27). This reduces the stress on the rear end of the car body when the boot is heavily laden, and also the stress on the lorry frame under full load (Fig. 1.20).

The longitudinal leaf springs can be fitted inclined, with the advantage that during cornering the rigid rear axle (viewed from above) is at a small angle to the vehicle longitudinal axis (Fig. 1.28). To be precise, the side of the wheel base on the outside of the bend shortens somewhat, while the side on the inside of the bend lengthens by the same amount. The rear axle steers into the bend and, in other words, it is forced to self-steer towards 'roll-understeering' (Fig. 1.29).

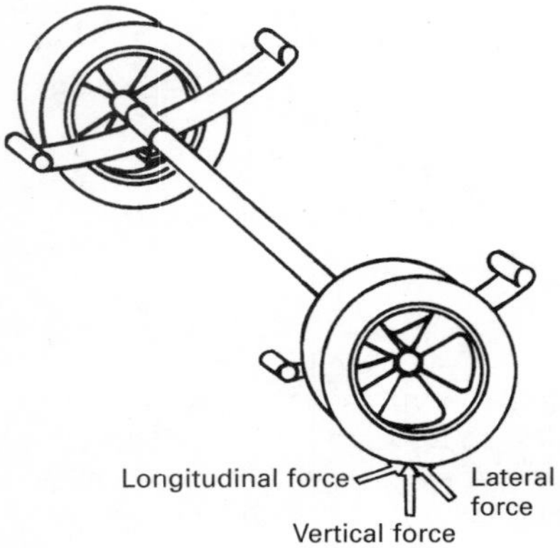

Fig. 1.26 Longitudinal leaf springs can absorb both forces in all directions and the drive-off, braking and lateral force moment. (See Section 6.2 in Ref. [3]).

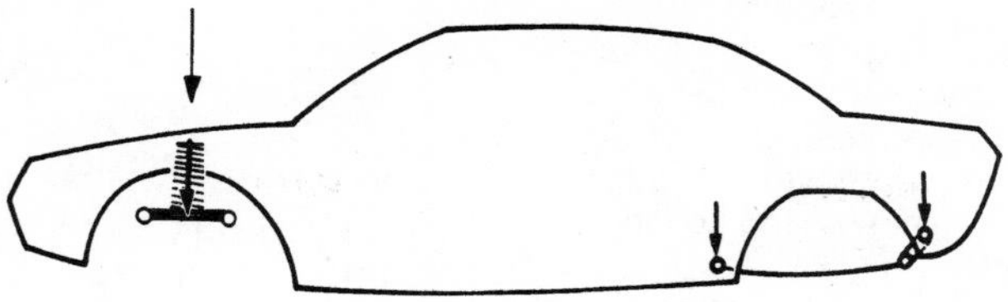

Fig. 1.27 Longitudinal rear leaf springs support the body of a car in two places – under the back seats and under the boot – with the advantage of reduced bodywork stress.

This measure can, of course, have an adverse effect when the vehicle is travelling on bad roads, but it does prevent the standard passenger car's tendency to oversteer when cornering. Even driven rigid axles exhibit – more or less irrespective of the type of suspension – a tendency towards the load alteration (torque steering) effect, but not to the same extent as semi-trailing link suspensions. Details can be found in Section 2.12.2 and in Ref. [2] and Ref. [9].

On front-wheel drive vehicles, the wheels of the trailing axle can take on a negative camber. This improves the lateral grip somewhat, but does not promote perfect tyre wear. This is also possible on the compound crank suspension (a

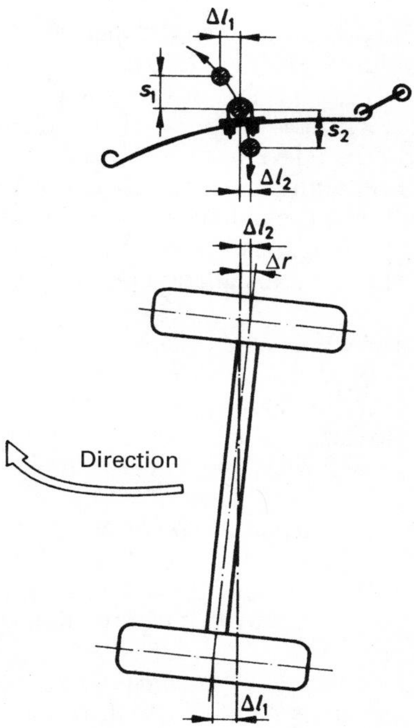

Fig. 1.28 Angled longitudinal leaf springs fixed lower to the body at the front than at the back cause the rigid rear axle to self-steer towards understeering (so-called roll pitch understeering). Where there is body roll, the wheel on the outside of the bend, which is compressing along the path s_1, is forced to accommodate a shortening of the wheel base Δl_1, whilst the wheel on the inside of the bend, which is extending by s_2, is forced to accommodate a lengthening of the wheel base by Δl_2. The axle is displaced at the steering angle Δr (see also Fig. 3.75).

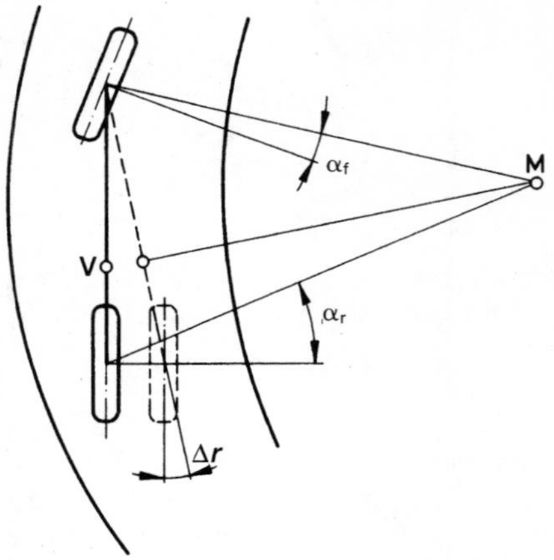

Fig. 1.29 If a rigid rear axle steers with the angle Δr towards understeer, the tail moves out less in the bend and the driver has the impression of more neutral behaviour. Moreover, there is increased safety when changing lanes quickly at speed.

The same occurs if the outside wheel of an independent wheel suspension goes into toe-in and the inside wheel goes into toe-out (see Fig. 3.79).

suspension-type halfway between a rigid axle and independent wheel suspension) which, up to now, has been fitted only on front-wheel drive vehicles. Details are given in Fig. 1.2 and Section 1.6.4.1.

1.3.2 Semi rigid crank axles

The compound crank suspension could be described as the new rear axle design of the 1970s (Figs 1.30 and 1.2) and it is still used in today's small and medium-sized front-wheel drive vehicles. It consists of two trailing arms that are welded to a twistable cross-member and fixed to the body via trailing links. This member absorbs all vertical and lateral force moments and, because of its offset to the wheel centre, must be less torsionally stiff and function simultaneously as an anti-roll bar. The axle has numerous advantages and is therefore found on a number of passenger cars which have come onto the market.

From an installation point of view:

- the whole axle is easy to assemble and dismantle;
- it needs little space;
- a spring damper unit or the shock absorber and springs are easy to fit;
- no need for any control arms and rods; and thus
- only few components to handle.

From a suspension point of view:

- there is a favourable wheel to spring damper ratio (See Section 5.3.5 in Ref. [3]);
- there are only two bearing points O_l and O_{rs}, which hardly affect the springing (Fig. 1.31);
- low weight of the unsprung masses (see Section 6.1.3); and
- the cross-member can also function as an anti-roll bar.

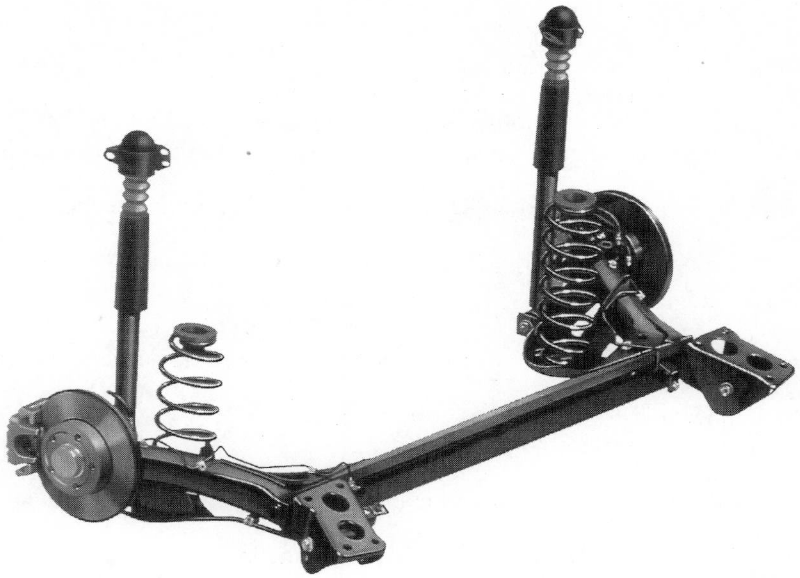

Fig. 1.30 Twist-beam suspension of the VW Golf IV (1997), VW Bora (1999) and Audi A3 (1996). The rubber–metal bearings of the axle body are set at 25° to the transverse suspension of the vehicle in order to improve the self-steering properties of the suspension together with the rigidity of the bearings which varies in three directions in space. Compared with the previous model, it was possible to reduce unwanted lateral-force toe-out steer resulting from link deformation by 30% to approximately 1 mm per 500 N of lateral force. Figure 1.72 shows the four-wheel drive version of the VW Golf IV.

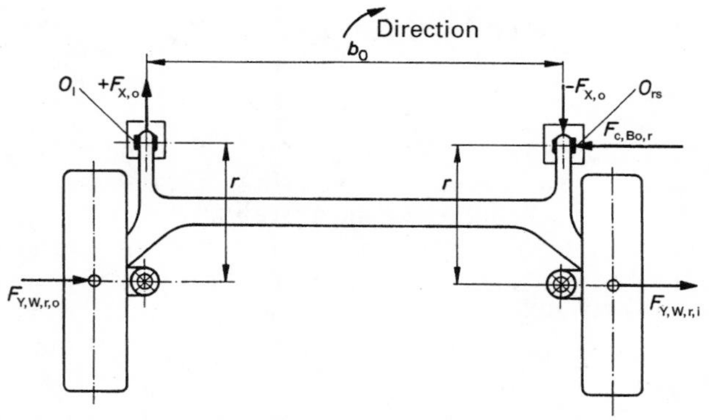

Fig. 1.31 The lateral forces $F_{Y,W,r,o}$ and $F_{y,W,r,i}$ occurring at the centres of tyre contact during cornering are absorbed at the bearing points O_l and O_{rs}. This results in a moment $M_y = (F_{Y,W,r,o} + F_{Y,W,r,i}) X r = F_{X,o}\, b_O$ which (depending on the elasticity of the rubber bearing) can cause 'lateral force oversteering'. The longer the control arms (distance r) and the closer the points O_l and O_{rs} (distance b_O), the greater the longitudinal forces $\pm F_{xO}$.

From a kinematic point of view:

- there is negligible toe-in and track width change on reciprocal and parallel springing;
- there is a low change of camber under lateral forces (Figs 3.54 and 3.57);
- there is low load-dependent body roll understeering of the whole axle (Fig. 3.38 and 3.78); and
- good radius-arm axis locations O_l and O_{rs} (Fig. 1.31), which reduce tail-lift during braking.

The disadvantages are:

- a tendency to lateral force oversteer due to control arm deformation (Fig. 3.72);
- torsion and shear stress in the cross-member;
- high stress in the weld seams; which means
- the permissible rear axle load is limited in terms of strength;
- the limited kinematic and elastokinematic opportunities for determining the wheel position;
- the establishment of the position of the instantaneous centre by means of the axle kinematics and rigidity of the twist-beam axle;
- the mutual effect on the wheel;
- the difficult decoupling of the vibration and noise caused by the road surface; and
- the considerable need for stability of the bodywork in the region of those points on the front bearings at which complex, superposed forces have to be transmitted.

1.4 Front-mounted engine, rear-mounted drive

In passenger cars and estate cars, the engine is approximately in the centre of the front axle and the rear wheels are driven (Fig. 1.32). To put more weight on the rear axle and obtain a more balanced weight distribution, Alfa Romeo, Porsche (928, 968 models) and Volvo integrated the manual transmission with the differential. This is also the case with the Chevrolet Corvette sports car (1998; Figs 1.33 and 1.34). With the exception of light commercial vehicles, all lorries have the engine at the front or centrally between the front and rear axles together with rear-wheel drive vehicles. The long load area gives hardly any other option. Articulated lorries, where a major part of the trailer weight – the trailer hitch load – is carried over the rear wheels, have the same configuration. On buses, however, the passengers are spread evenly throughout the whole interior of the vehicle, which is why there are models with front, central and rear engines.

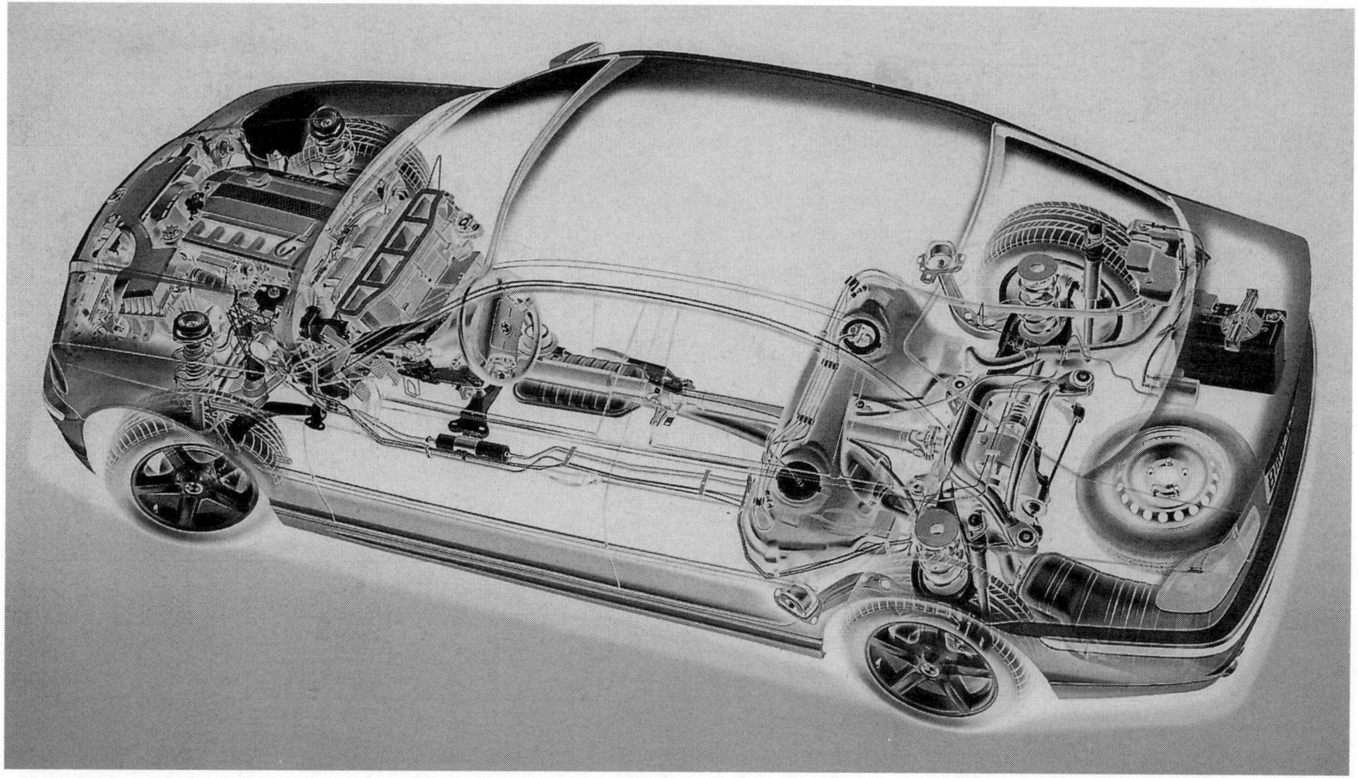

Fig. 1.32 Front-mounted engine, rear-mounted drive (BMW 3 series E46, 1998). The manual transmission is flange-mounted on the engine, which is longitudinally positioned over the front axle. The rear-axle differential is driven by means of a propshaft. The fuel tank is situated in front of the rear axle for safety in case of an accident. The battery was placed in the boot in order to achieve a balanced 50:50 axle-load distribution. Figure 1.1 shows the rear axle in detail.

Fig. 1.33 Chevrolet Corvette (1998). In order to achieve balanced axle-load distribution, a more rigid overall system (necessary on account of the greater flexibility of the plastic bodywork) and more leg room, the gearbox is integrated with the rear-axle differential. Compared with standard drives, the cardan shaft turns higher (with engine speed) but is subject to correspondingly less torque. The front and rear axles have plastic (fibreglass) transverse leaf springs.

Compared with the previous model, unwanted vibration, particularly on an uneven road surface, is reduced as a result of the shorter length of the wheel spindles of 63 mm and the small steering-axle angle of 8.8 degrees. Owing to the combination of a castor angle of 6.5 degrees with a castor trail of 36 mm (previous model: 5.9 degrees, 45 mm), a good compromise is achieved between high lateral rigidity of the axle and good feedback properties.

1.4.1 Advantages and disadvantages of the front-mounted engine, rear-mounted drive design

The standard design has a series of advantages on passenger cars and estate cars:

- There is hardly any restriction on engine length, making it particularly suitable for more powerful vehicles (in other words for engines with 8–12 cylinders).
- There is low load on the engine mounting, as only the maximum engine torque times the conversion of the lowest gear without differential transmission has to be absorbed.
- Insulation of engine noise is relatively easy.
- Under full load most of the vehicle mass is on the driven rear axle (important for estate cars and trailers (Figs 1.36 and 6.22)).
- A long exhaust system with good silencing and catalytic converter configuration.
- Good front crumple zone, together with the 'submarining' power plant unit, i.e. one that goes underneath the floor panel during frontal collision.

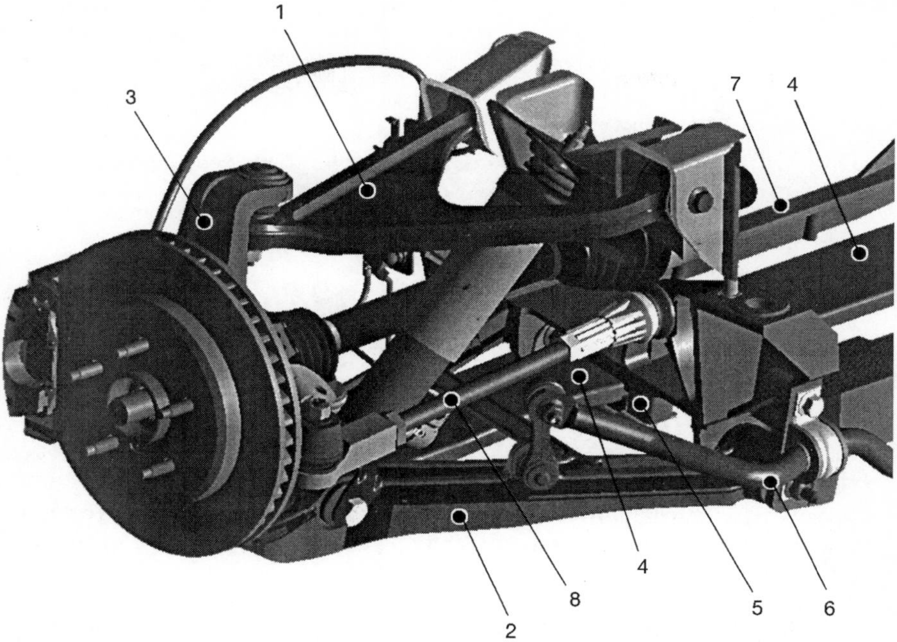

Fig. 1.34 Rear axle (left side of wheel) of the Chevrolet Corvette (1998). Links 1, 2 and wheel carrier 3 of the multi-link suspension are made from aluminium in order to reduce the unsprung masses. The plastic leaf spring 4 is mounted at two places on the right and left sides of the body (5) so that it also helps to make the body more resistant to roll. Roll spring stiffness is further increased by stabilizer 6. This is attached to subframe 7, which is also made of aluminium. The design of the wheel carrier 3 on the front and rear axles is the same, but not the wheel links 1 and 2. The toe-in control of the rear axle is exercised very stiffly and precisely, via tie rod 8.

- Simple and varied front axle designs are possible irrespective of drive forces.
- More even tyre wear thanks to function distribution of steering/drive.
- Uncomplicated gear shift mechanism.
- Optimum gearbox efficiency in direct gear because no force-transmitting bevel gear is in action (Fig. 6.19).
- Sufficient space for housing the steering system in the case of a recirculating ball steering gear.
- Good cooling because the engine and radiator are at the front; a power-saving fan can be fitted.
- Effective heating due to short hot-air and water paths.

The following disadvantages mean that, in recent years, only a few saloon cars under 2 l engine displacement have been launched internationally using this design, and performance cars also featured the front-mounted design:

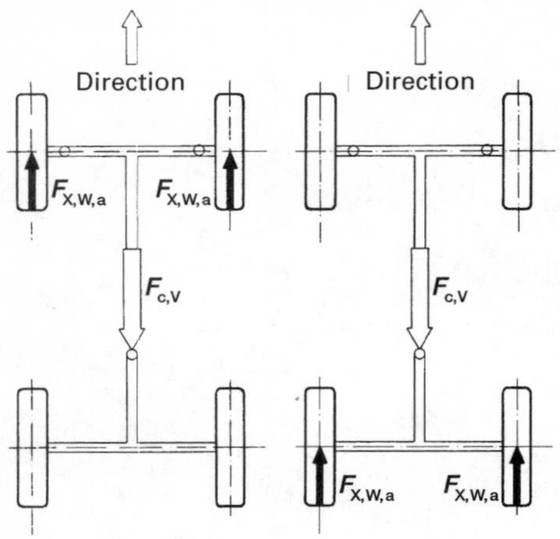

Fig. 1.35 On a front-wheel drive (left) the vehicle is pulled. The result is a more stable relationship between the driving forces $F_{X,W,a}$ and the inertia force $F_{c,V}$ Conversely, in the case of driven rear wheels an unstable condition is theoretically evident; front axle settings ensure the necessary stabilization.

- Unstable straight-running ability (Fig. 1.35), which can be fully corrected by special front suspension geometry settings, appropriate rear axle design and suitable tyres.
- The driven rear axle is slightly loaded when there are only two persons in the vehicle, leading to poor traction behaviour in wet and wintry road conditions – linked to the risk of the rear wheels spinning, particularly when tight bends are being negotiated at speed. This can be improved by setting the unladen axle load distribution at 50%/50% which, however, is not always possible (Figs 1.36 and 6.22). It can be prevented by means of drive-slip control (see Ref. [7]).
- A tendency towards the torque steer effect (Fig. 2.53) and, therefore,
- complex rear independent wheel suspension with chassis subframe, differential gear case and axle drive causing
- restrictions in boot size
- The need for a propshaft between the manual gearbox and differential (Fig. 1.32) and, therefore,
- a tunnel in the floor pan is inevitable, plus an unfavourable interior to vehicle –length ratio.

Fig. 1.36 Average proportional axle load distribution based on drive type and loading condition. With the standard design saloon, when the vehicle is fully laden, the driven rear wheels have to carry the largest load. With the front-wheel drive, however, with only two persons in the vehicle, the front wheels bear the greater load.

	Front-wheel drive		*Rear wheel drive*		*Rear engine*	
	front	*rear*	*front*	*rear*	*front*	*rear*
Empty	61	39	50	50	40	60
2 passengers at the front	60	40	50	50	42	58
4 passengers	55	45	47	53	40	60
5 passengers and luggage	49	51	44	56	41	59

1.4.2 Non-driven front axles

The standard design for passenger cars that have come onto the market in recent years have McPherson struts on the front axle, as well as double wishbone or multi-link suspensions. The latter type of suspension is becoming more and more popular because of its low friction levels and kinematic advantages. Even some light commercial vehicles have McPherson struts or double wishbone axles (Fig. 1.7). However, like almost all medium-sized and heavy commercial vehicles, most have rigid front axles. In order to be able to situate the engine lower, the axle subframe has to be offset downwards (Fig. 1.37).

The front wheels are steerable; to control the steering knuckle 5 (Fig. 1.38) on double wishbone suspensions, there are two ball joints that allow mobility in all directions, defined by full bump/rebound-travel of the wheels and the steering angle. The wishbone, which accepts the spring, must be carried on a supporting joint (item 7) in order to be able to transmit the vertical forces. A regular ball joint transferring longitudinal and lateral forces (item 8) is generally sufficient for the second suspension control arm. The greater the distance between the two joint points, the lower the forces in the components. Figure 1.39 shows a front axle with ball joints a long way apart.

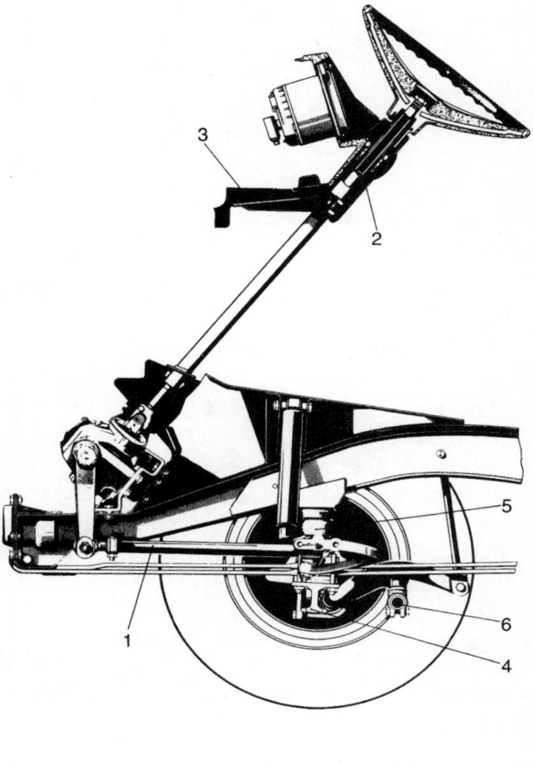

Fig. 1.37 The front rigid axle on the Mercedes-Benz light commercial vehicle of the 207 D/308 series with recirculating ball steering gear and steering rod 1 parallel to the two-layer parabolic spring. This rod has to be slightly shorter than the front side of the spring, so that both parts take on the same motion curve when the axle bottoms out (see also Fig. 4.6). The brace 3, running from the steering column jacket 2 to the body, bends on impact. The T-shaped axle casing 4, which is cranked downwards and to which the springs are fastened, can be seen in the section. The elastomer spring 5 sits on the longitudinal member of the frame and the two front wheels are joined by the tie rod 6. The safety steering wheel has additional padding.

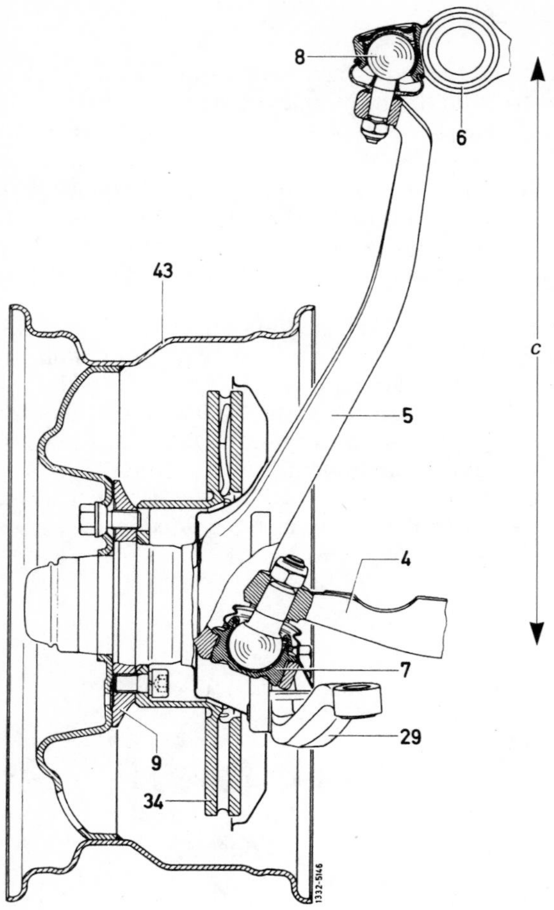

Fig. 1.38 Front hub carrier (steering knuckle) on the Mercedes-Benz S class (W40, 1997) with a large effective distance *c* (see also Fig. 1.4). The upper transverse control arm 6 forms the casing for the ball pivot of the guiding joint, whereas the lower supporting joint 7 is pressed into the hub carrier 5. The ventilated brake disc 34 (dished inwards), the wheel hub 9, the double-hump rim 43 with asymmetrical drop centre and the space for the brake caliper (not included in the picture) are clearly shown.

The base on McPherson struts is better because it is even longer. Figure 1.40 shows a standard design and Fig. 1.8 the details.

The coil spring is offset at an angle to reduce the friction between piston rod 2 and the rod guide. The lower guiding joint (point G) performs the same function as on double wishbones, whereas point E is fixed in the shock tower, which is welded to the wheel house panel. As the wheels reach full bump, piston rod 2 moves in the cylinder tube (which sits in the carrier or outer tube, see Fig. 5.53) and when there is a steering angle the rod and spring turn in an upper strut mount, which insulates noise and is located at point E (Fig. 1.9).

Wheel controlling damper struts do not require such a complex mount. The piston rod turns easily in the damping cylinder (Fig. 1.41). Only the rod needs noise insulation. The coil spring sits separately on the lower control arm, which must be joined to the steering knuckle via a supporting joint. The damper is

Fig. 1.39 Multi-link front suspension of the Mercedes-Benz model W220 (S class, 1998). Based on a double wishbone axle, two individual links (tension strut and spring link) are used instead of the lower transverse link in order to control the steering axle nearer to the middle of the wheel. As a result, the kingpin offset and disturbing force lever arm are reduced and vibrations are caused by tyre imbalances and brake-force fluctuations is consequently minimized. Crash performance is also improved by the more open design. The air-spring struts with integrated shock absorber (see Fig. 5.19) proceed directly from the spring link. The laterally rigid rack and pinion steering in front of the middle of the wheel leads to the desired elastokinematic understeer effect during cornering owing to the laterally elastic spring link bearings. The manufacturing tolerances are kept so small by means of punched holes that the adjustment of camber and camber angles in production is not necessary.

lighter than a shock-absorbing strut and allows a greater bearing span across the damping cylinder, permits a wider, flatter engine compartment (which is more streamlined) and is easier to repair. However, it is likely to be more costly and offsetting the spring from the damper (Figs 1.8 and 1.11) may cause slip-stick problems with a loss of ride comfort.

In the case of front-wheel drive vehicles, there may be a problem in the lack of space between the spring and the drive axle.

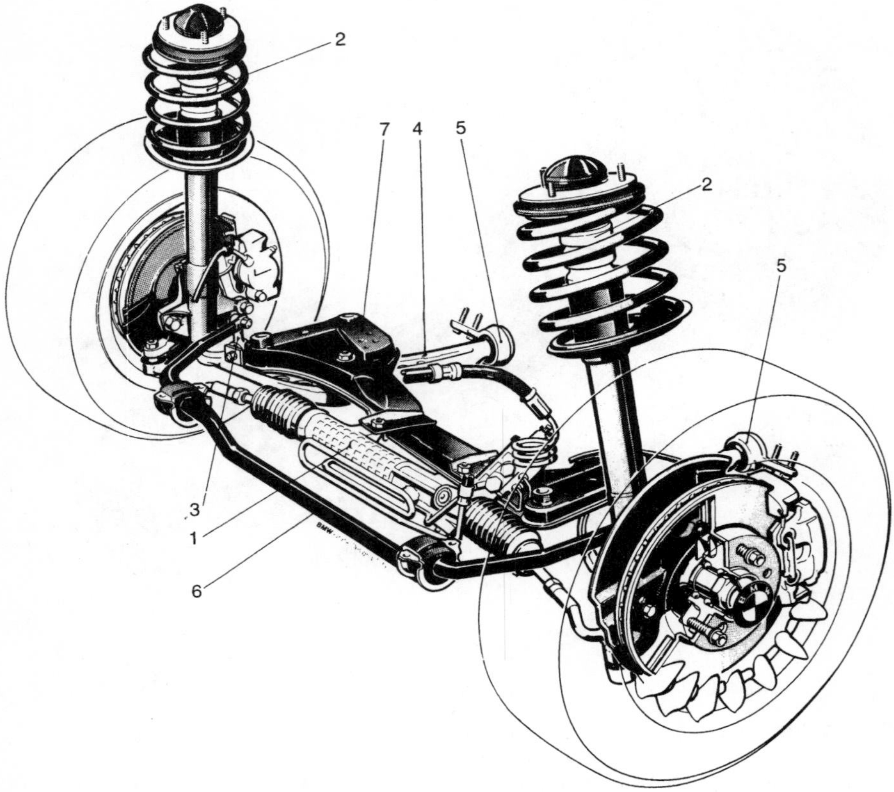

Fig. 1.40 Spring strut front axle of the BMW Roadster Z3, which Lemförder Fahrwerktechnik produce in the USA and supply directly to the assembly line there. The additional springs 2 are positioned in the coil springs (Fig. 1.11) which are offset at an angle in order to reduce friction. The stabilizer 6 is connected to the lower links by the struts 3.

The cross-member 7 which serves as the subframe takes the hydraulically supported rack and pinion steering 1 at the front and the transverse link 4 on its lower side. The L-shape of the transverse link makes good decoupling of the lateral rigidity and longitudinal elasticity possible: lateral forces are introduced directly into the rigid front bearing, while longitudinal forces produce a rotational movement about the front bearing as a result of the laterally elastic rear bearing 5. As shown in Fig. 3.84, these rubber elements ensure a defined lateral springing. The large-diameter internally ventilated brake discs (15″ rim) and the third-generation, two-row angular ball bearings, whose outside ring also acts as a wheel hub, are clearly shown.

The kingpin offset at ground (scrub radius) depends on the tyre width and thus the wheel offset (Fig. 3.10L); it is $r_\sigma = +10$ mm on 185/65 R 15 tyres and $r_2 = +5$ mm on 205/60 R 15 tyres.

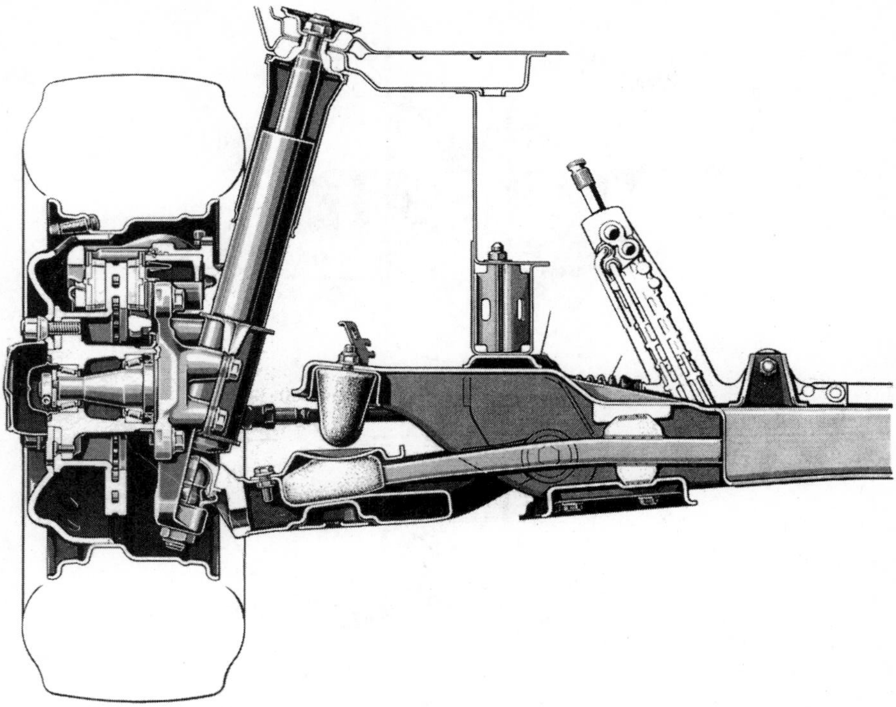

Fig. 1.41 Front axle of the Mercedes-Benz Sprinter series (1995). The wheel-controlling strut is screwed on to the wheel carrier, which is, in turn, connected to the lower cross-member by means of a ball joint. Both the vehicle suspension and roll stabilization are ensured by means of a transverse plastic leaf springmounted on rubber elements. Large rubber buffers with progressive rigidity act as additional springs and bump stops.

1.4.3 Driven rear axles

Because of their cost advantages, robustness and ease of repair rigid axles are fitted in practically all commercial and off-road vehicles (Fig. 1.43) in combination with leaf springs, coil springs or air springing (Figs 1.20 and 1.42). They are no longer found in saloons and coupés. In spite of the advantages described in Section 1.3, the weight of the axle is noticeable on this type of vehicle.

For independent suspension, the semi-trailing arm axle, shown in Figs 1.15 and 1.45, is used as independent wheel suspension in passenger and light commercial vehicles. This suspension has a chassis subframe to which the differential is either fixed or, to a limited degree, elastically joined to give additional noise and vibration insulation. The springs sit on the suspension control arms. This gives a flat, more spacious boot, but with the disadvantage that the forces in all components become higher.

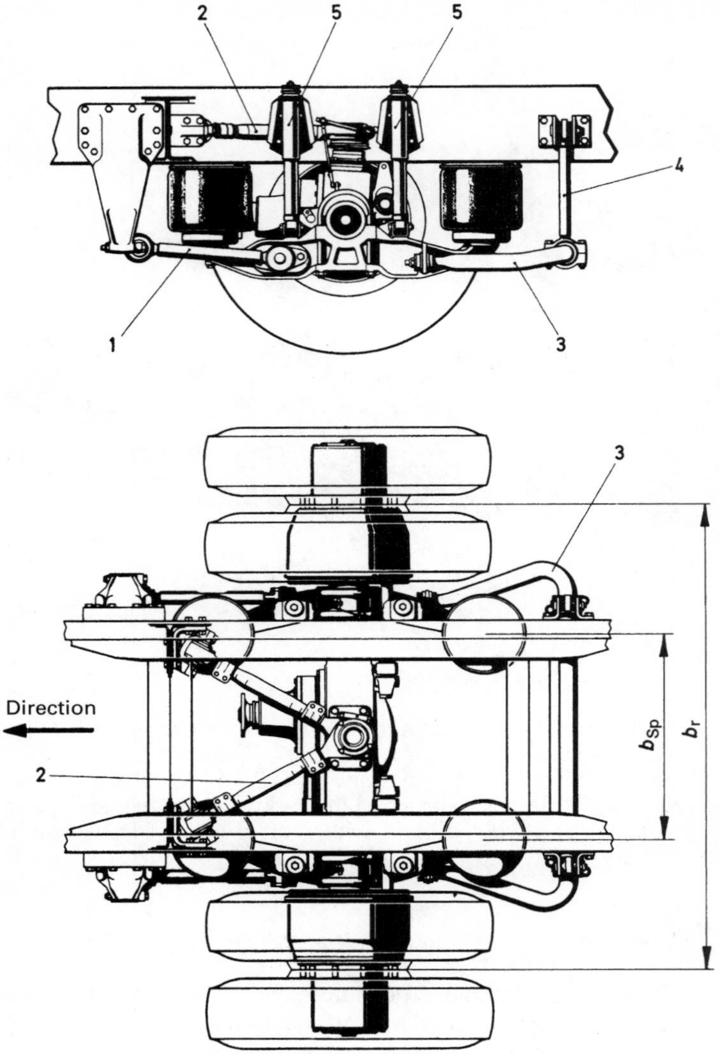

Fig. 1.42 Driven rear axle with air springs of the Mercedes-Benz lorry 1017 L to 2219 L 6 × 2. The axle is carried in the longitudinal and lateral directions by the two struts 1 and the upper wishbone type control arm 2. The four spring bellows sit under the longitudinal frame members and, because of the twin tyres, they have a relatively low effective b_{Sp}. The tracking width b_r divided by b_{Sp} yields approximately the ratio $i_\varphi = 2.2$. As shown in Equation 5.19, with reciprocal springing the rate is $c_{\varphi,r}$ which amounts to only 21% of the rate c_r with parallel springing.

To reduce body roll pitch the anti-roll bar 3 was placed behind the axle and is supported on the frame via the rod 4. The four shock absorbers 5 are almost vertical and are positioned close to the wheels to enable roll movements of the body to fade more quickly.

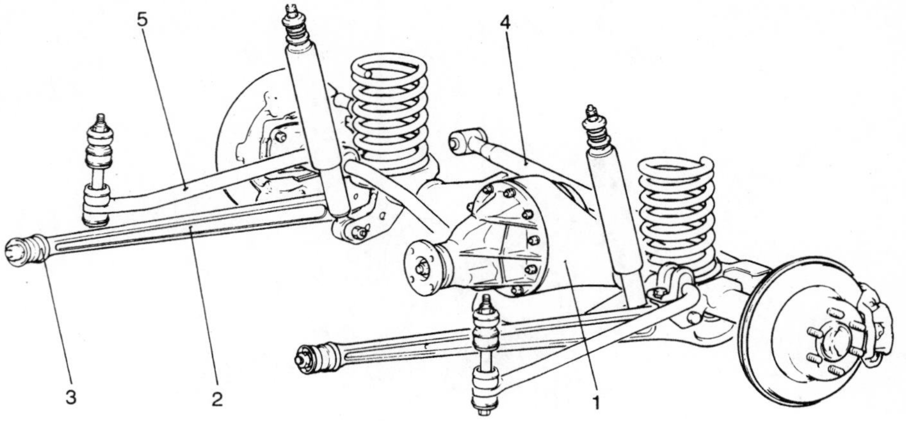

Fig. 1.43 The rear axle on the all-terrain, general-purpose passenger car, Mitsubishi Pajero. The rigid axle casing 1 is taken through the longitudinal control arms 2. These absorb the drive-off and braking forces (and the moments which arise) and transmit them to the frame. The rubber mountings 3 in the front fixing points 1, which also represent the vehicle pitch pole O_r (Fig. 3.160), are designed to be longitudinally elastic to keep the road harshness due to the dynamic rolling hardness of the radial tyre away from the body. The Panhard rod 4 absorbs lateral forces. The anti-roll bar 5 is (advantageously) fastened a long way out on the frame (Fig. 1.23). The disc brakes, coil springs and almost vertical shock absorbers can be clearly seen. Further details are contained in Section 3.5 in Ref. [2].

Because of its ride and handling advantages, more and more passenger cars have double wishbone suspension rear axles or so-called multi-link axles (Figs 1.1, 1.19, 1.34 and 1.72).

Most independent wheel suspensions have an easy-to-assemble chassis subframe for better wheel control and noise insulation. However, all configurations (regardless of the design) require drive shafts with length compensation. This is carried out by the sliding CV (constant velocity) joints fitted both at the wheel and the differential. Figure 1.17 shows a section through a joint of this type, and Fig. 1.44 shows a typical modern bearing of a driven rear wheel.

1.5 Rear and mid engine drive

The rear-mounted power plant consists of the engine and the differential and manual gearbox in one assembly unit, and it drives the rear wheels. The power plant can sit behind the axle (Fig. 1.45, rear-mounted engine) or in front of it (Fig. 1.46, central engine). This configuration makes it impossible to have a rear seat as the engine occupies this space. The resulting two-seater is only suitable as a sports or rally car.

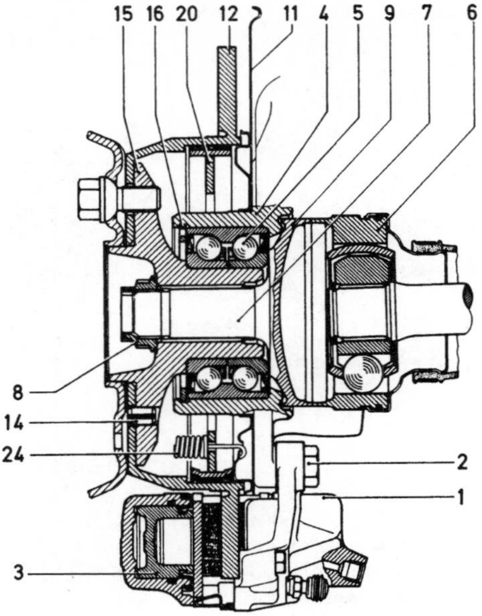

Fig. 1.44 Rear axle wheel hub carrier with wheel and brake. The drive shaft 7 is butt-welded to the CV slip joint 6. The drive shaft transmits the driving torque to the wheel hub 15 via a serrated profile. Part 15 is carried by the maintenance-free, two-row angular (contact) ball bearing 5. The one-part outer ring is held in the hub carrier 4 by the snap ring 16.

The seal rings on both sides sit in the permanently lubricated bearing unit. The covering panel 11 (that surrounds the brake disc 12) acts as additional dirt-protection outside, as does collar 9 of the CV joint on the inside. This grips into a cut-out in the wheel hub carrier 4 and creates a cavity. The centrifugal effect of the bell-shaped joint housing prevents ingress of dirt and water. The brake disc 12 is pulled from outside against the flange 15 and fixed by dowel 14 until the wheel is mounted. The jaws 20 of the drum brake acting as a handbrake act on the inside of part 12. At the lower end, the illustration shows the fixed calliper 1 of the disc brake. Two hexagonal bolts (item 2) fix it to the wheel hub carrier 4. Piston 3 and the outer brake pad are shown cut away (illustration: Mercedes-Benz).

The disadvantages of rear and central engine drive on passenger cars are:

- moderate straight running abilities (caster offset at ground angles of up to $\tau = 8°$ are factory set);
- sensitivity to side winds;
- indifferent cornering behaviour at the stability limit (central engine);
- oversteering behaviour on bends (rear-mounted engine, see Fig. 2.42);
- difficult to steer on ice because of low weight on the front wheels;
- uneven tyre wear front to rear (high rear axle load, see Fig. 1.36);
- the engine mounting must absorb the engine moment times the total gear ratio;

Fig. 1.45 VW Transporter, a light truck which could be used either as an eight-seater bus or for transporting goods, and which has the optimal axle load distribution of 50%/50% in almost all loading conditions. The double wishbone suspension at the front, the semi-trailing link rear axle and the rack and pinion steering, which is operated via an additional gear set in front, can be seen clearly. To achieve a flat load floor throughout, VW changed the Transporter to front-wheel drive in 1990.

Fig. 1.46 The Porsche Boxster (1996) has a water-cooled engine which is longitudinally installed in front of the rear axle. The front axle is designed as a spring strut-type axle. The transverse link is arranged almost in extension of the wheel axle; it is connected to the longitudinal link by a strut bush which is soft for reasons of comfort. This open design and link geometry make it possible to combine a high level of driving precision, a result of rigid wheel control, with riding comfort, owing to the longitudinal elasticity of the axle. At a camber angle of 8 degrees, good straight running results from the large castor displacement of 41 mm. The kingpin offset is –7 mm and the disturbing force lever arm is 83 mm. The pitch centre of the front axle was located near to the road to achieve kinematic wheel recession of the axle, which is important for riding comfort, with the result that braking-torque compensation is only 10%.
The rear axle is also a spring strut-type axle in an open link design; the wheel carrier, hub and bushes as well as the transverse link are the same as those found on the front axle. The open design makes it possible to have an inwardly inclined elastokinematic axis of rotation, so that a stabilizing toe-in position of the rear wheels is produced during braking. The axle can also be designed to understeer when subject to lateral forces.
The main disadvantages of the mid-engine design are apparent from the boot space: only 130 l are available at both the front and back.

- the exhaust system is difficult to design because of short paths;
- the engine noise suppression is problematic;
- complex gear shift mechanism;
- long water paths with front radiators (Fig. 1.46);
- high radiator performance requirement because of forced air cooling, the electric fan can only be used on the front radiator;

- the heating system has long paths for hot water or warm air;
- the fuel tank is difficult to house in safe zones;
- the boot size is very limited.

In the case of vehicles with a short wheel base and high centre of gravity with the engine on or behind the rear axle, there is a danger that the vehicle will overturn if it is rolling backwards down a steep slope and the parking brake, which acts upon the rear axle, is suddenly applied.

As a result of the logical further development of the kinematics and elastokinematics of the axles, Porsche have succeeded in improving straight running as well as cornering in the steady state (vehicles now understeer slightly up to high lateral accelerations) and transient state as well as when subject to torque steer effects. Even in the case of the Boxsters (with mid-engine, see Fig. 1.46, since 1996) and 911 (water-cooled since 1997), Porsche are adhering to rear-wheel drive (whereas the VW Transporter, Fig. 1.45, has not been built since 1991) and, in so doing, obtain the following benefits:

- very agile handling properties as a result of the small yawing moment;
- very good drive-off and climbing capacity, almost irrespective of load (Fig. 6.22);
- a short power flow because the engine, gearbox and differential form one compact unit;
- light steering due to low front axle load;
- good braking force distribution;
- simple front axle design;
- easy engine dismantling (only on rear engine);
- no tunnel or only a small tunnel in the floor pan;
- a small overhang to the front is possible.

1.6 Front-wheel drive

The engine, differential and gearbox form one unit, which can sit in front of, over, or behind the front axle. The design is very compact and, unlike the standard design, means that the vehicle can either be around 100–300 mm shorter, or the space for passengers and luggage can be larger. These are probably the main reasons why, worldwide, more and more car manufacturers have gone over to this design. In recent years only a few saloons of up to 2 l capacity without front-wheel drive have come onto the market. Nowadays, front-wheel drive vehicles are manufactured with V6 and V8 engines and performances in excess of 150 kW.

However, this type of drive is not suitable for commercial vehicles as the rear wheels are highly loaded and the front wheels only slightly. Nevertheless, some light commercial vehicle manufacturers accept this disadvantage so they can lower the load area and offer more space or better loading conditions (Fig. 1.47). The propshafts necessary on standard passenger cars would not allow this.

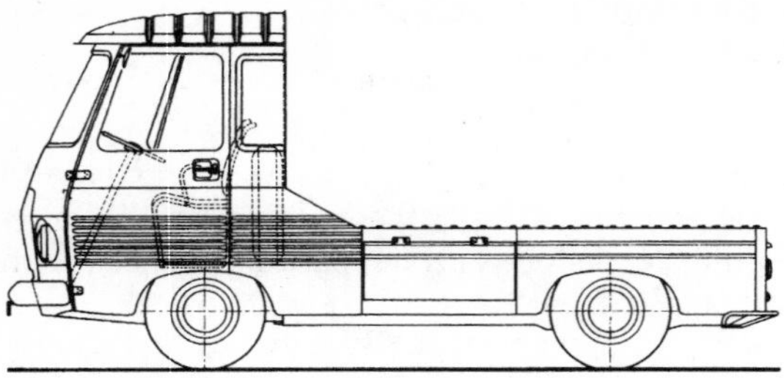

Fig. 1.47 The low cargo area on the Peugeot light commercial vehicle J 5/J 7 is achieved due to front-wheel drive and a semi-trailing link axle to the rear (similar to the one in Fig. 1.63).

1.6.1 Types of design

1.6.1.1 Engine mounted longitudinally 'North–South' in front of the axle

In-line or V engines mounted in front of the axle – regardless of the wheelbase – give a high front axle load, whereby the vehicle centre of gravity is pushed a long way forwards (Fig. 1.48). Good handling in side winds and good traction, especially in the winter, confirm the merits of a high front axle load, whereas the heavy steering from standing (which can be rectified by power-assisted steering), distinct understeering during cornering and poor braking force distribution would be evidence against it.

This type of design, as opposed to transverse mounting, is preferred in the larger saloons as it allows for relatively large in-line engines. The first vehicles of this type were the Audi 80 and 100. Inclining the in-line engine and placing the radiator beside it means the front overhang length can be reduced. Automatic gearboxes need more space because of the torque converter. This space is readily available with a longitudinally mounted engine.

A disadvantage of longitudinal engines is the unfavourable position of the steering gear: this should be situated over the gearbox. Depending on the axle design, this results in long tie rods with spring strut (McPherson) front axles (Fig. 1.57).

1.6.1.2 Transverse engine mounted in front of the axle

In spite of the advantage of the short front overhang, only limited space is available between the front wheel housings (Figs 1.49 and 1.50). This restriction means that engines larger than an in-line four cylinder or V6 cannot be fitted in a medium-sized passenger car. Transverse, asymmetric mounting of the engine and gearbox may also cause some performance problems. The unequal length of the drive shafts affects the steering. During acceleration the vehicle rises and the drive shafts take on different angular positions, causing uneven moments around the steering axes. The difference between these moments to the left and to the right causes unintentional steering movements resulting in a noticeable pull to one side (Fig. 3.88); drive shafts of equal length are therefore desirable. This also prevents different drilling angles in the drive shaft causing timing differences in drive torque build-up.

The large articulation angle of the short axle shaft can also limit the spring

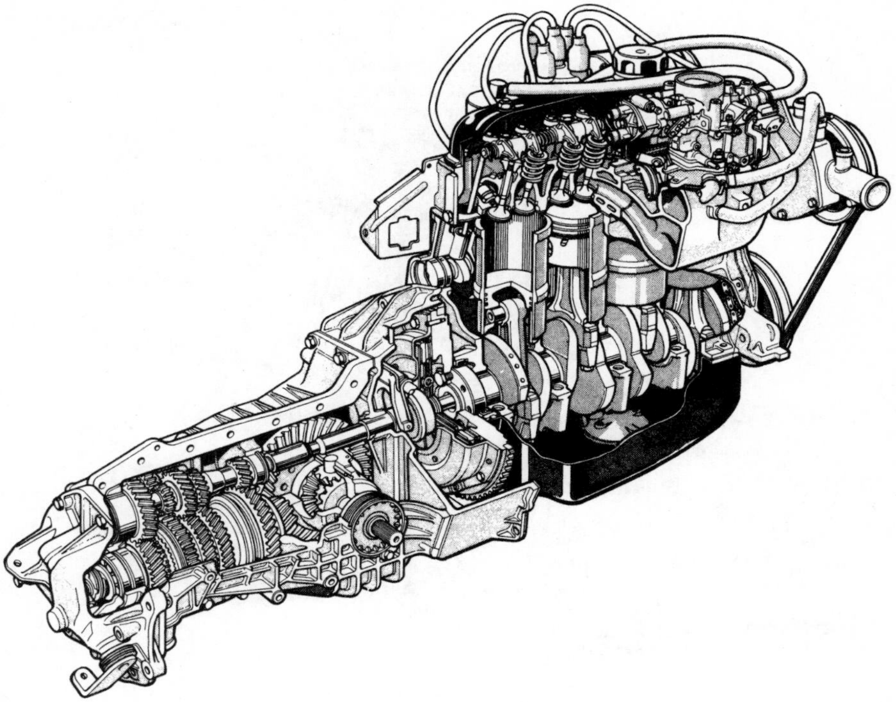

Fig. 1.48 In front-wheel drive vehicles the engine can be mounted longitudinally in front of the front axle with the manual gearbox behind. The shaft goes over the transverse differential (illustration: Renault).

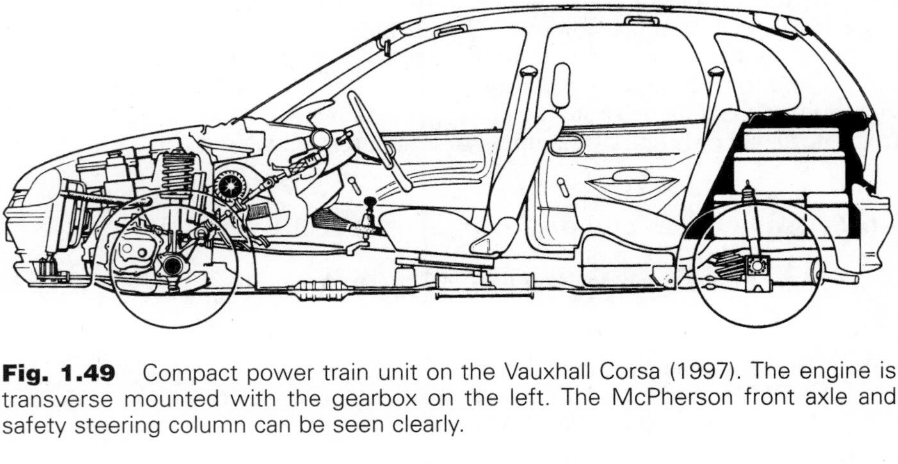

Fig. 1.49 Compact power train unit on the Vauxhall Corsa (1997). The engine is transverse mounted with the gearbox on the left. The McPherson front axle and safety steering column can be seen clearly.

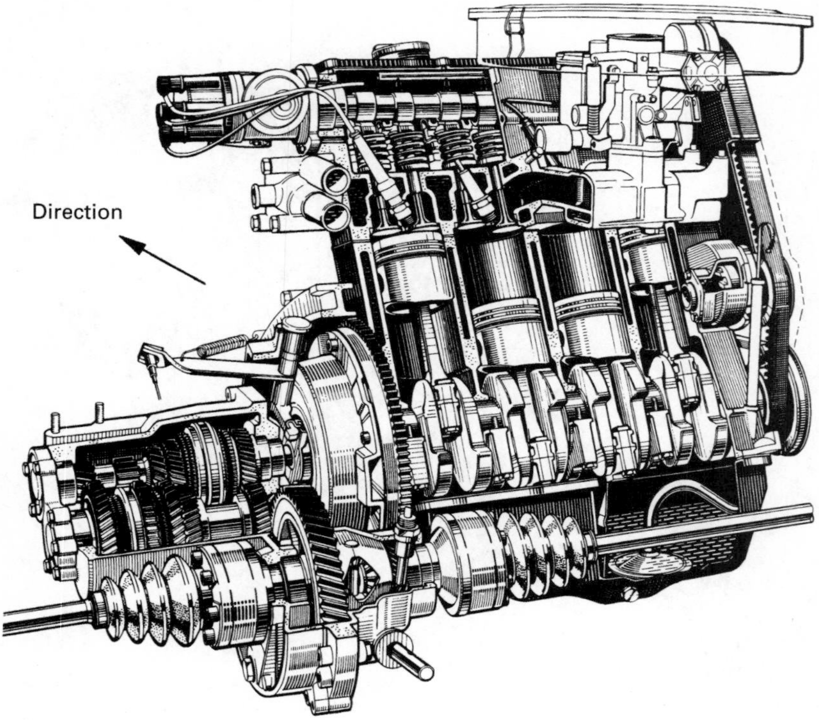

Fig. 1.50 Layout of transverse engine, manual gearbox and differential on the VW Polo. Because the arrangement is offset, the axle shaft leading to the left front wheel is shorter than that leading to the right one. The shifter shaft between the two can be seen clearly. The total mechanical efficiency should be around $n \approx 0.9$.

travel of the wheel. To eliminate the adverse effect of unequal length shafts, passenger cars with more powerful engines have an additional bearing next to the engine and an intermediate shaft, the ends of which take one of the two sliding CV joints with angular mobility (Figs 1.51 and 1.17). Moreover, 'flexing vibration' of the long drive shaft can occur in the main driving range. Its natural frequency can be shifted by clamping on a suppression weight (Fig. 4.1).

1.6.2 Advantages and disadvantages of front-wheel drive

Regardless of the engine position (see Fig. 1.52), front-wheel drive has numerous advantages:

- there is load on the steered and driven wheels;
- good road-holding, especially on wet roads and in wintry conditions – the car is pulled and not pushed (Fig. 1.35);

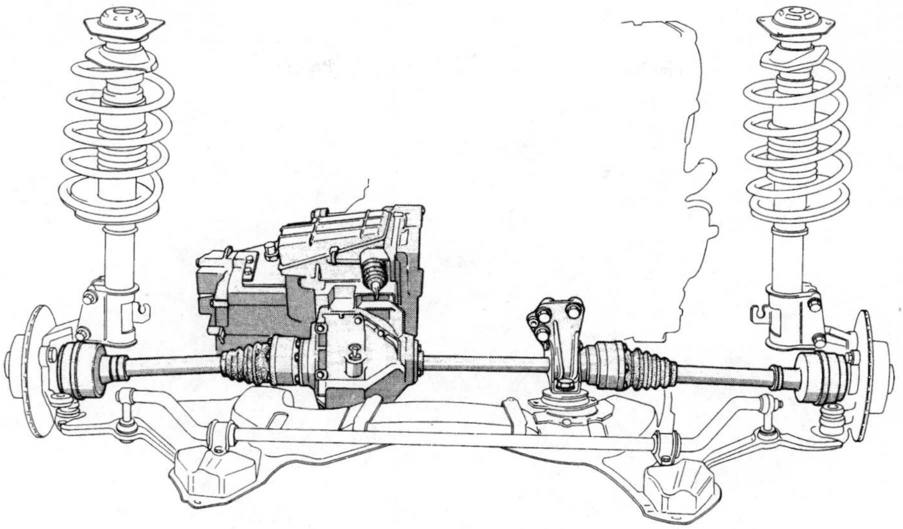

Fig. 1.51 Gearbox unit on the Lancia Thema, located beside the transverse engine and between the front axle McPherson struts. Owing to the high engine performance, the design features two equal-length axle shafts joined by an intermediate shaft. There are also internally ventilated disc brakes.

- good drive-off and sufficient climbing capacity with only few people in the vehicle (Fig. 6.22);
- tendency to understeer in cornering;
- insensitive to side wind (Fig. 3.125);
- although the front axle is loaded due to the weight of the drive unit, the steering is not necessarily heavier (in comparison with standard cars) during driving;
- axle adjustment values are required only to a limited degree for steering alignment (see Section 3.8);
- simple rear axle design – e.g. compound crank or rigid axles – possible;
- long wheelbase making high ride comfort possible;
- short power flow because the engine, gearbox and differential form a compact unit;
- good engine cooling (radiator in front), and an electric fan can be fitted;
- effective heating due to short paths;
- smooth car floor pan;
- exhaust system with long path (important on cars with catalytic converters);
- a large boot with a favourable crumple zone for rear end crash.

The disadvantages are:

- under full load, poorer drive-off capacity on wet and icy roads and on inclines (Figs 1.36 and 6.22);

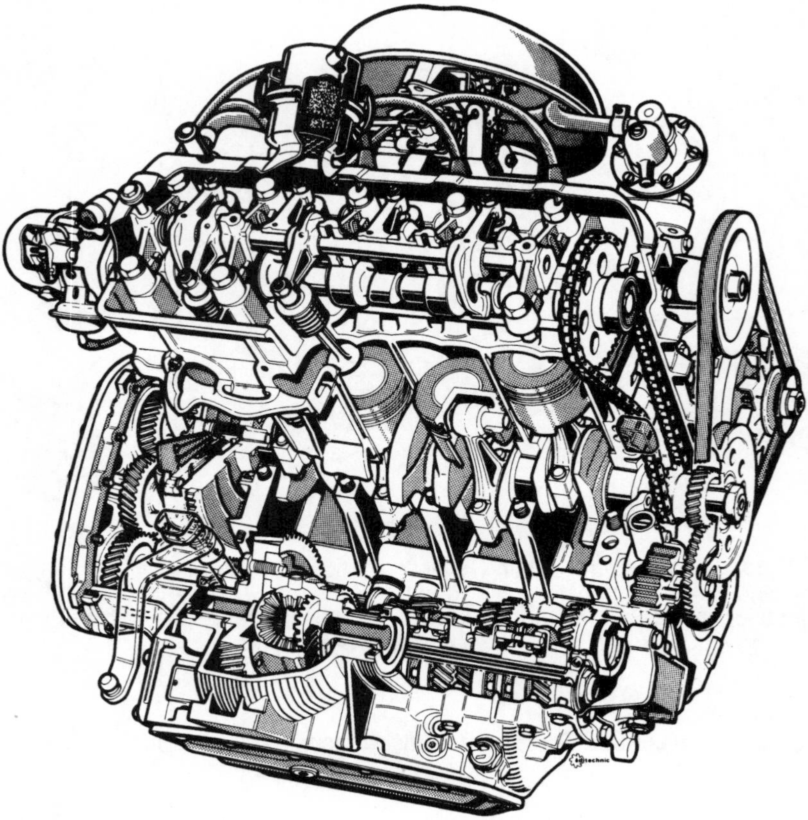

Fig. 1.52 Arrangement of the gearbox beneath the motor, which is inclined towards the rear, and the differential gear placed behind it. A single oil-economy undertakes the supply, in this case, of the driving unit, narrow in its design.
(Works Illustr. Fa Peugeot)

- with powerful engines, increasing influence on steering;
- engine length limited by available space;
- with high front axle load, high steering ratio or power steering is necessary;
- with high located, dash-panel mounted rack and pinion steering, centre take-off tie rods become necessary (Figs 1.57 and 4.39) or significant kinematic toe-in change practically inevitable (Fig. 3.67)
- geometrical difficult project definition of a favourable interference force lever arm and a favourable steering roll radius (scrub radius);
- engine gearbox unit renders more difficult the arrangement of the steering package;
- the power plant mounting has to absorb the engine moment times the total gear ratio (Figs 3.110, 6.20 and Equation 6.36)
- it is difficult to design the power plant mounting (see Ref. [5]) – booming

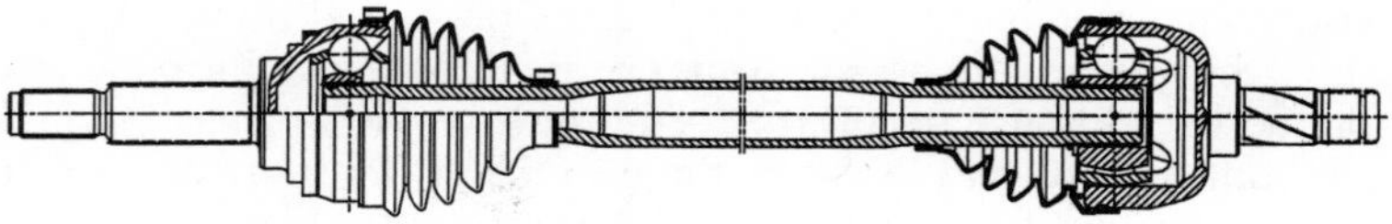

Fig. 1.53 Front-wheel output shaft of GKN Automotive. A constant-velocity sliding joint is used on the gearbox side and a constant-velocity fixed joint is used on the wheel side (Fig. 1.17). The maximum bending angles are 22° for the sliding joint and 47° for the fixed joint. For reasons of weight, the sliding joint is placed directly into the differential and fixed axially by a circlip. A central nut secures attachment on the wheel side. The intermediate shaft is designed as a carburized, shaped hollow shaft.

noises, resonant frequencies in conjunction with the suspension, tip in and let off torque effects etc., need to be suppressed;
- with soft mountings, wavy road surfaces excite the power plant to natural frequency oscillation (so-called 'front end shake', see Section 5.1.3);
- there is bending stress on the exhaust system from the power plant movements during drive-off and braking (with the engine);
- there is a complex front axle, so inner drive shafts need a sliding CV joint (Fig. 1.53);
- the turning and track circle is restricted due to the limited bending angle (up to 50°) of the drive joints (see Section 3.7.2);
- high sensitivity in the case of tyre imbalance and non-uniformity on the front wheels;
- higher tyre wear in front, because the highly loaded front wheels are both steered and driven;
- poor braking force distribution (about 75% to the front and 25% to the rear);
- complex gear shift mechanism which can also be influenced by power plant movements.

The disadvantage of the decreased climbing performance on wet roads and those with packed snow can be compensated with a drive slip control (ASR, see Chapter 6 in Ref. 7) or by shifting the weight to the front axle. On the XM models, Citroën moved the rear axle a long way to the rear resulting in an axle load distribution of about 65% to the front and 35% to the back. The greater the load on the front wheels, the more the car tends to understeer, causing adverse steering angles and heavy steering, which makes power steering mandatory (see Section 4.2.5).

1.6.3 Driven front axles

The following are fitted as front axles on passenger cars, estate cars and light commercial vehicles:

- double wishbone suspensions;
- multi-link axles;
- McPherson struts, and (only in very few cases);
- damper struts.

On double wishbone suspensions the drive shafts require free passage in those places where the coil springs are normally located on the lower suspension control arms. This means that the springs must be placed higher up with the disadvantage that (as on McPherson struts) vertical forces are introduced a long way up on the wheel house panel. It is better to leave the springs on the lower suspension control arms and to attach these to the stiffer body area where the upper control arms are fixed. Shock absorbers and springs can be positioned behind the drive shafts (see Fig. 1.54) or sit on split braces, which grip round the shafts and are jointed to the lower suspension control arms (Fig. 1.55). The axle is flatter and the front end (bonnet contour) can be positioned further down. The upper suspension control arms are relatively short and have mountings that are wide apart. This increases the width of the engine compartment and the spring shock absorber unit can also be taken through the suspension control arms; however, sufficient clearance to the axle shaft is a prerequisite.

Fig. 1.54 Double wishbone front axle assembly of the Audi A4. The Audi A6 of 1997, the Audi A8 (1996) and the VW Passat of 1996 are similar. Four individual transverse 'arms' on each side form what is effectively a double wishbone arrangement which provides lateral and longitudinal wheel location. The two upper members (1 and 2) are attached to the spheroidal graphite iron hollow-section stub-axle post (18) by low-friction ball and socket joints. The track rod (3) provides the steering input through a horizontal extension of this stub-axle post which forms a steering arm. The two lower suspension members consist of the radius arm (4) and the transverse arm (5). This latter must be capable of reacting high loads from the anti-roll bar (6) and spring/damper (7) attachment points. The co-axial spring/damper assembly incorporates a polyurethane rubber bump-stop, as well as the hydro-mechanical tension rod stop (Fig. 5.51). The spring/damper unit (7) and the inner bearings of the upper members (1) and (2) are mounted on the upper suspension bracket.

The inner ends of suspension members (4 and 5) are located by substantial rubber mountings on the inside of the sub-frame (10). The rear mounting (11) is hydraulically damped to absorb any harshness associated with radial tyres. The vehicle body is mounted on four rubber mountings (12 to 15) of specified elasticity to ensure a high standard of ride comfort.

The inner drive shafts are located to the rear of the spring dampers (7) and are connected to the drive-line by 'tripot' flexible couplings (16). The outer ends of the drive shafts transmit the drive to the wheels through double-row angular contact bearings. The inner races of these bearings are integral with the wheel hubs.

The hydraulically assisted steering rack is mounted on the vehicle's scuttle, with the steering damper (17) located on one side of the steering housing, and the other side attached to the steering rack.

The high location of the wheel-joint facilitates space saving and a consequent reduction of the lever-arm forces, and allows the inner valences of the mudguard to be located further outboard.

The advantages of this type of four-link suspension include the location of the points E and G of the paired arms 1 and 2, likewise 4 and 5 (Fig. 3.145), which are subjected to outward thrusts resulting from steering input to the steering-arm, which are thereby compressed through $r = 10$ mm (see para 3.9.3). Moreover the high location of the point E (Fig. 1.5) – together with the negative steering roll radius $r = -7$ mm (Fig. 3.106) – helps to reduce the loads in all components of the front suspension system.

Other design parameters of the suspension arrangement are:

King pin inclination	$\varepsilon = 30'$
Caster angle	$\tau = +3°50'$
Camber angle	$\sigma = +3°45'$
Caster linear trail	$\eta = +5.5$mm

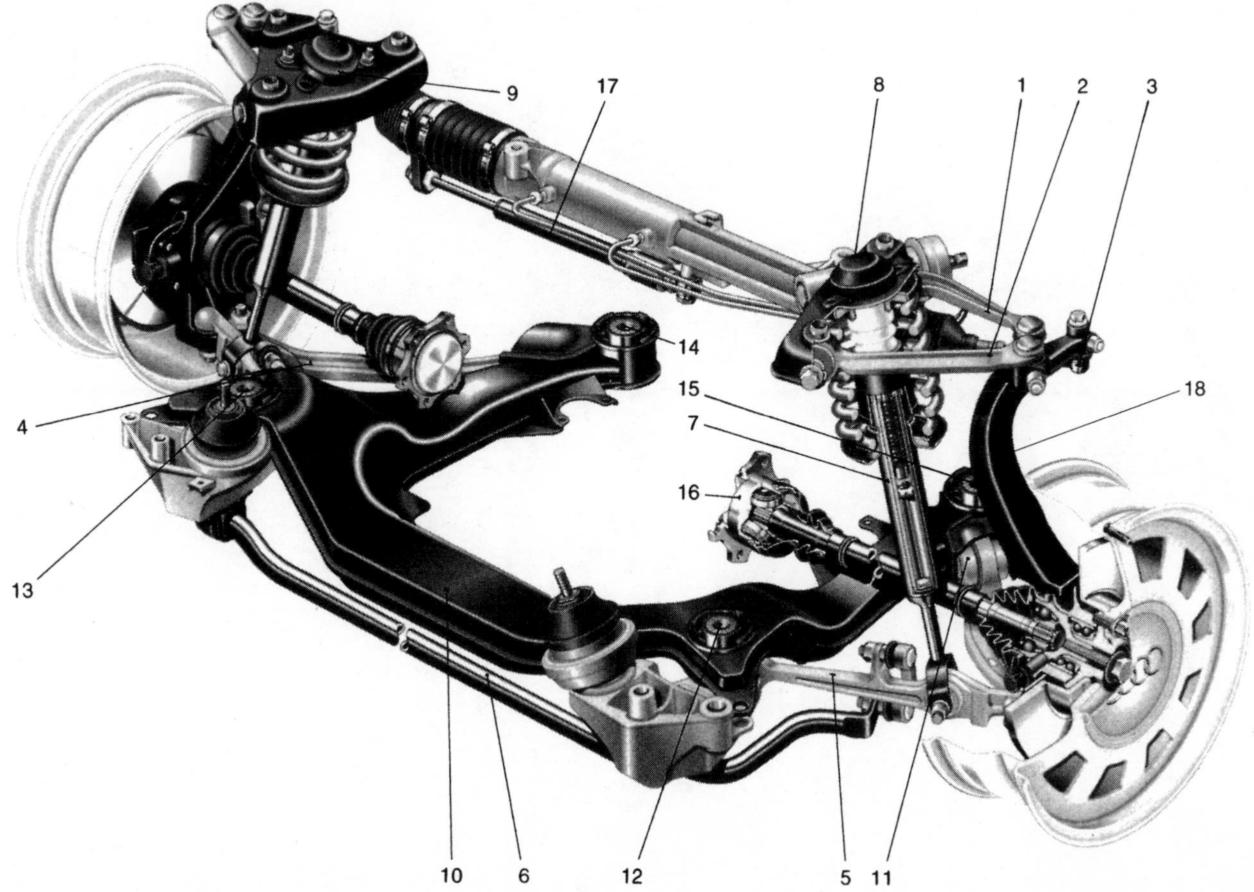

9
17
8
1
2
3
14
4
15
18
7
16
13
10
6
12
5
11

Fig. 1.55 Double wishbone front suspension on the Honda models Prelude and Accord with short upper wishbones with widely spaced bearings, lower transverse control arms and longitudinal rods whose front mounts absorb the dynamic rolling stiffness of the radial tyres. The spring shock absorbers are supported via fork-shaped struts on the transverse control arms and are fixed within the upper link mounts. This point is a good force input node. Despite the fact that the upper wheel carrier joint is located high, which gives favourable wheel kinematics, the suspension is compact and the bonnet can be low to give aerodynamic advantages. The large effective distance *c* between the upper and lower wheel hub carrier joints seen in Fig. 1.5 results in low forces in all mounts and therefore less elastic deflection and better wheel control.

Fig. 1.56 Lancia front axle. The McPherson strut consists of the wheel hub carrier 1 and the damping part 2; the two are connected by three screws. The lower spring seat 3 sits firmly on the outer tube and also acts as a buffer for the supplementary spring 4. This surrounds the outer tube 2 giving a longer bearing span (path *l–o*, Fig. 1.11). The supporting bearing 5 is arranged diagonally and thus matches the position of the coil spring which is offset to reduce damping friction. The rubber bearing 6 absorbs the spring forces, and the rubber bearing 7 absorbs the forces generated by the damping. Disc 8 acts as a compression buffer and plate 9 acts as a rebound buffer for this elastic bearing. Both parts come into play if the damping forces exceed certain values.

The centre of the CV joint 10 lies in the steering axis and the wheel hub 11 fits onto a two-row angular (contact) ball bearing. Guiding joint 12 sits in a cone of the wheel hub carrier 1 and is bolted to the lower transverse control arm 13. Inelastic ball joints provide the connection to the anti-roll bar 14. The steering axis inclination σ between the centre point of the upper strut mount and guiding joint 12 and the (here slightly positive) kingpin offset at ground (scrub radius) r_σ are included.

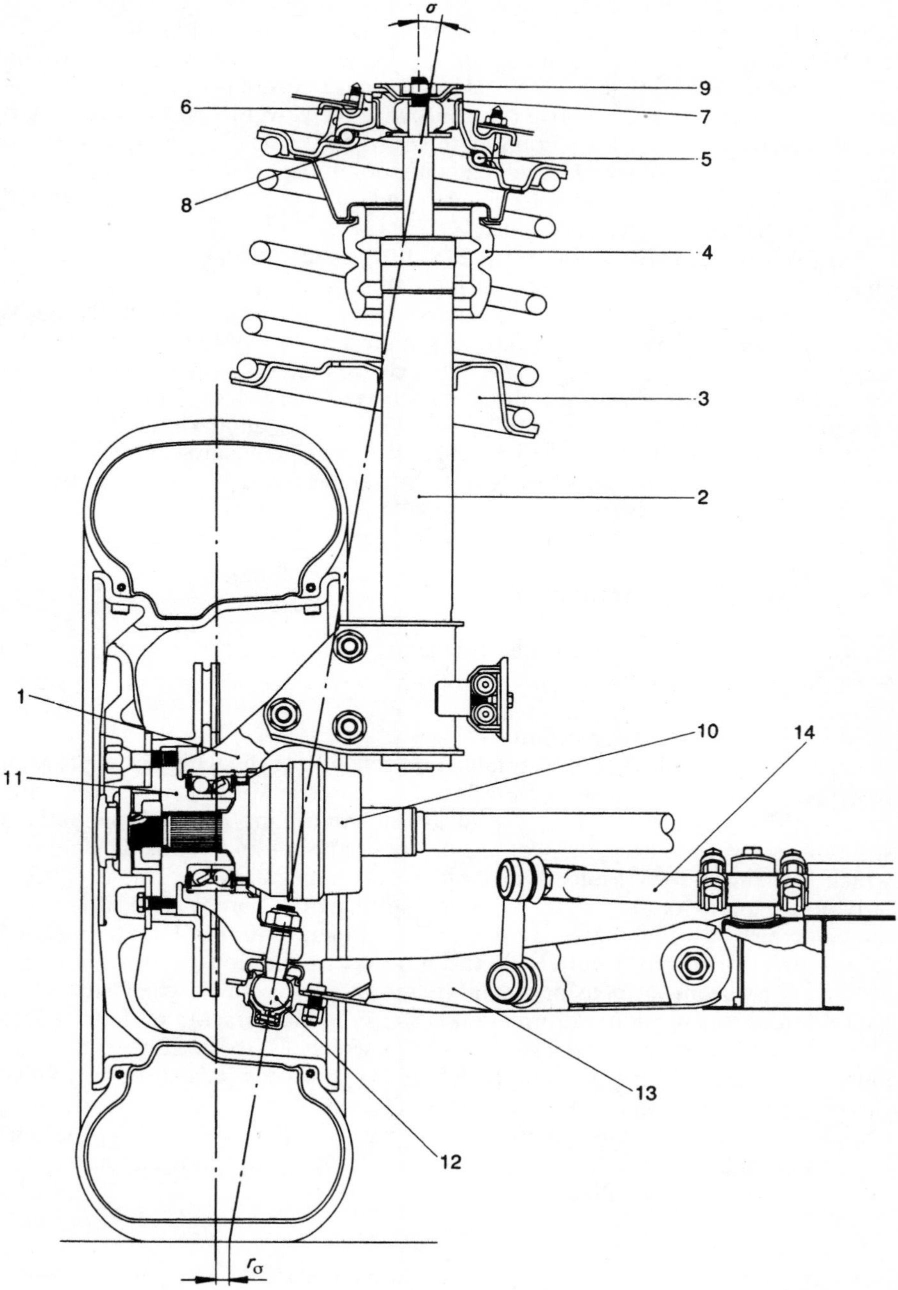
σ
9
6
7
5
8
4
3
2
1
10
14
11
13
12
r_σ

Due to the slight track width change, the change of camber becomes favourable. Furthermore, the inclination of the control arms provides an advantageous radius arm axis position and anti-dive when braking (see also Fig. 1.75).

Most front-wheel drives coming on to the market today have McPherson struts. It was a long time after their use in standard design cars that McPherson struts were used at the front axle on front-wheel drive vehicles. The drive shaft requires passage under the damping part (Fig. 1.56). This can lead to a shortening of the effective distance *l–o*, which is important for the axle (Fig. 1.11), with the result that larger transverse forces $F_{Y,C}$ and $F_{Y,K}$ occur on the piston and rod guide and therefore increase friction.

On front-wheel drive vehicles there is little space available to fit rack and pinion steering. If the vehicle has spring dampers or damper struts, and if the steering gear is housed with short outer take-off tie rods, a toe-in change is almost inevitable (Figs 4.4 and 3.67). A high steering system can readily be attached to the dash panel (Fig. 1.57), but a centre take-off is then necessary and the steering system becomes more expensive (Figs 4.1, 4.11 and 4.39). Moreover, the steering force applied to the strut is approximately halfway between mountings E and G (Fig. 1.11). The inevitable, greater yield in the transverse direction increases the steering loss angle and makes the steering less responsive and imprecise.

1.6.4 Non-driven rear axles

If rear axles are not driven, use can frequently be made of more simple designs of suspension such as twist-beam or rigid.

1.6.4.1 Twist-beam suspension

There are only two load paths available on each side of the wheel in the case of twist-beam axles. As a result of their design (superposed forces in the links, only two load paths), they suffer as a result of the conflicting aims of longitudinal springing – which is necessary for reasons of comfort – and high axle rigidity – which is required for reasons of driving precision and stability. This is particularly noticeable with the loss of comfort resulting from bumpy road surfaces. If the guide bearings of the axle are pivoted, the superposition of longitudinal and lateral forces should particularly be taken into consideration. As a result of the design, twist-beam suspensions exhibit unwanted oversteer when subject to lateral forces as a result of deformation of the swinging arms. In order to reduce the tendency to oversteer, large guide bearings which, as 'toe-in correcting' bearings, permit lateral movements of the whole axle body towards understeer when subject to a lateral force are provided. As the introduction of longitudinal and lateral forces into the body solely occurs by means of the guide bearings, it must be ensured that the structure of the bodywork is very rigid in these places (see Figs. 1.30 and 1.58, also Sections 3.6.3 and 3.6.4).

1.6.4.2 Rigid axle

Non-driven rigid axles can be lighter than comparable independent wheel suspensions. Their advantages outweigh the disadvantages because of the almost

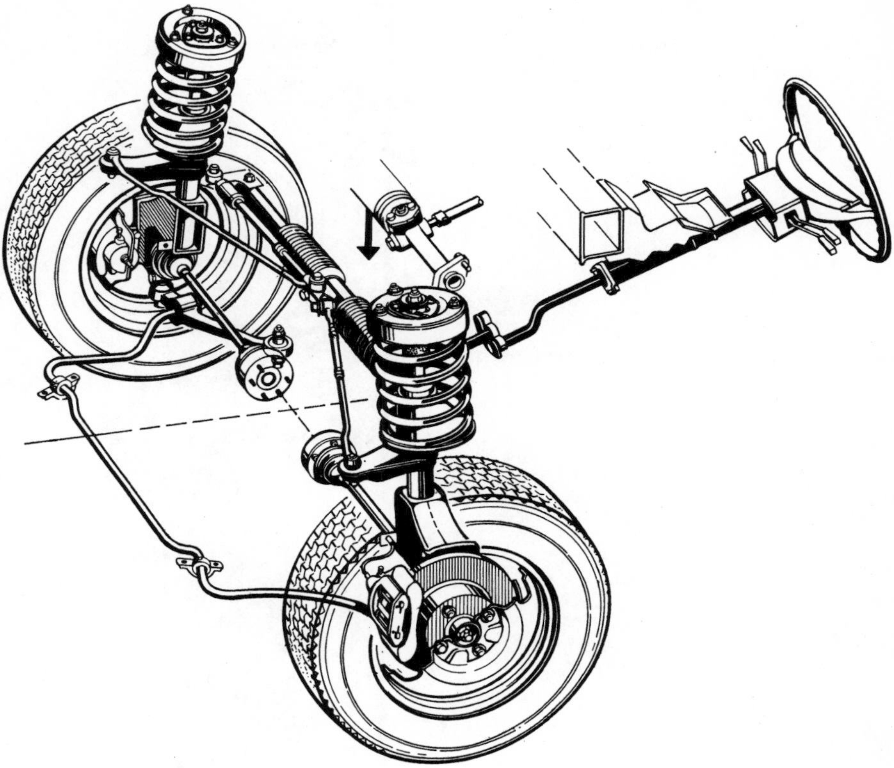

Fig. 1.57 Driven McPherson front axle on the Audi 6 (Audi 100, 1991). The dynamic rolling hardness of the radial tyre is absorbed by the rubber bearings shown in Fig. 3.85, which sits in the lower transverse control arms. The inner sleeves of these bearings take the arms of the anti-roll bar, which also act as a trailing link (classic McPherson construction). To avoid greater toe-in changes when the wheels are at full bump/rebound-travel, centre take-off tie rods are used on the rack and pinion steering higher up and in the centre (Fig. 4.39). Together with these rods, the steering damper located on the right is fastened to the end of the steering rack. The engine is mounted longitudinally, which means the drive shafts are of equal length (see Section 3.6.5.3).

The development of the axle since 1997 is shown in Fig. 1.54.

non-variable track and camber values during drive. Figure 1.24 illustrates an inexpensive yet effective design:

- axle casing in steel tubing;
- suspension on single leaf springs.

The lateral and longitudinal wheel control characteristics are sufficient for passenger cars in the medium to small vehicle range and delivery vehicles. The

Fig. 1.58 Twist-beam suspension of the Audi A6 (1997). An advantageously large support width of the guide on the links – important for force application – was chosen because of the overhung arrangement. The flexurally resistant, but torsionally soft V section profile of the axle is in an upright position in order to ensure that the suspension has roll understeer properties through the high position of the centre of thrust of the profile. The instantaneous centre height is 3.7 mm and the toe-in alteration is 0.21 min/mm. Braking-torque compensation of 73% is reached. The stabilizer situated in front of the axis of rotation increases the lateral rigidity of the axle design, because it accepts tension forces upon the occurrence of lateral wheel forces. The linear coil springs mounted on noise-insulating moulded rubber elements on both sides are separated from the shock absorbers to allow the maximum loading width of the boot as a result of their location under the side rails. The gas-pressure shock absorbers support additional springs made of cellular polyurethane which act softly, through specific rigidity balancing, to avoid uncomfortable changes in stiffness when reaching the limits of spring travel. Owing to the rigid attachment of the shock absorbers to the bodywork, these also work at low amplitudes; so-called 'parasitic' springing resulting from the unwanted flexibility of wheel suspension or bodywork components is thereby reduced.

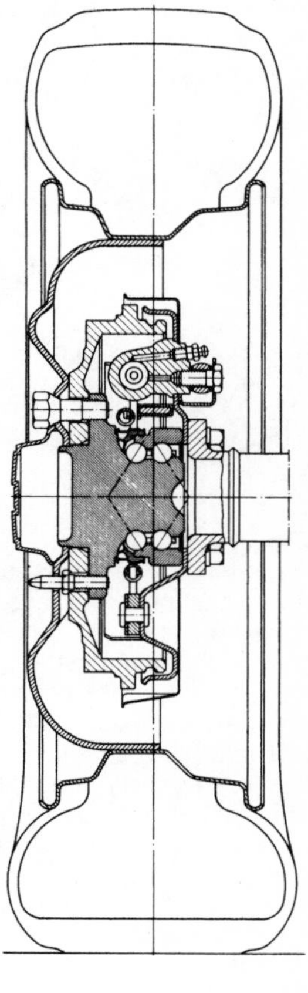

Fig. 1.59 Rear wheel bearing on the Fiat Panda with a third-generation, two-row angular (contact) ball bearing. The wheel hub and inner ball bearing ring are made of one part, and the square outer ring is fixed to the rigid axle casing with four bolts (picture: SKF).

resultant hard springing is acceptable and may even be necessary because of the load to be moved. The wheel bearing can be simple on such axles (Fig. 1.59). Faster, more comfortable vehicles, on the other hand, require coil springs and, for precise axle control, trailing links and a good central guide (Fig. 1.60) or Panhard rod. This is generally positioned behind the axle (Fig. 1.61).

1.6.4.3 Independent wheel suspension

An independent wheel suspension is not necessarily better than a rigid axle in terms of handling properties. The wheels may incline with the body and the lateral grip characteristics of the tyres decrease (Figs 3.53 and 3.54), and there are hardly any advantages in terms of weight (see Section 6.1.3). This suspension usually needs just as much space as a compound crank axle.

Among the various types, McPherson struts (Fig. 1.12), semi-trailing or trailing

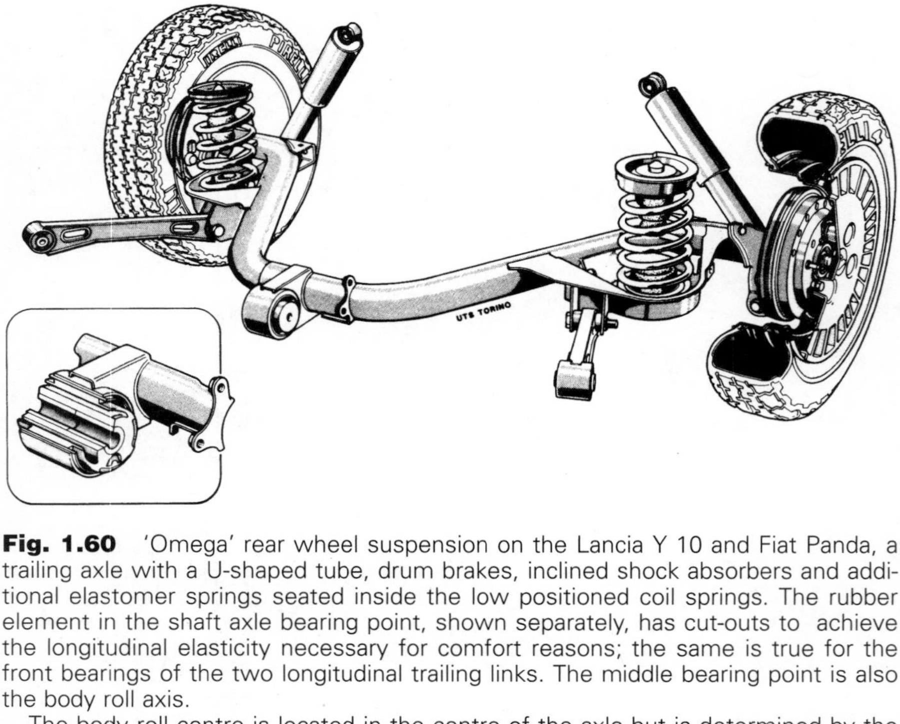

Fig. 1.60 'Omega' rear wheel suspension on the Lancia Y 10 and Fiat Panda, a trailing axle with a U-shaped tube, drum brakes, inclined shock absorbers and additional elastomer springs seated inside the low positioned coil springs. The rubber element in the shaft axle bearing point, shown separately, has cut-outs to achieve the longitudinal elasticity necessary for comfort reasons; the same is true for the front bearings of the two longitudinal trailing links. The middle bearing point is also the body roll axis.

The body roll centre is located in the centre of the axle but is determined by the level of the three mounting points on the body. The lateral forces are absorbed here. The angled position of the longitudinal trailing links is chosen to reduce the lateral force oversteering that would otherwise occur (shown in Figs 1.31 and 3.79). The coil springs are located in front of the axle centre and so have to be harder, with the advantage that the body is better supported on bends (for details, see Section 3.4 in Ref. [2]).

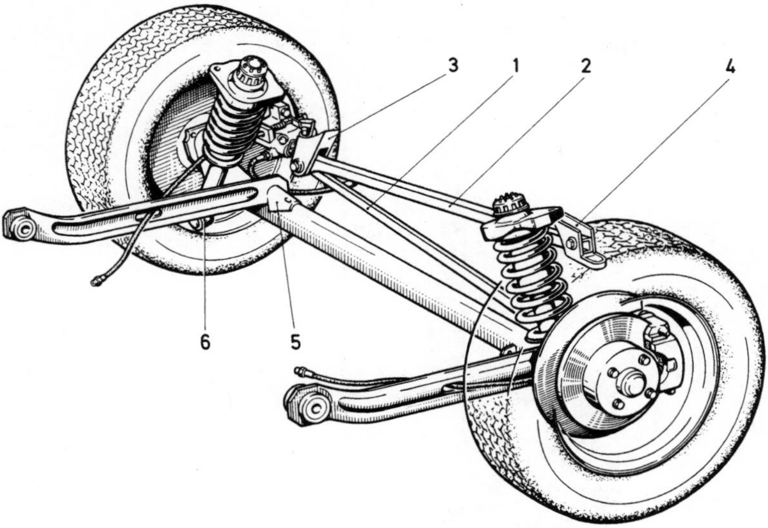

Fig. 1.61 Torsion crank axle on the Audi A6 (Audi 100, 1991) with spring dampers fixed a long way out at points 6 and which largely suppress body roll vibrations.

The longitudinal control arms therefore had to be welded further in to the U-profile acting as a cross-member and reinforced by shoe 5. The U-profile is also raised at the side to achieve higher torsional resistance. The anti-roll bar is located inside the U-profile.

Brace 2 distributes the lateral forces coming from the Panhard rod 1 to the two body-side fixing points 3 and 4. Bar 1 is located behind the axle, and the lateral force understeering thus caused and shown in Fig. 3.81 could be largely suppressed by the length of the longitudinal control arms. Furthermore, it was possible to increase the comfort and to house an 80 l fuel tank as well as the main muffler in front of the axle.

The only disadvantage is that the link fixing points, and therefore the body roll axis O_r, moves further forward and this reduces the 'anti-dive' described in Fig. 3.159, and that the suspension requires much space when assembled.

link axles (Figs 1.2, 1.13 and 1.63) and – having grown in popularity for some years now – double wishbone suspensions, mostly as so-called multi-link axles (Figs 1.1, 1.8 and 1.62; Section 5.3 in Ref. [2]) are all used. The latter are currently the best solution, due to:

- kinematic characteristics;
- elastokinematic behaviour;
- space requirements;
- axle weight;
- the possibility of being able to retrofit the differential on four-wheel drive (Figs 1.77 and 1.1).

(See also Section 1.4.3.).

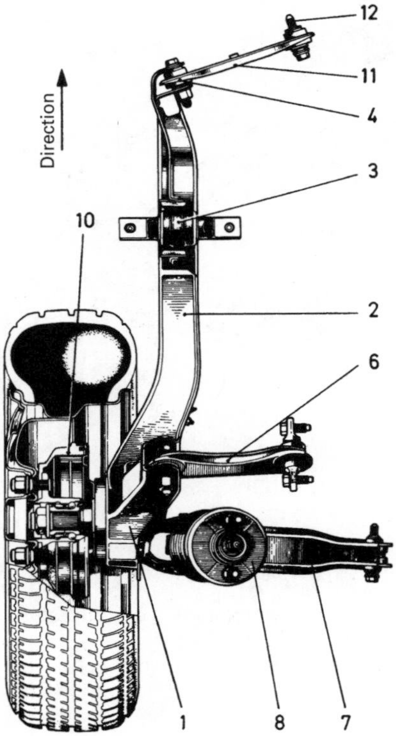

Fig. 1.62 Top view of the double wishbone rear axle on the Honda Civic. The trailing arm 2, which is stiff under flexure and torsion, and the wheel hub carrier 1 form a unit and, along with the two widely spaced lower transverse control arms 7 and 11, ensure precise wheel control and prevent unintentional toe-in changes. The rubber bearing in point 3, which represents the so-called 'vehicle roll axis' O_r, provides the real longitudinal wheel control of the axle (Fig. 3.159). The lateral control of wheel carrier 1 is performed by the short upper transverse control arm 6 and the longer lower one 7, which accepts the spring shock absorber 8 in point 9. The length difference in the control arms gives favourable camber and track width change (Figs 3.19 and 3.49).

During braking, bearing 3 yields in the longitudinal direction and, due to the angled position of the links 11 when viewed from the top, the front point 4 moves inwards (Fig. 3.82) and the wheel goes into toe-in. Behaviour during cornering is similar: the axle understeers due to lateral force and body roll (see Sections 3.6.3 and 3.6.4 and Figs 1.1 and 1.77). The wheel is carried by 'third-generation' angular (contact) ball bearings on which the outside ring is also designed as a wheel hub. In models with smaller engines, brake drums (item 10) are used, which are fixed to the wheel hub.

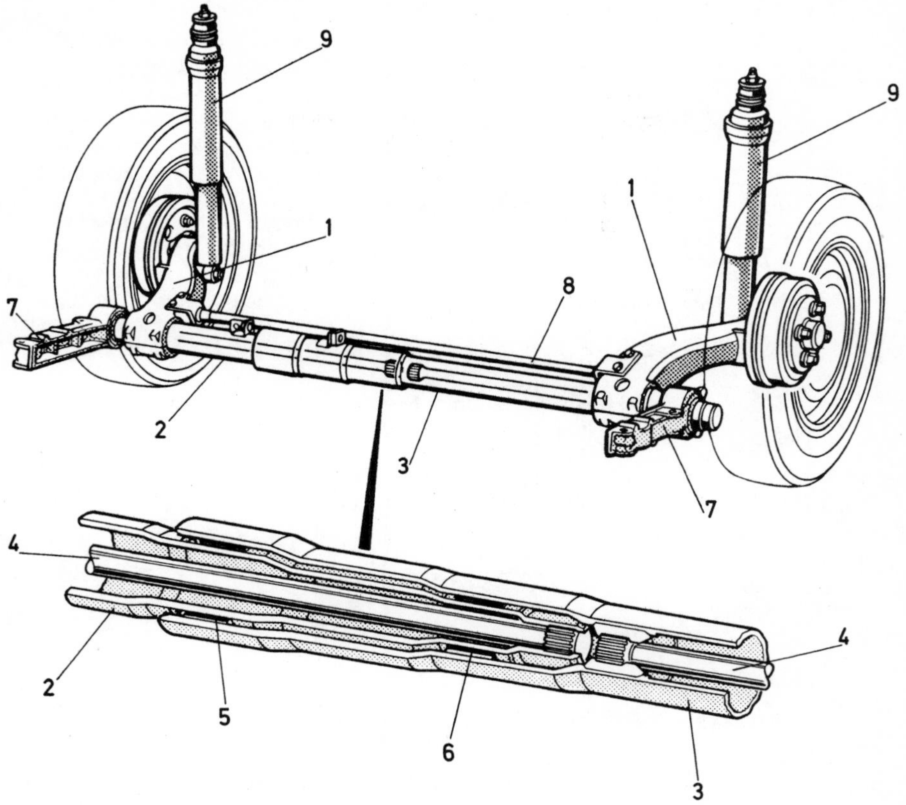

Fig. 1.63 Compact trailing arm rear axle, fitted by Renault to less powerful medium-sized vehicles. The short torsion bar springs grip into the guide tubes 2 and 3 in the centre of the vehicle. Parts 2, 3 and 4 are jointly subjected to torsional stresses and so the torsional stiffness of the transverse tubes contributes to the spring rate. On the outside, the cast trailing arms 1 are welded to the transverse tubes, which (pushed into each other) support each other on the torsionally elastic bearings 5 and 6. This creates a sufficiently long bearing basis, which largely prevents camber and toe-in changes when forces are generated.

The entire assembly is fixed by the brackets 7 which permits better force transfer on the body side sill. Guide tubes 2 and 3 are mounted in the brackets and can rotate, as well as the outer sides of the two torsion bars 4. The two arms thus transfer all vertical forces plus the entire springing moment to the body. The anti-roll bar 8 is connected to the two trailing arms via two U-shaped tabs. The two rubber bearings 5 and 6 located between the tubes 2 and 3 also contribute to the stabilizing effect.

The bump and rebound travel stops are fitted into the shock absorber 9 (see Section 5.6.8). As shown in Fig. 1.2, on the newer models the dampers would be inclined so that they can be fixed to the side members of the floor pan which also leads to more space between the wheel housings.

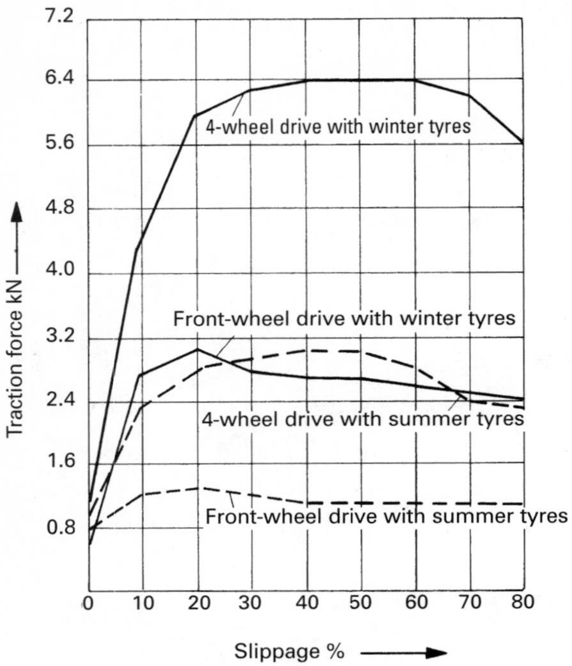

Fig. 1.64 With a loaded Vauxhall Cavalier on compacted snow ($\mu_{X,W} = 0.2$) driving forces are measured on the flat as a function of the slip (Fig. 2.33). The illustration shows the advantage of four-wheel drive, and the necessity, even with this type of drive, of fitting correct tyres. Regardless of the type of drive, winter tyres also give shorter braking (stopping) distances on these road surface conditions.

1.7 Four-wheel drive

In four-wheel drives, either all the wheels of a passenger car or commercial vehicle are continuously – in other words permanently – driven, or one of the two axles is always linked to the engine and the other can be selected manually or automatically. This is made possible by what is known as the 'centre differential lock'. If a middle differential is used to distribute the driving torque between the front and rear axles, the torque distribution can be established on the basis of the axle–load ratios, the design philosophy of the vehicle and the desired handling characteristics. That is why Audi choose a 50%:50% distribution for the V8 Quattro and Mercedes-Benz choose a 50%:50% distribution for M class off-road vehicles, whereas Mercedes-Benz transmits only 35% of the torque to the front axle and as much as 65% to the rear axle in vehicles belonging to the E class.

This section deals with the most current four-wheel drive designs. In spite of the advantages of four-wheel drive, suitable tyres – as shown in Fig. 1.64 – should be fitted in winter.

1.7.1 Advantages and disadvantages

In summary, the advantages of passenger cars with permanent four-wheel drive over those with only one driven axle are:

- better traction on surfaces in all road conditions, especially in wet and wintry weather (Figs 1.64, 1.65 and 1.66);
- an increase in the drive-off and climbing capacity regardless of load;
- better acceleration in low gear, especially with high engine performance;
- reduced sensitivity to side wind;
- stability reserves when driving on slush and compacted snow tracks;
- better aquaplaning behaviour;
- particularly suitable for towing trailers;
- balanced axle load distribution;
- reduced torque steer effect;
- even tyre wear.

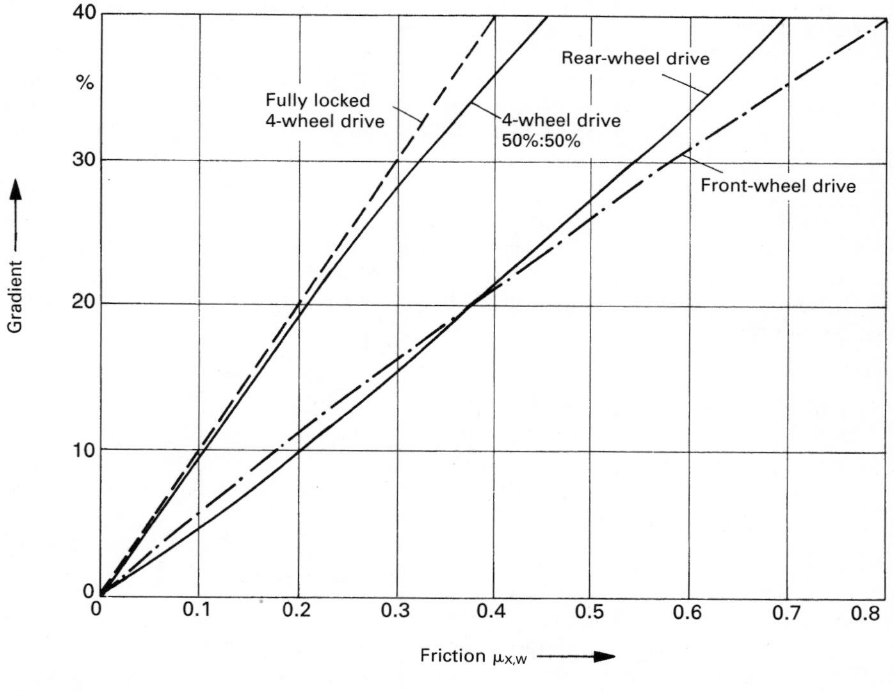

Fig. 1.65 Hill-climbing capacity on a homogeneous surface with front, rear-wheel and four-wheel drive, and with locked centre differential and a driving force distribution of 50%/50% on four-wheel drive. Of the cars studied, the front/rear axle load distribution was (Fig. 1.36):

front-wheel drive 57%/43%
rear-wheel drive 51%/49%
four-wheel drive 52%/48%
(see also Fig. 6.22).

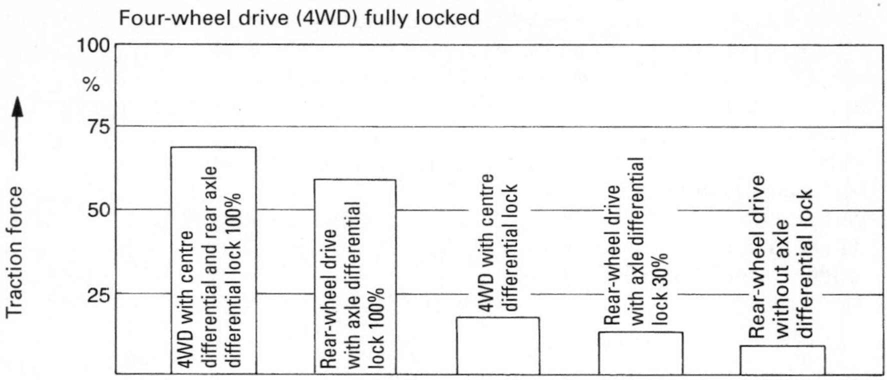

Fig. 1.66 Influence of the type of drive and differential lock on the propulsion force with 'μ split', in other words a slippery road surface with $\mu_{X,W}$ = 0.1/0.8 on one side only. 100% locking of the rear axle differential gives most benefits.

Some car manufacturers offer this option as *ASR* (or *EDS*) or using a hydraulic manual selection clutch (see details in Chapter 6 of Ref. [7]). However, only 25% to 40% locking is provided on the multi-disc limited-slip differentials that have usually been fitted on vehicles to date (see Section 5.3.2 in Ref. 9 and Section 6.4 in Ref. 8).

According to EU Directive 70/156/EWG, a 'towed trailer load' of 1.5 times the permissible total weight has been possible for multi-purpose passenger vehicles (four-wheel passenger vehicles) since 1994.

However, the system-dependent, obvious disadvantages given below should not be ignored:

- acquisition costs;
- around 6% to 10% higher kerb weight of the vehicle;
- generally somewhat lower maximum speed;
- 5% to 10% increased fuel consumption;
- in some systems, limited or no opportunity for using controlled brake gearing, for instance for anti-locking or ESP systems;
- not always clear cornering behaviour;
- smaller boot compared with front-wheel drive vehicles.

Predictability of self-steering properties even in variable driving situations, traction, toe-in stability and deceleration behaviour when braking, manoeuvrability, behaviour when reversing and interaction with wheel control systems are the principal characteristics of the vehicle movement dynamics which are taken into consideration for an assessment of four-wheel drive systems.

To transmit the available engine torque to all four wheels, interaxle differentials (such as cone, planet or Torsen differentials), which are manually or automatically lockable, or clutches (such as sprag, multi-disc or visco clutches) must be installed on the propshaft between the front and rear axles. Differentials must be present on both drive axles. However, on roads with different coefficients of friction on the left and right wheels, known as 'μ-split', and with traditional differentials, each driven axle can, at most, transmit double the propulsion force of the wheels running on the side with the lower coefficient of friction (μ-low).

Higher driving forces can be achieved with an 'axle differential lock' or controlled wheel brake gearing which creates the need for 'artificial' torque on the spinning wheel. Differential-locking can only be 100% effective on the rear axle as, at the front, there would no longer be problem-free steering control. The lock partially or completely stops equalization of the number of revolutions between the left and right wheel of the respective axle and prevents wheelspin on the μ-low-side.

In passenger cars, automatic locking differentials are used between front and rear axles. These can operate mechanically (multi-disc limited slip differentials (see Section 5.3.2 in Ref. [9], Torsen differential, Fig. 1.71) or based on fluid friction (visco lock, Fig. 1.74) and produce a locking degree of usually 25% to 40%. Higher values severely impede cornering due to the tensions in the power train (Fig. 1.69) Nevertheless, up to 80% locking action can be found in motor sport.

The locking action of uncontrolled or slip-dependent differential locks necessitates increased expenditure with the use of brake-power control systems (ABS, ESP). Thus in the case of the visco lock, a free-wheel clutch is required that is engaged during reversing. Here the advantage of controlled differentials (Haldex clutch, automatically controlled locking differential, see Sections 1.7.4 and 1.7.5) becomes apparent: They can be used to maximum effect in any operating conditions and with any brake-power control system, because the locking action is produced by an electronically controlled, hydraulically activated multi-disc clutch (Fig. 1.67).

Traditional differential locks are increasingly being replaced because of the

Fig. 1.67 The Mercedes G all-terrain vehicle, according to DIN 70, a so-called 'all-purpose passenger car', has high ground clearance and short overhangs both front and rear. This, together with the large ramp angles (α_f, α_r) and the overhang angle β, makes it particularly suitable for off-road driving.

use of wheel control systems in both front- and rear-wheel drive as well as four-wheel drive vehicles. In these systems, the wheel speed is measured, usually with the use of ABS sensors. If the speed of a wheel is established, this wheel is retarded by means of the wheel braking device. In the case of the differential, this corresponds to the build-up of torque on the side of the spinning wheel and it can now transmit torque at the higher coefficient of friction up to the adhesion limit of the wheel. Volkswagen AG calls this system electronic differential lock, as front-wheel drive forces which correspond to those of a driven axle with differential lock and 100% locking action, and which can even be exceeded in intelligent (slip-controlled) systems, such wheel gearing systems produce. The system can ensure that the driving torque that is to be applied to the side with the retarded wheel is equal to the torque on the side with the higher coefficient of friction. This 'lost torque' must be generated by the transmission, on the one hand, and retarded by the wheel brake, on the other, so that loss of engine power and heating of the wheel brakes are produced. The braking temperatures are calculated on the basis of the braking torque and period of application of the brakes. If the temperatures calculated exceed the permissible limits, application of the brakes is discontinued during the front-wheel drive phase until a calculated cooling of the system has taken place; the transmission then corresponds to that found in a conventional vehicle.

Another possibility for maximum utilization of grip is afforded by traction control systems in which engine power is reduced by means of the throttle, injection and ignition point so that the spinning wheels work in the region of lower slip and consequently higher adhesion. Both systems are used together, even without four-wheel drive. In models of the E and M class with an electronic traction system (ETS), Mercedes-Benz uses electronic locks instead of mechanical differential locks.

1.7.2 Four-wheel drive vehicles with overdrive

In four-wheel drive vehicles with overdrive the middle differential is not used. The engine torque is distributed to all four wheels by means of a clutch on the propshaft, as required. The clutch can be engaged manually, or automatically in response to slip. With the use of sprag clutches, which are usually engaged manually, the torque is transmitted in a fixed ratio between the front and rear axles; multi-disc or visco clutches permit variable torque distribution. As these systems have essential similarities with permanent four-wheel drive varieties, they are discussed in Section 1.7.4.

With sprag-clutch engaged transmissions, the design complexity, and therefore the costs, are lower than on permanent drive. Usually there is no rear axle differential lock, which is important on extremely slippery roads; while this results in price and weight advantages, it does lead to disadvantages in the traction.

Front-wheel drive is suitable as a basic version and the longitudinal engine has advantages here (Fig. 1.48). With the transverse engine, the force from the manual gearbox is transmitted via a bevel gear and a divided propshaft, to the rear axle with a differential (Fig. 1.68). There is relatively little additional

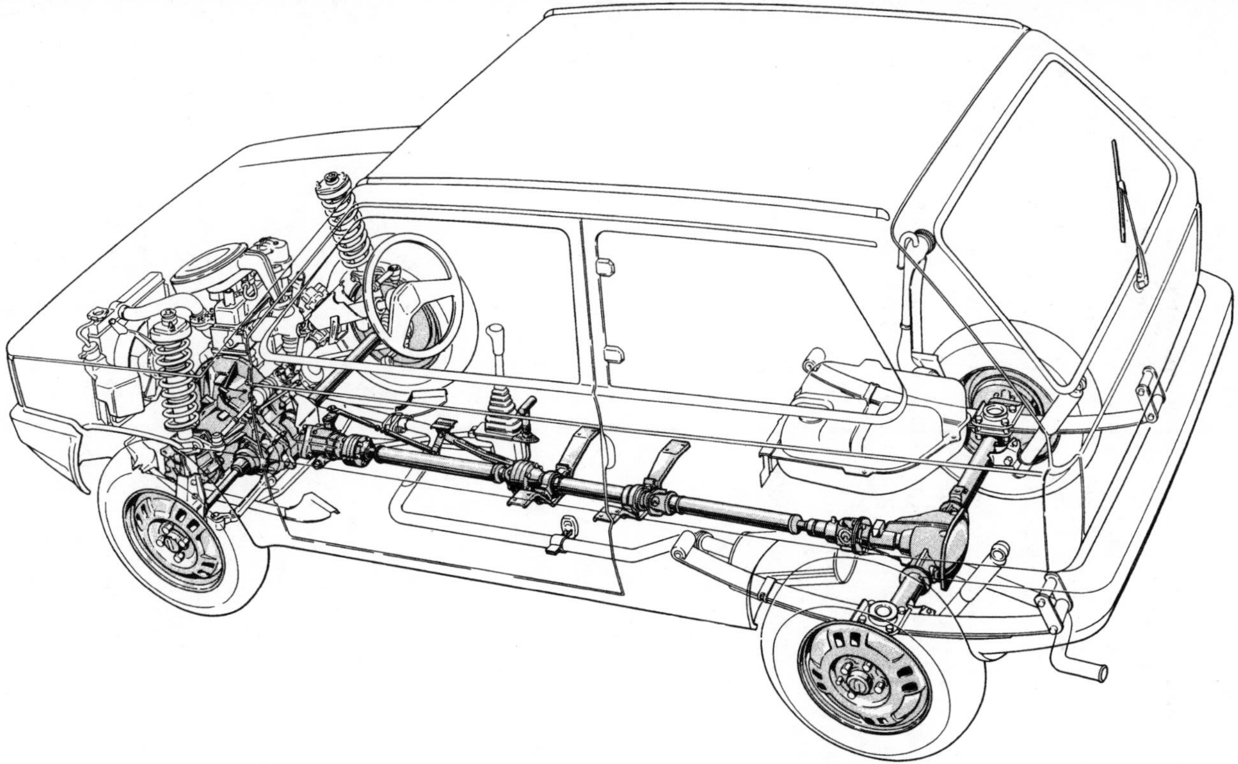

Fig. 1.68 The Fiat Panda Treking 4 × 4, a passenger car based on front-wheel drive with transverse engine. The vehicle has McPherson struts at the front and a rigid axle with longitudinal leaf springs at the back. The propshaft leading to it is divided into three to be able to take the rotational movements of the rigid axle around the transverse (y) axis during drive-off and braking and to absorb movements of the drive unit. The Fiat Panda is an estate car with the ratio

$$i_1 = \frac{2159\text{ mm}}{3689\text{ mm}} = 0.59 \text{ (see Equation 3.1)}$$

complexity compared with the front-wheel drive design, even if, on the Fiat Panda (Trecking 4 × 4), there is a weight increase of about 11% (90 kg), not least because of the heavy, driven, rigid axle. It is possible to select rear-wheel drive during a journey using a shift lever that is attached to the prop-shaft tunnel.

Manual selection on the Subaru Justy operates pneumatically at the touch of a button (even while travelling). This vehicle has independent rear-wheel suspension and weighs only 6% more than the basic vehicle with front-wheel drive. Traction is always improved considerably if the driver recognizes the need in time and switches the engine force onto all four wheels. In critical situations, this usually happens too late, and the abrupt change in drive behaviour becomes an additional disadvantage.

Conversely, if the driver forgets to switch to single axle drive on a dry road, tensions occur in the power train during cornering, as the front wheels travel larger arcs than the back ones (Figs 1.69 and 3.91). The tighter the bend, the greater the stress on the power train and the greater the tendency to unwanted tyre slip.

A further problem is the braking stability of these vehicles. If the front axle locks on a wet or wintry road during braking, the rear one is taken with it due to the rigid power train. All four wheels lock simultaneously and the car goes into an uncontrollable skid.

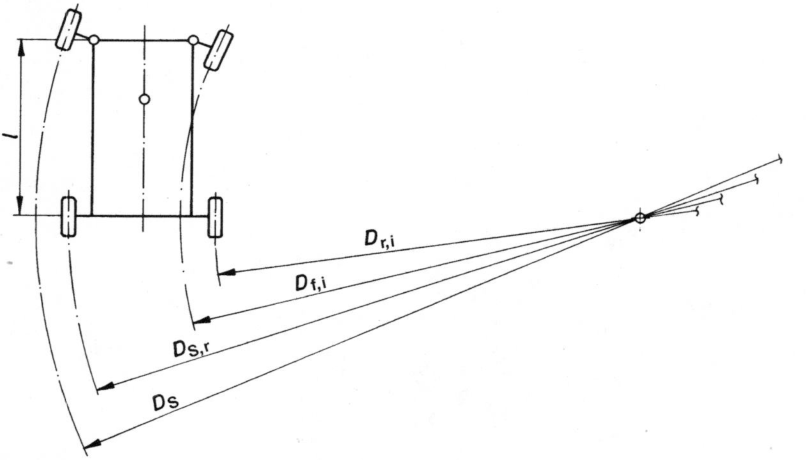

Fig. 1.69 The front wheel on the outside of the bend draws the largest arc during slow cornering, the track circle diameter D_S, while the inner wheel draws the considerably smaller arc $D_{f,i}$. This is the reason for the differential in the driven front axle of the front-wheel drive. The bend diameters $D_{S,r}$ and $D_{r,i}$ to the rear are even smaller, so the rolling distance of the two wheels of this axle decreases further and there can be tensions in the drive train if both axles are rigidly connected, a bend is being negotiated and when a dry road surface makes wheel slip more difficult because of high coefficients of friction.

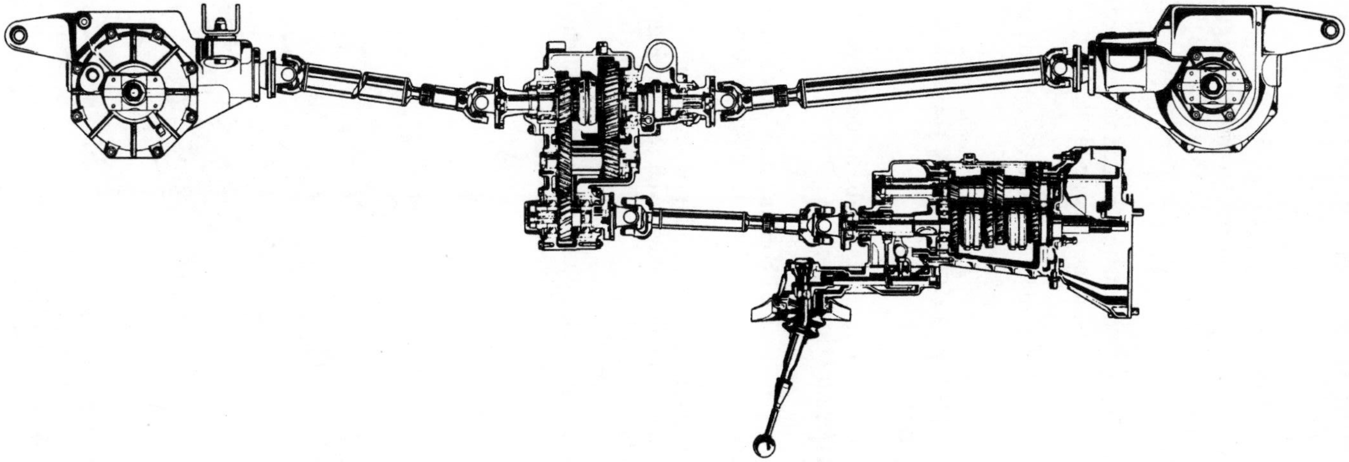

Fig. 1.70 Complex power distribution on the Fiat Campagnolo, a four-wheel drive, all-purpose passenger car. The drive moment is transferred from the manual gearbox via a centrally located two-gear power take-off gear to the differentials of the front and rear axles. Efficiency is not likely to be especially good.

1.7.3 Manual selection four-wheel drive on commercial and all-terrain vehicles

The basis for this type of vehicle is the standard design which, because of the larger ground clearance necessary in off-road vehicles (Fig. 1.67), has more space available between the engine and front axle differential and between the cargo area and the rear axle. Figure 1.70 shows the design details:

- a central power take-off gear with manual selection for the front axle, plus a larger ratio off-road gear, which can be engaged if desired;
- three propshafts;
- complex accommodation of the drive joints if there is a rigid front axle (Fig. 1.3).

1.7.4 Permanent four-wheel drive; basic passenger car with front-wheel drive

All four wheels are constantly driven; this can be achieved between the front and rear axle with different design principles:

- a bevel centre differential with or without manual lock selection;
- a Torsen centre differential with moment distribution, based on the traction requirement (Fig. 1.71);
- a planet gear central differential with fixed moment distribution and additional visco clutch, which automatically takes over the locking function when a difference in the number of revolutions occurs or a magnetic clutch (which is electronically controlled, Fig. 1.79);
- electronically controlled multi-disc clutches (Haldex clutch, Fig. 1.73);

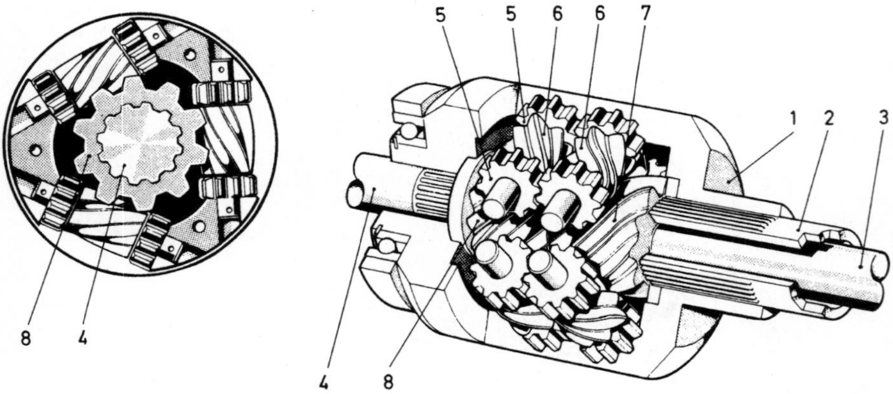

Fig. 1.71 Torsen central differential fitted in Quattro models (apart from the TT) by Audi. It consists of two worm gears, which are joined by spur gears and, depending on the traction requirement, can distribute the driving torque up to 75% to the front or rear axle. Under normal driving conditions 50% goes to each axle.

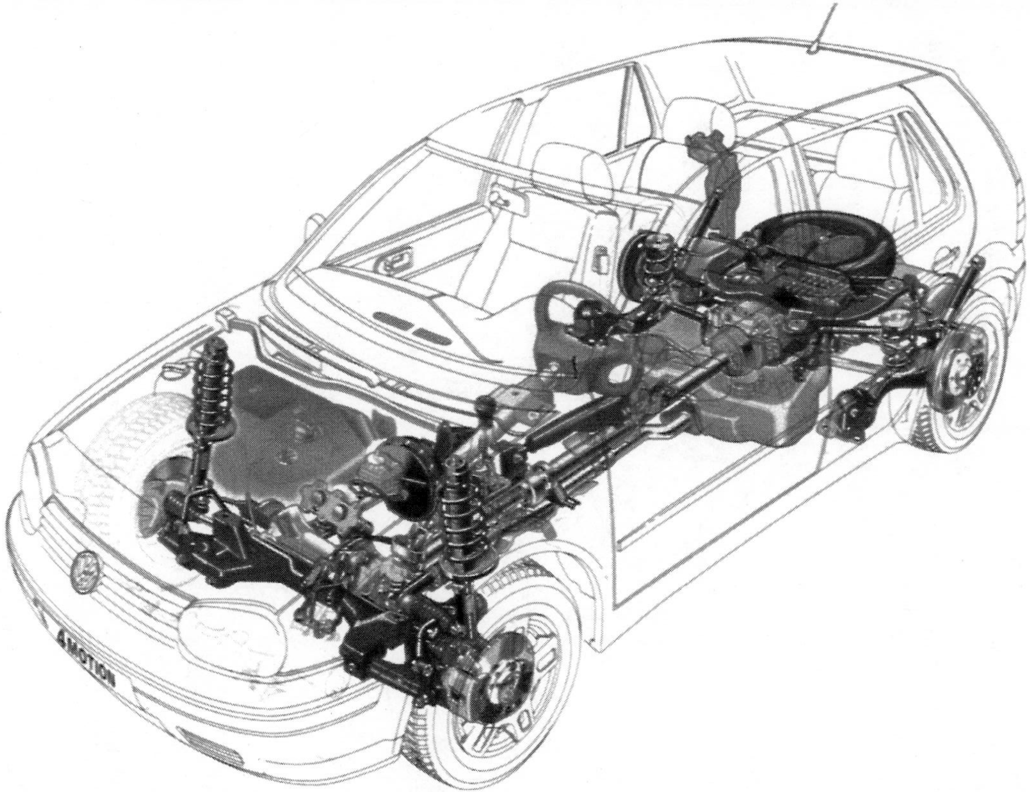

Fig. 1.72 Four-wheel drive Golf 4motion (1998). In the four-wheel drive vehicle, Volkswagen uses a multi-link suspension consisting of one longitudinal and two transverse links mounted on a subframe. The driving torque is transmitted to the rear axle via a wet multi-disc clutch by the Swedish company Haldex which is flange-mounted on the rear axle drive and runs in oil. This electronically controlled clutch can build up a coupling torque of up to 3200 Nm even at small cardan-shaft rotation angles of 45° and can be combined to good effect with brake-power control systems. The drive train of the Audi TT Quattro (1998) built on the same platform is built to almost the same design.

- a visco clutch in the propshaft power train, which selects the initially undriven axle depending on the tyre slip (Figs 1.72, 1.74 and 1.75).

Here too, the front-wheel drive passenger car is suitable as a basic vehicle. In 1979, Audi was the first company to bring out a car with permanent four-wheel drive, the Quattro, and today vehicles with this type of drive are available throughout the entire Audi range. On a longitudinally mounted engine, a Torsen centre differential distributes the moment according to the traction requirement (Fig. 1.71). The four-wheel drive increases the weight by around 100 kg.

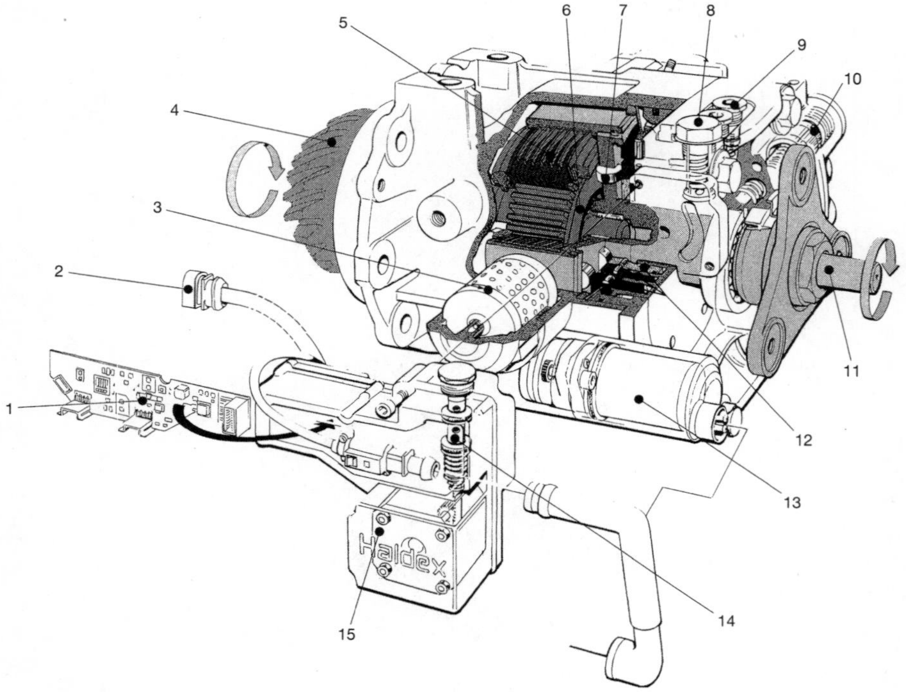

Fig. 1.73 Multi-disc clutch of the Swedish company Haldex, used in the Golf 4motion (1998) and Audi TT Quattro (1998). When there is a difference in speed between the front and rear axles, the disc cam 6 on the output shaft activates the working elements of the axial-piston pump 12 by means of the rollers 7. Via the control valve 14, the pressure produced activates the working piston which moves the discs. The torque transmitted is adjusted continuously by the control unit up to the maximum value, depending on the driving situation described by the wheel sensors, the signals from the slip and brake-power control systems, the position of the accelerator pedal, the engine speed etc. The clutch is disengaged when the ABS function is used. 1 electronic control unit, 2 connector vehicle (voltage, CAN, K leads), 3 oil filter, 4 shaft bevel wheel exit (rear axle gearing), 5 lamella, 6 cam plate, 7 coil, 8 relief valve, 9 pressure regulating valve, 10 accumulator, 11 input shaft, 12 axial piston pump, 13 pre-load pump, 14 control valve, 15 intermittent or step motor.

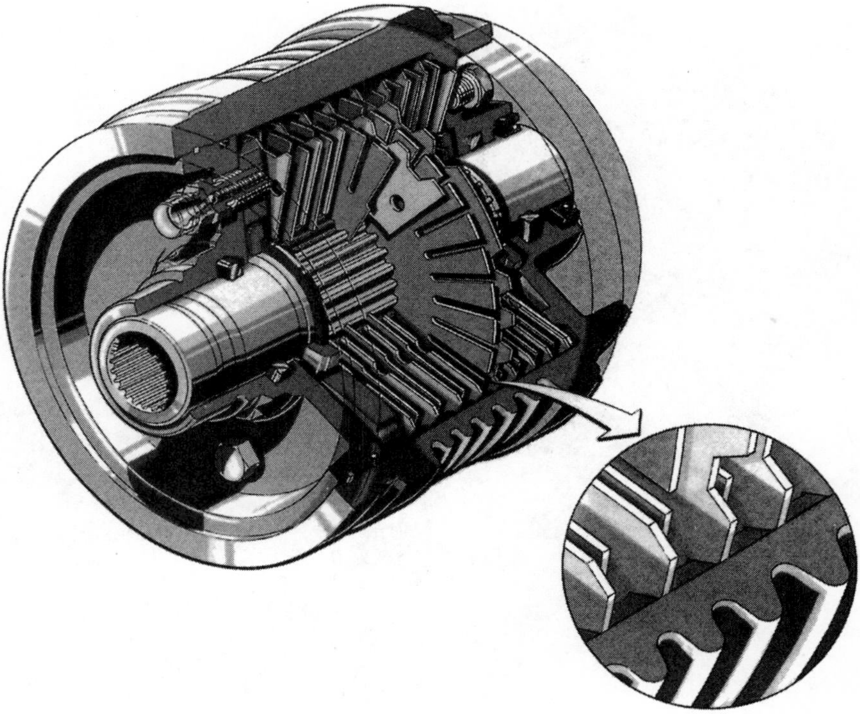

Fig. 1.74 Visco clutch with slip-dependent drive moment distribution. Two different packages sit in the closed drum-shaped housing: radially slit steel discs, which are moved by the serrated profile of the hollow shaft, and perforated discs which grip (as can be seen below) into housing keys. The shaft is joined with the differential and the casing with the propshaft going to the rear axle.

The discs are arranged in the casing so that a perforated disc alternates with a slit one. The individual parts have no definite spacing but can be slid against one another axially. The whole assembly is filled with viscous silicone fluid and the torque behaviour (therefore the locking effect) can be adjusted via the filling level.

If slip occurs between the front and rear axle, the sets of discs in the clutch rotate relative to one another and shearing forces are transferred via the silicone fluid. These increase with increasing slip and ensure a torque increase in the rear axle. The power consumed in the visco clutch leads to warming and thus to growing inner pressure. This causes an increase in the transferable torque which, under conditions of extreme torque requirement, ultimately leads to an almost slip-free torque transfer (rigid drive). With ABS braking, a free-wheeling device disengages the clutch; the latter must be engaged again when reversing.

Fig. 1.75 Driven front axle of the Porsche 911 Carrera 4 (1996, 1998). The visco clutch is flange mounted directly on the front axle to achieve a better distribution of axle load. With corresponding slip of the rear wheels, up to 40% of the driving torque is transmitted to the front axles. Particular attention was paid during the adjustment of the four-wheel drive to predictable self-steering properties independent of drive distribution and to controllability of the handling characteristics even at the stability limit. Instead of differential locks, specific wheel brake engagements are made in order to retard spinning wheels. Four-wheel drive is integrated into the Porsche Stability Management (PSM), a system for controlling the dynamics of vehicle movement with brake actuation.

Fig. 1.76 Double wishbone rear axle on the Audi A4 Quattro. The suspension subframe 1 is fixed to the body with four widely spaced rubber mountings (items 2 and 3) and houses the differential casing 8 and transverse control arms (items 4 and 5). The springs and shock absorbers are mounted next to the fixings for the upper control arms 7. The location 6 of the wheel hub carrier 5 was raised (long base *c*, Fig. 1.4) and drawn outwards. The lower transverse control arm 4 is fixed to part 1 with widely spaced mountings. These measures ensure a wide boot and low forces, making it easier to attain the desired kinematic characteristics.

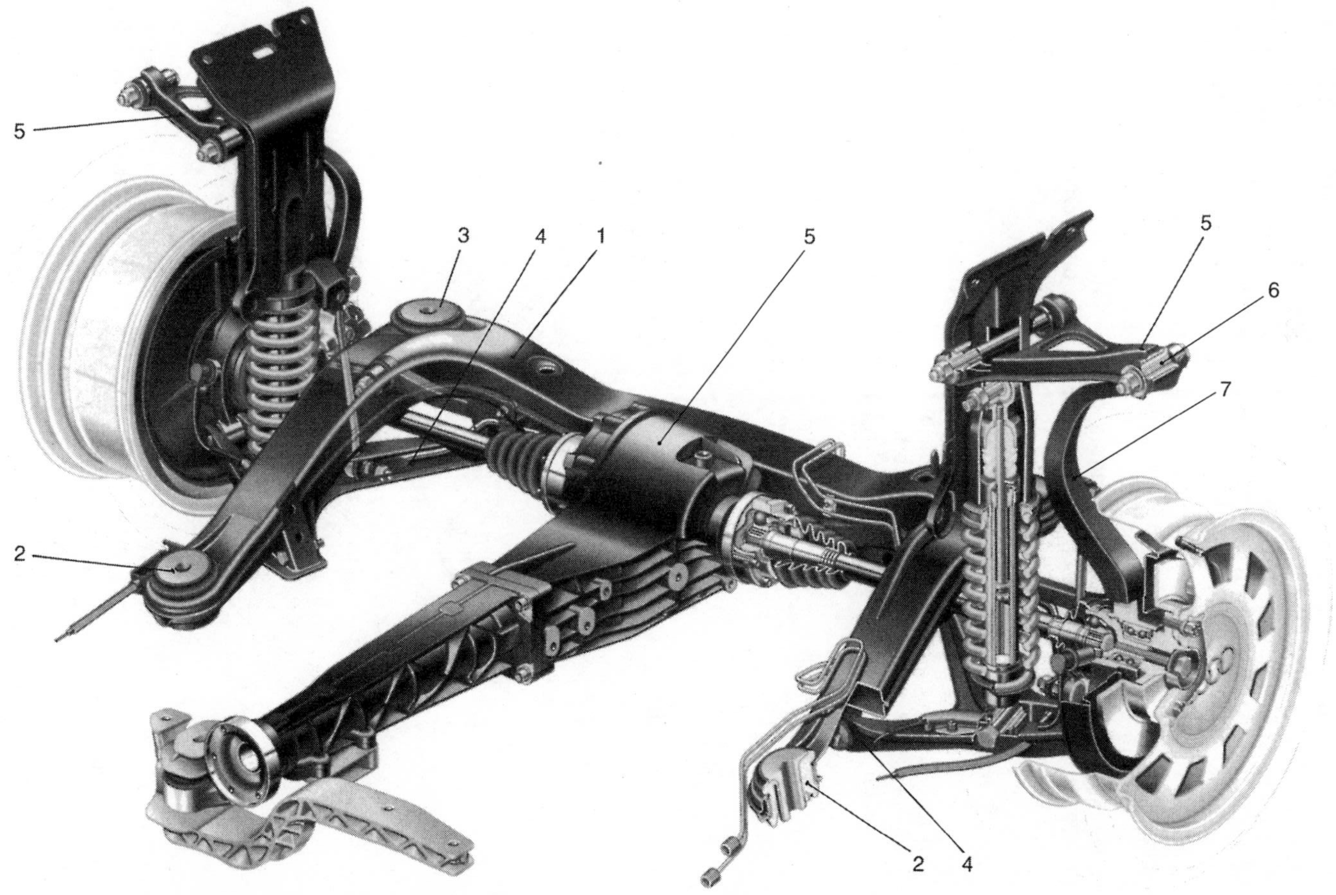

5
3
4
1
5
5
6
7
2
2
4

VW used a visco clutch in the power train (without centre differential) for the first time on the Transporter (Fig. 1.74) and then subsequently used it in the Golf syncro. The clutch has the advantage of the engine moment distribution being dependent on the tyre slip. If the slip on the front wheels, which are otherwise driven at the higher moment, increases on a wet or frozen surface or off-road, more drive is applied to the rear wheels. No action on the part of the driver is either necessary or possible. The transverse engine makes a bevel gear in front of the split propshaft necessary. The visco clutch sits in the rear differential casing and there is also an overrunning clutch, which ensures that the rear wheels are automatically disengaged from the drive, on overrun, to guarantee proper braking behaviour. This type of drive is fully ABS compatible. When reverse is engaged, a sliding sleeve is moved, which bridges the overrunning clutch to make it possible to drive backwards.

When selecting their rear axle design, manufacturers choose different paths. Audi fits a double wishbone suspension in the A4 and A6 Quattro (Fig. 1.76), Honda uses the requisite centre differential on the double wishbone standard suspension in the Civic Shuttle 4WD (Figs 1.77 and 1.62).

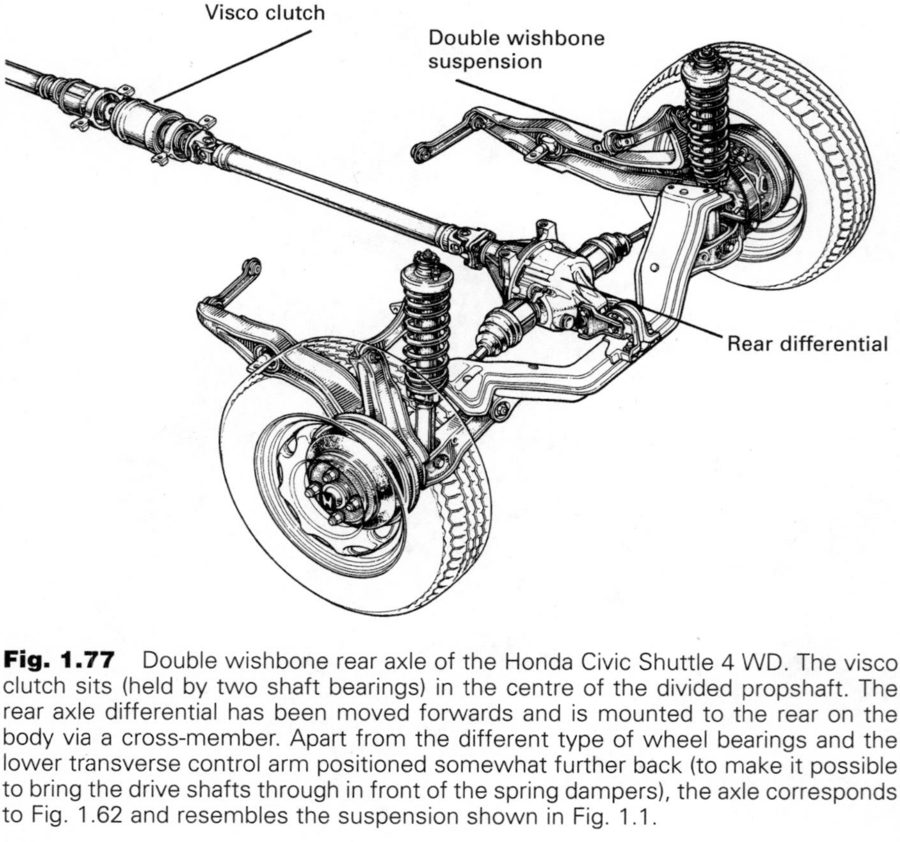

Fig. 1.77 Double wishbone rear axle of the Honda Civic Shuttle 4 WD. The visco clutch sits (held by two shaft bearings) in the centre of the divided propshaft. The rear axle differential has been moved forwards and is mounted to the rear on the body via a cross-member. Apart from the different type of wheel bearings and the lower transverse control arm positioned somewhat further back (to make it possible to bring the drive shafts through in front of the spring dampers), the axle corresponds to Fig. 1.62 and resembles the suspension shown in Fig. 1.1.

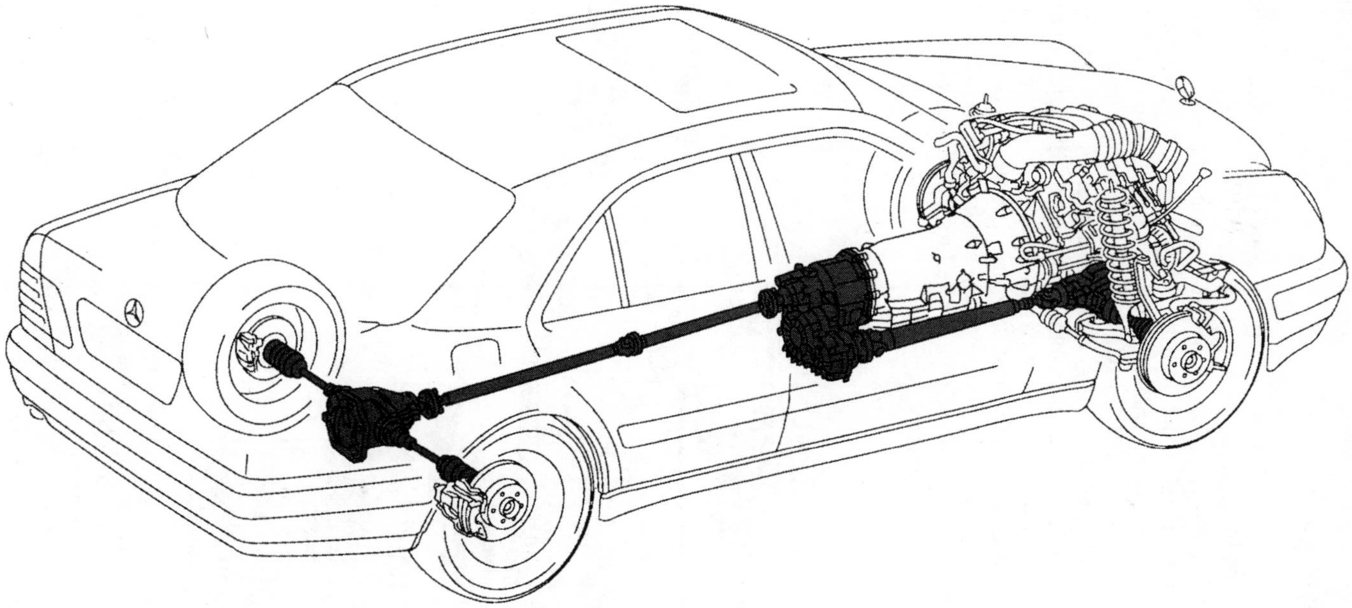

Fig. 1.78 Drive train of the four-wheel drive Mercedes-Benz E class 4MATIC (from 1997). In order to be able to control the drive shafts to the front wheels, an integrated spring-and-shock absorber strut in the shape of a fork on the lower transverse link is used. In the almost identical suspension design of other than off-road varieties, the springs and shock absorbers are separate.

1.7.5 Permanent four-wheel drive, basic standard design passenger car

Giving a standard design car four-wheel drive requires larger modifications, greater design complexity and makes the drive less efficient (Fig. 1.78). A power take-off gear is required, from which a short propshaft transmits the engine moment to the front differential. The lateral offset must be bridged, for example, with a toothed chain (Fig. 1.79). The ground clearance must not be affected and so changes in the engine oil pan are indispensable if the axle drive is to be accommodated (Fig. 1.80).

The power take-off gear (Fig. 1.79) contains a planet gear centre differential which facilitates a variable force distribution (based on the internal ratio); 36% of the drive moment normally goes to the front and 64% to the rear axle. A multi-disc clutch can also be installed that can lock the differential electromagnetically up to 100%, depending on the torque requirement (front to rear axle). Moreover, there is a further electrohydraulically controlled lock differential in the rear axle which is also up to 100% effective.

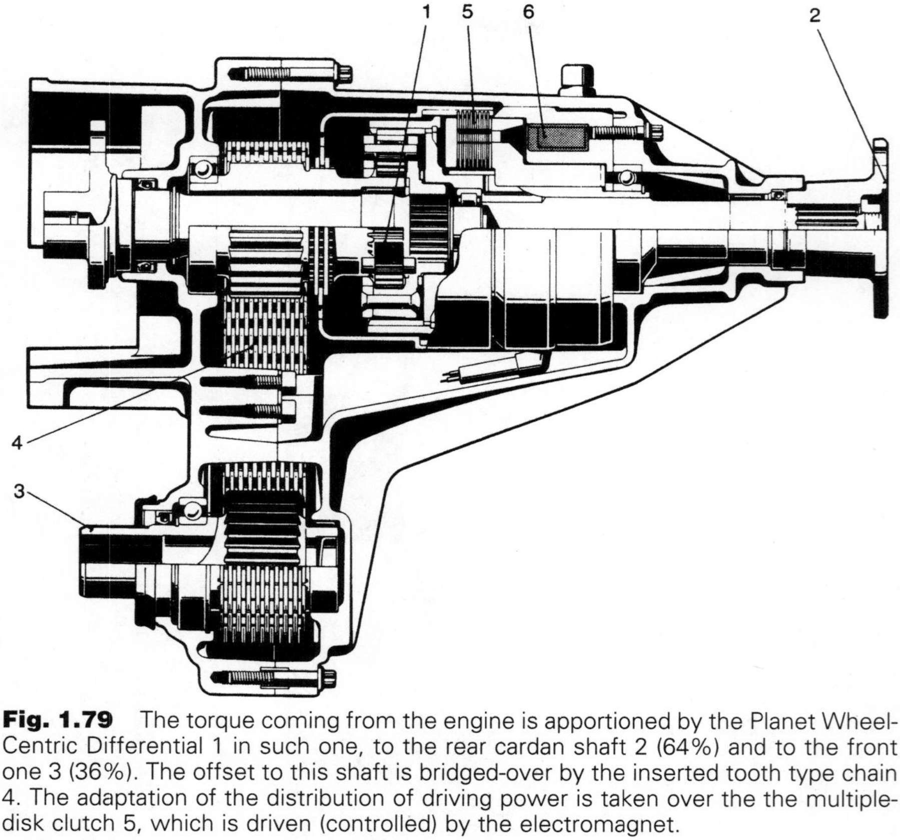

Fig. 1.79 The torque coming from the engine is apportioned by the Planet Wheel-Centric Differential 1 in such one, to the rear cardan shaft 2 (64%) and to the front one 3 (36%). The offset to this shaft is bridged-over by the inserted tooth type chain 4. The adaptation of the distribution of driving power is taken over the the multiple-disk clutch 5, which is driven (controlled) by the electromagnet.
Power Divider A110 of the Fa. ZF. (Zahnradfabrik Friedrichshafen)

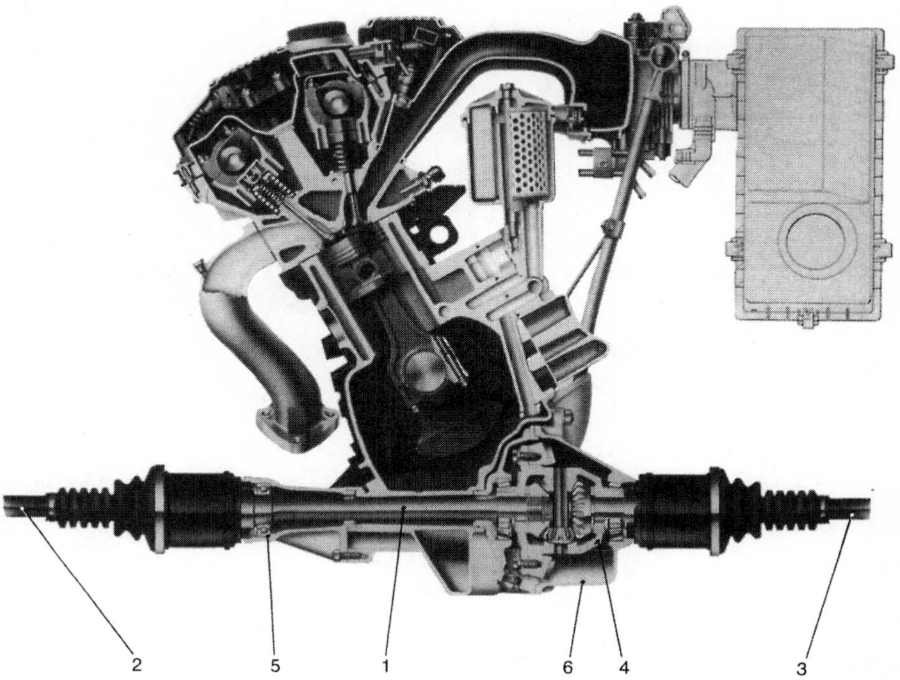

Fig. 1.80 Front cross-section view of the engine; and drive axle of a standard four-wheel drive vehicle (BMW assembly diagram). The basic vehicle has rear-wheel drive and, in order to also be able to drive the front wheels, the front axle power take-off 4 had to be moved into the space of the oil pan. The intermediate shaft 1 bridges the distance to the right inner CV joint and thus ensures drive shafts of equal length to both wheels (items 2 and 3 and Fig. 1.51). Part 1 is mounted on one side in the non-lockable differential 4 and on the other side in the outrigger 5. This, and the casing 6, are screwed to the oil pan.

The two differentials with variable degrees of lock offer decisive advantages:

- to reach optimal driving stability, they distribute the engine moments during overrun and traction according to the wheel slip on the drive axles;
- they allow maximum traction without loss of driving stability (Fig. 1.66).

The locks are open during normal driving. By including the front axle differential, they make it possible to equalize the number of revolutions between all wheels, so tight bends can be negotiated without stress in the power train and parking presents no problems. If the car is moved with locked differentials and the driver is forced to apply the brakes, the locks are released in a fraction of a second. The system is therefore fully ABS compatible.

In its four-wheel drive vehicles of the E class (Fig. 1.78), Mercedes-Benz uses

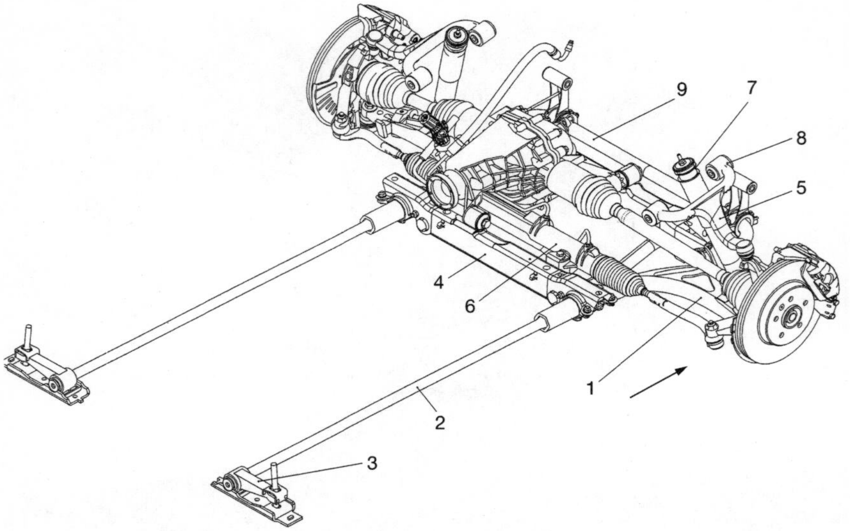

Fig. 1.81 Front suspension and drive axle of the Mercedes-Benz off-road vehicle of the M series. In off-road vehicles, rigid axles are mostly used. Instead of these, Mercedes-Benz installs double wishbone suspensions at the front and rear. In this way, the proportion of unsprung masses can be reduced by approximately 66%; driving safety and riding comfort are increased. For space reasons, torsion-bar springs are used for the suspension of the front axle.

1 lower transverse link in the form of a forged steel component because of the introduction of torque by the torsion bars (2) and notch insensitivity off road conditions; 2 torsion bars (spring rate of 50 Nm/degree); 3 vertically adjustable torque support which can be placed in any position in a transverse direction; 4 integral bearers (subframe) attached to the box-type frame by 4 bolts; 5 upper transverse link in the form of a forged aluminium component; 6 rack and pinion power steering, 7 twin-tube shock absorber with integrated rubber bump stop, 8 transverse link mounting points; 9 stabilizer application of force to lower transverse link.

a transfer gear with central differential situated on the gearbox outlet and a front axle gear integrated into the engine-oil pan. The (fixed) driving torque distribution is 35%:65%. Instead of traditional differential locks, the wheel brakes are activated on the spinning wheels as in off-road vehicles of the M class. This system permits maximum flexibility, its effect not only corresponds to differential locks on front and rear axles as well as on the central differential, but also makes it possible for other functions such as ABS and electronic yaw control (ESP) to be integrated without any problem. Design complexity – and thus cost – is considerable.

1.7.6 Summary of different kinds of four-wheel drive

The list in Fig. 1.83 shows the increasing use of slip-controlled clutches (visco and Haldex clutches) for the transmission of torque instead of an interaxle

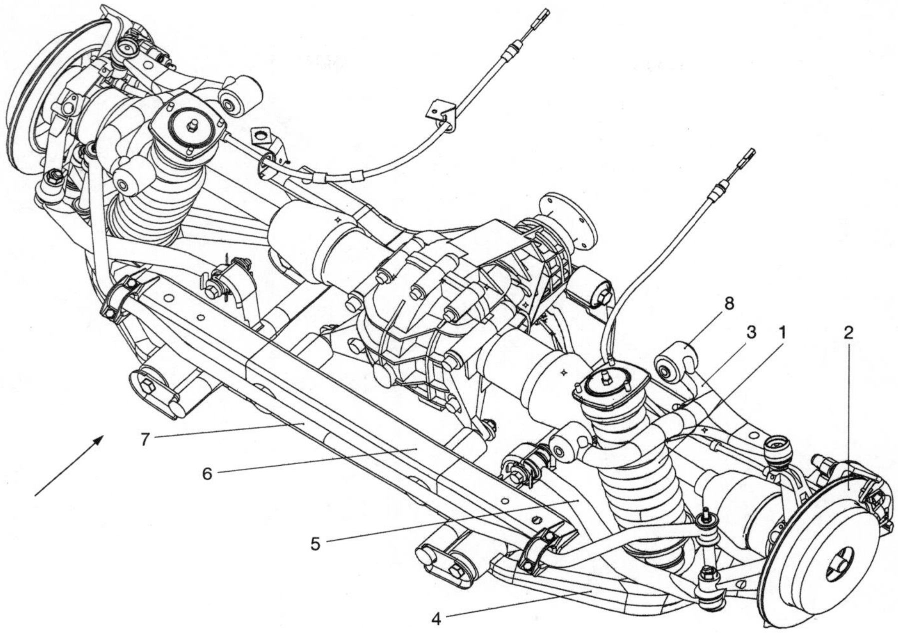

Fig. 1.82 Rear axle of the Mercedes-Benz off-road vehicle of the M series. Suspension and damping are ensured by the spring strut (1) whose spring is tapered for reasons of construction space (spring rate gradually increasing from 70 to 140 N/mm), 2 brake disc with integrated drum parking brake, 3 upper transverse link (forged aluminium component), 4 lower transverse link (forged aluminium component), 5 tie rod (forged steel component), 6 integral bearer (subframe), 7 stabilizer, 8 transverse link mounting points.

Common characteristics of front and rear axles: camber and castor are adjusted by positioning the transverse link mounting points (8) in long holes during assembly. Technical data: spring travel ±100 mm, kingpin offset –5 mm, disturbing force moment arm 56.7 mm, kingpin inclination 10.5°, camber angle –0.5°, castor for front axle/rear axle 7/–8.5°, castor trail for front axle/rear axle 37/–55 mm, wheel castor trail for front axle/rear axle 5/–4.5 mm, instantaneous centre height for front axle/rear axle 80/119 mm, braking-torque compensation for front axle/rear axle 38/21%, starting-torque compensation for front axle/rear axle –7/–3%. The axle concept was designed and developed by Mercedes-Benz. Mass production and assembly is undertaken by Zahnradfabrik Friedrichshafen AG who, via Lemförder Fahrwerktechnik AG, supply the complete subassemblies to the assembly line as required.

differential and the importance of electronic brake application systems which are used instead of lockable differential gears. Modern four-wheel varieties operate without functional restrictions with antilocking, slip and driving stabilization systems.

Fig. 1.83 Different kinds of four-wheel drive.

Motor position	Reduction drive	Drive on	Four-wheel drive			Middle differential locks via	Front axle differential		Rear axle differential		Example
			switched by	slip dependent	perm. by		locking	brake	locking	brake	
longit.	2.05:1	rear	sprag-clutch man.			N/A	n	n	n	n	Opel Frontera
longit.	1.425:1	rear	sprag-clutch man.			N/A	n	n	multi-disc clutch	n	Mitsubishi Pajero
longit.	2.15:1	rear	sprag-clutch man.			N/A	n	n	n	n	Suzuki Jimny Cross Country
longit.	2.43:1	rear	sprag-clutch man.			N/A	n	n	multi-disc clutch	n	Chevrolet Blazer
longit.	2.48:1	rear	multi-disc.	electron.		N/A	n	n	multi-disc clutch	n	Ford Explorer
transv.		front	multi-disc	electron.		N/A	n	n	n	n	Honda CR-V
longit.	2.72:1	rear		visco		N/A	n	n	multi-disc clutch	n	Jeep Grand Cherokee
transv.		front		visco		N/A	n	n	n	n	Chrysler Voyager 4WD
transv.		front		visco		N/A	n	y	n	y	Land Rover Freelander Discovery
longit.		rear		visco		N/A	n	y	n	y	Porsche Carrera 4
transv.		front		visco		N/A	n	y	multi-disc clutch	n	Volvo V70 R AWD
transv.		front		visco		N/A	n	y	multi-disc clutch	n	VW Multivan Synchro
transv.		front		Haldex multi-disc		N/A	n	y	n	n	Audi TT Quattro, Golf 4motion
longit.		f + r			longit. diff.	claw clutch	n	n	n	n	Daihatsu Terios
longit.	2.64:1	f + r			longit. diff.	N/A	n	y	n	y	Mercedes-Benz M series

longit.	y	f + r	longit. diff.	y	y	n	y	n	Mercedes-Benz G series
longit.		f + r	longit. diff.	N/A	n	y	n	y	Mercedes-Benz E series
longit.		f + r	longit. diff.	N/A	n	y	n	y	BMW Sports Activity Vehicle (SAV, E53)
longit.	1.21:1	f + r	longit. diff.	N/A	n	y	n	y	Land Rover Discovery
	y	f + r	longit. diff.	visco + man.					Toyoto LandCruiser
longit.	1.93:1	f + r	longit. diff.	visco + man.	n	n	sprag-clutch	n	Mitsubishi Pajero
longit.	1.20: 1	f + r	longit. diff.	visco lock	n	n	n	n	Subaru Legacy Outback
longit.	1.45:1	f + r	longit. diff.	visco lock	n	n	n	n	Subaru Forester
transv.		f + r	Torsen diff.	self-locking	n	y	n	y	Audi A4/A6 Quattro, VW Passat Synchro
transv.		f + r	Torsen diff.	self-locking	n	y	n	y	Audi A4/A6 Quattro, VW Passat Synchro

2

Tyres and wheels

2.1 Tyre requirements

The tyres are crucial functional elements for the transmission of longitudinal, lateral and vertical forces between the vehicle and road. The tyre properties should be as constant as possible and hence predictable by the driver. As well as their static and dynamic force transmission properties, the requirements described below – depending on the intended use of the vehicle – are also to be satisfied.

As tyres significantly affect the handling properties of vehicles, the properties of original tyres – the tyres with which the vehicle is supplied to the customer – are specified by the vehicle manufacturers in conjunction with the tyre manufacturers. However, spare tyres usually differ from the original tyres, despite their similar designation; hence handling characteristics can change. Individual vehicle manufacturers have therefore decided to identify tyres produced in accordance with their specifications by means of a symbol on the sidewall of the tyre or to sell tyres which meet the specifications of original tyres at their manufacturing branches.

2.1.1 Interchangeability

All tyres and rims are standardized to guarantee interchangeability, i.e. to guarantee the possibility of using tyres from different manufacturers but with the same designation on one vehicle and to restrict the variety of tyre types worldwide.

Within Europe, standardization is carried out by the European Tyre and Rim Technical Organization or ETRTO, which specifies the following:

- tyre and rim dimensions;
- the code for tyre type and size;
- the load index and speed symbol.

Passenger car tyres are governed by UNO regulation ECE-R 30, commercial vehicles by R 54, spare wheels by R 64, and type approval of tyres on the vehicle by EC directive 92/23/EC.

In the USA the Department of Transportation (or DOT, see item 9 in Fig. 2.18) is responsible for the safety standards. The standards relevant here are:

Standard 109	Passenger cars
Standard 119	Motor vehicles other than passenger cars.

The Tire and Rim Association, or TRA for short, is responsible for standardization.

In Australia, binding information is published by the Federal Office of Road Safety, Australian Motor Vehicle Certification Board.

ARD 23	Australian Design Rule 23/01: Passenger car tyres

is the applicable standard.

In Germany the DIN Standards (Deutsches Institut für Normung) and the WdK Guidelines (Wirtschaftsverband der Deutschen Kautschukindustrie Postfach 900360, D-60443, Frankfurt am Main) are responsible for specifying tyre data. All bodies recognize the publications of these two organizations.

At the international level, the ISO (International Organization for Standardization) also works in the field of tyre standardization and ISO Standards are translated into many languages.

2.1.2 Passenger car requirements

The requirements for tyres on passenger cars and light commercial vehicles can be subdivided into the following six groups:

- driving safety
- handling
- comfort
- service life
- economy
- environmental compatibility.

To ensure driving safety it is essential that the tyre sits firmly on the rim. This is achieved by a special tyre bead design (tyre foot) and the safety rim, which is the only type of rim in use today (Figs 2.5 and 2.21). Not only is as great a degree of tyre-on-rim retention as possible required, but the tyre must also be hermetically sealed; on the tubeless tyre this is the function of the inner lining. Its job is to prevent air escaping from the tyre, i.e. it stops the tyre from losing pressure. However, this pressure reduces by around 25–30% per year, which shows how important it is to check the tyre pressure regularly.

In order to guarantee driving safety, the aim is also to ensure that tyres are as insensitive to overloading and as puncture-proof as possible and that they have emergency running properties which make it possible for the driver to bring the vehicle safely to a halt in case of tyre failure.

Handling characteristics include the properties:

- high coefficients of friction in all operating conditions;
- steady build-up of lateral forces without sudden changes;
- good cornering stability;
- direct and immediate response to steering movements;
- guarantee requirement of sustained maximum speed;
- small fluctuations in wheel load.

Riding comfort includes the characteristics:

- good suspension and damping properties (little rolling hardness);
- high smoothness as a result of low radial tyre run-out and imbalances;
- little steering effort required during parking and driving;
- low running noise.

Durability refers to:

- long-term durability
- high-speed stability.

Both are tested on drum test stands and on the road.

Economic efficiency is essentially determined by the following:

- purchase cost;
- mileage (including the possibility of profile regrooving in the case of lorry tyres);
- wear (Fig. 3.46);
- rolling resistance;
- the necessary volume, which determines
- the amount of room required in the wheel houses and spare-wheel well;
- load rating.

Of increasing importance is environmental compatibility, which includes:

- tyre noise;
- raw material and energy consumption during manufacture and disposal;
- possibility of complete remoulding inherent in the construction.

The importance of

- tyre design, profile design and the 'radius–width appearance' must not be neglected either.

Further details are available in Refs [4], [6], [7] and [9].

2.1.3 Commercial vehicle requirements

In principle, the same requirements apply for commercial vehicles as for passenger cars, although the priority of the individual groups changes. After safety, economy is the main consideration for commercial vehicle tyres. The following properties are desirable:

- high mileage and even wear pattern
- low rolling resistance
- good traction
- low tyre weight
- ability to take chains
- remoulding/retreading possibilities.

Compared with passenger car tyres, the rolling resistance of commercial vehicle tyres has a greater influence on fuel consumption (20–30%) and is therefore an important point (Fig. 2.32).

2.2 Tyre designs

2.2.1 Diagonal ply tyres

In industrialized countries, cross-ply tyres are no longer used on passenger cars, either as original tyres or as replacement tyres, unlike areas with very poor roads where the less vulnerable sidewall has certain advantages. The same is true of commercial vehicles and vehicles that tow trailers, and here too radial tyres have swept the board because of their many advantages. Nowadays, cross-ply tyres are used only for:

- temporary use (emergency) spare tyres for passenger cars (due to the low durability requirements at speeds up to 80 or 100 km h^{-1});
- motor cycles (due to the inclination of the wheels against the lateral force);
- racing cars (due to the lower moment of inertia);
- agricultural vehicles (which do not reach high speeds).

Cross-ply tyres consist of the substructure (also known as the tyre carcass, Fig. 2.1) which, as the 'supporting framework' has at least two layers of rubberized cord fibres, which have a zenith or bias angle ξ of between 20° and 40° to the centre plane of the tyre (Fig. 2.2). Rayon (an artificial silk cord), nylon or even steel cord may be used, depending on the strength requirements. At the tyre feet the ends of the layers are wrapped around the core of the tyre bead on both sides; two wire rings, together with the folded ends of the plies, form the bead. This represents the frictional connection to the rim. The bead must thus provide the permanent seat and transfer drive-off and braking moments to the tyre. On tubeless tyres it must also provide the airtight seal.

The running tread, which is applied to the outer diameter of the substructure,

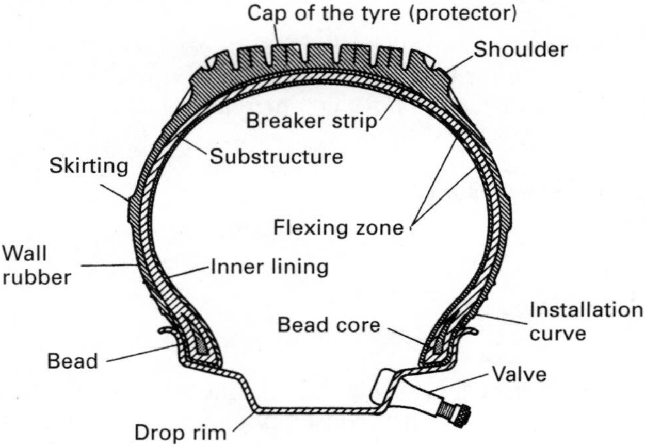

Fig. 2.1 Design of a diagonal ply tubeless car tyre with a normal drop rim and pressed-in inflating valve (see also Fig. 2.6).

Fig. 2.2 The diagonal ply tyre has crossed-bias layers; the zenith angle ξ was 30° to 40° for passenger cars. The 4 PR design should have two layers in each direction. Smaller angles ξ can be found in racing cars. Rolling resistance, lateral and suspension stiffness are significantly determined by the zenith angle.

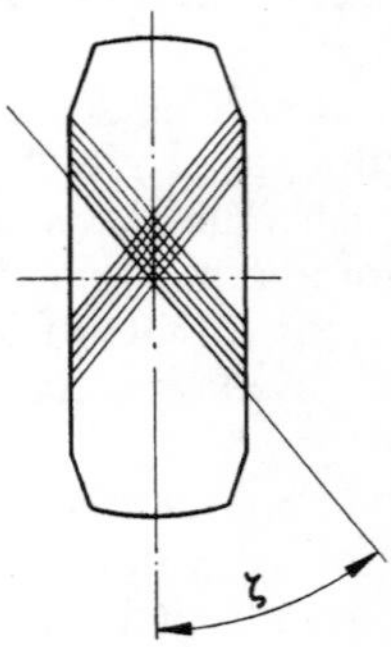

provides the contact to the road and is profiled. Some tyres also have an intermediate structure over the carcass as reinforcement.

At the side, the running tread blends into the shoulder, which connects to the sidewall (also known as the side rubber), and is a layer that protects the substructure. This layer and the shoulders consist of different rubber blends from the running tread because they are barely subjected to wear; they are simply deformed when the tyre rolls. This is known as flexing. Protective mouldings on the sides are designed to prevent the tyre from being damaged through contact with kerbstones. There are also GG grooves, which make it possible to see that the tyre is seated properly on the rim flange.

Cross-ply design and maximum authorized speed are indicated in the tyre marking by a dash (or a letter, Fig. 2.12) between the letters for width and rim diameter (both in inches) and a 'PR' (ply rating) suffix. This ply rating refers to the carcass strength and simply indicates the possible number of plies (Fig. 2.5). The marking convention is:

5.60-15/4 PR	(VW rear-engine passenger car, tyres authorized up to 150 km h^{-1})
7.00-14/8 PR	(VW Transporter, tyres authorized up to 150 km h^{-1})
9.00-20/14 PR	(reinforced design for a commercial vehicle)

and on the temporary use spare wheel of the VW Golf, which requires a tyre pressure of p_T = 4.2 bar and may only be driven at speeds up to 80 km h^{-1} (F symbol)

T 105/70 D 14 38 F

2.2.2 Radial ply tyres

The radial ply tyre consists of two bead cores joined together radially via the carcass (Fig. 2.3) – hence the name radial tyres. A belt of cords provides the necessary stiffness (Fig. 2.4), whereas the external part of the tyre consists of the tread and sidewall and the interior of the inner lining, which ensures the tyre is hermetically sealed (Figs 2.5 and 2.1). In passenger car tyres, the carcass is made of rayon or nylon, the belt of steel cord or a combination of steel, rayon or nylon cord, and the core exclusively of steel. Due to the predominance of steel as the material for the belt, these tyres are also known as 'steel radial tyres'. The materials used are indicated on the sidewall (Fig. 2.18, points 7 and

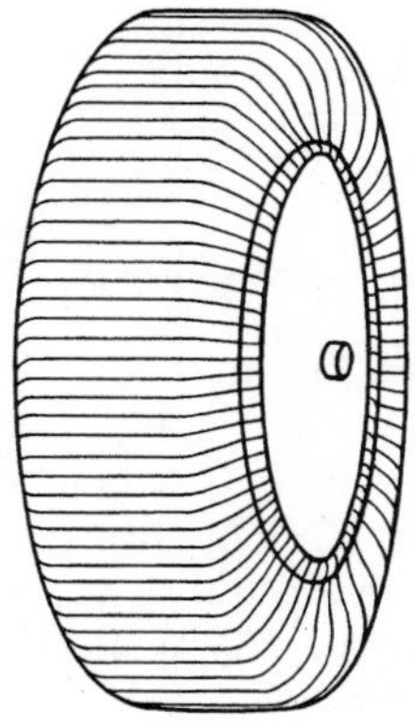

Fig. 2.3 Substructure of a radial tyre. The threads have a bias angle between 88° and 90°.

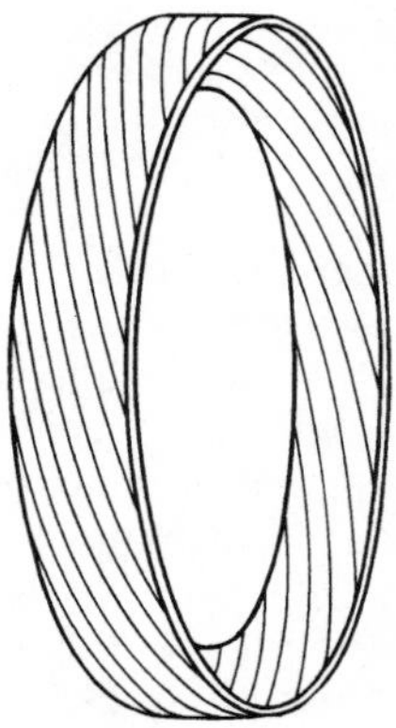

Fig. 2.4 The belt of the radial tyre sits on the substructure. The threads are at angles of between 15° and 25° to the plane of the tyre centre.

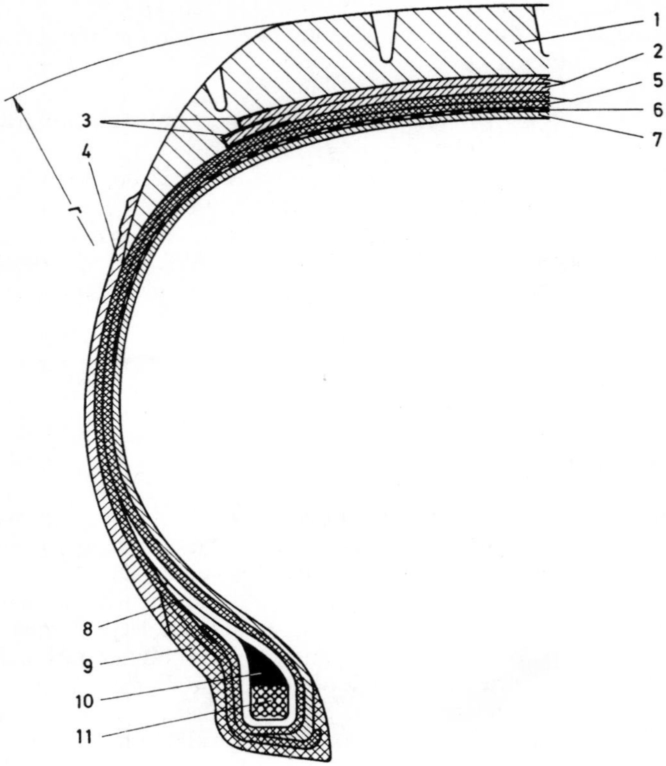

Fig. 2.5 Radial design passenger car tyres in speed category T (Fig. 2.12); the number of layers and the materials are indicated on the sidewall (see Fig. 2.18). The components are: 1 running tread; 2 steel belt; 3 edge protection for the belt, made of rayon or nylon; 4 sidewall; 5 substructure with two layers; 6 cap; 7 inner lining; 8 flipper; 9 bead profile; 10 core profile; 11 bead core.

8). In commercial vehicle designs this is particularly important and the carcass may also consist of steel.

The stiff belt causes longitudinal oscillation, which has to be kept away from the body by wheel suspensions with a defined longitudinal compliance, otherwise this would cause an unpleasant droning noise in the body, when on cobbles and poor road surfaces at speeds of less than 80 km h^{-1} (see Sections 3.6.5.2 and 5.1.2). The only other disadvantage is the greater susceptibility of the thinner sidewalls of the tyres to damage compared with diagonal ply tyres. The advantages over cross-ply tyres, which are especially important for today's passenger cars and commercial vehicles, are:

- significantly higher mileage
- greater load capacity at lower component weight

- lower rolling resistance
- better aquaplaning properties
- better wet-braking behaviour
- transferable, greater lateral forces at the same tyre pressure
- greater ride comfort when travelling at high speeds on motorways and trunk roads.

2.2.3 Tubeless or tubed

In passenger cars, the tubeless tyre has almost completely ousted the tubed tyre. The main reasons are that the tubeless tyre is

- easier and faster to fit
- the inner lining is able to self-seal small incisions in the tyre.

In tubeless tyres the inner lining performs the function of the tube, i.e. it prevents air escaping from the tyre. As it forms a unit with the carcass and (unlike the tube) is not under tensional stress, if the tyre is damaged the incision does not increase in size, rapidly causing loss of pressure and failure of the tyre. The use of tubeless tyres is linked to two conditions:

- safety contour on the rim (Fig. 2.21)
- its air-tightness.

Because this is not yet guaranteed worldwide, tubed tyres continue to be fitted in some countries. When choosing the tube, attention should be paid to ensuring the correct type for the tyre. If the tube is too big it will crease, and if it is too small it will be overstretched, both of which reduce durability. In order to avoid confusion, the tyres carry the following marking on the sidewall:

tubeless (Fig. 2.18, point 3)
tubed or tube type.

Valves are needed for inflating the tyre and maintaining the required pressure. Various designs are available for tubeless and tubed tyres (Figs 2.6 and 2.7). The most widely used valve is the so-called 'snap-in valve'. It comprises a metal foot valve body vulcanized into a rubber sheath, which provides the seal in the rim hole (Fig. 2.20). The functionality is achieved by a valve insert, while a cap closes the valve and protects it against ingress of dirt.

At high speeds, the valve can be subjected to bending stress and loss of air can occur. Hub caps and support areas on alloy wheels can help to alleviate this (see Fig. 2.24 and Section 7.2 in Ref. [4]).

2.2.4 Height-to-width ratio

The height-to-width ratio H/W – also known as the 'profile' (high or low) – influences the tyre properties and affects how much space the wheel requires

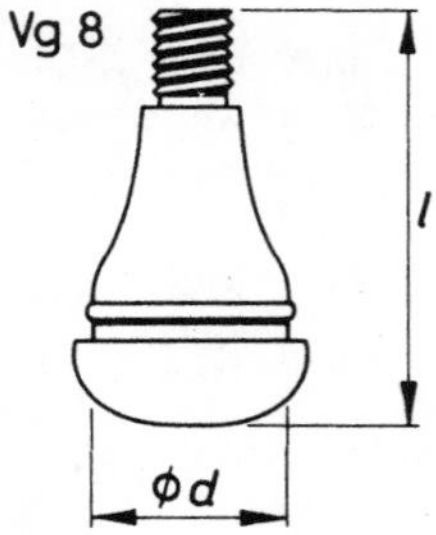

DIN	*l*	Diameter *d*
43 GS 11.5	43	15.2
43 GS 16	43	19.5

Fig. 2.6 Snap-in rubber valve for tubeless tyres, can be used on rims with the standard valve holes of 11.5 mm and 16 mm diameter. The numerical value 43 gives the total length in mm (dimension *l*). There is also the longer 49 GS 11.5 design.

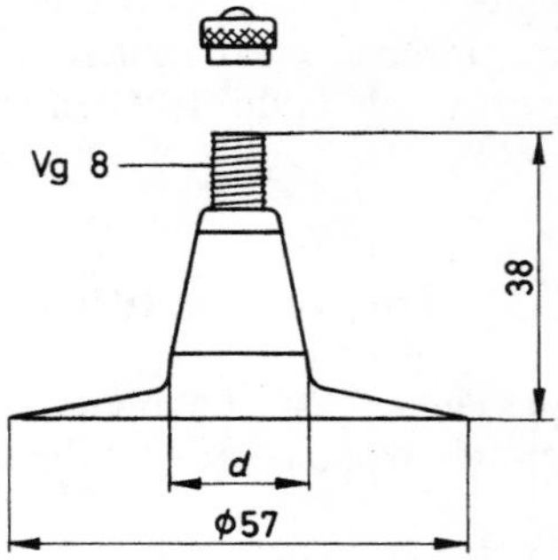

Valve specification	*d*
38/11.5	11.7
38/16	16.5

Fig. 2.7 Rubber valve vulcanized onto tubes. Designations are 38/11.5 or 38/16.

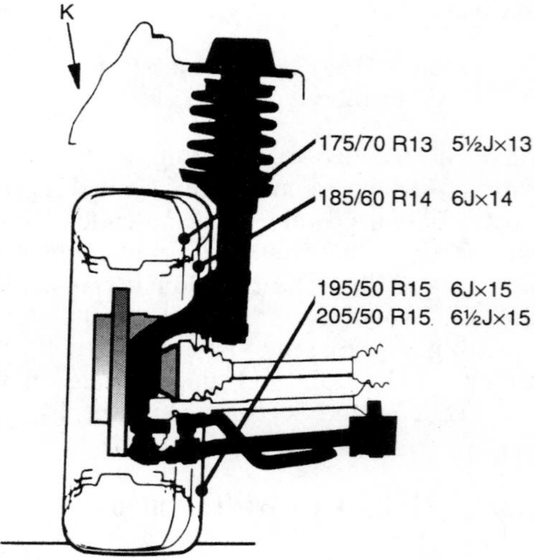

Fig. 2.8 Tyre sizes and associated rims used on the VW Golf III. All tyres fit flush up to the outer edge of the wing (wheel house outer panel) K. To achieve this, differing wheel offsets (depth of dishing) *e* are used on disc-type wheels (Fig. 2.23) with the advantage of a more negative rolling radius r_σ on wider tyres (Fig. 3.102). A disadvantage then is that snow chains can no longer be fitted and steering sensitivity changes very slightly.

(Fig. 2.8). As shown in Fig. 2.9, the narrower tyres with a H/W ratio = 0.70 have a reduced tread and therefore good aquaplaning behaviour (Fig. 2.35). Wide designs make it possible to have a larger diameter rim and bigger brake discs (Fig. 2.10) and can also transmit higher lateral and longitudinal forces.

W is the cross-sectional width of the new tyre (Fig. 2.11); the height H can easily be calculated from the rim diameter given in inches and the outside diameter of the tyre OD_T. The values OD_T and W are to be taken from the new tyre

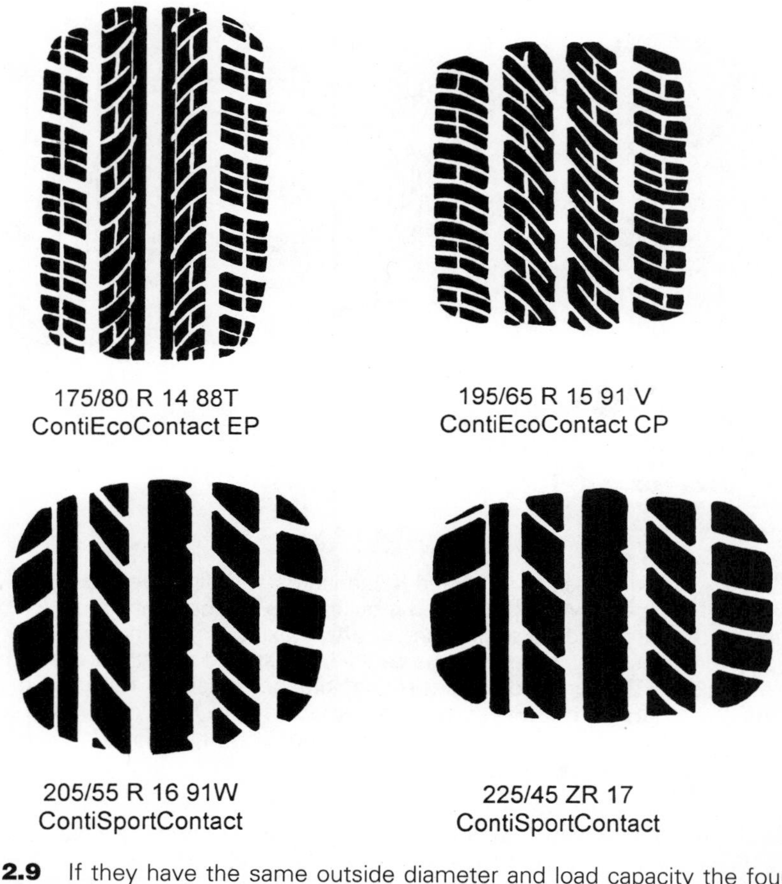

Fig. 2.9 If they have the same outside diameter and load capacity the four tyre sizes used on medium-sized passenger cars are interchangeable. The series 65, 55 and 45 wide tyres each allow a 1″ larger rim (and therefore larger brake discs). The different widths and lengths of the tyre contact patch, known as 'tyre print', are clearly shown (Fig. 3.119), as are the different designs of the standard road profile and the asymmetric design of the sports profile (see also Section 2.2.10). The 65 series is intended for commercial vehicles, and the 60, 55 and 45 series for sports cars. (Illustration: Continental; see also Fig. 2.19.)

Fig. 2.10 The flatter the tyre, i.e. the larger the rim diameter d (Fig. 2.11) in comparison with the outside diameter OD_T, the larger the brake discs or drums that can be accommodated, with the advantage of a better braking capacity and less tendency to fade. An asymmetric well-base rim is favourable (Figs 1.8 and 2.11).

Wheel rim diameter in inches	12	13	14	15	16	17
Brake disc outer diameter in mm	221	256	278	308	330	360
Brake drum inner diameter in mm	200	230	250	280	300	325

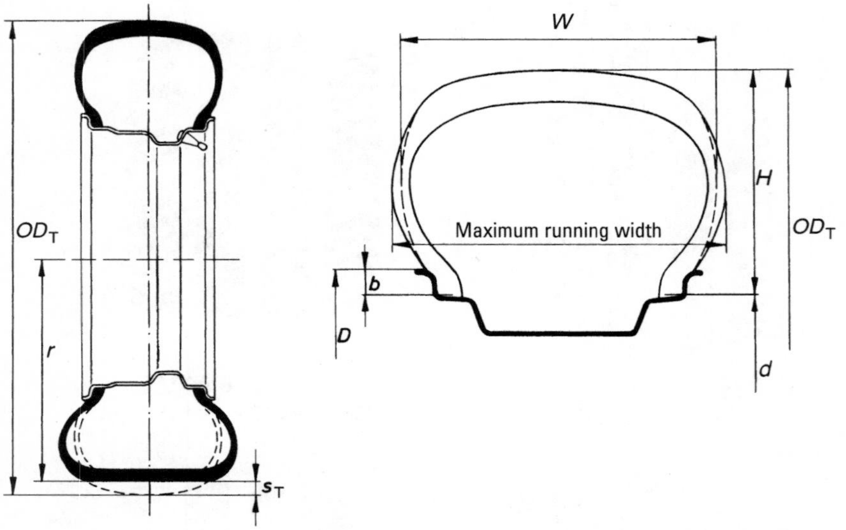

Fig. 2.11 Tyre dimensions specified in standards and directives. B is the cross-section width of the new tyre; the tread moulding (as can be seen in Fig. 2.1) is not included in the dimension. For clearances, the maximum running width with the respective rim must be taken into consideration, as should the snow chain contour for driven axles. The tyre radius, dependent on the speed, is designated r (see Section 2.2.8). Pictured on the left is an asymmetrical well-base rim, which creates more space for the brake caliper and allows a larger brake disc (Fig. 2.10).

mounted onto a measuring rim at a measuring tyre pressure of 1.8 bar or 2.3 bar on V-, W- or ZR tyres, Fig. 2.15):

$$H = 0.5\,(OD_T - d) \tag{2.1}$$

$$1'' = 1 \text{ in} = 25.4 \text{ mm} \tag{2.1a}$$

The 175/65 R 14 82 H tyre mounted on the measuring rim 5J × 14 can be taken as an example:

$OD_T = 584$ mm, $d = 14 \times 25.4 = 356$ mm and $W = 177$ mm

$H/W = [0.5 \times (OD_T - d)]/W = 114/177 = 0.644$

The cross-section ratio is rounded to two digits and given as a percentage. We talk of 'series', and here the ratio profile is 65% as shown in the tyre marking – in other words it is a 65 series tyre. A wider rim, e.g. 6J × 14 would give a smaller percentage.

2.2.5 Tyre dimensions and markings

2.2.5.1 Designations for passenger cars up to 270 km h^{-1}

The ETRTO standards manual of the European Tire and Rim Technical Organization includes all tyres for passenger cars and delivery vehicles up to 270 km h^{-1} and specifies the following data:

- tyre width in mm
- height-to-width ratio as a percentage
- code for tyre design
- rim diameter in inches or mm
- operational identification, comprising load index; LI (carrying capacity index) and speed symbol GSY.

The following applies to the type shown in Fig. 2.15:

175 / 65 R 14 82 H

- H – speed symbol (authorized up to 210 km h^{-1}, Fig. 2.12).
- 82 – load index (maximum load capacity 475 kg at 2.5 bar and 160 km h^{-1}, Figs 2.13 and 2.14).
- 14 – rim diameter in inches (Fig. 2.20).
- R – code for tyre design (R = radial, diagonal tyres have a dash '–' here (see Section 2.2.1 and Chapter 6 in Ref. 4).
- 65 – cross-section ratio profile as a % (can be omitted on 82 series or replaced by 80; see Section 2.2.5.2).
- 175 – width of the new tyre on the measuring rim and at measuring pressure of 1.8 bar.

Fig. 2.12 Standardized speed categories for radial tyres, expressed by means of a speed symbol and – in the case of discontinued sizes – by means of the former speed marking. Sizes marked VR or ZR may be used up to maximum speeds specified by the tyre manufacturer. The symbols F and M are intended for emergency (temporary use) spare wheels (see Chapter 6 in Ref. [5]).

v_{max} *in km/h*$^{-1}$	*Speed symbol*	*Identification*
80	F	
130	M	
150	P	
160	Q	
170	R	
180	S	
190	T	
210	H	
240	V	
270	W	
300	Y	
over 210	—	VR (old system)
over 240	—	ZR (old system)

The old markings can still be found on individual tyres:

155 S R 13

- 13 — rim diameter in inches
- R — radial tyre
- S — speed symbol (authorized up to 180 km h^{-1})
- 155 — width of the new tyre and 82 series, when details of the cross-section ratio missing

2.2.5.2 Designations of US tyres and discontinued sizes for passenger cars

Tyres manufactured in the USA and other non-European countries may also bear a 'P' for passenger car (see Fig. 2.17) and a reference to the cross-section ratio:

P 155/80 R 13 79 S

The old system applied up until 1992 for tyres which were authorized for speeds of over $V = 210$ km h^{-1} (or 240 km h^{-1}, Fig. 2.12); the size used by Porsche on the 928 S can be used as an example:

225/50 VR 16

- R — radial tyre
- V — speed symbol V (authorized over 210 km h^{-1})

Fig. 2.13 Load capacity/air pressure category specified in the directives. The load capacity on the left – also known as 'load index' (LI) – applies for all passenger cars up to the speed symbol W; they relate to the minimum load capacity values up to 160 km h^{-1} at tyre pressure 2.5 bar (see Section 2.2.6). Further criteria, such as maximum speed, handling etc., are important for the tyre pressures to be used on the vehicle. For LI values above 100, further load increases are in 25 kg increments:

LI = 101 corresponds to 825 kg,
LI = 102 corresponds to 850 kg etc. to
LI = 108 corresponds to 1000 kg.

Load index	*Wheel load capacity in kg with tyre pressure measured in bars* 1.5	1.6	1.7	1.8	1.9	2.0	2.1	2.2	2.3	2.4	2.5
69	215	225	240	250	260	270	285	295	305	315	325
70	225	235	245	260	270	280	290	300	315	325	335
71	230	240	255	265	275	290	300	310	325	335	345
72	235	250	260	275	285	295	310	320	330	345	355
73	245	255	270	280	295	305	315	330	340	355	365
74	250	260	275	290	300	315	325	340	350	365	375
75	255	270	285	300	310	325	335	350	360	375	387
76	265	280	295	310	320	335	350	360	375	385	400
77	275	290	305	315	330	345	360	370	385	400	412
78	280	295	310	325	340	355	370	385	400	410	425
79	290	305	320	335	350	365	380	395	410	425	437
80	300	315	330	345	360	375	390	405	420	435	450
81	305	325	340	355	370	385	400	415	430	445	462
82	315	330	350	365	380	395	415	430	445	460	475
83	325	340	360	375	390	405	425	440	455	470	487
84	330	350	365	385	400	420	435	450	470	485	500
85	340	360	380	395	415	430	450	465	480	500	515
86	350	370	390	410	425	445	460	480	495	515	530
87	360	380	400	420	440	455	475	490	510	525	545
88	370	390	410	430	450	470	485	505	525	540	560
89	385	405	425	445	465	485	505	525	545	560	580
90	400	420	440	460	480	500	520	540	560	580	600
91	410	430	450	475	495	515	535	555	575	595	615
92	420	440	465	485	505	525	550	570	590	610	630
93	430	455	475	500	520	545	565	585	610	630	650
94	445	470	490	515	540	560	585	605	625	650	670
95	460	485	505	530	555	575	600	625	645	670	690
96	470	495	520	545	570	595	620	640	665	685	710
97	485	510	535	560	585	610	635	660	685	705	730
98	500	525	550	575	600	625	650	675	700	725	750
99	515	540	570	595	620	650	675	700	725	750	775
100	530	560	590	615	640	670	695	720	750	775	800

Fig. 2.14 The tyre load capacity shown in the ETRTO standards manual in the form of the load index LI is valid for V tyres up to vehicle speeds of 210 km h^{-1}; for W tyres up to 240 km h^{-1} and for Y tyres up to 270 km h^{-1}. At higher speeds, lower percentages of the load capacity must be incurred; for VR and ZR tyres, which are no longer made, these values were determined by vehicle and tyre manufacturers.

	Tyre load capacity (%)		
Top speed of car		*Speed symbol*	
(km h^{-1})	*V*	*W*	*Y Tyres*
210	100	100	100
220	97	100	100
230	94	100	100
240	91	100	100
250	–	95	100
260	–	90	100
270	–	85	100
280	–	–	95
290	–	–	90
300	–	–	85

The following should be noted for VR tyres:

- over 210 km h^{-1} and up to 220 km h inclusive, the load may only be 90% of the otherwise authorized value;
- over 220 km h^{-1} the carrying capacity reduces by at least 5% per 10 km h^{-1} speed increment.

2.2.5.3 Designation of light commercial vehicle tyres

Tyres for light commercial vehicles have a reinforced substructure compared with those for passenger cars (Fig. 2.5), so they can take higher pressures, which means they have a higher load capacity. The suffix 'C' followed by information on the carcass strength (6, 8 or 10 PR) used to indicate suitability for use on light commercial vehicles, or the word 'reinforced' simply appeared at the end of the marking. The current marking (as for passenger cars) retains the speed symbol as well as the load index which, behind the slash, gives the reduced load capacity on twin tyres (Fig. 3.4). Compared with the previous marking, the new system is as follows:

Former	*Current*
–	205/65 R 15 98 S (Fig. 2.15)
185 SR 14	185 R 14 90 S
185 SR 14 reinforced	185 R 14 94 R
185 R 14 C 6 PR	185 R 14 99/97 M
185 R 14 C 8 PR	185 R 14 102/100 M

The 185 R 14 tyre is a passenger car size which is also fitted to light commercial vehicles.

2.2.5.4 Tyre dimensions

Figure 2.15 shows the important data for determining tyre size:

- size marking;
- authorized rims and measuring rim;
- tyre dimensions: width and outside diameter new and maximum during running;
- static rolling radius (Fig. 2.11);
- rolling circumference (at 60 km h^{-1}, Fig 2.16, see also Section 2.2.8);
- load capacity coefficient (load index LI, Fig. 2.13);
- tyre load capacity at 2.5 bar and up to 160 km h^{-1} (see Section 2.2.6).

2.2.6 Tyre load capacities and inflation pressures

The authorized axle loads $m_{V,f,max}$ and $m_{V,r,max}$ (see Section 5.3.5), and the maximum speed v_{max} of the vehicle, determine the minimum tyre pressure. However, the required tyre pressure may be higher to achieve optimum vehicle handling (see also Section 2.10.3.5 and Fig. 2.44).

2.2.6.1 Tyre load capacity designation

The load capacities indicated in the load index (item 6, Fig. 2.18) are the maximum loads per tyre permitted for all tyres up to the speed symbol 'H'. They are valid up to speeds of 210 km h^{-1} for tyres marked 'V' and up to 240 km h^{-1} for those marked 'R' 'W' or 'ZR'. For vehicles with a higher top speed, the load capacity has to be reduced accordingly.

Consequently, for tyres with speed symbol 'V', at a maximum speed of 240 km h^{-1} the load capacity is only 91% of the limit value (Fig. 2.14). Tyres designated 'W' on the sidewall are only authorized up to 85% at 270 km h^{-1}. In both cases the load capacity values between 210 km h^{-1} ('V' tyre) and 240 km h^{-1} ('W' tyre) and the maximum speed must be determined by linear interpolation.

For higher speeds (ZR tyres), the interpolation applies to the 240–270 km h^{-1} speed range. At higher speeds, the load capacity as well as the inflating pressure will be agreed between the car and tyre manufacturers. However, this approval does not necessarily apply to tyres which are specially produced for the US market and which bear the additional marking 'P' (Fig. 2.17 and Section 2.2.5.2).

2.2.6.2 Tyre pressure determination

For tyres with speed symbols 'R' to 'V' and standard road tyres the minimum pressures set out in the tables and corresponding with load capacities are valid up to 160 km h^{-1} (see Fig. 2.15 and Section 2.1.1).

Special operating conditions, the design of the vehicle or wheel suspension and expected handling properties can all be reasons for higher pressure specification by the vehicle manufacturer.

Further, for speeds up to 210 km h^{-1} the linear increase of basic pressure has to be by 0.3 bar (i.e. by 0.1 bar per Δv = 17 km h^{-1}; see also end of Section 2.84) and at speeds above 210 km h^{-1} the tyre load capacity has to be reduced

Fig. 2.15 Radial 65 series tyres, sizes, new and running dimensions, authorized rims and load capacity values (related to maximum 160 km h^{-1} and 2.5 bar); the necessary increase in pressures at higher speeds can be taken from Section 2.2.6. The tyre dimensions apply to tyres of a normal and increased load capacity design (see Section 2.2.5.3) and to all speed symbols and the speed marking ZR.

	Dimensions of new tyre				*Manufacturer's measurements*					
Tyre size	*Measuring rim*	*Width of cross-section*	*Outer diameter*	*Permissible rims according to DIN 7817 and DIN 7824*	*Max. width*	*Max. outer diameter*[4]	*Static radius ±2.0%*	*Circum-ference +1.5% −2.5%*	*Load index (LI)*	*Wheel load capacity*[5]
155/65 R 13	4.50 B × 13	157	532	4.00 B × 13[1]	158	540	244	1625	73	365
				4.50 B × 13[1]	164					
				5.00 B × 13[1]	169					
				5.50 B × 13[1]	174					
155/65 R 14	4½ J × 14	157	558	4 J × 14[2]	158	566	257	1700	74	375
				4½ J × 14[2]	164					
				5 J × 14[2]	169					
				5½ J × 14[2]	174					
165/65 R 13	5.00 B × 13	170	544	4.50 B × 13[1]	171	533	248	1660	76	400
				5.00 B × 13[1]	176					
				5.50 B × 13[1]	182					
				6.00 B × 13[1,3]	187					
165/65 R 14	5 J × 14	170	570	4½ J × 14[2]	171	579	261	1740	78	425
				5 J × 14[2]	176					
				5½ J × 14[2]	182					
				6 J × 14	187					
175/65 R 13	5.00 B × 13	177	558	5.00 B × 13[1]	184	567	254	1700	80	450
				5.50 B × 13[1]	189					
				6.00 B × 13[1,3]	194					
175/65 R 14	5 J × 13	177	584	5 J × 14[2]	184	593	267	1780	82	475
				5½ J × 14[2]	189					
				6 J × 14	194					
175/65 R 15	5 J × 15	177	609	5 J × 15[2]	184	618	279	1855	83	487
				5½ J × 15[2]	189					
				6 J × 15	194					
185/65 R 13	5.50 B × 14	189	570	5.50 B × 13[1]	191	580	259	1740	84	500
				5.50 B × 13[1]	197					
				6.00 B × 13[1,3]	202					
				6½ J × 13	207					
185/65 R 14	5½ J × 14	189	596	5 J × 14	191	606	272	1820	86	530
				5½ J × 14	197					
				6 J × 14	202					
				6½ J × 14	207					

185/65 R 15	5½ J × 15	189	621	5 J × 15	191	631	284	1895	88	560
				5½ J × 15	197					
				6 J × 15	202					
				6½ J × 15	207					
195/65 R 14	6 J × 14	201	610	5½ J × 14	204	620	277	1860	89	580
				6 J × 14	209					
				6½ J × 14	215					
				7 J × 14	220					
195/65 R 15	6 J × 15	201	635	5½ J × 15	204	645	290	1935	91	615
				6 J × 15	209					
				6½ J × 15	215					
				7 J × 15	220					
205/65 R 14	6 J × 14	209	622	5½ J × 14	212	633	282	1895	91	615
				6 J × 14	217					
				6½ J × 14	222					
				7 J × 14	227					
				7½ J × 14	233					
205/65 R 15	6 J × 15	209	647	5½ J × 15	212	658	294	1975	94[6]	670
				6 J × 15	217					
				6½ J × 15	222					
				7 J × 15	227					
				7½ J × 15	233					
215/65 R 15	6½ J × 15	221	661	6 J × 15	225	672	300	2015	96[7]	710
				6½ J × 15	230					
				7 J × 15	235					
				7½ J × 15	240					
215/65 R 16	6½ J × 16	221	686	6 J × 16	225	697	312	2090	98	750
				6½ J × 16	230					
				7 J × 16	235					
				7½ J × 16	240					
225/65 R 15	6½ J × 15	228	673	6 J × 15	232	685	304	2055	99	775
				6½ J × 15	237					
				7 J × 15	242					
				7½ J × 15	248					
				8 J × 15	253					

[1] Instead of wheel rims with the identification letter B, same-sized rims with the identification letter J may be used. For example 5½ J × 13 instead of 5.50 B × 13. (See Section 2.3.2.)

[2] Instead of wheel rims with the identification letter J, same-sized rims with the identification letter B may be used. For example 4.50 B × 14 instead of 4½ J × 14.

[3] The wheel rims without identification letters mentioned in the table are expected to be identified with DIN 7824 Part 1.

[4] The outer diameter of wheels with M & S – tread can be up to 1% bigger than the standard tread.

[5] Maximum in kg at 2.5 bar.

[6] Reinforced model, 750 kg at 3.0 bar (LI 98).

[7] Reinforced model, 800 kg at 3.0 bar (LI 100).

Fig. 2.16 Factor k_v, which expresses the speed dependence of the rolling circumference of passenger vehicle radial tyres above 60 km h^{-1} as a percentage. The permissible tolerances Δk_v have to be added (see Section 2.2.8), all taken from the German WDK Guideline 107, page 1.

v (km h^{-1})	60	90	120	150	180	210	240
Factor k_v (%)	–	+0.1	+0.2	+0.4	+0.7	+1.1	+1.6
Deviation Δk_v (%)	–	±0.1	±0.2	±0.4	±0.7	±1.1	±1.6

Fig. 2.17 ZR tyres manufactured specially for the American market and marked with a 'P' do not meet the European standard and are therefore not authorized here (photograph: Dunlop factory).

in accordance with item 2.2.6.1. If the tyre load is lower than the maximum load capacity, a lower additional safety pressure can be used in consultation with the tyre manufacturer.

For tyres with the speed symbol 'W', the pressures in Fig. 2.13 apply up to 190 km^{-1}. After this it has to be increased by 0.1 bar for every 10 km h^{-1} up to 240 km h^{-1}. For higher speeds, the load capacity must be reduced (see Section 2.2.6.1).

On vehicles, pressure should be tested on cold tyres, i.e. these must be adjusted to the ambient temperature. If the tyre pressure is set in a warm area in winter there will be an excessive pressure drop when the vehicle is taken outside.

On M & S winter tyres it has long been recommended that inflation pressures be increased by 0.2 bar compared with standard tyres. Newer brands of tyre no longer require this adjustment.

2.2.6.3 Influence of wheel camber

Wheel camber angles ε_W considerably influence tyre performance and service life. The camber angle should therefore not exceed 4° even in full wheel jounce condition. For angles above ±2° (see Section 3.5.1), the loadability of the tyres reduces at

$\varepsilon_W > 2°$ to 3° to 95%
$\varepsilon_W > 3°$ to 4° to 95%

Intermediate values have to be interpolated. Compensation can be achieved by increasing the inflation pressure. The values are as follows:

Camber angle	2°20′	2°40′	3°	3°20′	3°40′	4°
Pressure increase	2.1%	4.3%	6.6%	9.0%	11.5%	14.1%

Taking all the influences into account, such as top speed, wheel camber and axle load, the minimum tyre pressure required can be calculated for each tyre category (size and speed symbol). Formulas are shown in the 'WdK 99' guidelines from the Wirtschaftsverband der Deutschen Kautschukindustrie.

2.2.6.4 Tyre pressure limit values

Tyre pressure limit values should be adhered to. These values are

Q and T tyres	3.2 bar
H to W and ZR tyres	3.5 bar
M & S tyres (Q and T tyres)	3.5 bar

2.2.7 Tyre sidewall markings

All tyres used in Europe should be marked in accordance with the ETRTO standards (see Section 2.1.1).

In the USA, Japan and Australia, additional markings are required to indicate the design of the tyre and its characteristics. The characters must also bear the import sizes – the reason why these can be found on all tyres manufactured in Europe (Fig. 2.18).

2.2.8 Rolling circumference and driving speed

The driving speed is:

$$v = 0.006(1 - S_{X,W,a})\frac{C_{R,dyn} \times n_M}{i_D \times i_G} \text{ (km/h)} \tag{2.1b}$$

This includes:

$S_{X,W,a}$	the absolute traction slip (Equation 2.4f)
$C_{R,dyn}$	the dynamic rolling circumference in m (Equation 2.1d)
n_M	the engine speed in rpm
i_D	the ratio in the axle drive (differential)
i_G	the ratio of the gear engaged (Equation 6.36)

The following can be assumed for slip $S_{X,W,a}$:

1st gear	0.08	4th gear	0.035
2nd gear	0.065	5th gear	0.02
3rd gear	0.05		

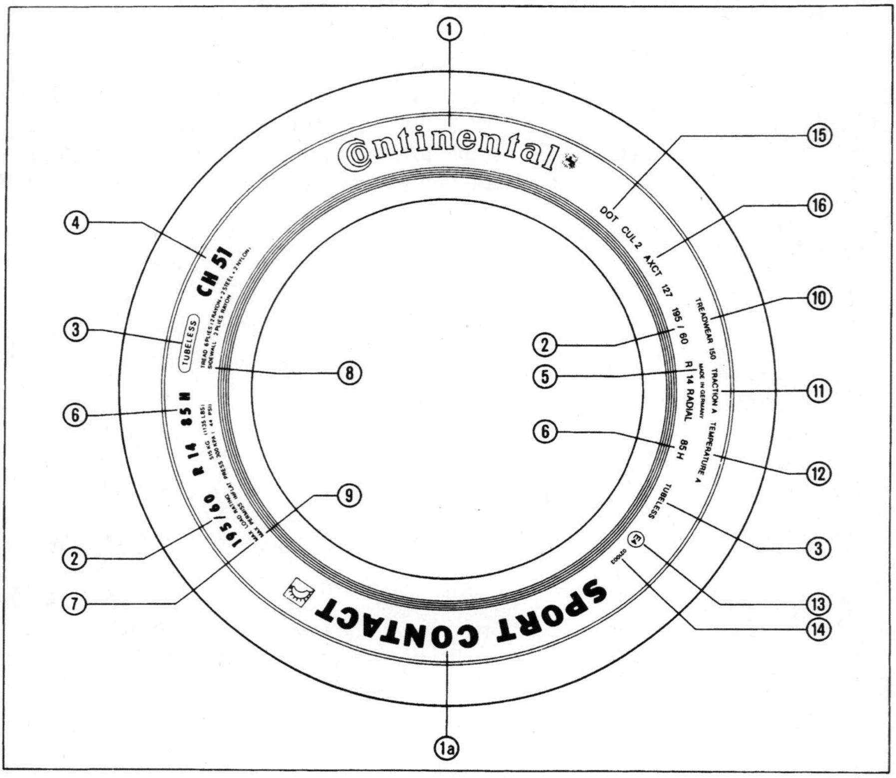

Fig. 2.18 Explanation of the marking on the sidewall of a tyre manufactured by Pneumatiques Kléber SA:

Legal and industry standard markings on the sidewalls of tyres according to:
FMVSS and CIR 104
UTQG (USA)
CSA Standard (Canada)
ADR 23B (Australia)
ECE–R30 (Europe)

1 Manufacturer (brand)
1a Product name
2 Size marking
195 = nominal tyre wideth in mm
60 = height–width ratio (60%)
radial type construction
14 rim diameter in inches
3 Tubeless
4 Trade code
5 Country of manufacture
6 Load capacity index (LI)
7 Maximum load capacity for the USA
8 Tread: under the tread are 6 plies carcass rayon, 2 plies steel belt, 2 plies nylon) Sidewall: the substructure consists of 2 plies rayon
9 Maximum tyre pressure for the USA
10, 11, 12 USA: manufacturer's guarantee of compliance with the Uniform Tire Quality Grade (UTQG) which specifies: 10 tread wear: relative life expectancy compared with US-specific standard test values; 11 traction: A, B, C = braking performance on wet surfaces 12 temperature resistance: A, B or C = temperature resistance at higher test stand speeds; C fulfills the legal requirement in the USA
13 E 4 = tyre fulfils the ECE R30 value requirements
4 = country in which approval was carried out
(4 = The Netherlands)
14 identity number according to ECE R-30
15 DOT = tyre fulfils the requirements according to FMVSS 109 (DOT = Department of Transportation)
16 Manufacturer's code:
CU = factory (Continental)
L2 = tyre size
AXCT = model
127 = date of manufacture: production week 12, 1987

According to DIN 75020 Part 5, the rolling circumference C_R given in the tyre tables relates to 60 km/h and operating pressure of 1.8 bar. At lower speeds it goes down to $C_{R,stat}$:

$$C_{R,stat} = r_{stat}\, 2\pi \tag{2.1c}$$

The values for r_{stat} are also given in the tables. At higher speeds, C_R increases due to the increasing centrifugal force. The dynamic rolling circumference $C_{R,dyn}$ at speeds over 60 km h^{-1} can be determined using the speed factor k_v. Figure 2.16 shows the details for k_v as a percentage, increasing by increments of 30 km h^{-1}. Intermediate values must be interpolated. The circumference would then be:

$$C_{R,dyn} = C_R\,(1 + 0.01 \times k_v)\ \text{(mm)} \tag{2.1d}$$

The dynamic rolling radius can be calculated from $C_{R,dyn}$ as

$$r_{dyn} = C_R/2\pi$$

or, at speeds of more than 60 km h^{-1},

$$r_{dyn} = C_{R,dyn}/2\pi \tag{2.2}$$

Taking as an example the tyre 175/65 R 14 82 H at v = 200 km h^{-1} (Fig. 2.15) gives

$$k_{v180} = 0.7\% \text{ and } k_{v210} = 1.1\%$$

and interpolation gives:

$$k_{v200} = 0.007 + 0.0027 = 0.0097$$
$$k_{v200} = 0.97\%$$

The rolling circumference C_R taken from Fig. 2.15, according to Equation 2.1d, gives

$$C_{R,dyn200} = 1780 \times (1 + 0.0097) = 1797 \text{ mm}$$

and thus the dynamic radius in accordance with Equation 2.2 is:

$$r_{dyn60} = 283 \text{ mm and } r_{dyn200} = 286 \text{ mm}$$

The outside diameter (construction measure) is

$$OD_T = 584 \text{ mm and thus } OD_T/2 = 292 \text{ mm}$$

a value which shows the extent to which the tyre becomes upright when the vehicle is being driven: r_{dyn} is only 9 mm or 6 mm less than $OD_T/2$. Chapter 3 of Ref. [3] gives further details.

2.2.9 Influence of the tyre on the speedometer

The speedometer is designed to show slightly more than, and under no circumstances less than, the actual speed. Tyres influence the degree of advance, whereby the following play a role:

- the degree of wear
- the tolerances of the rolling circumference
- the profile design
- associated slip.

The EC Council directive 75/443, in force since 1991, specifies an almost linear advance Δv,

$$+\Delta v \leqslant 0.1 \times v + 4 \ (\text{km h}^{-1}) \tag{2.2a}$$

On vehicles registered from 1991 onwards the values displayed may only be as follows:

Actual speed (km h^{-1})	30	60	120	180	240
Max displayed value (km h^{-1})	37	70	136	202	268

As Fig. 2.15 indicates, at 60 km h^{-1} the rolling circumference C_R has a tolerance range of $\Delta C_R = +1.5\%$ to -2.5%, and according to Fig. 2.16 with a speed factor of k_v, deviations of up to $\Delta k_v = \pm 1.6\%$ are possible. When related to the dynamic rolling circumference $C_{R,dyn}$ (Equation 2.1d), the following tolerance limits (rounded to the nearest figure) may prevail and result in the displayed values when only the minus tolerances are considered, and if the speedometer has the maximum authorized advance:

Actual speed (km h^{-1})	60	120	180	240
Possible overall tolerance (%)	+1.5	+1.7	+2.2	+3.1
	−2.5	−2.7	−3.2	−4.1
Max display value at minus tolerance (km h^{-1})	72	140	208	279

The slip should be added directly to this, which in direct gear amounts to around 2% (see equations 2.1b and 2.4f), in other words

$$S_{X,W,a} = 0.02$$

If the manufacturer fully utilizes the advance specified in Equation 2.2a, it is possible that although the speedometer indicates 140 km h^{-1}, the vehicle is only moving at 120 km h^{-1}. This occurs, in particular, when the tyres are worn:

3 mm wear gives an advance of around 1%

Fig. 2.19 Designs of Continental tyre. (Top) Summer tyre (tyre foot prints see Fig. 2.9) EcoContact EP (size 185/65 R14T) and Sport Contact (size 205/55 R16W). (Below) Winter tyre WinterContact TS760 (size 185/65 R14T) and WinterContact TS770 (size 235/60 R16H).

Tyres with an M & S winter profile can, however, have a 1% larger outside diameter so that the profile can be deeper (Fig. 2.15, note 5 and Fig. 2.19). They would therefore reduce the degree by which the speedometer is advanced if the tyres are not yet worn. The same applies where the positive tolerances given in the above table are used. In this instance it is also possible that even a very precise speedometer could display too low a speed.

2.2.10 Tyre profiles

The design of tyre profiles (Fig. 2.19) depends on the intended use, taking into account the parameters of height-to-width ratio, construction and mixture and design. The aquaplaning properties are improved by increasing the negative proportion (light places in the tyre impression, Fig. 2.9). The shoulder region with its transverse water-drainage grooves is particularly important for its properties in a lateral direction and the middle region with straight longitudinal grooves is important for its properties in a longitudinal direction. An asymmetrical profile design ('sports' profile) is chosen for wide tyres, tread lugs in the outside shoulder, which are subject to greater stress during cornering, can be designed to be more rigid. By adjusting the correct balance between profile rigidity and belt rigidity, it must be ensured that no conical forces are produced. Profiled bands around the middle region increase noise reduction and improve the steering response properties and, via the increase in circular rigidity, the brake response properties.

Winter tyre profiles are improved, in terms of their force transmission properties in the wet, snow and ice, by a higher negative profile component, transverse grooves and a large number of sipes. Directional profiles (TS770) can be used to increase water dispersal, the longitudinal force coefficient and self-cleaning by means of transverse grooves which run diagonally outwards. Noise control is improved by variation in block length, sipes cut up to under the groove base or ventilation grooves running around the tyre.

2.3 Wheels

2.3.1 Concepts

Tyres are differentiated according to the loads to be carried, the possible maximum speed of the vehicle, and whether a tubed or tubeless tyre is driven. In the case of a tubeless tyre, the air-tightness of the rim is extremely important. The wheel also plays a role as a 'styling element'. It must permit good brake ventilation and a secure connection to the hub flange (see Chapter 9, in Ref. [6]). Figure 2.20 shows a passenger car rim fitted with a tubeless tyre.

2.3.2 Rims for passenger cars, light commercial vehicles and trailers

For these types of vehicle only well-base rims are provided. The dimensions of the smallest size, at 12″ and 13″ diameter and rim width up to 5.0″, are contained in the standard DIN 7824. The designation for a standard rim, suitable for the 145 R 13 tyre (Fig. 2.1) for example is:

DIN 7824 – drop base rim 4.00 B $\times$ 13

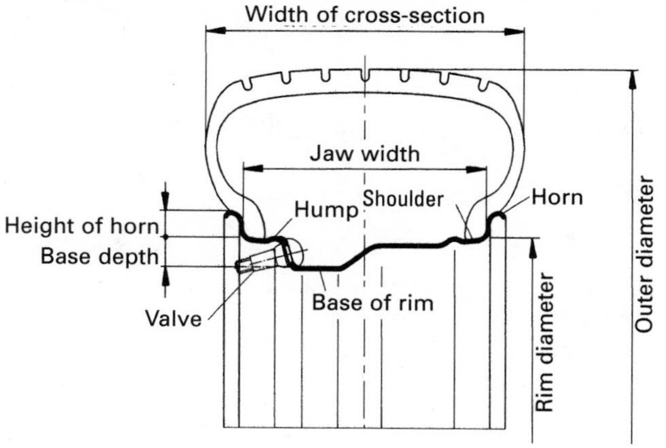

Fig. 2.20 Series 55 wide tyre designs, mounted on a double hump rim with the inflating valve shown in Fig. 2.6. The actual rim consists of the following:

- rim horns, which form the lateral seat for the tyre bead (the distance between the two rims is the jaw width a);
- rim shoulders, the seat of the beads, generally inclined at 5° ± 1° to the centre where the force transfer occurs around the circumference (Fig. 2.5);
- well base (also known as the inner base), designed as a drop rim to allow tyre fitting, and mostly shifted to the outside (diagram: Hayes Lemmerz).

This type of rim used on passenger cars up to around 66 kW (90 PS) has only a 14 mm high rim flange and is identified with the letter B. The DIN standard can generally be dropped.

In order to make it possible to fit bigger brakes (Fig. 2.10), more powerful vehicles have larger diameter rims as follows:

- series production passenger cars: 14″ to 17″ rims
- sports cars: 16″ to 18″ rims.

The J rim flange applied here is used on rims from 13″ upwards and is 17.3 mm high. The rim base can (as shown in Fig. 2.1) be arranged symmetrically or shifted outwards. The rim diameter, which is larger on the inside, creates more space for the brake (Figs 1.8, 1.56, 2.10, 2.11 and 2.20). DIN 7817 specifies the rim widths from $3\frac{1}{2}$″ to $8\frac{1}{2}$″. The definition of a normal asymmetrical rim with a 5″ width, J rim flange and 14″ diameter is:

DIN 7817 drop base rim – 5 J × 14

The symmetrical design is identified by the suffix 'S'. The standards also contain precise details on the design and position of the valve hole (see also Figs 2.20 and 2.24).

C tyres for light commercial vehicles require a broader shoulder (22 mm

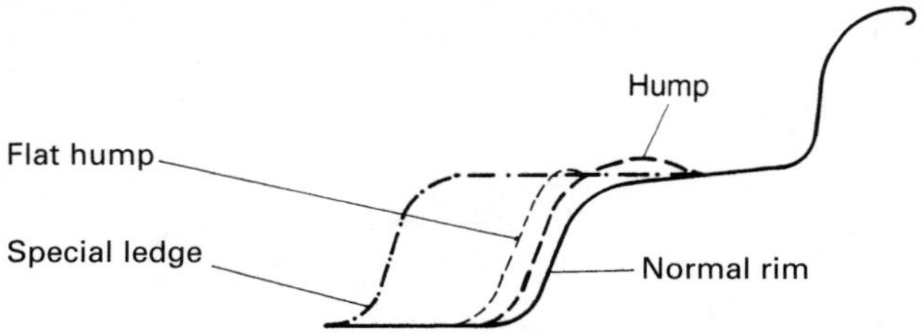

Fig. 2.21 Standard rim and contours of the safety shoulders which can be used on passenger cars and light commercial vehicles.

instead of 19.8 mm), which can be referred to by adding the letters LT (light truck) at the end of the marking:

DIN 7817 drop base rim – $5\frac{1}{2}$ J × 15 – LT

There is a preference worldwide for using tubeless radial tyres on passenger cars and light commercial vehicles. Where these tyres are used, it is essential to have a 'safety contour' at least on the outer rim shoulder. This stops air suddenly escaping if the vehicle is cornering at reduced tyre pressure.

The three different contours mainly used are (Fig. 2.21):

Hump	(H, previously H1)
Flat-hump	(FH, previously FHA)
Contre Pente	(CP)

Sheets 2 and 3 of DIN 7817 specify the dimensions of the first two designs. The 'hump' runs around the rim, which is rounded in H designs, whereas a flat hump rim is simply given a small radius towards the tyre foot. The fact that the bead sits firmly between the hump and rim flange is advantageous on both contours. An arrangement on both the outside and inside also prevents the tyre feet sliding into the drop bases in the event of all the air escaping from the tyre when travelling at low speeds, which could otherwise cause the vehicle to swerve. The disadvantage of hump rims is that changing the tyre is difficult and requires special tools.

A French design, intended only for passenger car rims, is the 'Contre Pente' rim, known as the CP for short. This has an inclined shoulder towards the rim base, which for rim widths between 4″ and 6″ is provided on one or both sides.

For years, the rims of most passenger cars have had safety shoulders on both sides, either a double hump (Figs 2.20 and 2.24) or the sharp-edged flat-hump on the outside and the rounder design on the inside (Fig. 2.23). The desired contour must be specified in the rim designation. Figure 2.22 gives the possible combinations and abbreviations which must appear after the rim diameter data. A complete designation for an asymmetrical rim would then be as follows:

Fig. 2.22 Marking of the various safety shoulders when used only on the outside of the rim or on both the inside and outside. Normal means there is no safety contour (Fig. 2.1). Further details are contained in standard DIN 7817.

	Nature of safety shoulder		
Denomination	*Outside of rim*	*Inside of rim*	*Identification letters*
One-sided hump	Hump	Normal	H
Double hump	Hump	Hump	H2
One-sided flat hump	Flat hump	Normal	FH[1]
Double-sided flat hump	Flat hump	Flat hump	FH2[1]
Combination hump	Flat hump	Hump	CH[2]

[1] In place of the identification letters FH the identification letters FHA were also permitted.
[2] In place of the identification letters CH the identification letters FH1-H were also permitted.

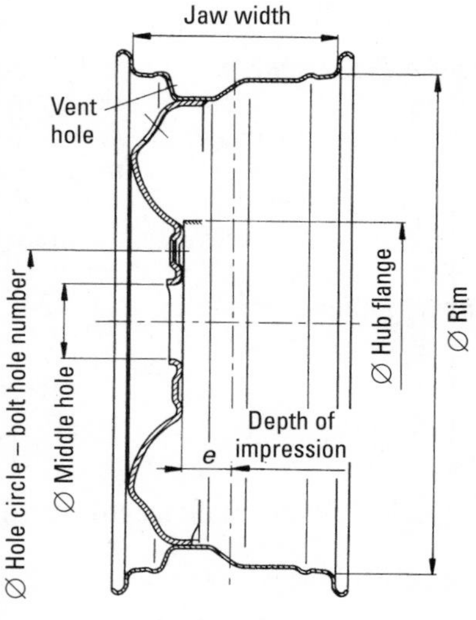

Fig. 2.23 The sheet metal disc-type wheel used in series production vehicles consists of a rim and disc. To avoid fatigue fractures, the wheel hub flange diameter should be greater than the dish contact surface. Wheel offset *e* (depth of impression) and kingpin offset at ground r_σ are directly correlated. A change in *e* can lead to an increase or a reduction in r_σ.

The dome-shaped dish leading to the negative kingpin offset at ground is clearly shown (diagram: Hayes Lemmerz).

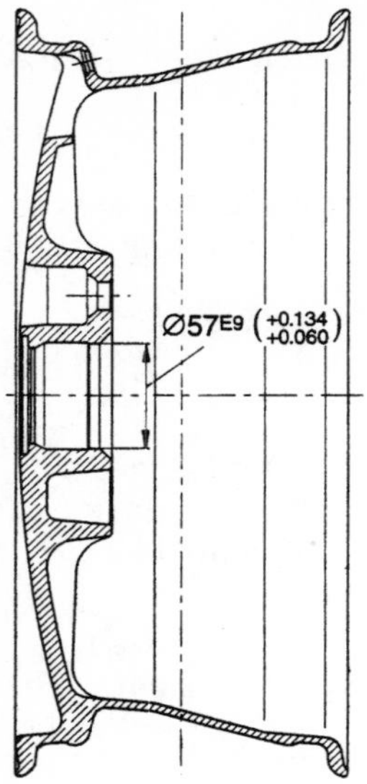

Fig. 2.24 Hayes Lemmerz alloy wheel for the Audi 80, made of the aluminium alloy GK-Al Si 7 Mg wa. The wheel has a double-hump rim (H2) and middle centring and is fixed with four spherical collar bolts. The different wall thicknesses, which are important for the strength, the shape of the bolt hole, the different shape of the drop-rim and the position of the valve hole are clearly shown. At high speeds the snap-fit valve (Fig. 2.6) is pressed outwards by the centrifugal force and supported below the rim base.

2.3.3 Wheels for passenger cars, light commercial vehicles and trailers

Most passenger cars and light commercial vehicles are fitted with sheet metal disc wheels, because these are economic, have high stress limits and can be readily serviced. They consist of a rim and a welded-on wheel disc (also known as an attachment face, Fig. 2.23). Cold-formable sheet metal, or band steel with a high elongation, can be used (e.g. RSt37-2 to European standard 20) depending on the wheel load, in thicknesses from 1.8 to 4.0 mm for the rim and 3.0 to 6.5 mm for the attachment faces.

There is a direct correlation between wheel offset e and 'kingpin offset at ground' r_σ; the more positive r_σ, the smaller can be the depth dimension e. However, a negative kingpin offset $-r_\sigma$, especially on front-wheel drive, results in a significant depth e and severe bowing of the attachment faces (as can be seen in Figs 2.8, 2.23, 2.25 and 3.102 and Section 7.3 in Ref. [6]).

The wheel disc can be perforated to save weight and achieve better brake cooling. Despite the fact that they cost almost four times as much as sheet metal designs, alloy wheels are becoming increasingly popular (Figs 1.56 and 2.24). Their advantages are:

- lower masses;
- extensive styling options; and therefore
- better appearance;
- processing allows precise centring and limitation of the radial and lateral runout (see Section 2.5);
- good heat transfer for brake-cooling (see Chapter 9 in Ref. [6]).

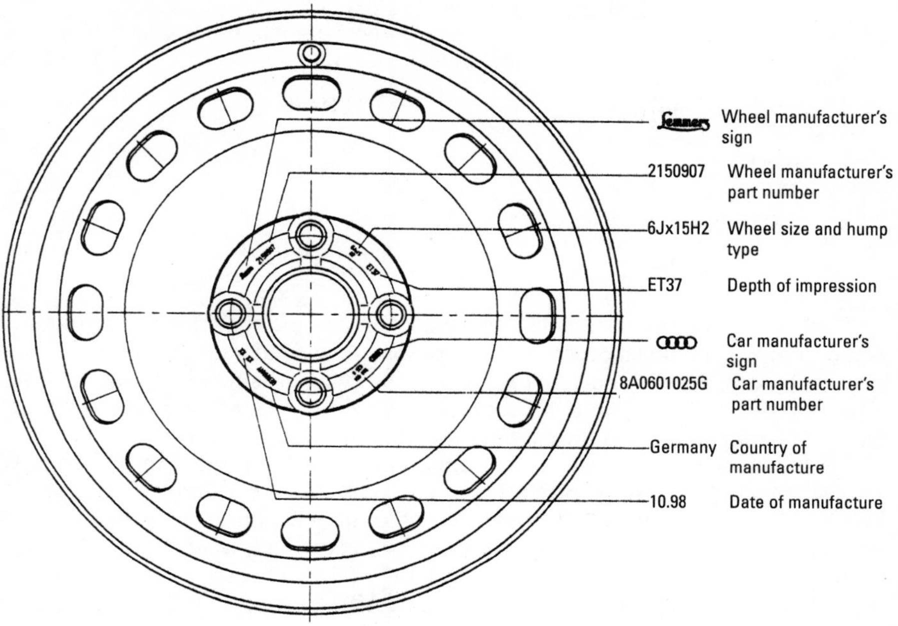

Fig. 2.25 Double-hump sheet metal disc-type wheel with openings for cooling the brakes. Also pictured is the stamp in accordance with the German standard DIN 7829, indicating manufacturer code, rim type and date of manufacture (week or month and year).

Also specified is the wheel offset (ET37) and, in the case of special wheels with their own ABE (General operating approval), the allocation number of the KBA, the German Federal Vehicle Licensing Office. If there is not much space the stamp may be found on the inside of the dish. The date of manufacture also points to when the vehicle was manufactured (diagram: Hayes Lemmerz).

Often incorrectly called aluminium rims, alloy wheels are mainly manufactured using low-pressure chill casting, occasionally forging or aluminium plate, and generally consist of aluminium alloys with a silicon content (which are sometimes heat hardenable), such as GK-Al Si 11 Mg, GK-Al Si 7 Mg T (T = tempered after casting) etc.

Regardless of the material, the wheels must be stamped with a marking containing the most important data (Fig. 2.25).

2.3.4 Wheel mountings

Many strength requirements are placed on the wheel disc sitting in the rim (or the wheel spider on alloy wheels); it has to absorb vertical, lateral and longitudinal forces coming from the road and transfer them to the wheel hub via the fixing bolts.

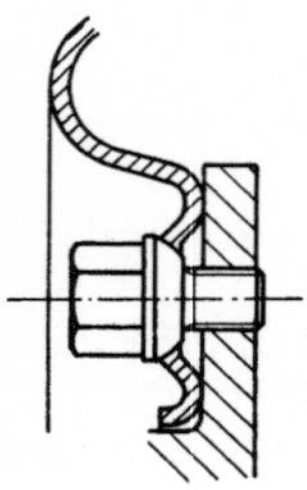

Fig. 2.26 Depression design with special springing characteristics on a passenger car sheet metal disc-type wheel. The wheel can be centred using the fixing bolts or by fitting into the toleranced hole (Fig. 2.24).

The important thing here is that the contact area of the attachment faces, known as the 'mirror', should sit evenly and, for passenger cars, that the hub flange should have a slightly larger diameter (Fig. 2.23), otherwise it is possible that the outer edge of the hub will dig into the contact area, with a loss of torque on the bolts. The notch effect can also cause a fatigue fracture leading to an accident.

The number of holes and their circle diameter are important in this context. This should be as large as possible to introduce less force into the flange and fixing bolts. If the brake discs are placed onto the wheel hub from the outside – which is easier from a fitting point of view – it is difficult to create a hole larger than 100 mm on 13″ wheels, and using a 14″ or 15″ wheel should make for the best compromise (Figs 1.8, 1.41, 1.44 and 2.10). German standard DIN 74361 contains further details.

The brake disc can also be fixed to the wheel hub from the inside (Fig. 1.38). However, the disadvantage of this is that the hub has to be removed before the disc can be changed. This is easy on the non-driven axle, but time-consuming on the driven axle (see Section 2.5 in Ref. 2 and Chapter 9 in Ref. 6). This brief look shows that even the brakes play a role in the problems of fixing wheels.

Nowadays, wheels are almost always fixed with four or five metric M12 × 1.5 or M14 × 1.5 DIN 74361 spherical collar bolts. The high friction between the spherical collar and the stud hole prevents the bolts from coming loose while the vehicle is in motion. For this reason, some car manufacturers keep the contact surface free of paint. On sheet metal disc wheels with attachment faces up to 6.5 mm thick, the spring action of the hole surround (Fig. 2.26) is an additional safety feature, which also reduces the stress on the wheel bolts as a result of its design elasticity. Sheet metal rings are often inserted in the alloy wheels to withstand high stresses underneath the bolt head.

Generally, the spherical collar nuts also do the job of centring the wheels on the hub. Hub centring has become increasingly popular because of a possible hub or radial run-out and the associated steering vibrations. A toleranced collar placed on the hub fits into the dimensioned hole which can be seen in Fig. 2.24.

2.4 Springing behaviour

The static tyre spring rate c_T – frequently also known as spring stiffness or (in the case of a linear curve) spring constant – is the quotient of the change in vertical force $\Delta F_{Z,W}$ in Newtons and the resultant change Δs_T – the compression in mm within a load capacity range corresponding to the tyre pressure p_T (Fig. 2.27; see also Section 2.2.5.4):

$$c_T = \Delta F_{Z,W}/\Delta s_T \text{ (N/mm}^{-1}\text{)} \tag{2.3}$$

The parameter c_T forms part of the vibration and damping calculation and has a critical influence on the wheel load impact factor (see Section 5.2 in Ref. [3], Section 4.1). The stiffer the tyre, the higher the damping must be set and the greater the stress experienced by the chassis components. The following parameters influence the spring rate:

- vertical force
- tyre pressure
- driving speed
- slip angle
- camber angle
- rim width

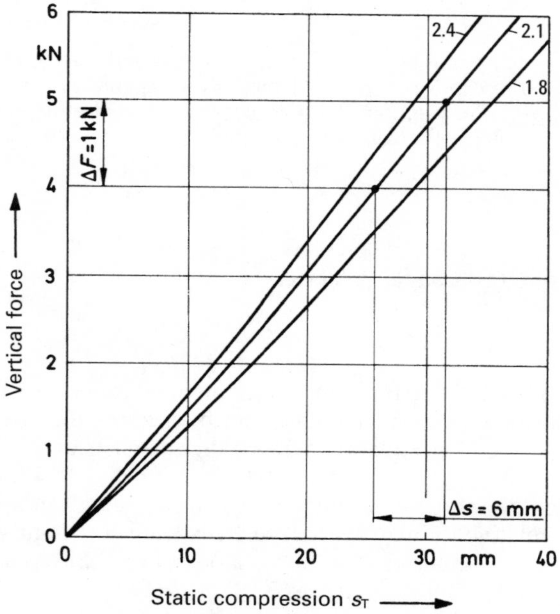

Fig. 2.27 The static tyre spring rate c_T is the quotient of the force and the deflection travel shown on the radial tyre 175/70 R 13 80 S at p_T = 1.8 bar, 2.1 bar and 2.4 bar; the example shown gives:

$$C_T = \frac{\Delta F_{Z,W}}{\Delta S_T} = \frac{1000 \text{ N}}{6 \text{ mm}} = 167 \text{ N/mm}$$

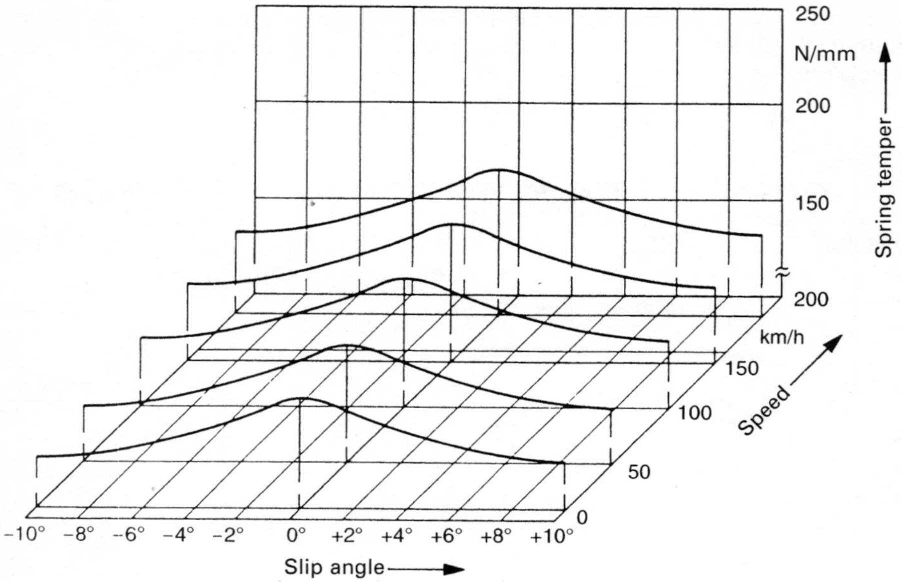

Fig. 2.28 Tyre springing rate as a function of slip angle and road speed, measured on a radial tyre 185/70 R 13 86 S at p_T = 2.1 bar. Speed increases the springing rate as the belt stands up due to the centrifugal force. However, the slip angle makes it softer because the belt is pushed away to the side and the shoulders take over part of the springing effect.

- height-to-width ratio
- construction of tyre (bias angle, material)
- tyre wear and tear
- wheel load frequency.

As can be seen in Fig. 2.27, apart from in the low load range, the spring rate is independent of the load. A linear increase can be seen as the speed increases (Figs 2.16 and 2.28; see also Equation 5.5a), which persists even when the tyre pressure changes.

During cornering, the force $F_{Y,W}$ (Fig. 3.119) shifts the belt in a lateral direction, and so it tips relative to the wheel plane. This leads to a highly asymmetrical distribution of pressure and (as can be seen from Fig. 2.28) to a reduction in the spring rate as the slip angles increase.

2.5 Non-uniformity

The tyre consists of a number of individual parts, e.g. carcass layers, belt layers, running tread, sidewall stock and inner lining, which – put together on a tyre rolling machine – give the tyre blank (Fig. 2.5). In the area where it is put together, variations in thickness and stiffness occur, which can lead to non-uniformity.

Owing to the irregularities caused during manufacture, the following occur around the circumference and width of the tyre:

- thickness variations
- mass variations
- stiffness variations.

These cause various effects when the tyre rolls:

- imbalance
- radial tyre runout
- lateral tyre runout
- variation in vertical and/or radial force
- lateral force variations
- longitudinal force variation
- ply steer (angle) force
- conicity force.

Imbalance U occurs when an uneven distribution of mass and the resulting centrifugal forces are not equalized. Because the uneven distribution occurs not only around the circumference, but also laterally, we have to differentiate between static and dynamic imbalance (Fig. 2.29). This is calculated in size and direction on balancing machines and eliminated with balancing weights on the rim bead outside and inside the wheel.

Radial and lateral runout are the geometrical variations in the running tread and the sidewalls. They are measured with distance sensors on a tyre-uniformity machine. The German WdK Guideline 109 contains full details.

The most important of the three force variations is the radial force variation. For greater clarity, it is shown on the model in Fig. 2.30, where the tyre consists of different springs whose rates fluctuate between c_1 and c_8. The resulting phenom-

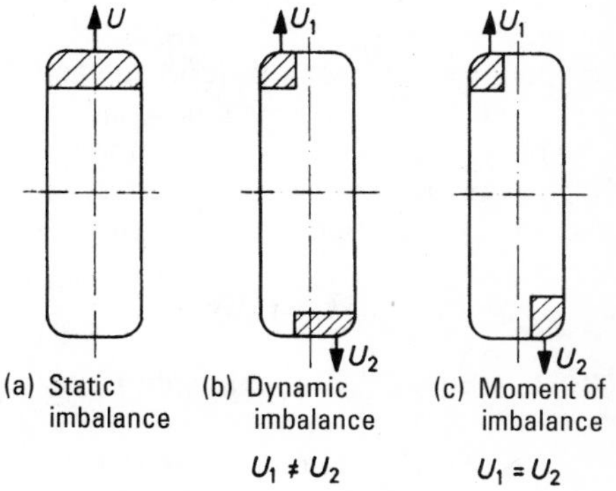

Fig. 2.29 Different forms of imbalance U: (a) static, (b) dynamic. The imbalance is equalized in (c).

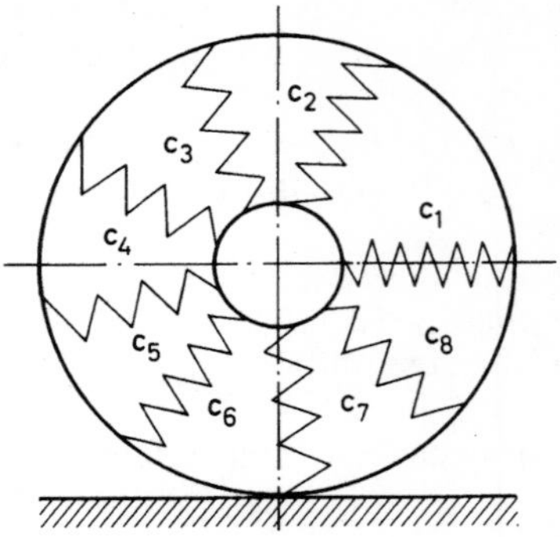

Fig. 2.30 The tyre spring rate can fluctuate depending on the manufacturing process, shown as c_1 to c_8.

ena should be indicated on the 175 R 14 88 S steel radial tyre, loaded at $F_{Z,W}$ = 4.5 kN and pressurized to p_T = 1.9 bar. Assuming this had a mean spring rate c_T = 186 N m^{-1}, which fluctuates by ±5%, the upper limit would be $c_{T,max}$ = 195 N mm^{-1} and the lower limit would be $c_{T,min}$ = 177 N mm^{-1}. Under vertical force $F_{Z,W}$ = 4.5 kN = 4500 N the tyre would, according to Equation 2.3a, have as its smallest jounce travel

$$s_{T,min} = \frac{F_{Z,W}}{C_{T,max}} = \frac{4500}{195}; \qquad s_{T,min} = 23.1 \text{ mm} \tag{2.3a}$$

and

$$s_{T,max} = 25.4 \text{ mm}$$

as the greatest travel. The difference is

$$\Delta s_T = s_{T,max} - s_{T,min} = 2.3 \text{ mm}$$

This difference in the dynamic rolling radius of Δs_T = 2.3 mm would cause variations in vertical force $\Delta F_{Z,W}$, which nevertheless is still smaller than the friction in the wheel suspension bearings. At a speed of perhaps 120 km/h and travelling on a completely smooth road surface, this would nevertheless lead to vibration that would be particularly noticeable on the front axle.

The vehicle used as an example should have a body spring rate of c_f = 15 N/mm per front axle side. The travel Δs_T would then give a vertical force difference, in accordance with Equation 5.0a, of:

$$\Delta F_{Z,W,f} = c_f \, \Delta s_T = 15 \times 2.3; \qquad \Delta F_{Z,W,f} = 34.5 \text{ N}$$

The friction per front axle side is, however, not generally below

$$F_{fr} = \pm 100 \text{ N (Fig. 5.6)}$$

so it can only be overcome if greater variations in vertical force occur as a result of non-uniformity in the road surface. The more softly sprung the vehicle, the more the variations in radial force in the tyre make themselves felt (see Section 5.1.2).

The lateral force variations of the tyre influence the straight-running ability of the vehicle. Even with a tyre that is running straight, i.e. where the slip angle is zero, lateral forces occur, which also depend on the direction of travel (see Chapter 11 in Ref. [4]).

The variations in longitudinal force that occur must be absorbed on the chassis side by the rubber bearings described in Section 3.6.5.2.

The ply steer force dependent on the rolling angle results from the belt design because of the lateral drift of the tyre contact area as a consequence of flat spotting. In contrast, the conicity force, resulting from a change in diameter across the width of the tyre, is not dependent on the rolling angle. Both forces disturb the straight running of the vehicle (see Chapter 11 in Ref. [4]).

2.6 Rolling resistance

2.6.1 Rolling resistance in straight-line driving

Rolling resistance is a result of energy loss in the tyre, which can be traced back to the deformation of the area of tyre contact and the damping properties of the rubber. These lead to the transformation of mechanical into thermal energy, contributing to warming of the tyre.

Sixty to 70% of the rolling resistance is generated in the running tread (Fig. 2.5) and its level is mainly dependent on the rubber mixture. Low damping running tread mixtures improve the rolling resistance, but at the same time reduce the coefficient of friction on a wet road surface. It can be said that the ratio is approximately 1:1, which means a 10% reduction in the rolling resistance leads to a 10% longer braking distance on a wet road surface. The use of new combinations of materials in the running tread (use of silica) has led to partial reduction of the conflict between these aims.

Rolling resistance is either expressed as a rolling resistance force F_R or as the rolling resistance factor k_R – also known as the coefficient of rolling resistance:

$$F_R = k_R \times F_{Z,W} \text{ (N)} \tag{2.4}$$

The factor k_R is important for calculating the driving performance diagram and depends on the vertical force $F_{Z,W}$ and the tyre pressure p_T. Figure 2.31 shows the theoretical k_R curve of tyres of different speed classes as a function of the speed. Although the coefficient of rolling friction of the T tyre increases disproportionally from around 120 km h^{-1}, this increase does not occur in H and V tyres until 160 to 170 km h^{-1}. The reason for this behaviour is the shape of the rolling hump that occurs at different speeds depending on the speed class, and is dependent on the stiffness of the belt, in other words on its design. The lower k_R values for the T tyres result from the usually poorer wet skidding behaviour of this speed class.

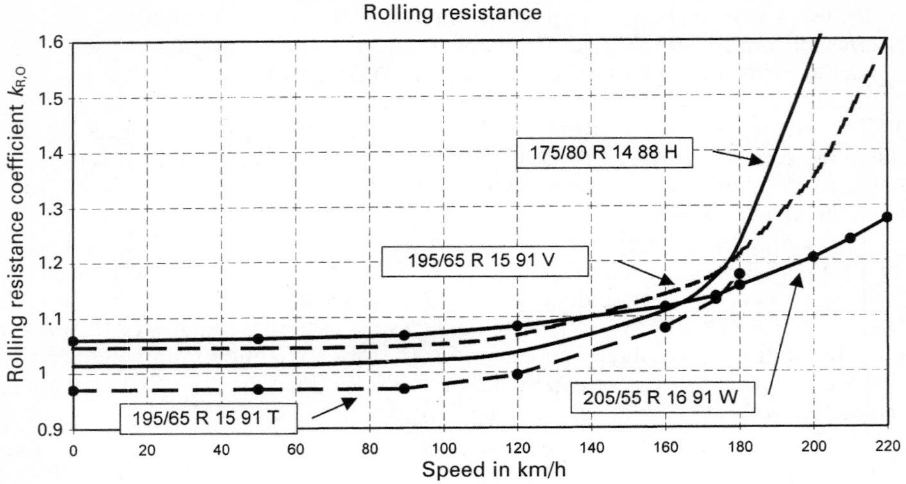

Fig. 2.31 Rolling resistance coefficients $k_{R,0}$, average values of radial tyres as a function of the speed, measured on a drum test rig. Tyres authorized up to 210 km h^{-1} have a lower rolling resistance below 160 km h^{-1} (than the V and W designs) whilst the value rises sharply above this speed (measurements: Continental).

Asphalted roads cause $k_{R,0}$ to increase by around 20% as k_R and rough concrete to at least 30%. The ratios i_R are then 1.2 or 1.3 to 1.4 and the actual value of k_R is:

$$k_R = i_R \times k_{R,0} \tag{2.4a}$$

The difference is due to the different design emphases during development of the tyres. The design priorities for H, V and W tyres are high-speed road holding and good wet skidding and aquaplaning behaviour, whereas T tyres are designed more for economy, i.e. lower rolling resistance (which plays an important role at lower speeds and influences urban driving fuel consumption, Fig. 2.32) and long service life.

2.6.2 Rolling resistance during cornering

Rolling resistance can change dramatically during cornering; its value depends on the speed and the rolling radius R, in other words on $\mu_{Y,W}$ (see Equations 2.9 and 2.11 and Fig. 2.43) and $\alpha_{f \text{ or } r}$. The rolling resistance $k_{R,co}$, which is included in some calculations (see Equation 3.35), comprises the coefficient k_R for straight running and the increase Δk_R:

$$k_{R,co} = k_R + \Delta k_R$$

$$\Delta k_R \approx \mu_{Y,W} \times \sin \alpha \tag{2.4b}$$

The following data can provide an example:

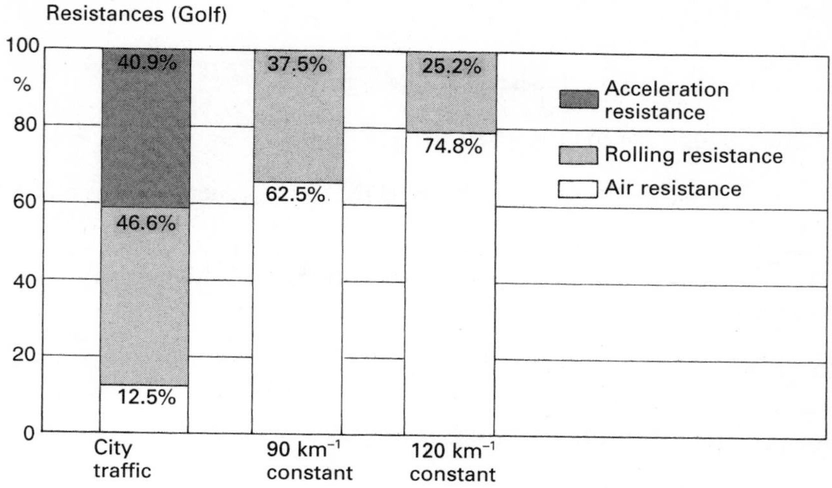

Fig. 2.32 In town and when the vehicle is travelling at low speeds on rural roads, fuel consumption is determined up to 40% by the rolling resistance, whereas at higher speeds the air drag is the determining factor see Section 2.1 and Section 2.2 in Ref. [3]). The figure shows a study carried out by VW on the Golf.

Front axle force $F_{Z,V,f} = 7\,\text{kN}$; $\mu_{Y,W} = 0.7$ (asphalted road)
Tyres 155 R 13 78 S $\quad p_T = 1.8$ bar, $v \leqslant 120$ km h

In accordance with Equation 2.11 related to one wheel:

$$F_{Y,W,f} = \mu_{Y,W}\, F_{Z,W,f} = \mu_{Y,W}\, F_{Z,V,f}/2 = 0.7 \times 3.5\ \text{kN}$$
$$F_{Y,W,f} = 2.45\ \text{kN}$$

The slip angle read off at $F_{Y,W,f}$ in Fig. 2.44 is 4° and corresponds to the values in Fig. 2.43.

However, the dynamic wheel load transfer seen in Fig. 1.5 plays a role during cornering, leading to a greater slip angle on the wheel on the outside of the curve (and thus also on the inner wheel), than resulted from test rig measurements. On '82' series tyres, α is about 5°, in accordance with Fig. 2.38:

$$\alpha \approx 7\,\mu_{Y,W} \tag{2.4c}$$

With sin 5° in accordance with Equation 2.4b there is an increase of

$$\Delta k_R \approx 0.7 \times 0.087 = 0.061$$

Assuming a value of $k_{R,0} = 0.012$, in accordance with Equation 2.4a, on asphalted road

$$k_R = i_R \, k_{R,0} = 1.2 \times 0.010 = 0.012$$

and therefore the rolling resistance during cornering is

$$k_{R,co} = 0.012 + 0.061 \approx 0.073$$

In the case of the understeering vehicles (Fig. 2.41) $k_{R,co}$ increases as a result of the additional steering input and – if the wheels are driven – μ_{rsl} should be inserted for $\mu_{Y,W}$ (see Equation 2.18); the slip angle increases further. '65 Series' tyres, on the other hand, require a smaller steering input and thus make the vehicle easier to handle:

$$\alpha = 3 \times \mu_{Y,W} \qquad (2.4d)$$

2.6.3 Other influencing variables

The rolling resistance increases in certain situations:

- in the case of a large negative or positive camber (the influence can be ignored up to $\pm 2°$);
- due to a change to track width (Fig. 3.6);
- in the case of deviations in zero toe-in around 1% per $\delta = 10'$ or $v = 1$ mm;
- on uneven ground.

In general it can be said that the ratio i_R (see Fig. 2.31) will take the following values:

- around 1.5 on cobbles
- around 3 on potholed roads
- around 4 on compacted sand
- up to 20 on loose sand.

2.7 Rolling force coefficients and sliding friction

2.7.1 Slip

If a tyre transfers drive or braking forces, a relative movement occurs between the road and tyre, i.e. the rolling speed of the wheel is greater or less than the vehicle speed (see Equation 2.1b). The ratio of the two speeds goes almost to ∞ when the wheel is spinning, and is 0 when it locks. Slip is usually given as a percentage. The following equation applies during braking:

$$S_{X,W,b} = \frac{\text{vehicle speed} - \text{circumferential speed of wheel}}{\text{vehicle speed}}$$

$$S_{X,W,b} = \frac{v - v_W}{v} \times 100\ (\%) \tag{2.4e}$$

Drive slip is governed by:

$$S_{X,W,a} = \frac{v_W - v}{v_W} \times 100\ (\%) \tag{2.4f}$$

The different expressions have the advantage that, in both cases where the wheel is spinning or locked, the value is 100% and is positive.

Further details can be found in Section 2.2.8, in Ref. 6 (Section 1.2), Ref. 7 (Chapter 1) and in Ref. 9 (Section 2.2).

2.7.2 Friction coefficients and factors

The higher the braking force or traction to be transmitted, the greater the slip becomes. Depending on the road condition, the transferable longitudinal force reaches its highest value between 10% and 30% slip and then reduces until the wheel locks (100% slip). The quotient from longitudinal force F_x and vertical force $F_{Z,W}$ is the coefficient of friction, also known as the circumferential force coefficient

$$\mu_{X,W} = F_{X,W}/F_{Z,W} \tag{2.5}$$

when it relates to the maximum value, and the coefficient of sliding friction, also called sliding friction factor

$$\mu_{X,W,lo} = F_{X,W}/F_{Z,W} \tag{2.5a}$$

when it is the minimal value (100% slip) (Fig. 2.33). F_x is designated $F_{X,W,b}$ during braking and $F_{X,W,a}$ during traction.

In all cases $\mu_{X,W}$ is greater than $\mu_{X,W,lo}$; in general it can be said that

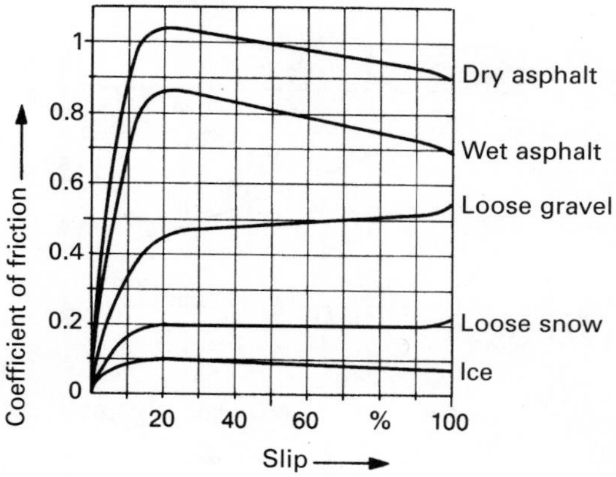

Fig. 2.33 Coefficient of friction $\mu_{X,W}$ of a summer tyre with 80 to 90% deep profile, measured at around 60 km/h and shown in relation to the slip on road surfaces in different conditions (see also Fig. 1.64). Wide tyres in the '65 series' and below have the greatest friction at around 10% slip, which is important for the ABS function (see Chapter 1 in Ref. [7]).

on a dry road $\mu_{X,W} \approx 1.2\ \mu_{X,W,lo}$ (2.6)

on a wet road $\mu_{X,W} \approx 1.3\ \mu_{X,W,lo}$ (2.6a)

2.7.3 Road influences

2.7.3.1 Dry and wet roads

On a dry road, the coefficient of friction is relatively independent of the speed (Fig. 2.34), but a slight increase can be determined below 20 km/h. The reason lies in the transition from dynamic to static rolling radius (see the example in Section 2.2.5.4) and is therefore linked to an increasing area of tyre contact. At speeds a little over zero, on a rough surface, a toothing cogging effect can occur, which causes a further increase in the coefficient of friction, then:

$$\mu_{X,W} \geqslant 1.3 \tag{2.6b}$$

When the road is wet, the coefficient of friction reduces, but is still independent of the speed. This situation changes as the amount of water increases and also with shallower profile depth. The water can no longer be moved out of the profile grooves and the μ value falls as speed increases.

2.7.3.2 Aquaplaning

The higher the water level, the greater the risk of aquaplaning. Three principal factors influence when this occurs:

- road
- tyres
- speed.

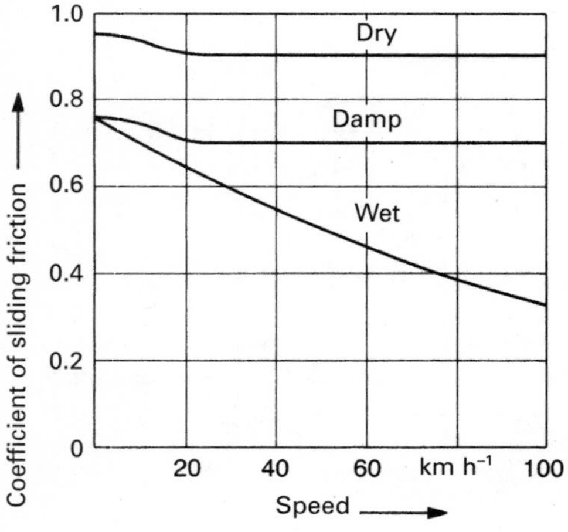

Fig. 2.34 Dependency of the coefficient of sliding friction $\mu_{X,W,lo}$ on speed on different road conditions.

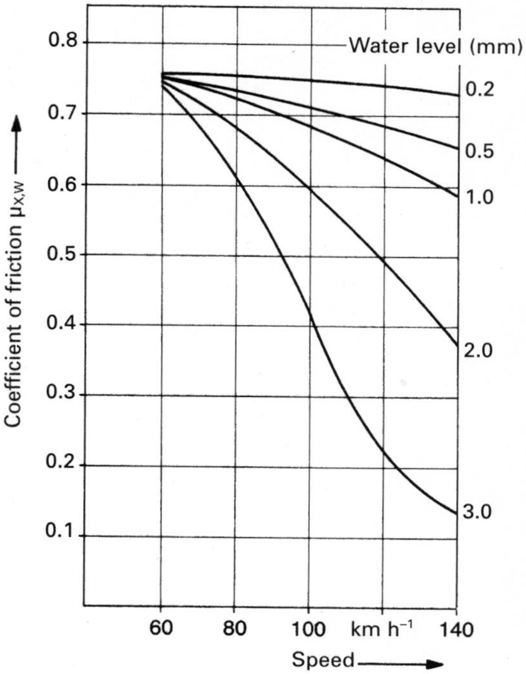

Fig. 2.35 Coefficients of friction $\mu_{X,W}$ of a summer tyre with an 8 mm deep profile dependent on speed at different water levels. Hardly any influence can be detected under 60 km h^{-1}; at higher speeds and 3 mm water depth, the curve shows a lowering of $\mu_{X,W}$ which indicates the aquaplaning effect.

With regard to the road, the water level is the critical factor (Fig. 2.35). As the level rises, there is a disproportionate increase in the tendency towards aquaplaning. When the level is low, the road surface continues to play a role because the coarseness of the surface absorbs a large part of the volume of water and carries it to the edge of the road. Following rainfall, the water levels on roads are generally up to 2 mm; greater depths can also be found where it has been raining for a long time, during storms or in puddles.

On the tyre, the tread depth has the greatest influence (Fig. 2.47). There can be up to a 25 km h^{-1} difference in speed between a full tread and the legal minimum tread depth of 1.4 mm. High tyre pressure and low running surface radius r (Fig. 2.5) lead to the area of contact becoming narrower, giving the advantage of improved aquaplaning behaviour as the distribution of ground pressure becomes more even (Fig. 2.9). Lower tyre pressure and contours with larger radii make aquaplaning more likely; this also applies to wider tyres (Fig. 2.19) particularly when tread depths are low. However, the greatest influence by far is the speed, especially when the water level increases and tread depths are low. This is why reducing speed is the best way to lessen the risk of aquaplaning, and is a decision drivers can make for themselves.

2.7.3.3 Snow and ice

Similar to aquaplaning, low coefficients of friction occur on icy roads, although these are highly dependent on the temperature of the ice. At close to 0°C, special

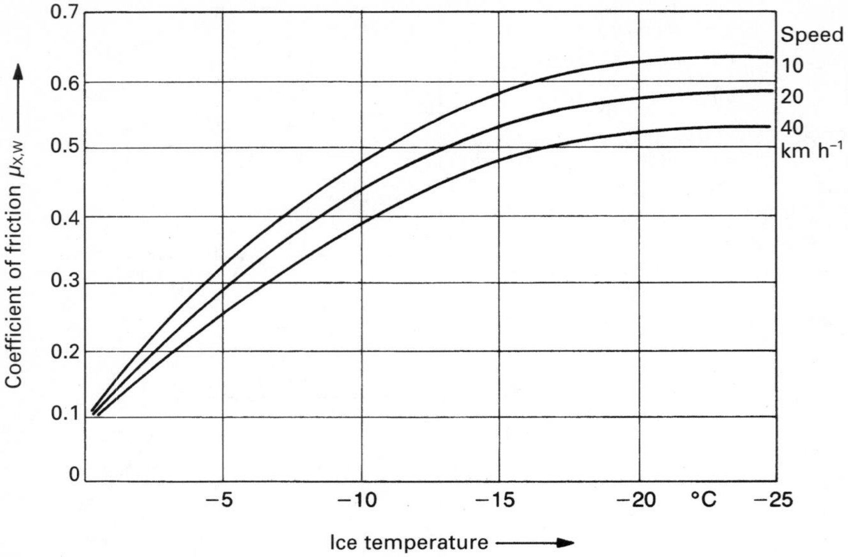

Fig. 2.36 Influence of ice temperature and car speed on the coefficient of friction $\mu_{X,W}$ of an 82 series winter tyre; the extremely low values at 0°C can be seen clearly.

conditions occur; compression of the surface can lead to the formation of water which has a lubricating effect and reduces the coefficient of friction to $\mu_{X,W} \leqslant 0.08$ (Fig. 2.36). At −25°C, a temperature that is by no means rare in the Nordic countries, values of around $\mu_{X,W} = 0.6$ can be reached. At low temperatures, coefficients of friction and sliding friction are further apart:

$$\mu_{X,W} \sim 2\,\mu_{X,W,lo} \tag{2.7}$$

2.8 Lateral force and friction coefficients

2.8.1 Lateral forces, slip angle and coefficient of friction

Lateral forces on a rolling tyre can be caused by the tyre rolling diagonal to the direction of travel (so-called slip), the tendency of a tyre to move from its position vertical to the road, camber or conical effects. The build-up of lateral forces as a result of slip will be discussed next.

If a disturbing force $F_{c,V}$ acts at the centre of gravity of the vehicle (e.g. a wind or side negative lift force), lateral wheel forces $F_{Y,W,f,o}$, $F_{Y,W,f,i}$, $F_{Y,W,r,o}$ and $F_{Y,W,r,i}$ are needed to balance the forces (Fig. 2.37). To build up these forces, the vehicle must alter its direction of travel about the angle α, the slip angle. The size of the slip angle depends on the force transmission properties of the tyre and the disturbing force (Fig. 2.38).

When cornering, the interference force should be equal to the centrifugal force $F_{c,V}$, which results from the speed v in m/s and the radius of the bend R in m, on which the vehicle centre of gravity V (Fig. 2.29a) moves. With the total weight $m_{V,t}$ of the vehicle the equation is:

$$F_{c,V} = m_{V,t} \times v^2/R = m_{V,t} \times a_y = F_{Y,V} \text{ (N)} \tag{2.8}$$

The centrifugal or disturbance force is just as large as the lateral forces on the wheels (Fig. 2.37):

$$F_{Y,V} = F_{Y,W,f,o} + F_{Y,W,f,i} + F_{Y,W,r,o} + F_{Y,W,r,i} = \Sigma F_{Y,W} \tag{2.8a}$$

and

$$\Sigma F_{Y,W} = \mu_{Y,W} \times \Sigma F_{Z,W} = \mu_{Y,W} \times F_{Z,V,t}$$

Together the two equations give

$$\mu_{Y,W} F_{Z,V,t} = \mu_{Y,W} \times m_{V,t}\, g = m_{V,t} \times a_y \tag{2.9}$$

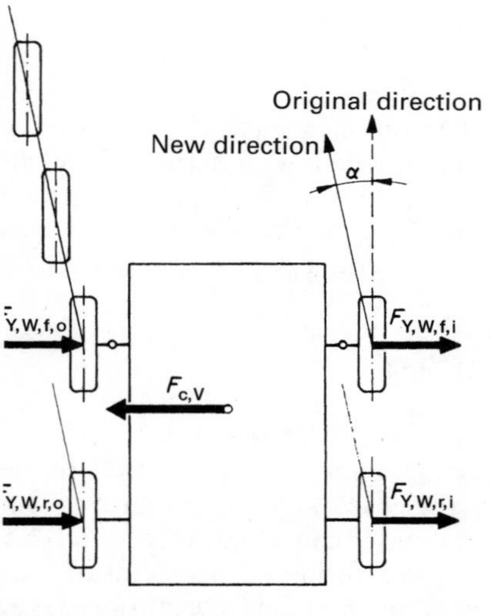

Fig. 2.37 Tyres are only able to transfer a lateral force $F_{Y,V}$ acting on the vehicle if they are rolling at an angle to the vehicle. Regardless of whether these are $F_{Y,V}$ or the centrifugal force $F_{C,Y}$ during cornering, the lateral forces $F_{Y,W}$ should be regarded as being perpendicular to the wheel centre plane.

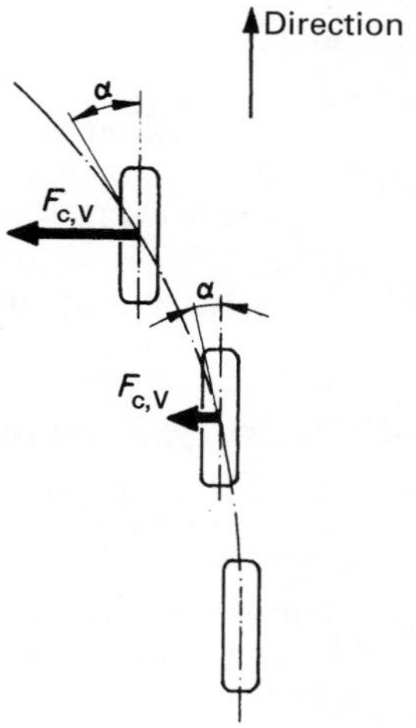

Fig. 2.38 The higher the lateral force $F_{Y,W}$, the greater the tyre slip angle α.

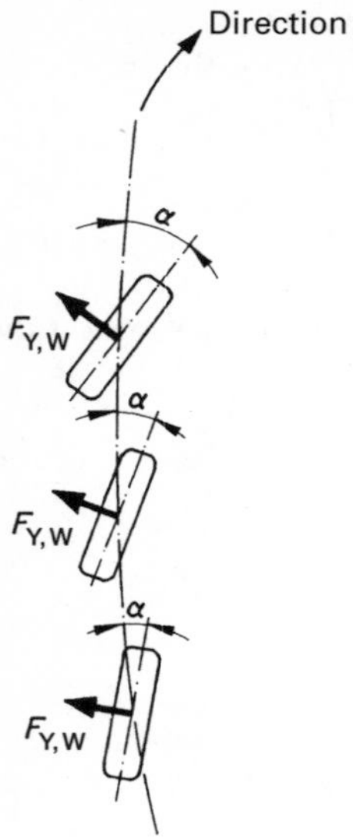

Fig. 2.39 Increasing lateral forces $F_{Y,W}$ during cornering caused by the centrifugal force $F_{c,V}$ leads to increasing slip angles α.

and

$$\mu_{Y,W} = g/a_y$$

The coefficient of friction $\mu_{Y,W}$ is not dependent on the radius of the curve and driving speed and is therefore more suitable for calculating cornering behaviour (see also Equation 6.13a).

The faster the vehicle negotiates a bend, the higher the coefficient of friction used and the greater the slip angles (Fig. 2.39).

2.8.2 Self-steering properties of vehicles

The self-steering properties of a vehicle describe the lateral force and hence slip angle ratios produced during steady-state cornering (radius and driving speed constant; no external disturbances). In the case of an understeering vehicle, a larger slip angle is required on the front axle than at the rear axle ($\alpha_f > \alpha_r$, Fig. 2.41). During cornering with an increase in lateral acceleration, the driver must force the vehicle into the bend by increasing the steering angle (see Fig. 5.2). If the necessary slip angles on the front and rear axles are the same ($\alpha_f = \alpha_r$, Fig. 2.40), one speaks of neutral handling characteristics. Over-steering behaviour is present if the tail of the vehicle moves outwards during cornering and the slip angle on the rear axle is greater than on the front axle ($\alpha_f < \alpha_r$, Fig. 2.42). The driver must respond to this by reducing the steering angle.

As understeering behaviour is consistent with the expectations and experience of the driver, it is this which needs to be aimed for. In normal driving conditions

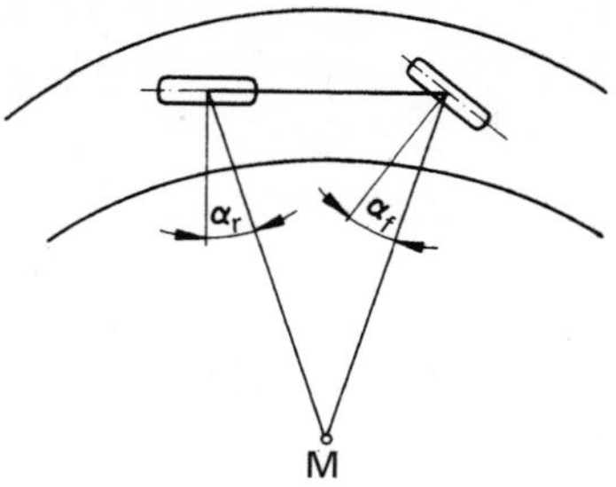

Fig. 2.40 If, during cornering, $\alpha_f \sim \alpha_r$, the handling of a vehicle can be described as neutral.

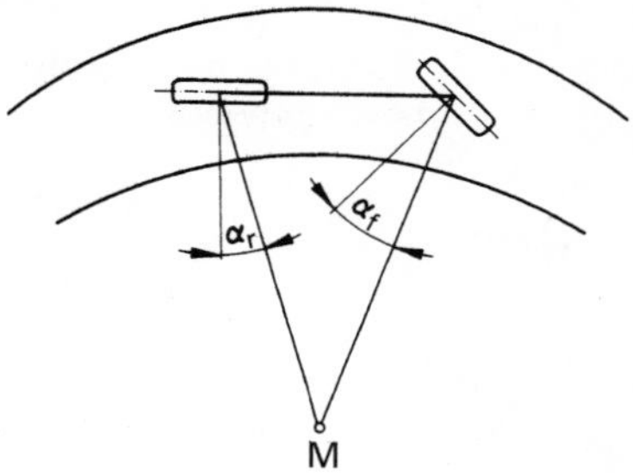

Fig. 2.41 If there is a greater slip angle α_f on the front wheels than α_r on the rear, the vehicle understeers.

Fig. 2.42 If there is a greater slip angle α_r on the rear wheels than on the front (α_f), the vehicle oversteers. The positive angle describes the angle between the vehicle longitudinal axis and its speed at the centre of gravity.

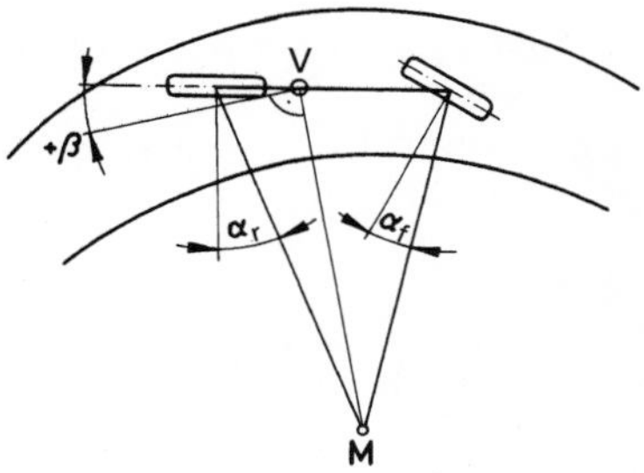

(anti-skid roadway, lateral acceleration of less than 6 m/s), all vehicles, therefore, are now designed to understeer. With increasing lateral acceleration, the understeering behaviour should be as linear as possible and then, also as a warning to the driver that the stability limit is about to be reached, increase progressively. If the handling characteristics change to oversteer at the stability limit, for instance with very high acceleration, this is an unpredictable driving situation which the untrained driver can only control with difficulty. For active riding safety, the predictability of self-steering properties in all kinds of conditions (vehicle loading, the distribution of driving torque in four-wheel drive vehicles, different coefficients of friction, acceleration or braking procedures, changes in tyre pressure, etc.) is of paramount importance.

For a simplified representation of the relationships described, the so-called single-track model is used, in which the wheels of the vehicle are drawn together in the middle of the vehicle, without taking into account the height of the centre of gravity (flat model).

Since in greater bend radii the average steering angle δ_m is less than 5°, it can be assumed that the sine and radius values of the angle are equal, and the angles δ_o and δ_i correspond to this (Fig. 3.91 and Equation 3.17):

$$\sin \delta_m \approx \delta_m \approx \delta_o \approx \delta_i \text{ (rad)}$$

Using Equation 3.12 it is now possible to determine the relationship between steering angle, turning circle diameter D_S (Figs 1.69 and 3.89) and slip angles at a constant cornering speed:

$$\delta_m = \frac{2 \times l}{D_S} + \alpha_f - \alpha_r \tag{2.10}$$

The kingpin offset at ground r_σ is so negligable in comparison to D_S that it can be ignored.

2.8.3 Coefficients of friction and slip

To determine the cornering behaviour, the chassis engineer needs the lateral forces (or the coefficient of friction) based on the slip angle and the parameters:

- vertical force (or wheel load) in the centre of tyre contact
- tyre pressure
- wheel camber
- tyre type.

The measurements are generally taken on test rigs, up to slip angles of $\alpha = 10°$. The drum surface with its friction values of $\mu_0 = 0.8$–0.9 sets limits here, and larger angles hardly give increasing lateral coefficients of friction:

$$\mu_{Y,W} = F_{Y,W}/F_{Z,W} \tag{2.11}$$

Conditions on the road are very different from those on the test rig; the type of road surface and its condition play a role here. As can be seen in Fig. 2.43, the coefficient of friction on rough, dry concrete increases to $\alpha = 20°$ and then falls. In precisely the same way as with the longitudinal force the slip $S_{Y,W}$ (in the lateral direction) is also taken into consideration; this is as a percentage of the sine of the slip angle times 100:

$$S_{Y,W} = \sin \alpha \times 100\ (\%) \tag{2.12}$$

In conjunction with the drum value $\alpha = 10°$, this would give a slip of $S_{Y,W} = 17\%$, and on the street at $\alpha = 20°$ slip values of up to $S_{Y,W} = 34\%$. If the tyre is further twisted to $\alpha = 90°$, it slides at an angle of 90° to the direction of travel; sin α would then be equal to one and $S_{Y,W} = 100\%$. The coefficient of friction then becomes the coefficient of lateral sliding friction $\mu_{Y,W,lo}$, which on average is around 30% lower:

$$\mu_{Y,W,lo} \approx 0.7 \times \mu_{Y,W} \tag{2.13}$$

In contrast to dry concrete (as also shown in Fig. 2.43) on asphalt and, in particular on wet and icy road surfaces, no further increase in the lateral cornering forces can be determined above $\alpha = 10°$ (i.e. $S_{Y,W} \approx 17\%$).

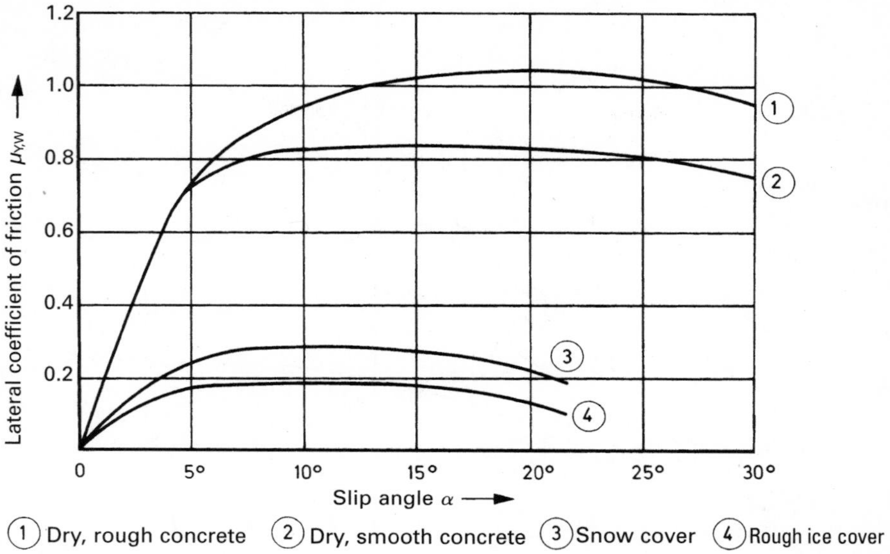

Fig. 2.43 Lateral coefficients of friction $\mu_{Y,W}$ as a function of slip angle and road condition, shown for an '82 series' summer tyre with around 90% deep profile. The ice temperature is around –4°C. The vertical force $F_{Z,W}$ was kept constant during the measurements to obtain the dimensionless values of $\mu_{Y,W}$. The maximum at $\alpha = 20°$ on a very skid-resistant road can be seen clearly. The further $\mu_{Y,W}$ sinks, the further it moves towards smaller angles.

2.8.4 Lateral cornering force properties on dry road

Figure 2.44 shows the usual way in which a measurement is carried out for a series 82 tyre. The lateral force appears as a function of the vertical force in kilonewtons and the slip angle α serves as a parameter. A second possibility can be seen in Fig. 2.45; here, for the corresponding series 70 tyre, $\mu_{Y,W} = F_{Y,W}/F_{Z,W}$ is plotted against α and $F_{Z,W}$ serves as a parameter. The degree of curvature of the graphs in both figures shows that slope at any point changes as a function of $F_{Z,W}$ or $\mu_{Y,W}$. The maximum occurs with large angles and small vertical forces. A less stressed tyre in relation to its load capacity therefore permits greater coefficients of friction and higher cornering speeds than one whose capacity is fully used.

This result, which has been used for a long time in racing and sports cars, has also become popular in modern cars, A mid-range standard car can be taken as an example. The car manufacturer specifies $p_T = 2.2$ bar/2.5 bar under full load for the front and rear wheels 185/65 R 15 88H. At these pressures, the load capacity, in accordance with Figs 2.13 and 2.15, is:

front 505 kg and rear 560 kg

Figure 5.10 contains the authorized axle loads from which the wheel load (divided by two) results:

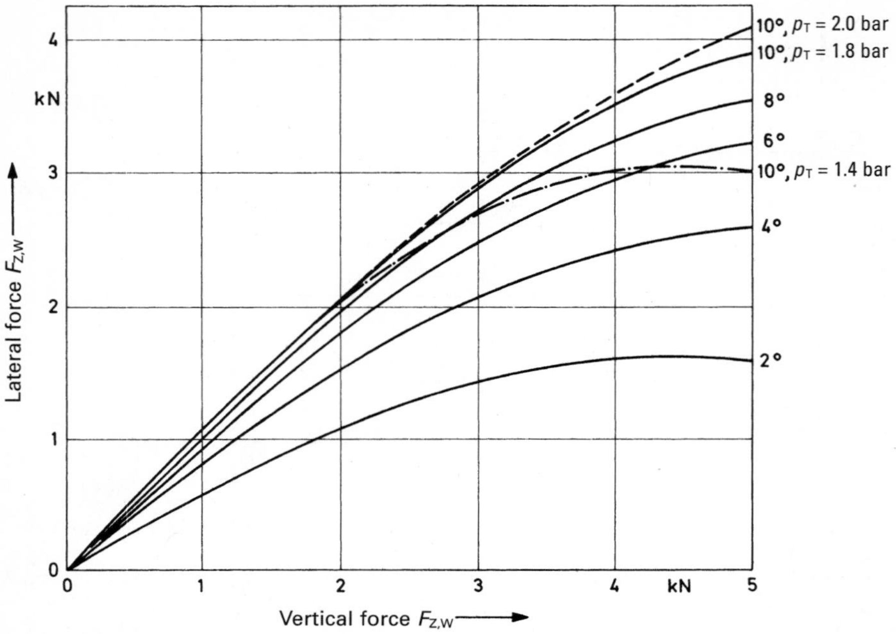

Fig. 2.44 Lateral cornering forces of the 155 R 13 78 S '82 series' steel radial tyre, measured on a dry drum at p_T = 1.8 bar. The load capacity at this pressure is around 360 kg, corresponding to a vertical force $F_{Z,W}$ = 3.53 kN. Also shown are the forces at α = 10° and P_T = 1.4 bar and 2.0 bar to indicate the influence of the tyre pressure on the lateral cornering properties.

front 375 kg and rear 425 kg

As described in Section 2.2.6, at speeds up to 210 km h^{-1} (H tyres), an increase in tyre pressure of 0.3 bar is necessary or there is only a correspondingly lower load capacity. This then is, with p_T = 1.9 bar at the front or 2.2 bar at the back,

450 kg and 505 kg

Thus, the actual load factor k_m at 210 km/h becomes:

$$\begin{aligned} &\text{front } k_{m,f} = (375/450) \times 100 = 83\% \\ &\text{back } k_{m,r} = (425/505) \times 100 = 84\% \end{aligned} \tag{2.14}$$

2.8.5 Influencing variables

2.8.5.1 Cross-section ratio *H/W*

The 185/65 R 15 88H size used as an example in the previous section is a 65 series wide tyre; the 15″ diameter also allows a good sized brake disc diameter (Fig. 2.10).

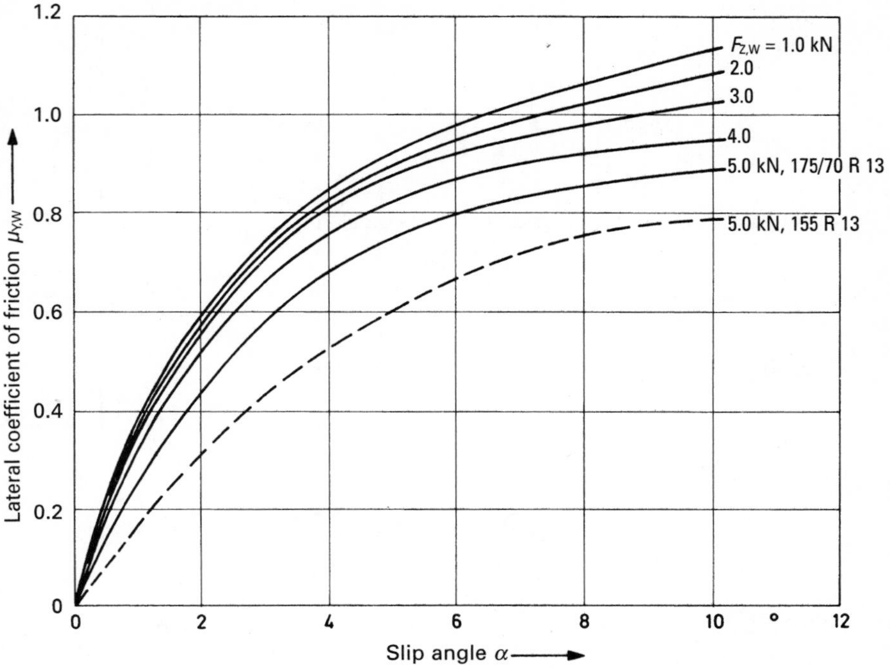

Fig. 2.45 Lateral coefficients of friction $\mu_{Y,W}$ as a function of the slip angle α and the vertical force $F_{Z,W}$, measured on a dry drum on a 175/70 R 13 82 S tyre at p_T = 2.0 bar. The tyre, which has been inflated in such a manner, carries 395 kg or $F_{Z,W}$ = 3.87 kN. In order to indicate the influence of the cross-section on the transferable lateral forces the 82 series 155 R 13 78 S tyre was also included.

In contrast to the 82 series standard tyre, the sizes of the 70 series and wide tyres (H/W = 0.65 and below) generate higher lateral cornering forces at the same slip angles (Figs 2.9, 2.45 and 2.46). As can be seen in Fig. 1.6, these, as $F_{Y,W,o} = \mu_{Y,W}\,(F_{Z,W} + \Delta F_{Z,W})$, are all the greater, the faster the vehicle takes a bend.

2.8.5.2 Road condition

The force transmission ratios between the tyres and road are determined by the state of the road (see construction, surface roughness and condition; Figs 2.43 and 2.47).

2.8.5.3 Track width change

The track width change that exists, in particular on independent wheel suspensions described in Section 3.3, causes undesirable lateral forces at the centres of tyre contact on both wheels when the vehicle is moving unimpeded in a straight line. Figures 3.5 and 3.6 show this, and also what lateral forces can occur if a series 82 radial tyre rolling in a straight line is brought out of its direction by an

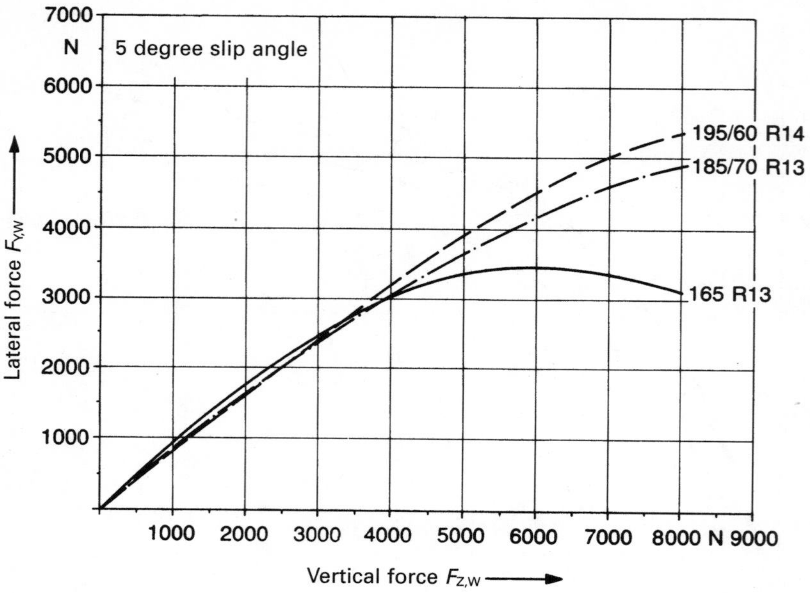

Fig. 2.46 Lateral force $F_{Y,W}$ dependent on vertical force $F_{Z,W}$ and tyre sizes of different *H/W* ratios: 165 R 13 82 H, 185/70 R 13 85 H and 195/60 R 14 85 H.

Up to $F_{Z,W}$ = 4000 N the curves are more or less the same, but at higher loads the more favourable lateral cornering properties of the wide tyre are evident.

suspension-kinematic dependent change. This effect is magnified by an increase in slip rigidity, as, for example, in wide tyres.

2.8.5.4 Variations in vertical force

During cornering, vertical force variations $\pm\Delta F_{Z,W}$ in the centre of tyre contact cause a reduction in the transferable lateral forces $F_{Y,W}$ as the tyre requires a certain amount of time and distance for the build-up of lateral forces. The loss of lateral force $\Delta F_{Y,W,4}$ depends on the effectiveness of the shock absorbers, the tyre pressure p_T (which can enhance the 'springing' of the wheels, see Equation 5.6) and the type of wheel suspension link mountings. Further influences are wheel load and driving speed. To calculate cornering behaviour, an average loss of lateral force $\Delta F_{Y,W,4}$ due to variations in vertical force and dependent only on tyre design and slip angle α, should be considered:

$$\Delta F_{Y,W,4} \approx 40 \text{ N per degree } \alpha \tag{2.15}$$

2.8.5.5 Camber change

Wheels that incline with the body during cornering have a similar, detrimental influence on the transferability of lateral forces. As can be seen from Fig. 1.6, positive angle ($+\varepsilon_W$) camber changes occur on the outside of the bend and negative

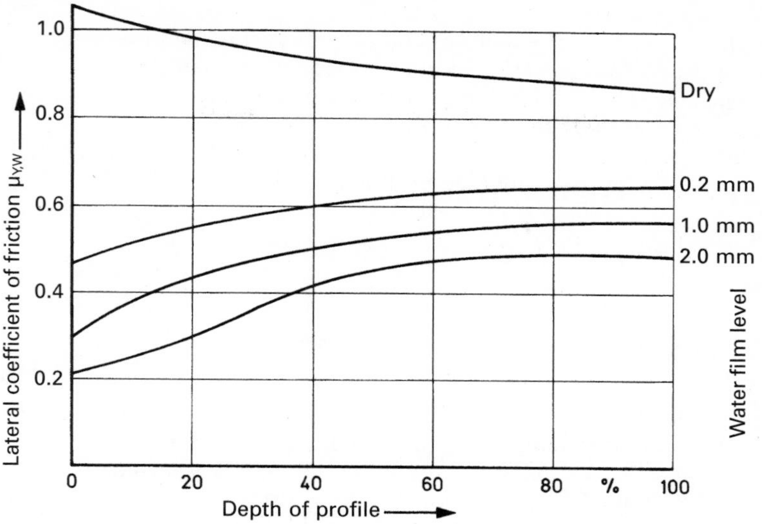

Fig. 2.47 Possible lateral friction coefficients $\mu_{Y,W}$ of a steel radial tyre 155 R 13 78 S depending on the depth of the tyre profile as a percentage (starting from 8 mm = 100%) at p_T = 1.8 bar, α = 10°, v = 60 km/h and varying water film levels in mm.

The improved grip of the treadless tyre on a dry road can be seen clearly as can its significantly poorer grip in the wet; a fact which also applies to the coefficient of friction in the longitudinal direction (see Section 2.7.2).

angles ($-E_W$)on the inside of the bend as a consequence of the body roll. The lateral forces are directed to the centre point of the bend (Fig. 3.13). If a wheel is 'cambered' against this, in other words inclined at the top towards the outside of the bend, the possibility of transferring lateral forces reduces; on a dry road surface, depending on the tyre size, the change is

$$\Delta F_{Y,W,3} = 40 \text{ N to } 70 \text{ N per degree of camber} \tag{2.16}$$

To counteract this, a greater slip angle must occur and greater steering input becomes necessary for the front wheels. This makes the vehicle understeer more (Fig. 2.41) and appear less easy to handle. Furthermore, the steering aligning moment (see Section 3.10.3) also increases. If this effect occurs on the rear axles – as is the case with longitudinal link axles (Fig. 1.14) – the vehicle has a tendency to oversteer. Negative camber $-\varepsilon_W$ on the outside of the bend and positive $+\varepsilon_W$ on the inside would have exactly the opposite effect. Wheels set in this manner would increase the lateral forces that can be absorbed by the amount stated previously for $\Delta F_{Y,W,3}$ and cause a reduction in the tyre slip angle.

2.8.5.6 Lateral force due to camber

Wheels according to the body roll inclined towards the outside edge of the bend (Fig. 1.6) try to roll outwards against the steering direction, so that additional camber forces are required in the tyre contact patches to force the wheels in the desired steering direction. As these camber forces act in the same direction as the centrifugal force $F_{c,Bo\ or\ V}$ in the case described, greater lateral slip forces $F_{Y,W,f,o}$, $F_{Y,W,f,i}$, $F_{Y,W,r,o}$ and $F_{Y,W,r,i}$ and hence greater slip angles must be applied to maintain the balance of forces on the part of the tyres.

The average force F_{ε_W} with the standard camber values for individual wheel suspensions on a dry road are (see Section 2.2.3 in Ref. 9):

$$F_{\varepsilon_W} \approx F_{Z,W} \times \sin \varepsilon_W \tag{2.17}$$

2.9 Resulting force coefficient

Rolling resistance increases when negotiating a bend (see Equation 2.4a), and the vehicle would decelerate if an increased traction force $F_{X,W,A}$ did not create the equilibrium needed to retain the cornering speed selected. In accordance with Equation 6.36, $F_{X,W,A}$ is dependent on a series of factors and the type of drive system (front- or rear-wheel drive); on single-axle drive (see Sections 1.4 to 1.6), the traction force on the ground stresses the force coefficient of friction (the coefficient of)

$$\mu_{X,W} = F_{X,W,A,f\ or\ r}/F_{Z,V,f\ or\ r} \tag{2.15}$$

and thus greater slip angles at the driven wheels. With given values for cornering speed and radius (see Equation 2.8) the resulting force coefficient μ_{rsl} can be determined:

$$\mu_{rsl} = (\mu_{Y,W}^2 + \mu_{X,W}^2)^{\frac{1}{2}} \tag{2.18}$$

μ_{rsl} cannot be exceeded because the level depends on the road's surface and the condition.

When braking on a bend, additional longitudinal forces $F_{X,W,b}$ occur on all wheels (see Section 6.3.1), and act against the direction of travel. In this case Equation 2.18 also applies.

On standard vehicles and front-wheel drives, the front wheels take 70–80% of the braking force and the rear wheels only 20–30%. This means that the slip angles increase on both axles, but more at the front than the rear and the vehicle tends to understeer (Fig. 2.41 and Equation 6.20). If the wheels of an axle lock, the friction becomes sliding friction and the vehicle pushes with this pair of wheels towards the outside of the bend (Figs 6.8 to 6.10).

Taking into consideration the maximum possible values in the longitudinal and lateral direction of the road – known respectively as $\mu_{X,W,max}$ and $\mu_{X,W,min}$ – the increasing force coefficient can be calculated:

$$\mu_{X,W} = \mu_{X,W,max}\left[1 - \left(\frac{\mu_{Y,W}}{\mu_{Y,W,max}}\right)^2\right]^{\frac{1}{2}} \tag{2.19}$$

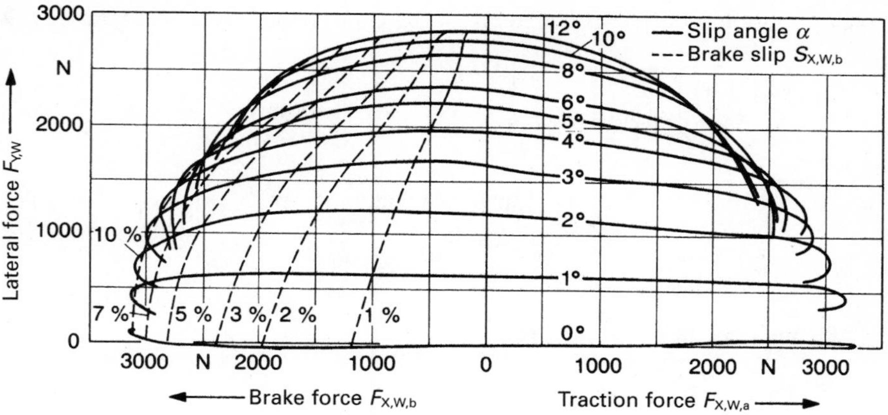

Fig. 2.48 Tyre-tangential lateral force performance characteristics with slip angles and brake slip as parameters. The study was carried out on a 18565 R 14 86 S radial tyre loaded at 300 kg at p_T = 1.5 bar. The shape of the curves indicates that, with increasing longitudinal forces, those which can be absorbed laterally reduce. At 1.5 bar, the tyre carries a weight of 350 kg, i.e. it is only operating at 86% capacity.

Consider as an example a braking process on a dry road at 100 km/h on a bend with R = 156 m. Using Equation 2.9 the calculation gives $\mu_{Y,W} = 0.5$.

Figure 2.48 shows a measurement on the tyre in question where the greatest coefficient of friction in the lateral direction at $F_{Z,W}$ = 2490 N, ε_W = 10% and α = 4° (see Equation 2.11) amounts to

$$\mu_{Y,W,max} = F_{Y,W}/F_{Z,W} = 2850/2940 \text{ (N/N)} \qquad \mu_{Y,W,max} = 0.97$$

In the longitudinal direction the possible braking force $F_{X,W,b}$ = 3130 N is at α = 0° and therefore (see Equation 2.5),

$$\mu_{X,W,max} = F_{X,W,b}/F_{Z,W} = 3130/2940 \text{ (N/N)}$$
$$= 1.06$$

and

$$\mu_{X,W} = 1.06\left[1 - \left(\frac{0.5}{0.97}\right)^2\right]^{\frac{1}{2}}$$
$$= 0.91$$

The lateral forces that the tyre can absorb during braking can also be calculated:

$$\mu_{Y,W} = \mu_{Y,W,max}\left[1 - \left(\frac{\mu_{X,W}}{\mu_{X,W,max}}\right)^2\right]^{\frac{1}{2}} \qquad (2.19a)$$

$\mu_{X,W} = 0.7$ should be given. The lateral force coefficient (which can be used) is:

$$\mu_{Y,W} = 0.97\left[1 - \left(\frac{0.7}{1.06}\right)^2\right]^{\frac{1}{2}}$$
$$= 0.73$$

At $S_{X,W,b} = 10\%$ and $\alpha = 4°$ the transferable lateral force is

$$F_{Y,W} = \mu_{Y,W} \times F_{Z,W} = 0.73 \times 2940$$
$$= 2146 \text{ N}$$

and the available braking force is

$$F_{X,W,b} = \mu_{X,W} \times F_{Z,W} = 0.7 \times 2940$$
$$= 2058 \text{ N}$$

2.10 Tyre self-aligning torque and caster offset

2.10.1 Tyre self-aligning torque in general

The focal point of the force of the tyre contact patch lies behind the middle of the wheel because of its load- and lateral-force-related deformation. As a result, the point of application of the lateral force alters by the amount $r_{\tau,T}$, known as the caster offset, and comes to lie behind the centre of the wheel (Fig. 3.119). On the front wheels, the lateral cornering force $F_{Y,W,f}$ together with $r_{\tau,T}$ (as the force lever) gives the self-aligning moment $M_{Z,T,Y}$ which superimposes the kinematic alignment torque and seeks to bring the input wheels back to a straight position (Section 3.8).

The self-aligning torque, lateral force and slip angle are measured in one process on the test rig. $M_{Z,T,Y}$ is plotted as a function of the slip angle (Fig. 2.49), the vertical force $F_{Z,W}$ serves as a parameter. The higher $F_{Z,W}$, the greater the self-alignment and, just like the lateral force, the moment increases to a maximum and then falls again. $M_{Z,T,Y,max}$ is, however, already at $\alpha \approx 4°$ (as can be seen in Fig. 2.43) and not, on a dry road, at $\alpha \geqslant 10°$.

2.10.2 Caster offset

Caster offset, $r_{\tau,T}$, is included in practically all calculations of the self-aligning moment during cornering (see Section 3.10.3). The length of this can easily be calculated from the lateral force and moment:

$$r_{\tau,T} = M_{Z,T,Y}/F_{Y,W} \text{ (m)} \tag{2.20}$$

This requires two images, one which represents $F_{Y,W} = f(F_{Z,W}$ and $\alpha)$ or $\mu_{Y,W} = f(F_{Z,W}$ and $\alpha)$, and another with $M_{Z,T,Y} = f(F_{Z,W}$ and $\alpha)$. The values of the 175/70R

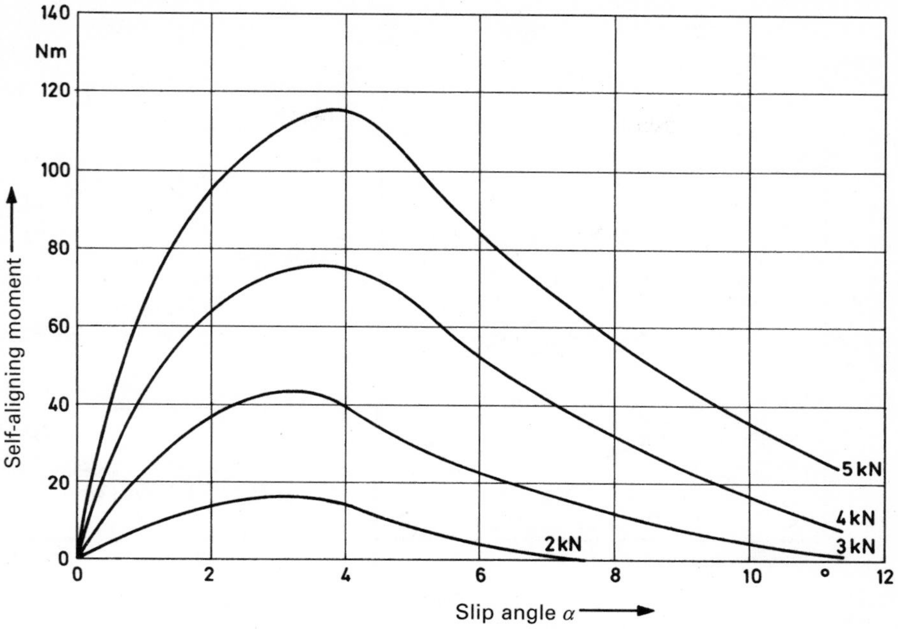

Fig. 2.49 Self-aligning torques of a 175/70 R 13 82 S steel radial tyre measured on a dry drum as a function of the slip angle at p_T = 2.0 bar. The vertical force $F_{Z,W}$ in kilonewtons is used as a parameter. The torques increase sharply at low angles, reach a maximum at α = 3° to 4° and then reduce slowly. As the cornering speed increases, the tyre self-aligning torque decreases, while the kinematically determined torque increases (see Section 3.8).

13 82 S steel radial tyre shown in Figs 2.45 and 2.49 and measured at $p_T = 2.0$ bar serve as an example. At $\alpha = 2°$ and $F_{Z,W} = 5.0$ kN the coefficient of friction $\mu_{Y,W} = 0.44$ and therefore:

$$F_{Y,W} = \mu_{Y,W} \times F_{Z,W} = 0.44 \times 5.0 = 2.2 \text{ kN}$$
$$= 2200 \text{ N}$$

At the same angle and with the same wheel force, the self-aligning torque is $M_{Z,T,Y} = 95$ Nm and therefore

$$r_{\tau,T} = M_{Z,T,Y}/F_{Y,W} = 95/2200 = 0.043 \text{ m}$$
$$= 43 \text{ mm}$$

Figure 2.50 shows the caster (caster offset trail) calculated in this manner. Higher lateral forces necessitate greater slip angles, and the latter result in smaller self-aligning moments and a reduced caster offset. The explanation for this fact is that, at low slip angles, only the tyre profile is deformed at the area

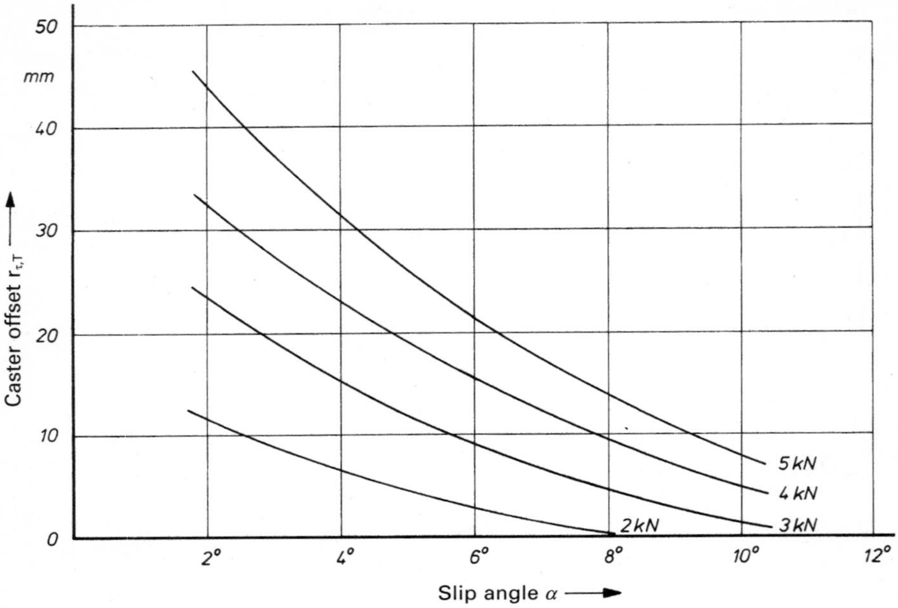

Fig. 2.50 Caster offset of tyre $r_{\tau,T}$ calculated from Figs 2.45 and 2.49 for 175/70 R 13 82 S steel radial tyres at p_T = 2.0 bar. The higher the vertical force $F_{Z,W}$ (in kN) and the smaller the angle α, the longer is $r_{\tau,T}$.

of contact. The point of application of the lateral force can therefore move further back, unlike large angles where, principally, the carcass is deformed. High vertical wheel forces cause the tyre to be severely compressed and therefore an increase both in the area of tyre contact and also in the caster offset occur.

2.10.3 Influences on the front wheels

The tyre self-aligning torque is one of the causes for the steering forces during cornering; its level depends on various factors.

2.10.3.1 Dry roads

The self-aligning torque is usually measured on a roller test bench with the drum allowing a coefficient of friction of μ_0 = 0.8 to 0.9 between its surface and the tyre. If the resultant self-aligning torque on the open road is required, it is possible to approximate the value $M_{Z,T,Y,\mu}$ using a correction factor:

$$k_\mu = \mu_{Y,W}/\mu_0 \qquad (2.21)$$

A cement block with $\mu_{Y,W} \sim 1.05$ (Fig. 2.43) and the 175/70 R 13 82 S radial tyre can be used as an example. In accordance with Fig. 2.49,

$$M_{Z,T,Y} = 40 \text{ N m with } F_{Z,W} = 3 \text{ kN and } \alpha = 4°$$

As a correction factor this gives

$$k_\mu = \mu \frac{\text{road}}{\text{roller}} = \frac{\mu_{Y,W}}{\mu_0} = \frac{1.05}{0.80}$$
$$= 1.31$$

and thus

$$M_{Z,T,Y,\mu} = k_\mu \times M_{Z,T,Y} = 1.31 \times 40$$
$$= 52.4 \text{ N m}$$

2.10.3.2 Wet roads

Provided that k_μ is independent of tyre construction and profile, the approximate value for a wet road can also be determined. In accordance with Fig. 2.47, with 1 mm of water on the surface and full profile depth the $\mu_{Y,W}$ value reduces from 0.86 to 0.55. Owing to the reduced coefficient of friction, only a smaller value $M_{Z,T,Y,\mu}$, can be assumed; in other words,

$$k_\mu = \mu_{Y,W} \frac{\text{wet}}{\text{roller}} = \frac{0.55}{0.86} = 0.64, \text{ and}$$

$$M_{Z,T,Y,\mu} = 0.64 \times 40 \text{ Nm}$$
$$= 25.6 \text{ Nm}$$

A greater water film thickness may cause the coefficient of friction to reduce but the self-aligning moment increases and the water turns the wheel back into the straight position. Furthermore, the self-aligning maximum shifts towards smaller slip angles when the road is wet.

2.10.3.3 Icy roads

Only with greater vertical forces and small slip angles is the smoothness of the ice able to deform the area of tyre contact and generate an extremely small moment, which is nevertheless sufficient to align the tyre. Low front axle loads or greater angles α arising as a result of steering corrections would result in a negative moment $-M_{Z,T,Y}$ (in other words in a 'further steering input' of the tyres). The wheel loads at the front, which were only low, were already a problem on rear-engine passenger vehicles.

2.10.3.4 Longitudinal forces

As shown in Fig. 3.119, traction forces increase the self-aligning torque; the equation for one wheel is

$$M_{Z,W,a} = F_{Y,W} \cdot r_{\tau,T} + F_{X,W,a} \cdot r_T = F_{z,W} \, (\mu_{Y,W} \cdot r_{\tau,T} + \mu_{X,W} \cdot r_T) \quad (2.22)$$

During braking the moment fades and reduces to such an extent that it even becomes negative and seeks to input the wheels further. The formula for one wheel is

$$\begin{aligned} M_{Z,W,b} &= F_{Y,W} \cdot r_{\tau,T} - F_{X,W,b} \cdot r_T \\ &= F_{Z,W} (\mu_{Y,W} \cdot r_{\tau,T} - \mu_{X,W} \cdot r_T) \end{aligned} \tag{2.23}$$

The length of the paths $r_{\tau,T}$ and r_T can be found in the details of Fig. 3.117.

2.10.3.5 Tyre pressure

When the tyre pressure is increased the self-aligning torque reduces by 6–8% per 0.1 bar, and increases accordingly when the pressure reduces, by 9–12% per 0.1 bar.

A reduction in pressure of, for example, 0.5 bar could thus result in over a 50% increase in the moment, a value which the driver would actually be able to feel.

2.10.3.6 Further influences

The following have only a slight influence:

- positive camber values increase the torque slightly, whereas negative ones reduce it;
- $M_{Z,T,Y}$ falls as speeds increase because the centrifugal force tensions the steel belt which becomes more difficult to deform (Fig. 2.16);
- widening the wheel rim width slightly reduces self-alignment.

2.11 Tyre overturning moment and displacement of point of application of force

A tyre which runs subject to lateral forces on the tyre contact patch is subject to deformation; there is a lateral displacement between the point of application of the normal force (wheel load; Fig. 3.119) and the centre plane of the wheel. Figure 2.51 shows the lateral drift of the normal (wheel load) point of application which is dependent on the size of the tyre, the lateral force and the camber angle and to a large extent on the construction of the tyre. Low section tyres with a small height-to-width ratio and a high level of sidewall rigidity exhibit greater lateral displacement. The rollover resistance of the vehicle is considerably reduced, as there is a decrease in the distance between the point of contact of the wheel and the centre of gravity of the vehicle.

This displacement results in the emergence of tyre overturning moments $M_{X,T,\alpha}$ about the longitudinal axis of the tyre (Fig. 2.52).

Both the lateral displacement of the point of application of the normal force and the tyre overturning moments must be taken into account when considering the overturning behaviour of vehicles, as they can considerably reduce rollover resistance, if, for example, a vehicle has a high centre of gravity and a small track dimension.

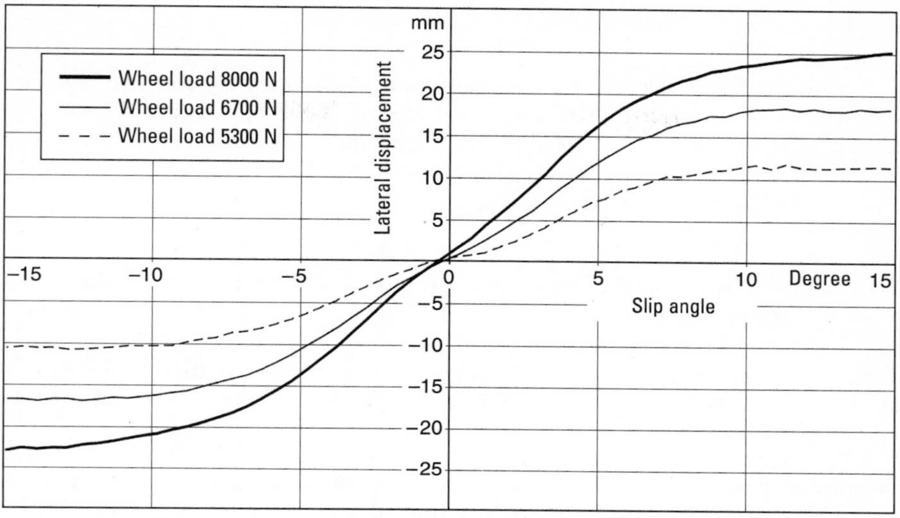

Fig. 2.51 Lateral displacement of normal (wheel load) point of application depending on slip angle and wheel load; measurements by Continental on a tyre of type 205/65 R 15 94 V ContiEcoContact CP.

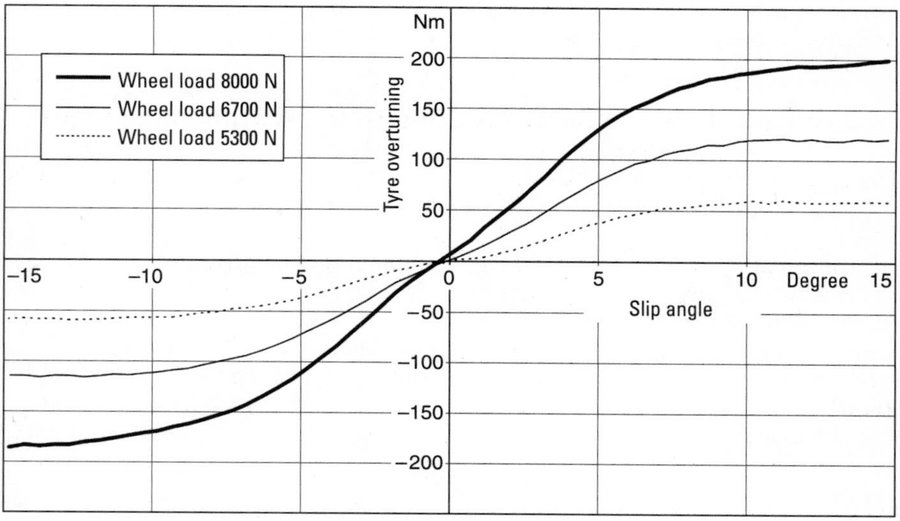

Fig. 2.52 Tyre overturning moments $M_{X,T,\alpha}$ on the wheel as a result of the build-up of lateral forces at different slip angles and wheel loads $F_{Z,W}$; measurements by Continental on a tyre of type 205/65 R 15 94 V ContiEcoContact CP.

2.12 Torque steer effects

Torque steer effects, i.e. changes in longitudinal forces during cornering, are an important criterion for the definition of transient handling characteristics. The torque steer effects depend on the size of the change in the longitudinal force, the adherence potential between the tyres and the road, the tyres and the kinematic and elastokinematic chassis design.

2.12.1 Torque steer effects as a result of changes in normal force

Torque steer effects usually occur during cornering when a driver has to slow down on a wrongly assessed bend by reducing the amount of acceleration or applying the brake.

The reaction force acting at the centre of gravity of the vehicle causes an increase in front axle load with a simultaneous reduction in the load on the rear axle. At an initially unchanged slip angle, the distribution of lateral forces changes as a result. If the force coefficient relating to the simultaneous transfer of longitudinal and transverse forces is sufficient, e.g. in the case of torque steer effects owing to reduction in acceleration or gentle braking (cf. Fig. 2.48), the increased lateral force corresponding to the increase in normal force on the front axle results in a yawing moment which allows the vehicle to turn into the bend.

If the adhesion potential is exceeded as a result of fierce braking or a low force coefficient, the tyres are no longer able to build up the necessary lateral forces. This results in an over- or understeering vehicle response depending on the specific case, be it a loss of lateral force on the front axle or rear axle or both.

2.12.2 Torque steer effects resulting from tyre aligning torque

The lateral displacement of the tyre contact area as a result of lateral forces leads to longitudinal forces being applied outside the centre plane of the wheel (Fig. 2.53).

This effect causes an increase in tyre aligning torque in driven wheels. In rear-wheel drive vehicles, this torque has an understeering effect with tractive forces, whereas it has an oversteering effect where there is a change in braking power.

In front-wheel drive vehicles, the resultant tractive force vector applies about lever arm $l_f \times \sin \delta_f$ offset from the centre of gravity of the vehicle (Fig. 2.54), so that an oversteering yawing moment is produced during driving which alters with application of a braking force to a (small) understeering yawing moment.

2.12.3 Effect of kinematics and elastokinematics

An attempt is made to keep the torque steer effects of a vehicle low by means of specific chassis design. The above-mentioned changes in forces produce

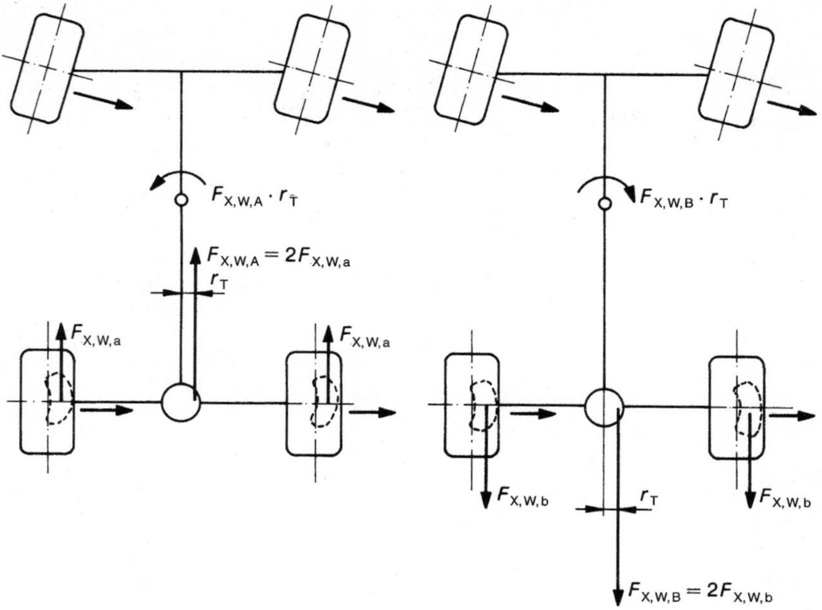

Fig. 2.53 The deformation of the tyre contact area during cornering results in aligning torque of the lateral forces which is further intensified by tractive forces and produces an understeering yawing moment. If there is a change in load, the braking forces produce an oversteering yawing moment.

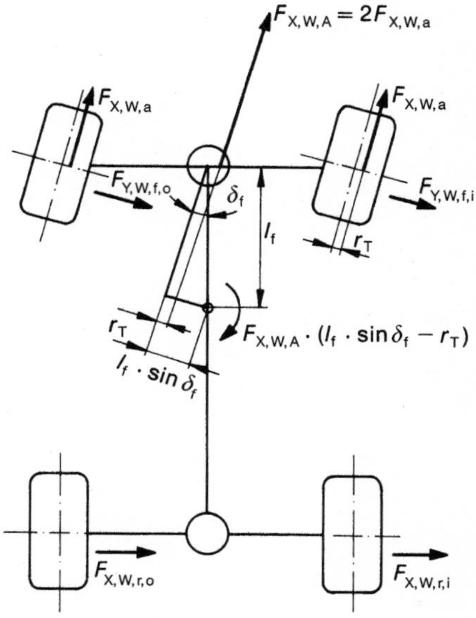

Fig. 2.54 With front-wheel drive, an oversteering yawing moment is produced, because the resultant tractive force vector is applied about lever arm $l_f \times \sin \delta_f$ displaced to the centre of gravity of the vehicle.

bump and rebound travel movements on the axles. The results, depending on the design of the chassis, in kinematic and elastokinematic toe-in and camber changes which can be used to compensate for unwanted changes in lateral forces, particularly in the case of multi-link suspensions. With unfavourable axle design and construction, there is, however, also the possibility of an increase in the torque steer effects.

3

Wheel travel and elastokinematics

'Kinematics' – wheel travel, according to DIN often also called wheel (or steering/suspension) geometry – describes the movement caused in the wheels during vertical suspension travel and steering, whereas 'elastokinematics' defines the alterations in the position of the wheels caused by forces and moments between the tyres and the road (Fig. 3.1 and Section 3.6.5), or the longitudinal movement of the wheel, against suspension anchorage required to prevent compliance, kinematic changes (Fig. 3.2). The changes are the result of the elasticity in the suspension parts. The coordinate directions (within which everything is to be considered) and the kinematic formulas are laid down in the German Standards DIN 70 000 and DIN 74 250 (Figs 3.3 and 3.101), as well as in the International Standards ISO 4130 and ISO 8855.

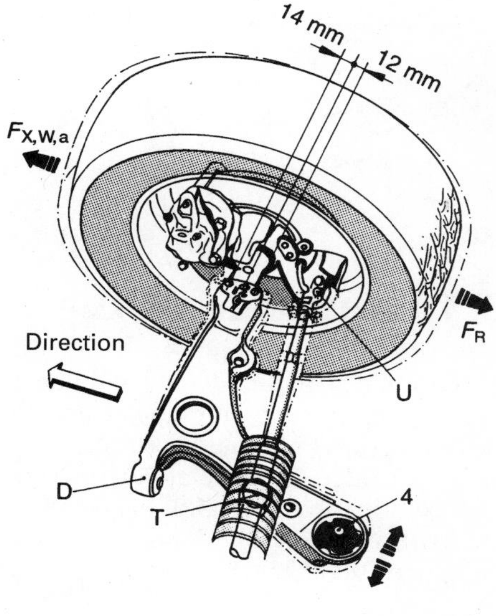

Fig. 3.1 Spring strut type front axle of the VW Passat (1995). As well as the vertical springing, the longitudinal springing shown is required in order to reduce the rolling hardness of the tyres and short-stroke movements caused by the road surface. This longitudinal springing is achieved by the lateral flexibility of the rear bearing 4; unwanted steering effects are corrected by the appropriate arrangement of steering tie rod points U and T (also see Fig. 3.83). The suspension arm is L-shaped in order to enable introduction of lateral wheel forces directly into the rigid bearing D to achieve a high level of lateral rigidity without a force component acting on the bearing 4.

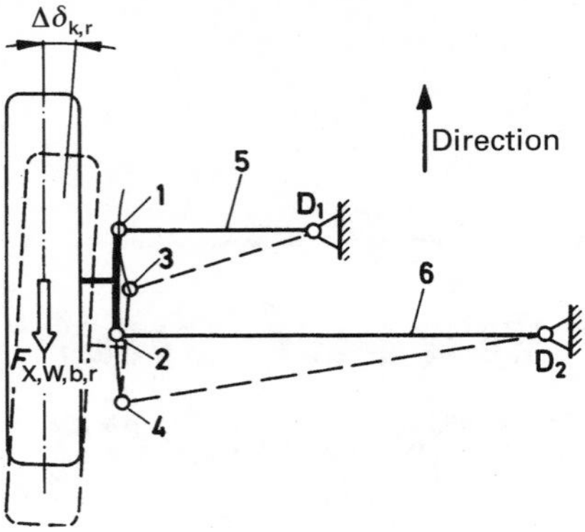

Fig. 3.2 If the front transverse link 5 on the bottom pair of suspension control arms of a rear McPherson axle is shorter than the rear one 6, and if the longitudinal forces are absorbed by a trailing link (not illustrated), its front bearing, which is fixed to the underbody, can comply in a defined manner when braking forces $F_{X,W,b,r}$ occur. The outer point 1 of the link 5 then moves in an arc around D_1 to 3 and point 2 of the link 6 around D_2 to 4. Due to the different radii of the two arcs, a toe-in angle $\Delta\delta_{k,r}$ occurs which opposes the returning moment $M_b = F_{X,W,b,r}\, r_b$ (Fig. 3.109) and produces braking force understeering effects in the handling.

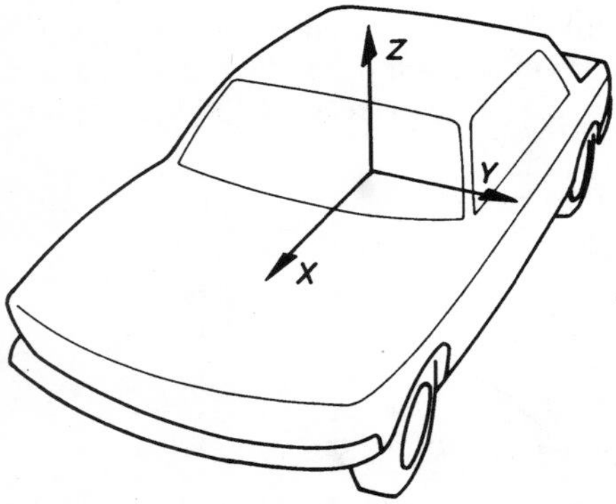

Fig. 3.3 Axis of coordinates in accordance with ISO 4130 and DIN 70000. The positive *Z* direction points upwards and, when viewed into the direction of travel (*X* direction), the *Y* arrow points left (see Fig. 3.101).

3.1 Purpose of the axle settings

To ensure the required road holding and directional stability and to prevent excessive tyre wear, automobile manufacturers specify certain settings, including the permissible tolerances for the front axles of all models and for the rear axles, provided these are not driven rigid axles. Toe-in can be set via the tie rods or eccentric discs (Fig. 3.62) and camber and caster angles can also be adjusted on some vehicles. The remaining manufacturers' data for kingpin inclination, kingpin offset at ground (scrub radius), caster offset and differential toe angle are design data and not easy to measure and are actually only used for roadworthiness of a vehicle which has been damaged in an ac reached a given age.

As shown in the figures in the following sections, the axle setti load and load distribution. In order to make the measurements easier for garages to carry out, only the curb weight, in accordance with recommendation DIN 70 020 (see Section 5.3.1.1) should be used as the basis for measurements.

3.2 Wheelbase

The wheelbase l, measured from the centre of the front to the centre of the rear axle (Fig. 6.1), is an important variable in the vehicle's ride and handling properties. A long wheelbase relative to the overall length of the vehicle makes it possible to accommodate the passengers easily between the axles and reduces the influence of the load on the axle load distribution (see Section 5.3.6). The short body overhangs to the front and rear reduce the tendency to pitch oscillations and make it possible to fit soft springing, normally associated with a high level of ride comfort. A short wheelbase, on the other hand, makes cornering easier, i.e. gives a smaller swept turning circle for the same steering input (see Section 3.7.2).

Vehicle designers seek to achieve a long wheelbase on both front-wheel drive passenger cars and on conventional designs. However, this depends on the body shape. (See Section 1.1 in Ref. [8] and Ref. [20]). A hatchback estate saloon (Figs 1.68 and 1.72) can be of a more compact design, giving a longer wheelbase relative to the vehicle length than notchback saloons and the estate cars developed from them. The ratio

$$i = \frac{\text{wheel base}}{\text{vehicle length}} \qquad (3.1)$$

can be used as a reference and should be as large as possible:

i_l = 0.57–0.67 on estate saloons, and
i_l = 0.56–0.61 on notchback saloons

In coupés the value can be below 0.56 and on small cars it is up to 0.72.

The wheelbase is quoted in the manufacturers' brochures and the trade press and lies between:

l = 2160 and 3040 mm

3.3 Track

The size of the track b_f at the front and b_r at the rear (Figs 3.4 and 3.90) has a decisive influence on the vehicle's cornering behaviour and its tendency to body

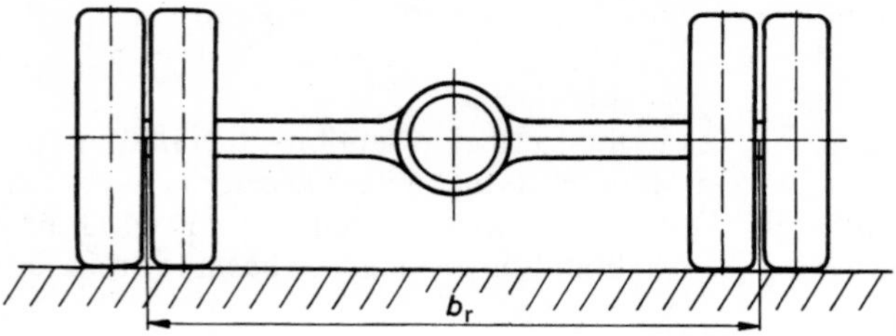

Fig. 3.4 On twin tyres the track specification b_r relates to the mean distance; the lower load capacity of the tyres should be noted here (see Section 2.2.5.3).

roll (see Section 5.4.3.1). It should be as large as possible but cannot exceed a certain value relative to the vehicle width. On the front axle the compressing, fully turned wheel may not come into contact with the wheel house (arch) (Fig. 2.8) and on the driven axle (regardless of whether front, rear or both) there has to be enough space for snow chains to be fitted. When the wheels compress or rebound, they must not come into contact with any part of the chassis or the bodywork.

The tread width on passenger cars is normally:

$$b_{\mathrm{f\ or\ r}} = 1210 \text{ to } 1602 \text{ mm}$$

and i_b can be used as a ratio for the width utilization and should be as large as possible:

$$i_b = \frac{\text{tread width}}{\text{vehicle width}} = 0.84 \text{ to } 0.87 \tag{3.1a}$$

When the wheels travel in bump and rebound-travel direction, the track changes on almost all independent wheel suspensions, which may be the result of functional factors or, as the following section shows, unavoidable if a higher body roll centre is necessary. However, the track size alteration causes the rolling tyre to slip (Figs 3.5 and 3.6) and, on flat cross-sections in particular, causes lateral forces, higher rolling resistance and a deterioration in the directional stability of the vehicle, and may even influence the steering.

Track variation on the front and rear axle must be checked on the drawing when the vehicle is at an early design stage. On a double wishbone suspension, arcs with suspension control arm lengths *c* and *f* must be drawn around points C and D (i.e. the suspension control arm axes of rotation), and the centres of the outer ball joints marked as points 1 and 2 (Fig. 3.7). A template can be prepared to show the steering knuckle and wheel (Fig. 3.8) and, in addition to points 1 and 2, must also have holes indicating the centre of tyre contact W and, if necessary, the central point U of the outer tie rod joint (see Section 4.6.3).

As shown in Fig. 3.7, points 1 and 2 of this template must be drawn upwards

Fig. 3.5 On independent wheel suspensions, the bump and rebound-travel of the wheels as they go over a bump can lead to a track alteration and this, in turn, to the tyres running at the slip angle α. This causes disturbing lateral forces, particularly if bump travel occurs on one side; directional stability and rolling resistance deteriorate.

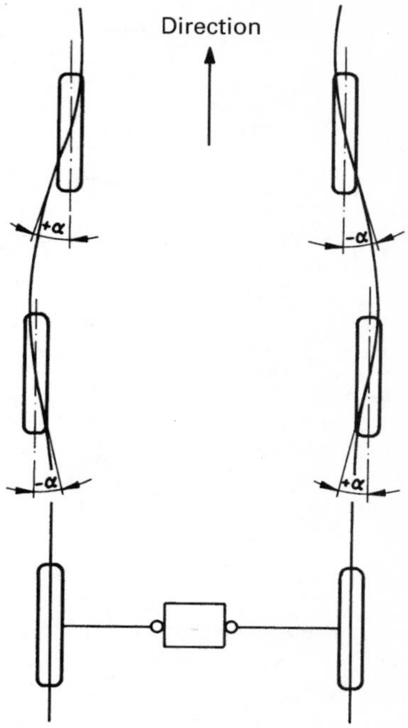

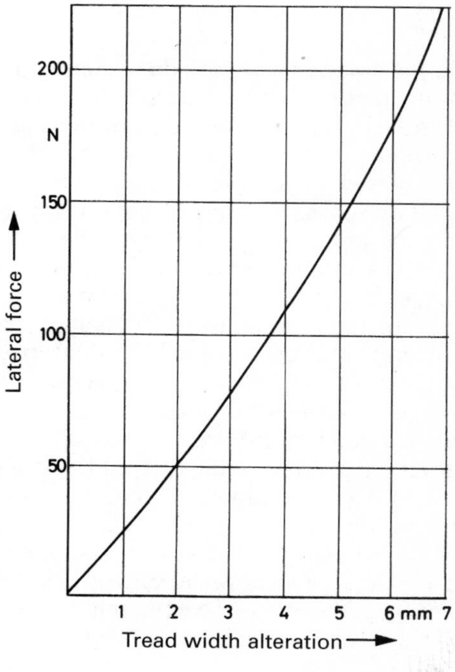

Fig. 3.6 Lateral forces $F_{Y,W}$ from the tyre to the road resulting from an alteration in track – shown on a radial 175/65 R 14 82 H tyre inflated to 1.9 bar under a load of 380 kg and at a speed of 80 km h^{-1}.

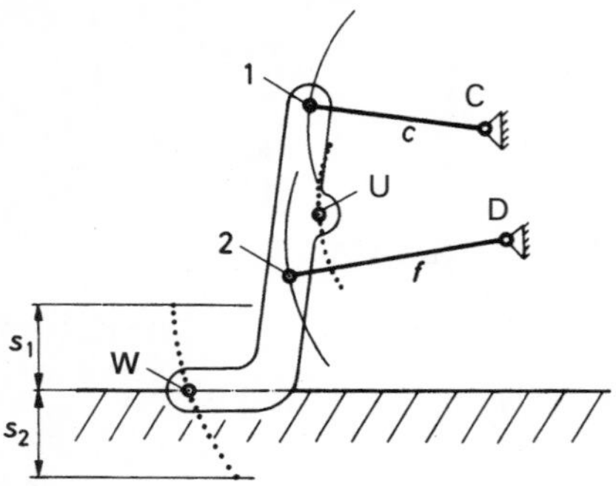

Fig. 3.7 Calculation by drawing of the alteration in track of a wheel (in the centre of tyre contact W) and the path of the outer tie rod joint U on the double wishbone suspension, using the template shown in Fig. 3.8.

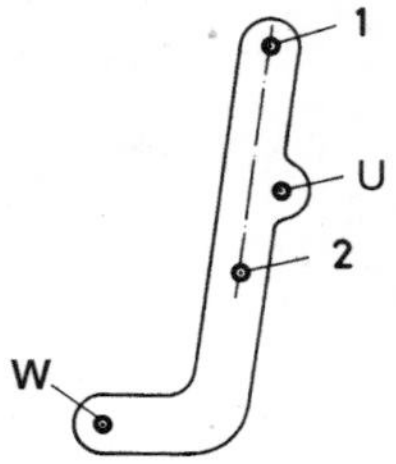

Fig. 3.8 Template for easy calculation of alteration in track, can be used on double wishbone suspensions (Fig. 3.7) and longitudinal link axles (Fig. 3.9).

along the arcs around C and D until point W of the template has reached the end of the bump travel s_1, previously indicated by a parallel to the ground, and downwards over the rebound travel s_2. The motions of W and U are then filled in step by step with a pencil. The line linking the points, which have been found in this way, gives the alteration of the track and the travel of the tie rod joint, but takes no account of any elasticity in the suspension control arm bearings (see Fig. 3.18).

In the case of the longitudinal control arm axle an arc must be drawn around D at the bottom, whilst a vertical line must be drawn on the suspension control arm axis of rotation (Fig. 3.9) and must go through point 1. At the same time a template as per Fig. 3.8 is moved along the arc and the vertical line to determine the tread width alteration.

McPherson struts have a mounting point E (Fig. 1.7) in the wheel house. When the wheel is in bump travel, the distance of the lower ball joint 2 to point C shortens and then lengthens again when the wheel rebounds (Fig. 3.11). The template has to take this length alteration into account (Fig. 3.10) and it has a slot in the direction of the strut damper centre line EE (only in the direction of the steering axis E2 if point 2 lies in its extension, see Figs 3.29, 3.30 and 4.46). Using point 2, which also has to appear on the template, a movement is made along the arc around D, whilst the slit is shifted over point C. A needle should mark this point on the drawing board.

If an arc is drawn around poles P, the track alteration of the dual joint swing axle can easily be drawn. Figure 3.12 shows both this and the advantages of lowering the tail end of the vehicle, i.e. achieving a smaller and thus more favourable camber angle and a higher lateral camber force on bends.

On all-independent wheel suspensions the position of the pole P determines the momentary alteration $\pm \Delta b$ (present in a small springing range, Fig. 3.14). Tread width alteration is avoided completely if P is at ground level and the lengths of the

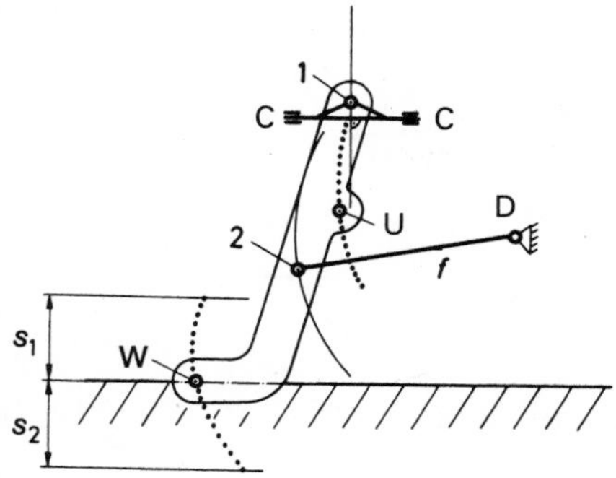

Fig. 3.9 Determination of the track alteration and the track of the outer tie rod joint U using the template shown in Fig. 3.8 on the longitudinal link axle. The description of this wheel suspension can be seen in Figs 3.32 and 3.157 and Section 9.4 in Ref. [2].

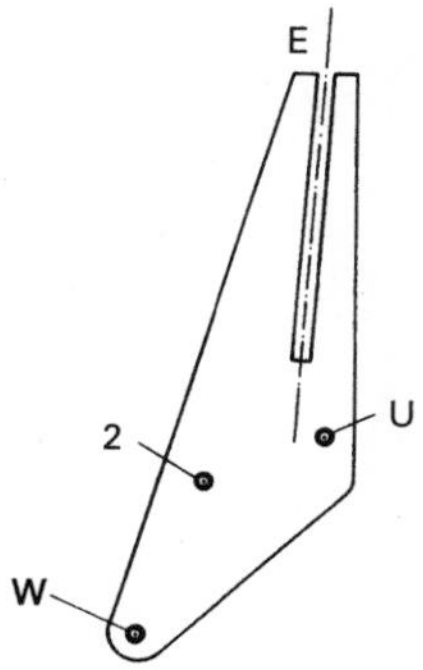

Fig. 3.10 The template needed to calculate, by drawing, the tread width alteration on the McPherson strut and strut damper must have a slot in the direction of the damper centre line E.

suspension control arms on a double wishbone suspension have been determined so that the pole moves horizontally from side to side on it when the wheels compress and rebound (Fig. 3.13). This can be demonstrated up to wheel travel $s = \pm 70$ mm using a drawing, calculation or models whereby any elasticity has been ignored (Fig. 3.18).

The tread width alteration can be measured as a function of the bump and rebound travel (s_1 and s_2) on the finished vehicle by determining the lateral shift of two parallel plates on which the two wheels of an axle are standing. It is necessary to run them parallel because a kinematic toe-in alteration when the wheels reach full bump/rebound travel (see Section 3.6.2) could turn the plates slightly and distort the measurement results.

Represented as a graph, the wheel travel should be plotted on the y-axis (Fig.

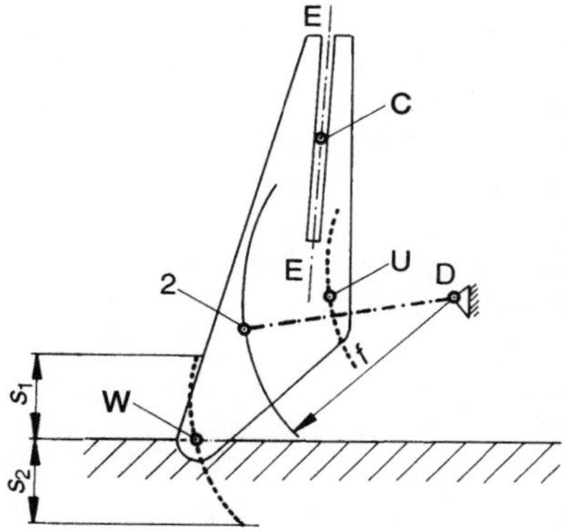

Fig. 3.11 Calculation by drawing of the alteration in track of one wheel and the path of the outer tie rod joint U on the McPherson strut and strut damper using the template shown in Fig. 3.10. C is the centre of the upper strut mount; this point is marked as E in Figs 1.8 and 3.139.

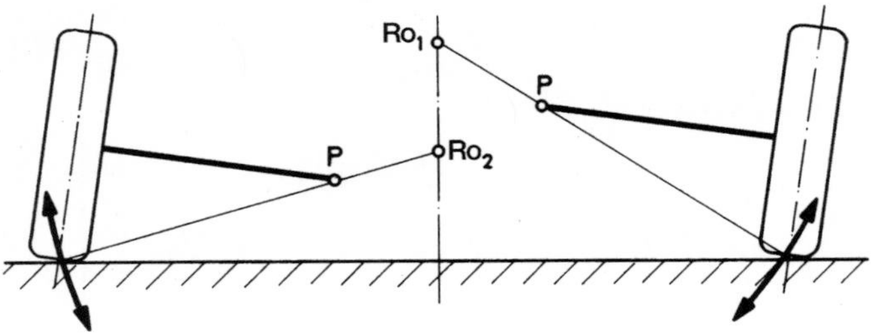

Fig. 3.12 Lowering the suspension control arm pivots P reduces the alteration in track on the dual swing axle, causes the body roll centre to drop from Ro_1 to Ro_2 and a wider track. With two people in the vehicle, there is already negative camber on the wheels – giving the advantage of accepting more of the lateral forces by the tyres, but the disadvantage of reduced bump travel (see description of swing axle in Ref. 2, Section 9.1).

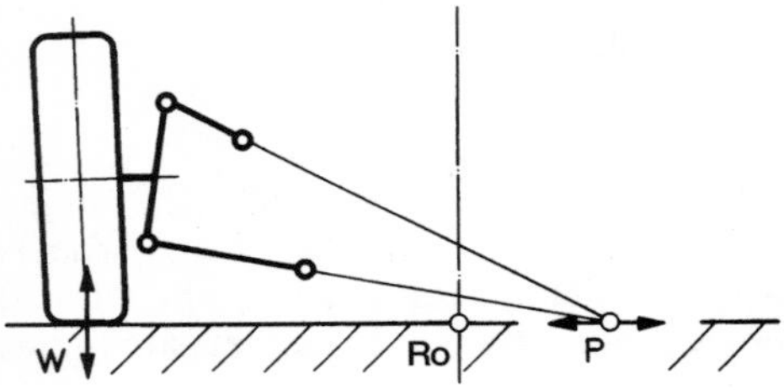

Fig. 3.13 An almost zero alteration in track requires a body roll centre at ground level (or at infinity, Fig. 3.25). Better kinematic properties are also obtained if the roll-centre axis is on the ground.

3.14) and – in accordance with the direction in which the axle is moving – bump travel can be shown as positive and upward (s_1), and rebound travel as downward (s_2). The zero position should correspond to the design weight (see Section 5.3.4), in other words the weight when three (or even two) people, each weighing 68 kg, are in the vehicle. An empty vehicle would be unrealistic.

The track alteration Δb of the two wheels (or $\Delta b/2$ of one wheel) appears on the x-axis, with the increase (as a positive value) entered to the right and the reduction (as a negative value) to the left. The existing track dimension $b_{\text{f or r}}$ in the zero position is an important dimension that should be stated. The tread width difference Δb to fully laden (or empty) can be determined using the spring-rate characteristic. The spring travel Δs_1 from the zero position to the permissible axle load (or the bump travel Δs_2 to the 'empty status') can be read off this to obtain the track alteration curve Δb as a function of Δs.

Figure 5.9 shows the front wheel springing of a front-wheel drive vehicle, where the dimension 80 mm must be deducted from 115 mm to get the rebound travel, Δs_2 = 35 mm starting from the zero position (here, two people each weighing 68 kg). The vehicle moves in bump-travel (from the zero position) by $\Delta s_1 = 92 - 50 = 42$ mm at the permissible wheel load (half the axle load). The paths are marked in Fig. 3.14; Δs_1 gives Δb_1 = +4 mm and Δs_2 gives Δb_2 = −8 mm. The track should be specified for the kerb weight: b_f = 1286 mm.

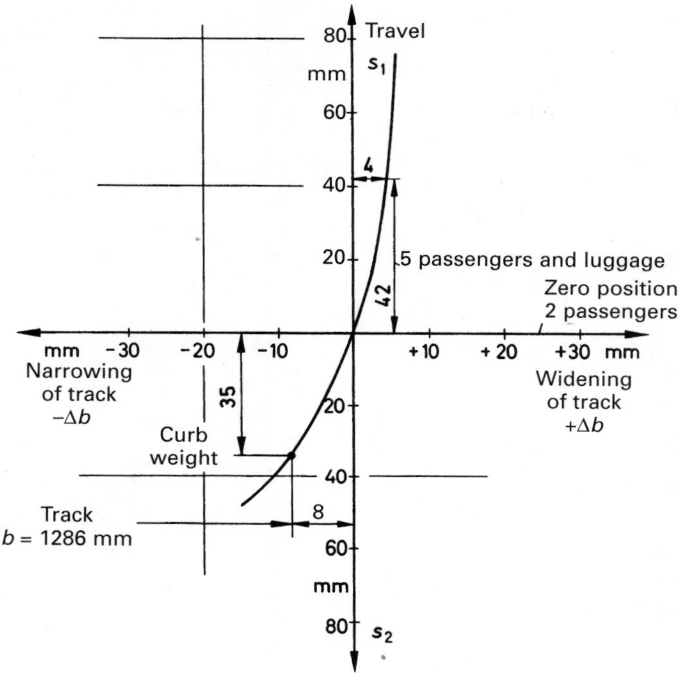

Fig. 3.14 The track (b) between the two wheels of an independent wheel suspension depends on the loading.

Figures 3.7, 3.15 and 3.18 show the track alteration of double wishbone suspensions and McPherson struts and the lower alteration values in bump travel can be clearly seen. As described in more detail in Section 3.4.1, the shape of the curve determines the level of the body roll centre. On all three passenger vehicle body configuration Ro_f is above the ground and falls perceptively (with the exception of the Honda, Fig. 3.15) when the vehicle is laden.

If the vehicle manufacturer has designed it at ground level as standard and the vehicle is subsequently lowered (Fig. 3.16) the body roll centre then moves into an adverse position; Ro_f drops below ground level and directional stability is likely to be impaired, particularly with wide tyres.

In double wishbone suspensions, the springs sit on the upper or lower suspension control arms and, in both cases, a moment arises (Figs 3.17 and 1.6) which, as a result of the elasticity in the suspension control arm bearings, causes the tread width alteration curve to take on a slightly different shape, thereby slightly altering the position of the roll centre (Fig. 3.18). The alteration curve, determined by measurements on the vehicle (with springs), gives the correct height in any case.

The tread width alteration curves of typical rear wheel suspensions are shown

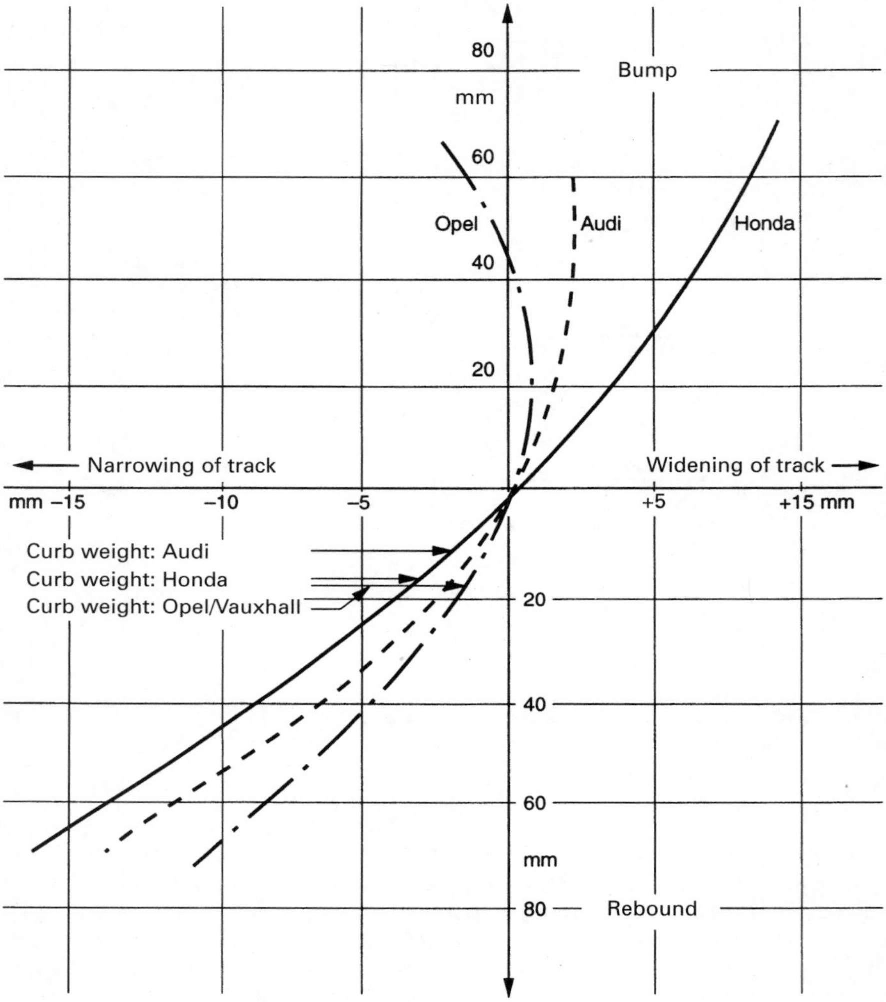

Fig. 3.15 Alteration in track of one wheel measured on the front axles of front-wheel drive Audi A6 (1996), Opel Astra (1996) and Honda Accord (1996) (Figs 1.57, 5.52 and 1.55). The Honda is the only passenger car to have double wishbone suspension; the kinematic advantages can be seen clearly.

The 'body roll centre height' $h_{Ro,f}$ in mm is:

Vehicle	Design position	Permissible axle load
Opel/Vauxhall	40	15
Audi	77	30
Honda	138	111

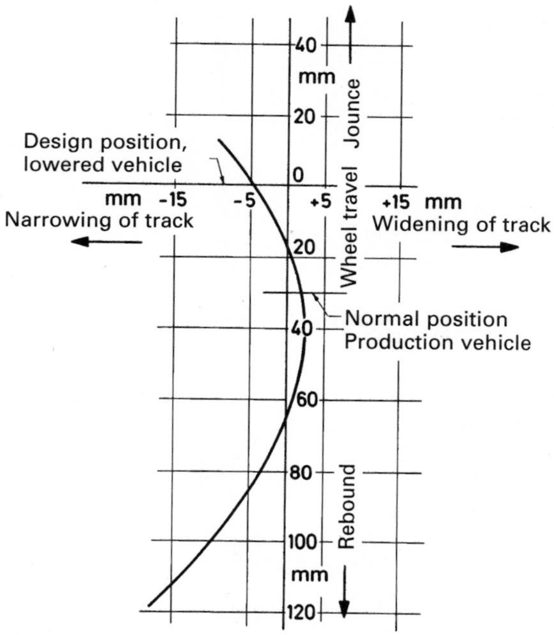

Fig. 3.16 Track alteration of both wheels measured on the front axle of a lowered VW Golf II GTi. In the normal position, specified by the manufacturer, the body roll centre is around road level. Lowering the vehicle by 30 mm means the body roll centre moves 115 mm below ground, resulting in a longer body roll lever and a theoretically increased roll. However, due to the early acting jounce bumper and virtually non-existent bump travel, the cornering inclination is greatly reduced (see Fig. 5.16 and Section 5.5.3).

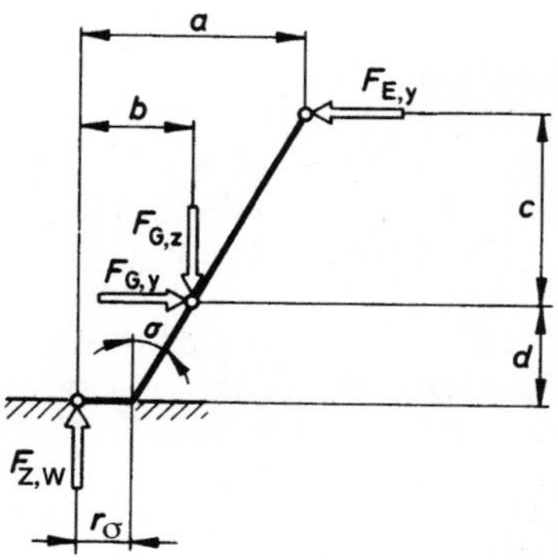

Fig. 3.17 The force $F_{Z,W}$ at the centre of the tyre contact and $F_{G,z}$ on the lower supporting ball joint form a moment, which is absorbed laterally on the suspension control arms causing the force pair $+F_{E,y}$ and $-F_{G,y}$ here. For reasons of simplification, upper and lower suspension control arms are assumed to be horizontal.

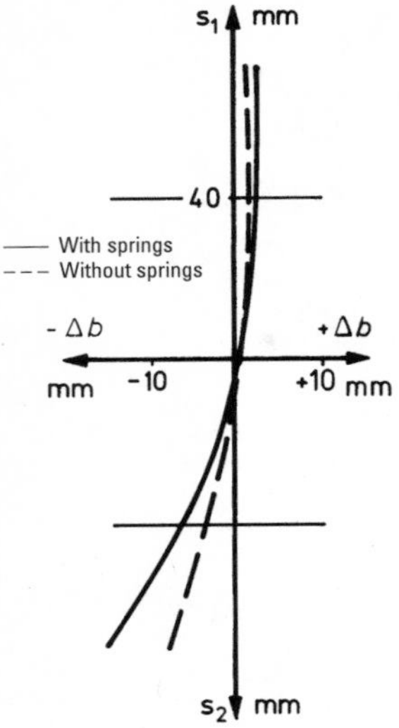

Fig. 3.18 Alteration in track of both wheels, measured with and without springs as a function of the spring travel on a double wishbone suspension. The curvature differs, being equivalent to a higher body roll centre on the drivable vehicle than the theoretical value (without opposing spring force) calculated or drawn on the drawing board (see Fig. 3.7).

in Figs 3.12, 3.19, 3.20 and 3.74. Non-driven rigid and twist-axle suspensions experience an increase or decrease in track as a result of the elastic camber alteration (Fig. 3.55).

3.4 Roll centre and roll axis

In all independent wheel suspensions, there is a direct correlation between the alteration in track and the height of the roll centre, so the two should be examined together. See Refs [2] and [9] for details.

3.4.1 Definitions

According to the German standard DIN 70 000, the body roll centre Ro is the point in the vertical plane which passes through the wheel centre points (Fig. 3.21), and in which transverse forces (y-direction) can be exerted on the sprung mass, in other words the body, without kinematic roll angles occurring.

The body roll centre is therefore the point in the centre of the vehicle (from the front), and in the centre of the axle (when viewed from the side), around which the body begins to roll when a lateral force acts, and at which reaction forces are absorbed between axle and body. Based on the existing track alteration curve of a wheel, the body roll centre is the point Ro in the centre of the vehicle (Fig. 3.22), which is intersected by a vertical, drawn on the tangent AB laid on the alteration curve in the centre of tyre contact. The height $h_{\mathrm{Ro,f}}$ of point Ro at the front (or $h_{\mathrm{Ro,r}}$ at the rear) can be determined in this way using the paths

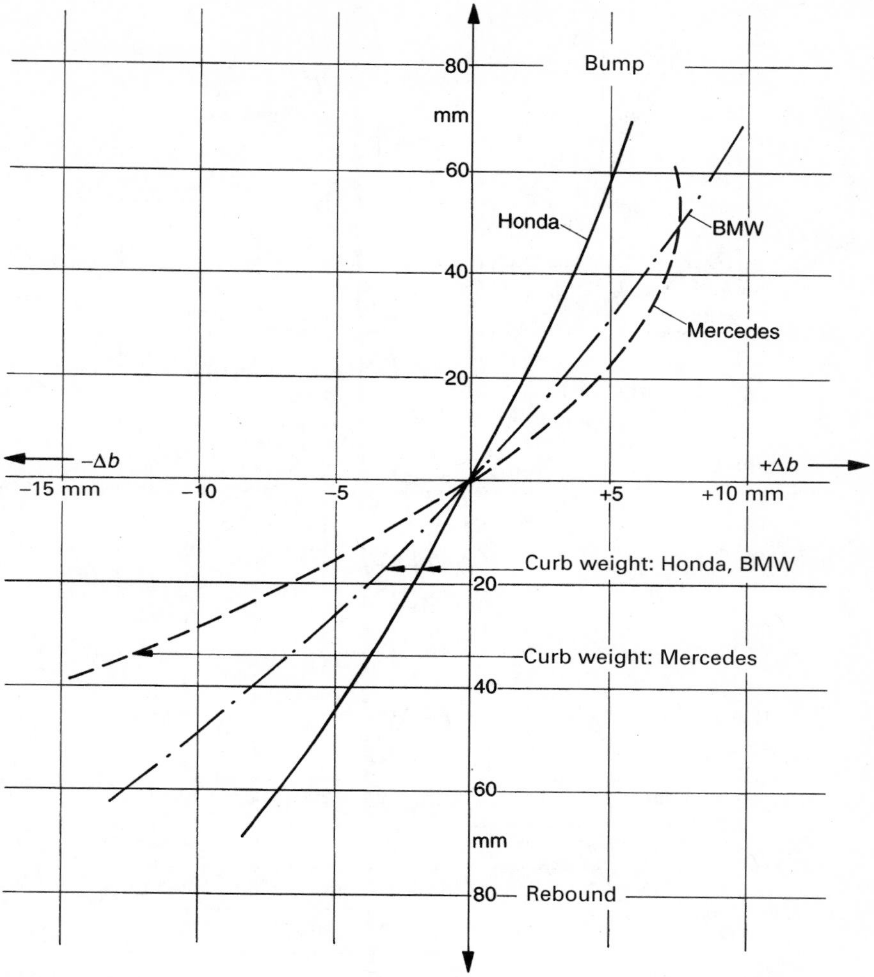

Fig. 3.19 Track alteration of one wheel, measured on the driven rear axle of a Mercedes (see Section 5.3.4 in Ref. 2), a BMW 3-series (Fig. 1.1) and the non-driven axle of a Honda Accord (Fig. 1.55). The shape of the curve indicates that, with the multi-link axle of the Mercedes, the body roll centre falls under load (Fig. 3.22). The levels $h_{Ro,r}$ (in mm) are as follows:

Vehicle	Design position	Permissible axle load
BMW	122	92
Honda	74	58
Mercedes	65	–

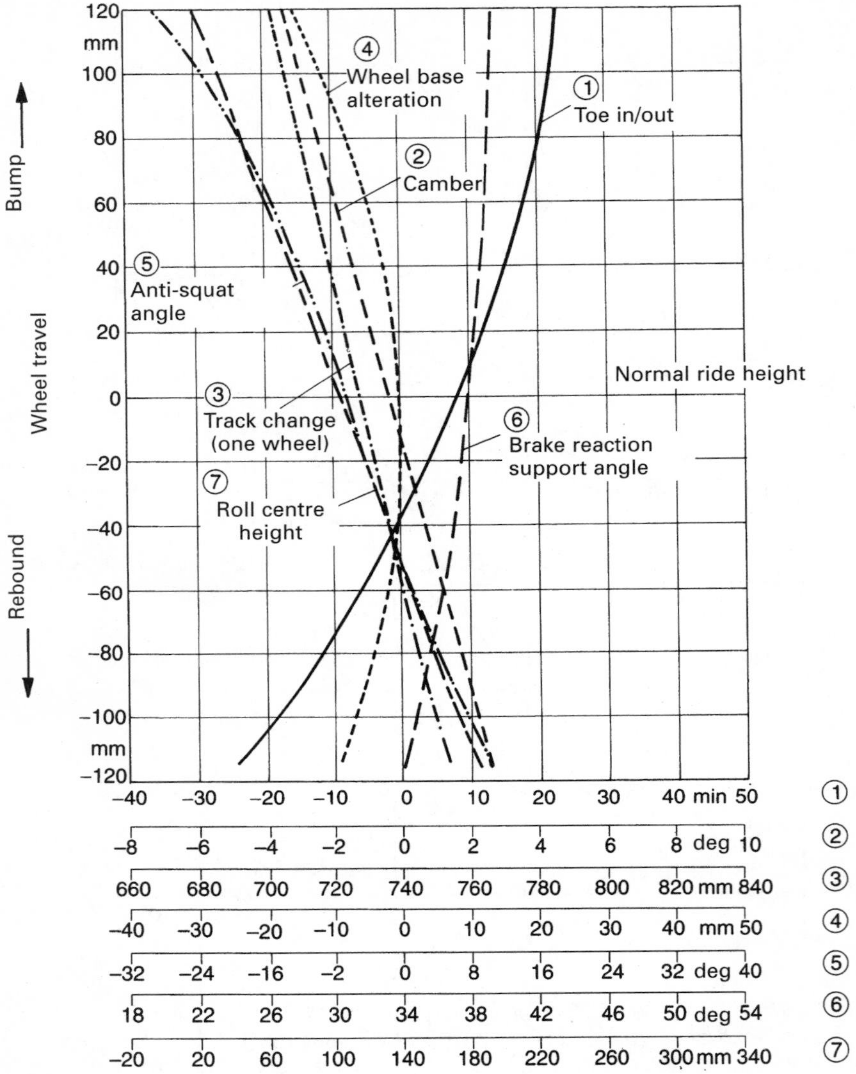

Fig. 3.20 Kinematics of the semi-trailing rear axle of an Opel Omega (1996). This measurement shows the change in track of one wheel only. The variation of toe or steer with suspension vertical deflection curve indicates a roll-steer effect on the rear axle tending towards understeering. This was achieved by the addition of a 'toe control link' on each side. The lowering of the rear body roll centre under load favourably reduces the dynamic wheel load transfer on the bend at permissible axle load (relative to that on the front): it allows the vehicle to understeer more.

Brake reaction support angle ε and anti-squat (diagonal springing) angle χ are shown in Fig. 3.160. The axle is shown in Fig. 1.15.

Fig. 3.21 The body roll centre is in the centre of the vehicle (viewed from the front) and in the centre of the axle (viewed from the side).

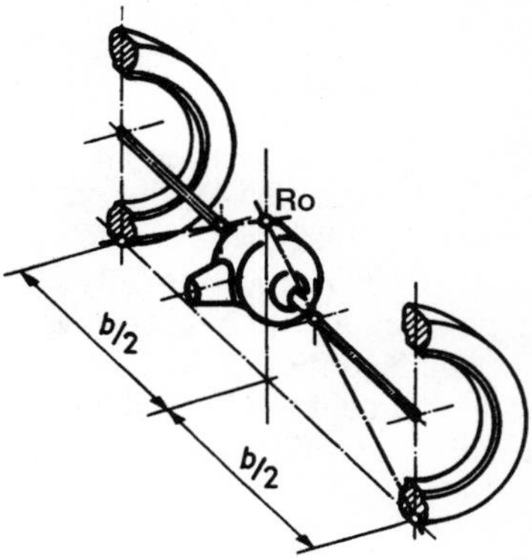

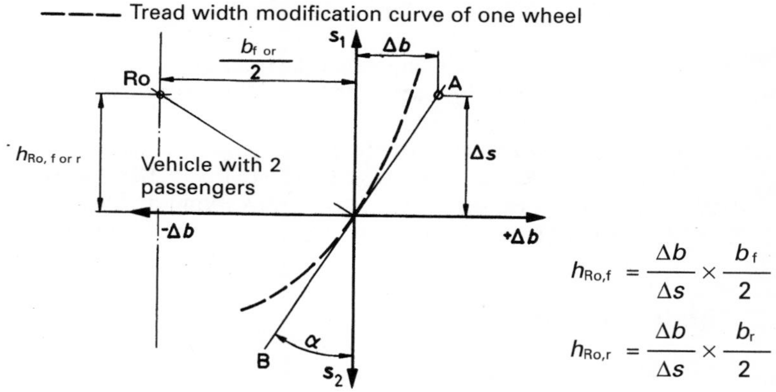

Fig. 3.22 The height $h_{Ro,f\ or\ r}$ of the body roll centre can be determined using a tangent from the measured track alteration curve in the respective load condition.

Δs and Δb drawn at the tangents, considering all elasticities in the suspension control arm bearings (Fig. 3.18). It behaves as follows:

$$\frac{\Delta b}{\Delta s} = \frac{h_{Ro,f\ or\ r}}{0.5 b_{f\ or\ r}} = \tan\alpha \tag{3.2}$$

and therefore the height of the body roll centre related to one wheel is

$$\text{front } h_{Ro,f} = \frac{\Delta b}{\Delta s}\frac{b_f}{2} \text{ and rear } h_{Ro,r} = \frac{\Delta b}{\Delta s}\frac{b_r}{2} \tag{3.2a}$$

Where $b_f = 1400$ mm, $\Delta b = 6$ mm per wheel and $\Delta s = 40$ mm,

$$h_{\mathrm{Ro,f}} = \frac{6}{40}\,\frac{1400}{2} = 105\ \mathrm{mm}$$

The greater the tread width alteration in the point corresponding to the respective load (Fig. 3.14), the steeper the vertical on the tangent becomes and the higher the body roll centre lies above ground. However, in the case of small track alterations, Ro is only slightly above, or on, the ground if the tangent **AB** is parallel to the *y*-axis (Fig. 3.13). If (as partly shown in some figures in Section 3.3) the track alteration due to both wheels is entered, the height of the body roll centre can be determined in the same way but only half the alteration travel, i.e. $\Delta b/2$, has to be considered. The equation is therefore related to both wheels:

$$h_{\mathrm{Ro,f\ or\ r}} = \frac{\Delta b\ b_{\mathrm{f\ or\ r}}}{\Delta s\ 4} \qquad (3.3)$$

In Fig. 3.15, in the Audi and Opel, tangents drawn on the upper curve are always parallel to the *y*-axis when the wheels compress, this being the equivalent of a drop in the body roll centre under load, a characteristic of McPherson struts. However, on the double wishbone suspension the tangent angle, and therefore the height of point Ro, alters less under load (Honda and Fig. 3.18). The same applies to this type of rear axle (Figs 3.19 and 3.20). With varying deflections to left and right, the body roll centre is generally no longer located at the vehicle centre.

3.4.2 Body roll axis

The position of the roll centres at the front and back and the course of the direction line joining these – the roll axis C (Fig. 3.23) – is of decisive importance for the handling properties: the height of the roll centres determines both the wheel load differences of an axle and hence the self-steering properties of the vehicle through the tyre properties, as well as the necessary roll suspension, which is again crucial to comfort in the case of unilateral deflection where a high level of roll rigidity is required and a stabilizer is used. The position of the roll centre also depends on the instantaneous position of the wheel links, i.e. the roll centre usually only lies in the centre plane of the vehicle if there is symmetrical wheel displacement and alters its position both horizontally and as vertically with unilateral displacement (cornering), resulting in the unwanted support effects of the wheel link forces on the body. A roll centre which decreases with symmetrical displacement helps to remedy this.

The height of the roll centre and the change in the roll centre with wheel travel is consequently a compromise between the following requirements:

- defined changes in wheel load during cornering to achieve the required (understeering) self-steering properties;
- track changes with wheel travel which are not critical for the dynamics of vehicle movement;

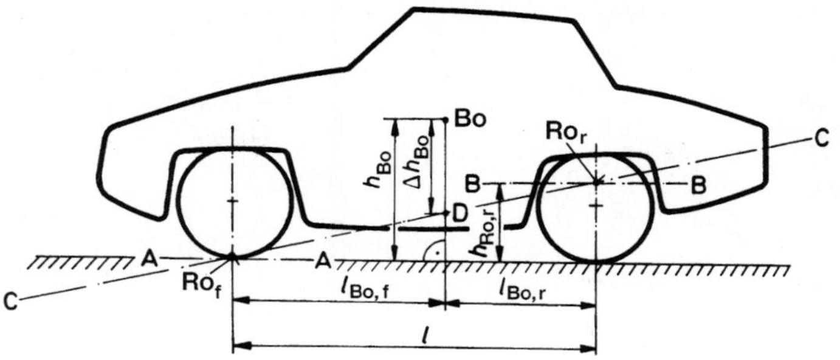

Fig. 3.23 Line C joining the front and rear body roll centre represents the theoretical roll axis (here at an angle). The path Δh_{Bo} is the body roll lever pointing vertical to the ground between this line and the body centre of gravity Bo. If the passenger car has a rigid rear axle, this angled disposition is beneficial. The body roll axis of a vehicle with independent wheel suspensions front and rear should only be at a slight angle (h_{Bo} see Equations 6.7 and 6.24).

- roll spring stiffness which is not crucial to comfort;
- desired – or permissable – camber change;
- as small as possible reaction forces acting on the body;
- the position of the roll axis.

The roll axis should rise slightly towards the rear in order to make use of fractions of the body damping to damp the yawing movements of the vehicle. Roll centre heights in the design of independent wheel suspensions are

h = 30 to 100 mm at the front
h = 60 to 130 mm at the rear.

Particular attention has to be paid to the superposition of high wheel loads with traction forces and hence a reduction in lateral force potential.

Depending on the curvature of the track alteration curve, the body roll centres fall under load to a greater or lesser degree (Figs 3.15, 3.19, 3.20 and 3.22).

The design of a chassis firstly requires the determination of the height $h_{Ro,f}$ of the front body roll centre (dependent on the track alteration) so that, in a second step, an appropriate rear axle can be provided; in the case of independent wheel suspensions with a slightly higher $h_{Ro,r}$.

If the vehicle is fitted with a rigid axle, the body enjoys less anti-roll support on bends ($i_\varphi = b_{Sp}/b_r$, Fig. 1.23) as a result of the shorter effective distance b_{Sp} of the springs relative to the track b_r. To balance this out, it is recommended that the body roll centre be designed slightly higher at the rear (as shown in Fig. 3.23). The possibilities for this can be taken from Ref. [2].

The additional lines A and B drawn in Fig. 3.23, are the actual body roll axes, which are mostly parallel to the ground. The precise location depends on the

The following dimensions have to be known: c, d, b_f, r_σ, α, β, σ

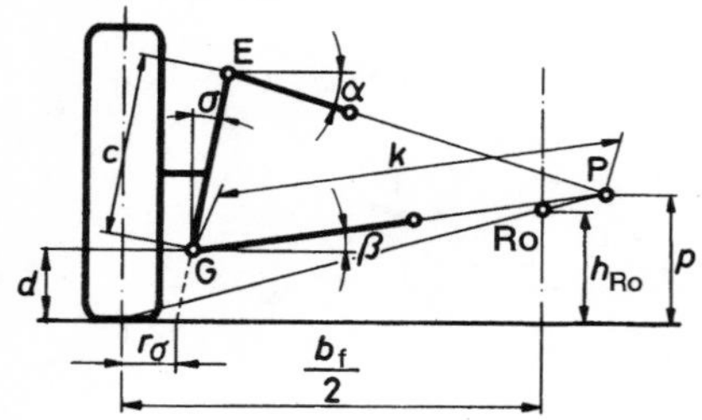

$$h_{Ro} = \frac{b_f}{2} \cdot \frac{p}{k \cos \beta + d \tan \sigma + r_\sigma} \qquad p = k \sin \beta + d$$

$$k = c \, \frac{\sin (90° + \sigma - \alpha)}{\sin (\alpha + \beta)}$$

Fig. 3.24 Determination by drawing and calculation of the paths h_{Ro} and p on double wishbone suspensions and a multi-link as well as longitudinal transverse axes (Figs 1.1 and 3.32).

angular position of the steering control arms. The body inclines around A and B under the influence of a lateral force.

3.4.3 Body roll centre on independent wheel suspensions

The height of the (instantaneous centre of rotation) P determines the position of the body roll centre Ro (Fig. 3.24). If P is above ground level, Ro will also be above ground. As can be seen in Fig. 3.22, the tangent drawn at the zero point on the track alteration curve varies by the angle α from the vertical. However, the shape of the curve at this point depends on the distance between virtual centre of rotation P and the centre of tyre contact W. The further the two are apart (i.e. the longer the path q, Fig. 3.30), the less pronounced the curvature and the lower the camber alteration (see Section 3.5.2). The following figures show the determination of height h_{Ro} of the body roll centre and path p by drawing. The virtual centre of rotation distance q from virtual centre of rotation to tyre contact-patch centre can be measured or calculated simply:

$$q = \frac{p \, b_{f \text{ or } r}}{h_{Ro} 2} \tag{3.4}$$

As can be seen in Figs 3.24 and 3.7, on the double wishbone suspension only the position of the steering control arms is important (i.e. the sizes of the angles α and β). The lines connecting the inner and outer steering control arm pivots need to be extended to fix virtual centre of rotation P and, at the same time, its height p. P linked with the centre of tyre contact W gives the body roll centre Ro in the intersection with the vehicle centre plane. In the case of parallel control arms, P is at ∞, and a line parallel to them needs to be drawn through W (Fig. 3.25).

Where the virtual centre of rotation is a long way from the wheel centre of contact, it is recommended that the distances p and h_{Ro} be calculated using the formulae in Fig. 3.24. Steering control arm axes of rotation, which are sloped

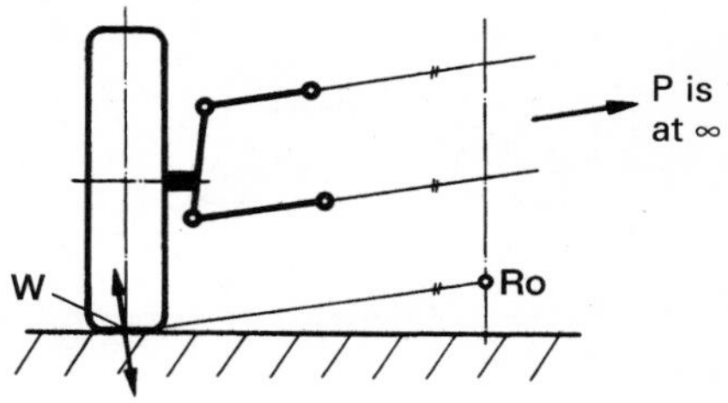

Fig. 3.25 Determination of the body roll centre on parallel double wishbones; the virtual centre of rotation is at infinity.

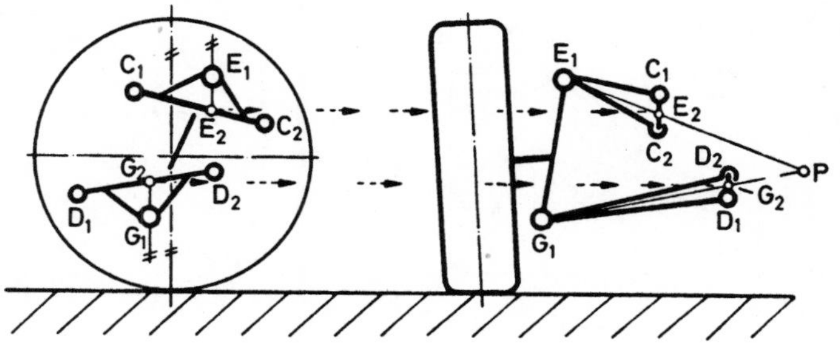

Fig. 3.26 If the suspension control arm axes of rotation are at an angle to one another when viewed from the side, a vertical should first be drawn to the ground through the points E_1 and G_1; the intersections with the axes of rotation C_1C_2 and D_1D_2 yield the points E_2 and G_2, needed for determining the virtual centre of rotation when viewed from the rear.

when viewed from the side (designed this way to obtain a vehicle pitch axis – Fig. 3.155), need E_1 and G_1 to be moved perpendicularly up or down (Fig. 3.26). The points E_2 and G_2 obtained in this way – linked with E_1 and G_1 when viewed from the rear – give the virtual centre of rotation P, and the line from this axis to the centre of tyre contact (as shown in Fig. 3.24) gives the body roll centre. If the axle is controlled by transverse leaf springs, where these are held in the middle (Fig. 3.27), the kinematic lever L_3 is important for calculating the body roll centre and,

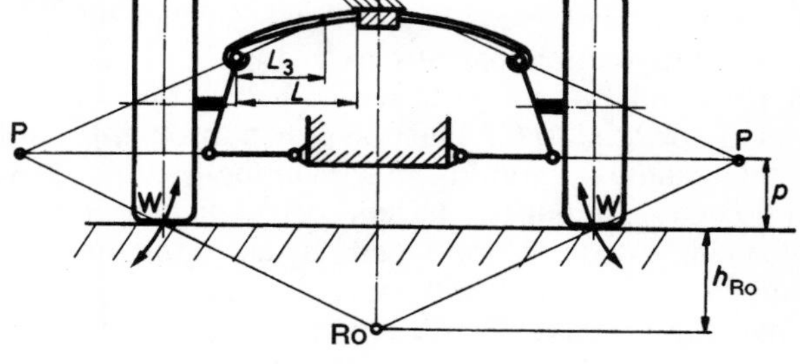

Fig. 3.27 Determination of Ro and P on a high, centrally anchored transverse leaf spring.

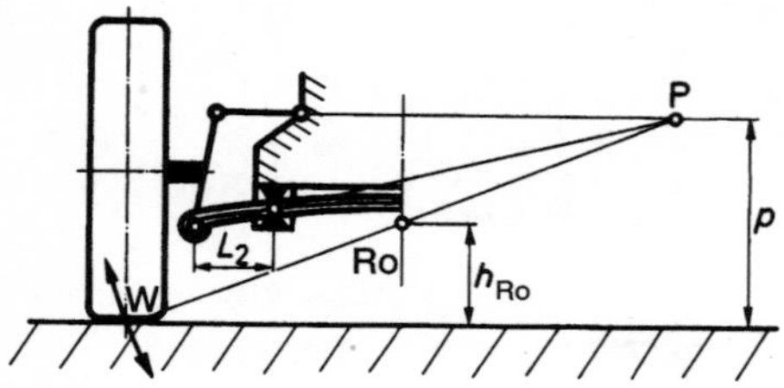

Fig. 3.28 Determination of Ro and P on a low transverse leaf spring supported in two places.

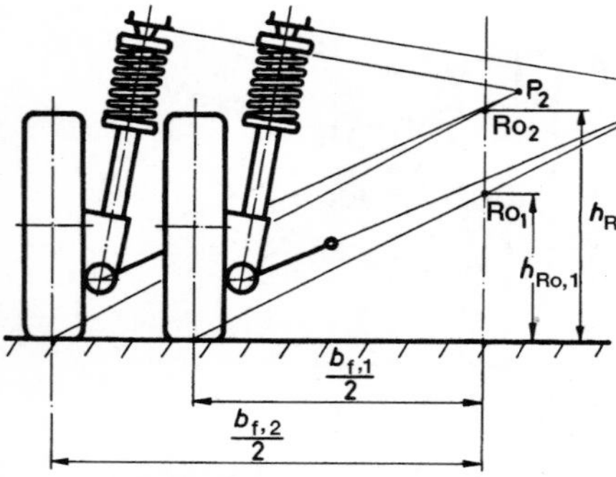

$b_{f,1} < b_{f,2}$ therefore $h_{Ro,1} < h_{Ro,2}$

Fig. 3.29 The greater the tread width b_f, the higher the body roll centre Ro, shown using the example of a McPherson strut (Fig. 1.56).

if the springs are attached at two points, the distance L_2 to the spring attachment point is important (Fig. 3.28). Further details are given in Section 4.7.3.1.

On McPherson struts, or strut dampers, a vertical must be created in the body side fixing point E to the centre line of the shock absorber piston rod, and the lower steering control arm must be extended. The intersection of the two lines will then give P (Fig. 3.29). The illustration also shows how increasing the track from $b_{f,1}$ to $b_{f,2}$ results in the body roll centre being raised from Ro_1 to Ro_2. A negative kingpin offset at ground makes it necessary to shift the lower swivel joint in to the wheel (Fig. 3.102) which separates the kingpin axis from the shock absorber centre line. Figure 3.30 shows the path EP, which is then vertical to the shock absorber centre line and also that h_{Ro} is not dependent on the steering control arm length, which is the decisive factor for the kinematic properties. Where the suspension control arm lies flat, it is recommended that the heights h_{Ro} and p be calculated because, if drawn, the virtual centre of rotation would be too far outside the drawing board (Fig. 3.31). Section 4.7.3.2 contains further details.

On the longitudinal link axle (Fig. 3.32), the direction of movement of the upper point E (vertical to the suspension control arm axis of rotation) plays a role. A parallel to CF must be drawn through E to obtain P and Ro. The calculation can be seen in Fig. 3.24. On the McPherson strut, the height of the body roll centre can only be influenced by placing the lower suspension control arm at an angle and only marginally by changing the angle between steering axis EG and the McPherson strut centre line (Fig. 3.30), which is a disadvantage of this type of suspension. On the longitudinal control arm axle it is possible to increase

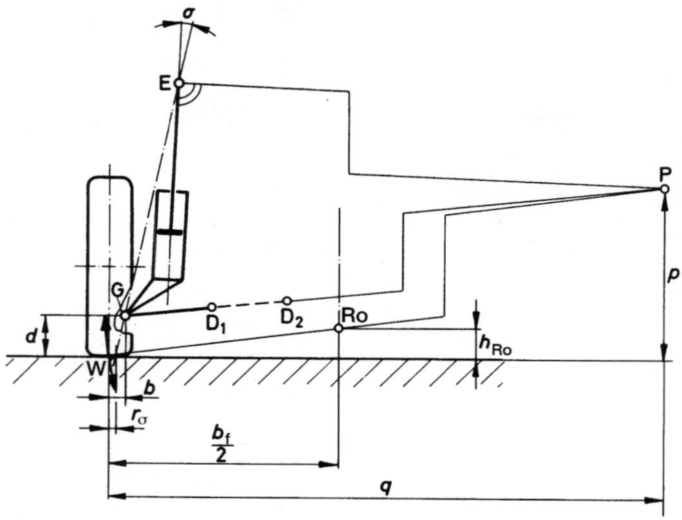

Fig. 3.30 The more vertical the McPherson struts and dampers and the more horizontal the lower control arm GD_1, the closer the body roll centre Ro is to the ground. This results in an adverse camber alteration when the wheels are in bump travel. Lengthening the lower suspension control arm (point D_1 to D_2) improves the kinematic properties.

To achieve a small or negative kingpin offset at ground r_σ, point G must be drawn outwards into the wheel, giving the benefit of a shorter lever b for the vertical force $F_{Z,W}$. The shorter can be path b, the less friction occurs between the piston rod and rod guide, as well as at the piston, and the smaller the forces in bearing points D, E and G (see also Fig. 1.11). A long path q means tread width alteration can be restricted. Fig. 1.8 shows the precise position of points E and G.

The lever b is easy to calculate:

$$b = r_\sigma + d \tan \sigma \tag{3.4a}$$

Depending on the design, either $+r_\sigma$ or $-r_\sigma$ has to be included in the equation (see Section 7.2 in Ref. [3]).

the angle of the axis of rotation CF further and therefore to raise Ro. At the same time, the virtual centre of rotation moves closer to the wheel, giving the additional advantage that the compressing wheels move more strongly into negative camber.

The heights $h_{Ro,f}$ of the front body roll centres determined in accordance with Figs 3.24 to 3.32 only agree in the case of bearings which, although they can be rotated, are otherwise not flexible, and only at body roll angles up to $\varphi = 2°$. The elasticity of the rubber elements used slightly alters the height available on the vehicle (Fig. 3.18). Furthermore, calculations and studies have both shown that, in the case of larger body roll angles, the left and right pivot axes take on a different position, but that the body roll centre in the centre of the vehicle experiences an alteration of only $\Delta h_{Ro} = \pm 10$ mm. Parallel measurements carried out on passenger cars showed a deviation of up to $\Delta h_{Ro} = 20$ mm.

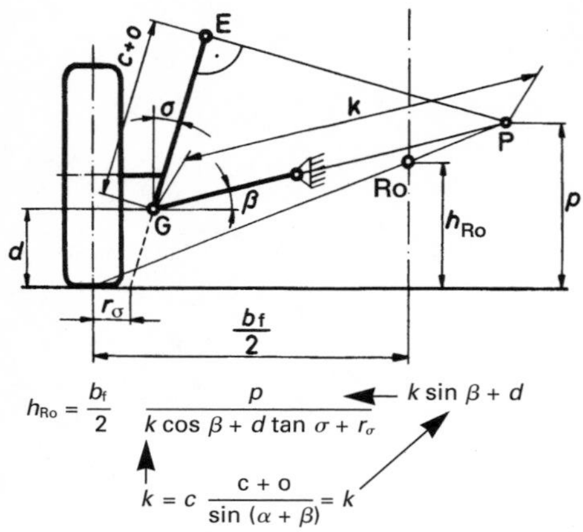

Fig. 3.31 Calculation of the paths h_{Ro} and p in the standard configuration of a McPherson strut and strut damper.

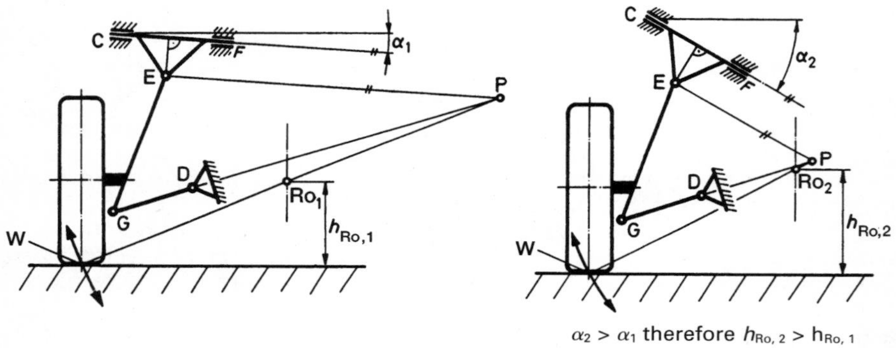

Fig. 3.32 With the longitudinal transverse suspension, a parallel to CF should be drawn through E and this made to intersect with the extension of the path GD to determine the roll centre Ro. Pole P is then connected to W to give Ro in the vehicle centre plane. The greater the angle of the upper suspension control arms, when viewed from the rear (α_2 right), the closer P moves to the vehicle centre; tread width and camber alteration increase and Ro_1 becomes Ro_2 at a higher level (see also Fig. 4.49).

In contrast to the front independent wheel suspensions, rear ones sometimes have only one control arm on each side; here, too the position of the virtual centre of rotation determines the height of the body roll centre, with the direction of movement of the wheel providing additional information. If the axis of rotation lies horizontal (Fig. 3.33) on the link axle, the wheel moves vertically and the roll centre Ro is at ground level. If the axis of rotation is inclined (Fig. 3.34), Ro moves above ground or, if the angle is in the other direction, below ground.

The single joint swing axle (Fig. 3.35) has its point of rotation in the centre of the vehicle. The pole is, at the same time, the body roll centre, unlike the dual joint swing axle on which point P is to the side next to the differential and Ro is

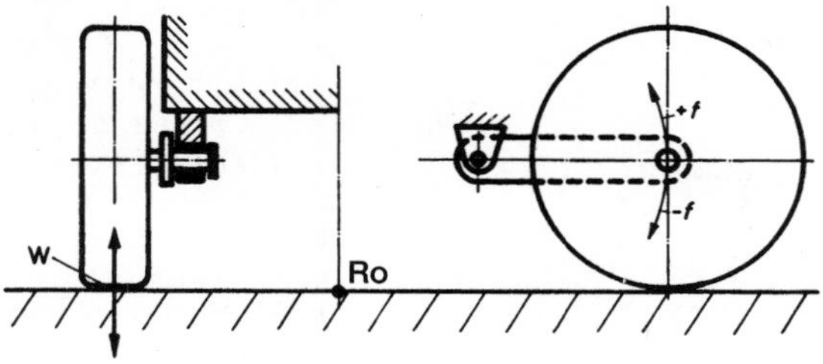

Fig. 3.33 If, with longitudinal links, the axis of rotation is horizontal, the body roll centre is at ground level and P is at ∞; the magnitude of torsional springing ±*f* depends on the suspension control arm length (diagonal springing angle χ, see Fig. 3.58).

therefore disproportionally high. Figure 3.12 shows how Ro is calculated, with the fall in the body roll centre in the case of negative camber $-\varepsilon_W$ (left) clearly indicated.

In the case of the semi-trailing link axle, the movement of the wheel vertical to the three-dimensional axis of rotation EG plays a role (Fig. 3.36). The point at which the extension of the axis of rotation intersects a vertical plane in the centre of the axle gives the virtual centre of rotation P_1 (= P_2), from which the height h_{Ro} of the body roll centre in the middle of the vehicle can be determined. To find this, first draw the top view, taking into account the angle α, and in it the extension of the suspension control arm axis of rotation made to intersect with the axle centre. The pole P_1 obtained in this way is moved perpendicularly down in to the rear view and made to intercept with the extension of the axis of rotation – this time using the angle β. Finally, the pole P_2 found in the rear view must be linked with W. With small angles α and β, it may be sensible to calculate h_{Ro} and p as a function of the dimensions specified by the designer. Figure 3.36 also contains the formulas for these relationships.

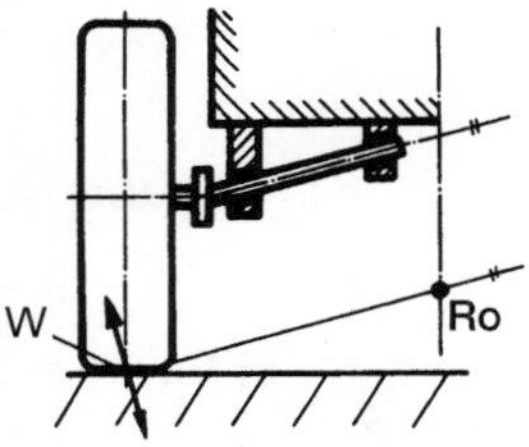

Fig. 3.34 If, with longitudinal links, the axis of rotation is at an angle, the body roll centre will lie above ground (or below it, if the angle is reversed); P is at ∞ in both cases (see Fig. 3.158).

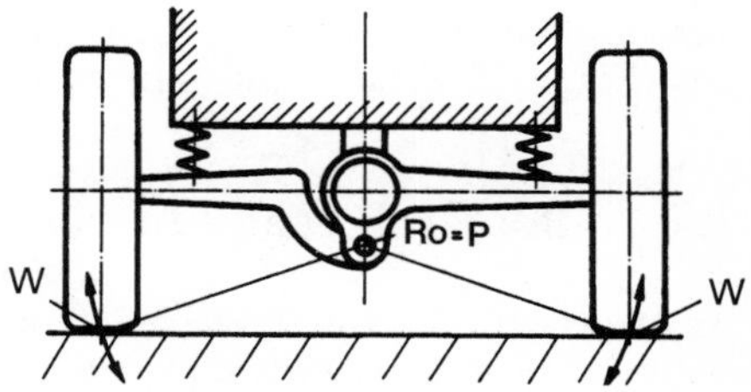

Fig. 3.35 On the single-joint swing axle, the suspension control arm pivot, which is approximately at the centre of the vehicle, is both the rear pivot axis and roll centre (see Section 9.2 in Ref. [2]).

The following dimensions have to be known: $e, f, k, b, \alpha, \beta$

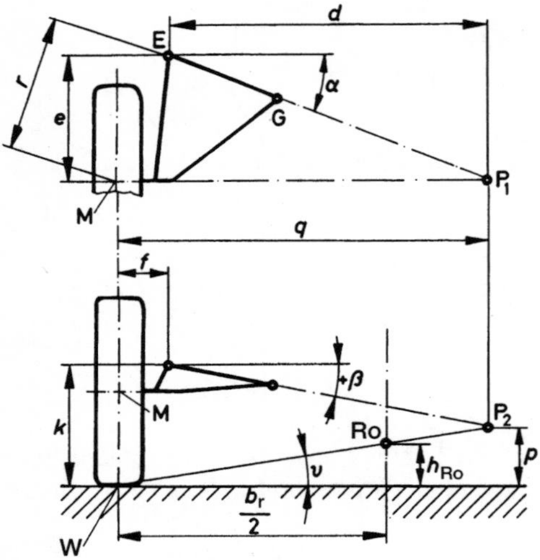

$$p = k - \tan\beta \, d$$

$$h_{Ro} = \frac{b}{2} \times \frac{p}{f + d} \qquad d = e \cot \alpha$$

Fig. 3.36 On the semi-trailing suspension, the positions of the virtual centre of rotation P and roll centre Ro are determined by the length r of the suspension control arms and the top view angle α and rear view angle β. The equations are used for calculating the height h_{Ro} in the vehicle centre. When the vehicle is laden, points E and G (and therefore also P and Ro) move down. The momentary tread width alteration results from an arc around P_2 (see also Figs 3.20 and 3.160).

3.4.4 Body roll centre on twist-beam suspensions

The kinematic or static body roll centres of this suspension are the bearing points O (Figs 3.37 and 1.31) at which – as specified in DIN 70 000 and described in Section 3.4.1 – the lateral forces are absorbed. The elastokinematic body roll centre, on the other hand, determines the alteration to toe-in and camber on reciprocal springing. Owing to the low torsion resistance of the transverse members the wheels swing (precess) during cornering, as on the semi-trailing link suspension, around the line connecting the points O_l and O_{rs} with the thrust centre point SM (Fig. 3.38). Toe-in and camber alteration are shown in Figs 3.54 to 3.57.

3.4.5 Body roll centre on rigid axles

As shown in Figs 1.25 and 1.26, on rigid axle suspension the lateral forces are absorbed in only one or two places. The body roll centre can therefore only occasionally be determined using the theory of transmission kinematics. It is the laws of statics which mainly apply, and the spring axle mounting point – at which the forces are transferred between body and axle – which should be observed.

If longitudinal leaf springs are used as the suspension, the lateral force is concentrated on the main leaves, and Ro is at their centre within the clamp (Fig.

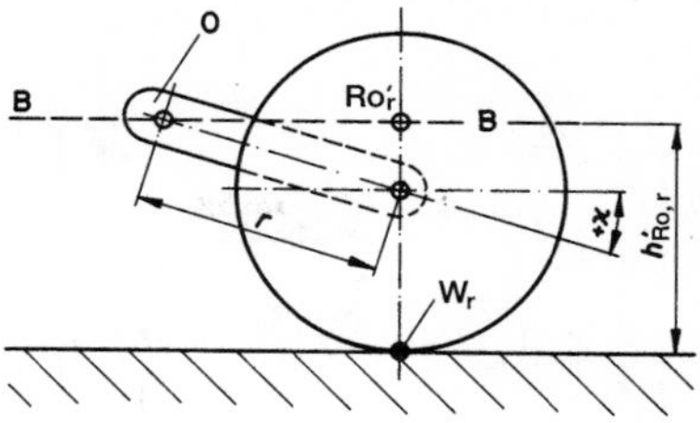

Fig. 3.37 On what is sometimes called the 'compound crank axle' (also called the torsion or twist beam axle) lateral loads are reacted by the two trailing links, which are stiff in torsion and bending. The height above ground of pivot points O determines the roll centre location. The position of O is dependent on the arm length *r* and its angle $\pm\chi$ to the horizontal. On the linked trailing arm or torsion beam (sometimes called 'twist beam') rear suspension the lateral forces are reacted by the two trailing arms, which are stiff in torsion and bending. The roll centre position is determined by the height above ground of the pivot points O. This is itself determined by the arm length *r* and its angle to the horizontal plane $\pm\chi$.

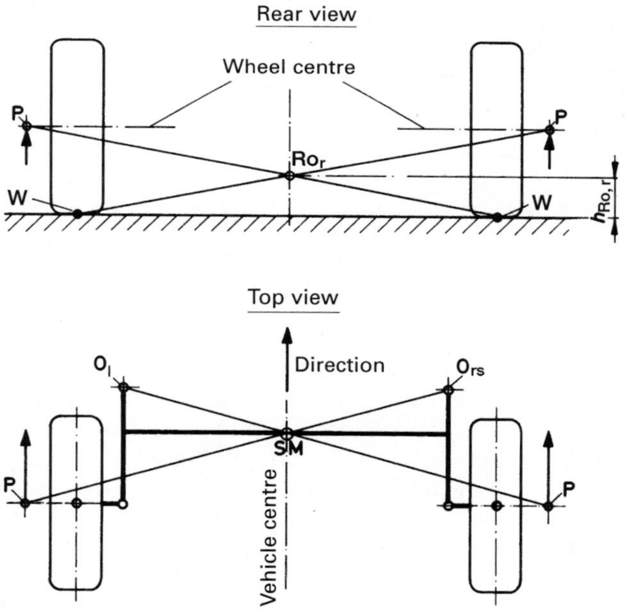

Fig. 3.38 Determination of the height $h_{Ro,r}$ of the elastokinematic roll centre Ro_r around which the body inclines under the influence of the centrifugal force acting on the body centre of gravity for the twist-beam suspension. The thrust centre point SM of the cross-member, which must be linked in the top view with the bearing points O and intersected with a straight line through the wheel centres, must be known. The resulting centres of rotation must be moved vertically upwards to the wheel centre axis in the rear view and linked with the centres of tyre contact W to obtain point Ro_r in the vehicle centre.

The position of the 'thrust centre point' also determines camber and caster alteration on counteracting bump/rebound-travel springing (Figs 3.54 and 3.55) as well as the lever arm ratio between the spring and shock absorber. (For more details see Ref. [2], Section 4.3.)

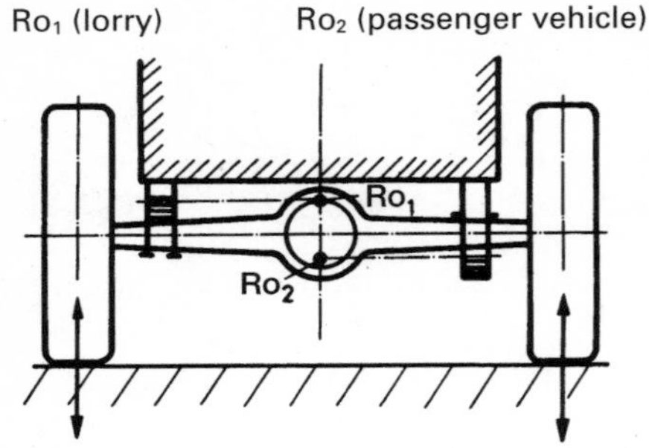

Fig. 3.39 If the rigid axle is carried by longitudinal leaf springs, the lateral forces are concentrated in its main bearings. The body roll centre is on the axle mounting in the middle of the main leaf, regardless of whether the spring is fixed above (left side and Figs 1.24 and 1.37) or below the axle (right and Fig. 1.26).

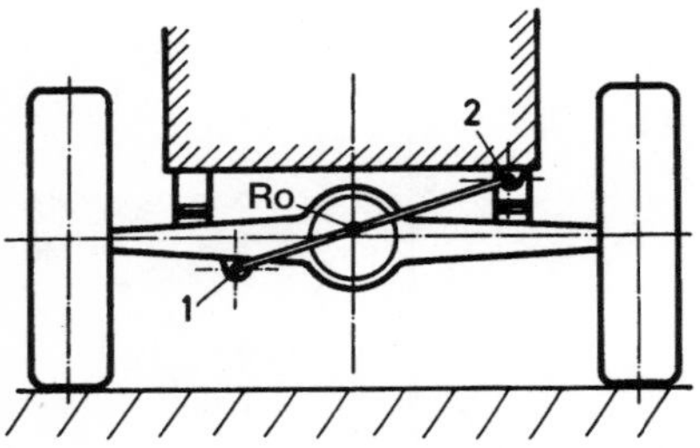

Fig. 3.40 If a panhard rod provides lateral force reaction support, the body roll centre is at the intersection of the rod with the vehicle centre line.

3.39). To keep it flat for a low underbody-ground clearance on a passenger car, the spring is underslung below the axle (right-hand side of the picture), whereas commercial vehicles need a high body roll centre to reduce the body inclination. The spring is then above the axle (left-hand side, see also Fig. 1.37) with the advantage that the fixing bolts are not subject to further tensile forces.

If the lateral force is supported by a panhard rod (Fig. 3.40), the body roll centre will be at the intersection of the panhard rod with the vehicle centre line (and not, as sometimes thought, in the centre of the bar). During cornering, the position of the bar changes and therefore so does the height of Ro. However, if a watt linkage supports the forces in a lateral direction, the point at which it is fixed to the axle housing is the decisive point of reference (Fig. 3.41).

The upper pair of longitudinal control arms and the panhard rod can be replaced by an A-arm (Fig. 3.42), which transfers lateral and longitudinal forces

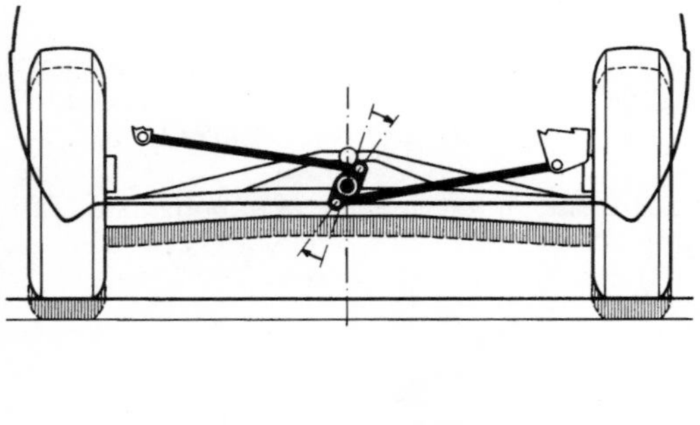

Fig. 3.41 Watt linkage on a passenger car rear axle. This allows the axle to be carried without any lateral deviation. When the springs deflect in bump and rebound-travel, the linkage turns around the mounting point on the axle housing, which is also the roll centre.

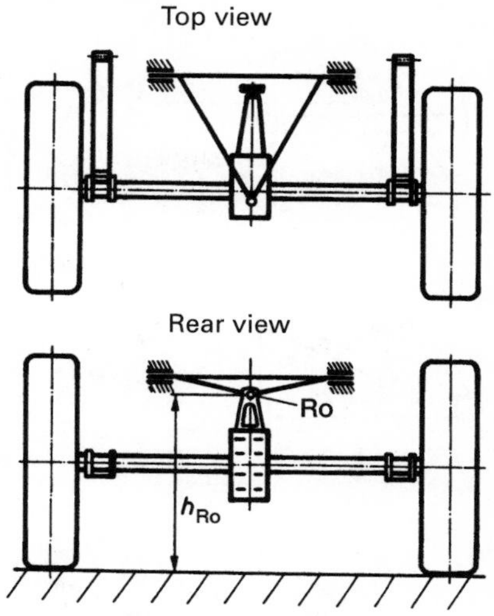

Fig. 3.42 If a longitudinal A-arm supports the rigid axle, its fixing position on the axle housing is also point Ro.

to the body. The body roll centre Ro is then the fixing point on the axle. In contrast to the panhard rod, point Ro maintains its height h_{Ro} when subjected to load.

Instead of the upper A-arm, two suspension control arms at an angle to one another can be used (Fig. 3.43). In this case, the intersection of the extension of the suspension control arm from the top view gives the virtual centre of rotation P_1 which must be brought down perpendicularly in the side view. In the case of parallel lower suspension control arms, a line drawn in the same direction as the arms intersects with the axle centre in the body roll centre Ro.

Unlike the rigid axle suspensions discussed so far, on the drawbar (longitudinal-pivot) axle (also known as the A-bracket axle) lateral forces can be absorbed jointly on the front bearing point O_r and two lateral struts (Fig. 1.60). The body roll centre is then at the height at which these three parts are attached to the body. If, instead of the two struts, there is a panhard rod, the forces are supported on this and point O_r. The side view shown in Fig. 3.44 next to the top view clearly shows both reaction forces $F_{O,y}$ and $F_{T,y}$. The body roll centre is therefore on the line linking the two points, which can be seen in the side view. If (as shown in Fig. 3.40) the panhard rod is at an angle, the mean height of the rod in the rear view must be determined and then transferred to the side view.

3.5 Camber

3.5.1 Camber values and data

In accordance with the standards DIN 70 000, camber is the angle between the wheel centre plane and a vertical to the plane of the road. It is positive if the

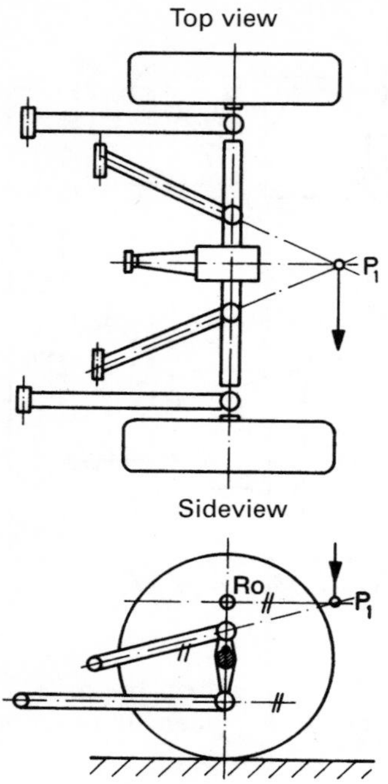

Fig. 3.43 If the two upper suspension links, which lie at an angle to one another in the top view, absorb the lateral forces, their extensions give virtual centre of reflection P_1. To determine Ro in the side view, a parallel must be drawn to the lower suspension control arms through P_1. As these two suspension links point in the same direction, as can be seen in the top view, their virtual centre is at ∞.

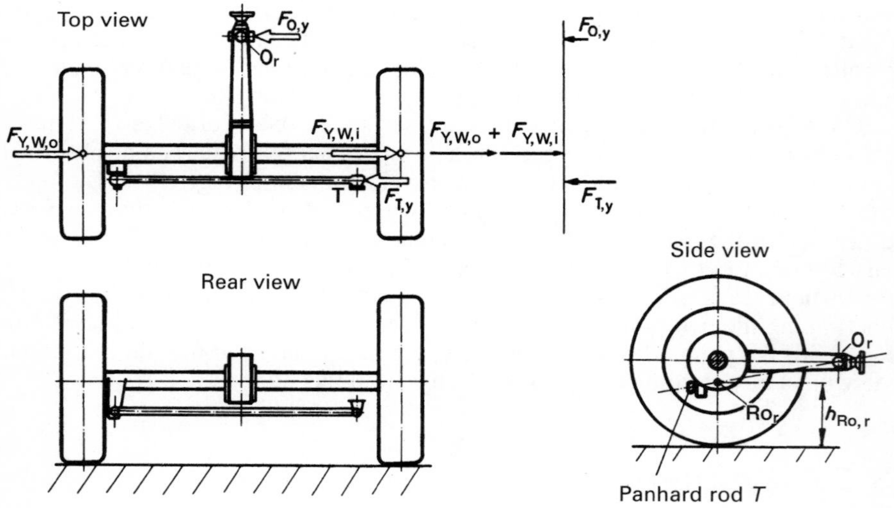

Fig. 3.44 The lateral forces $F_{Y,W,o}$ and $F_{Y,W,i}$ are transferred from the axle to the body at the foreward differential-housing exension mounting and the rear panhard rod. The reaction forces $F_{O,y}$ and $F_{T,y}$ occur. The body roll centre Ro_r must therefore lie on the line connecting points T and O_r from the side view. (The "drawbar" mounting is described in Ref. [2], Section 3.4.)

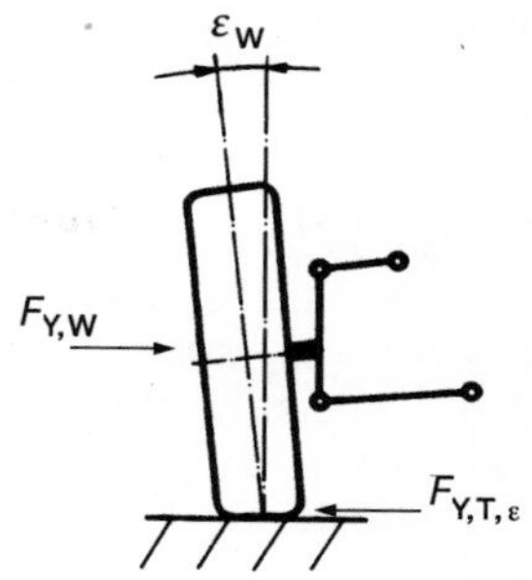

Fig. 3.45 Positive camber $+\varepsilon_W$ is the inclination of the wheel plane outwards from the vertical. The wheel shown would roll to the left because of the $F_{Y,T,\varepsilon}$ 'lateral camber force', if a right-hand counterweight did not restore the balance (i.e the direction straight ahead).

wheel is inclined outwards (Fig. 3.45) and negative, as $-\varepsilon_W$, when inclined inwards.

When a vehicle is loaded with two or three persons (design weight, see Section 5.3.4), a slightly positive camber would be useful on passenger cars to make the tyres roll as upright as possible on the slightly transverse-curved road surface and give more even wear and lower rolling resistance. As Fig. 3.46 shows, the optimum value for this purpose would be

$$\varepsilon_W = 5' \text{ to } 10', \text{ i.e. around } 0.1°$$

To give better lateral tyre grip on bends and improve handling, nowadays this rule is generally no longer adhered to and, on passenger cars, the setting is negative even when the vehicle is empty. Front axle values are as follows on newer production vehicles:

$$\varepsilon_{W,f,ul} = 0° \text{ to } -1° \, 20'$$

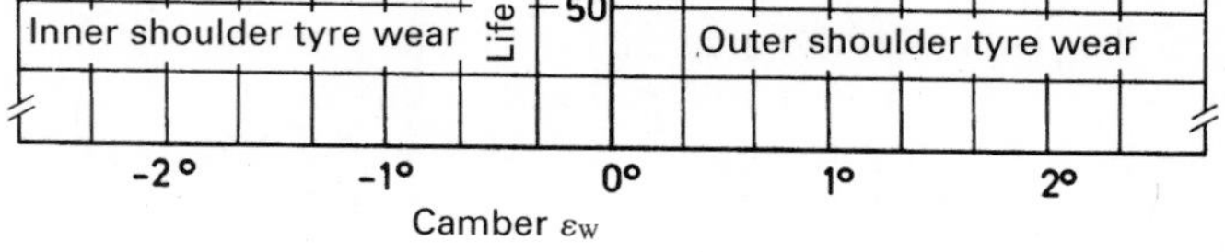

Fig. 3.46 Studies have shown that a camber of ε_W = +5′ to 10′ leads to the most even tyre wear; more positive camber would lead to more pronounced wear on the outer shoulder and negative camber to more pronounced wear on the inside of the tyre tread.

In addition to the absolute camber, the tolerance values are important, i.e. both the deviation from the permitted value and also the difference between the left and right wheel. A ±30′ deviation is usual to enable the components of the front axle to be manufactured economically. This is why it is not always possible to adjust the camber on front wheel suspensions. The various designs are described in Ref. [2].

To avoid the steering pulling to one side when the vehicle is moving in a straight line, the difference in the kingpin inclination angle between left and right wheels should not exceed $\Delta\sigma = 30'$. As can be seen in Fig. 3.103, camber and kingpin inclination are directly related, i.e. if the camber deviation is too great, so is the kingpin inclination angle. This is why no camber difference greater than 30' should be allowed as a factory setting. The information in the subassembly drawing of the front axle would then be as follows, for example:

Camber – 40′ ± 30′;
maximum difference between left and right 30′. (3.4b)

The measurement condition, which must relate to the kerb weight (i.e. the unoccupied vehicle, see DIN 70 020), must also be added. In the case of rear independent wheel suspensions and compound crank axles, designers prefer to use negative camber to increase lateral tyre grip; the mean value for the kerb weight can then be:

Camber – 1° 30′ ± 20′;
maximum difference between left and right 20′. (3.4c)

The existing setting options allow tighter tolerances here. On semi-trailing link axles there is a danger of too negative a value in the fully laden condition (Fig. 3.49); this could lead to the risk of the tyres becoming excessively warm and the protective cover coming free. This is the reason why passenger car manufacturers have reduced the kinematic camber alteration on this type of suspension by means of the angles α and β of the control arm axis of rotation (see Fig. 3.36 and Section 2.2.6.5).

3.5.2 Kinematic camber alteration

As described in Section 1.2.1, one disadvantage of independent wheel suspension is that the wheels incline with the body on a bend, i.e. the wheel on the outside of the bend goes into positive camber relative to the ground, and the lateral grip of the tyre under the greatest load (unlike the one on the inside of the bend) reduces (Figs 3.54 and 3.55). To balance this out, manufacturers tend to design the suspension on passenger cars such that the wheels go into negative camber as they travel in bump and into positive camber as they rebound (Figs 3.47 and 3.48).

On the x-axis, negative camber is given in degrees on the left and positive camber on the right, whereas wheel travel is plotted on the y-axis; wheel bump travel s_1 is plotted in mm upwards and rebound travel s_2 downwards. The curve for the double wishbone suspension, which bends sharply into the negative

Fig. 3.47 In independent wheel suspensions, the wheels incline with the body when the vehicle is cornering (Fig. 1.6). To even this out, the wheels, in bump travel, should go into negative camber and the rebounding ones into positive camber.

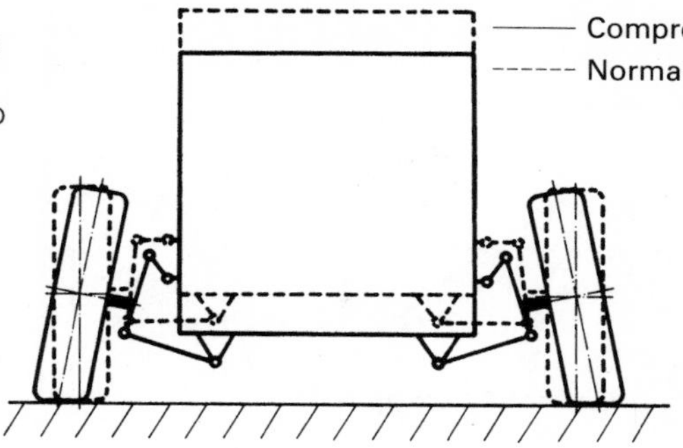

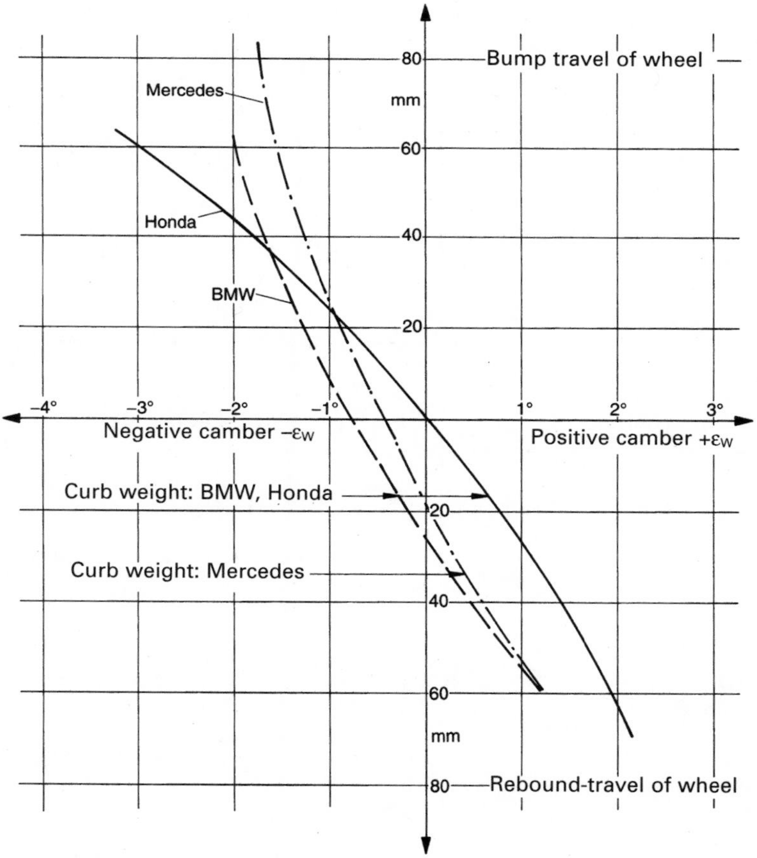

Fig. 3.48 Camber alteration on the front double wishbone suspension of a Honda Accord (Fig. 1.55) as a function of the wheel jounce travel s_1 and rebound travel s_2 in comparison with the McPherson suspension of a 3-series BMW (Fig. 1.40) and the strut damper axle of a Mercedes.

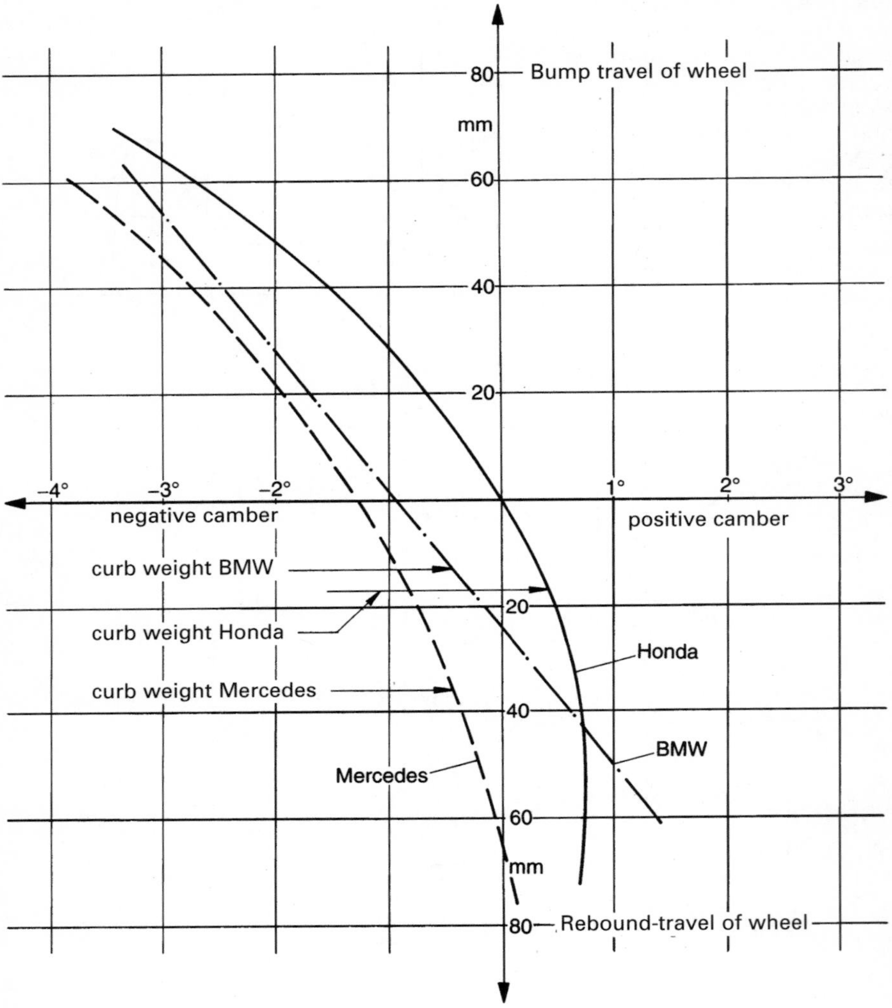

Fig. 3.49 Camber alteration on the rear wheels of a Mercedes, a 3-series BMW and a Honda Accord. The multi-link independent suspension of the Mercedes has a fairly precise camber setting. In the empty condition this was $\varepsilon_{W,O,l}$ = –55′ and $\varepsilon_{W,O,rs}$ = –35′ and increased to around –1°30′ when there were three people in the vehicle. When the springs compress, the curve shape is slightly progressive. The manufacturer's specification for the empty condition is ε_W = –50′ ± 30′ (see Ref. [2], Section 5.3.4).

The multi-link axle of the BMW (Fig. 1.1) exhibits a straight-line curve; when the springs deflect in bump travel, the negative camber is less than on the Mercedes.

The double wishbone suspension of the Honda (Fig. 1.55) has zero camber in the design position, but the wheels take on higher alterations (negative values) when the springs deflect in bump travel than on the two other suspensions.

during the compression, shows the advantage of this axle. For the McPherson strut or strut damper the curve bends (unfavourably) in the other direction. However, the wheel on the strut dampers takes on more positive camber during rebound, this being the equivalent of better lateral force absorption on the (less loaded) wheel on the inside of the bend.

The camber alteration curves for rear independent wheel suspensions are shown in Figs 3.20, 3.49 and 3.74, where improved properties can be seen than on the front ones. As there is no steering input to be considered, the semi-trailing links or transverse links can adopt an improved position. From the zero position shown, as can be seen in Fig. 5.14, the Mercedes compresses by 53 mm under full load. The camber is then $\varepsilon_{W,t} = -2°50'$ and remains above the critical value $\varepsilon W_{max} = -4°$, which should not be exceeded.

3.5.3 Camber alteration calculation by drawing

From a construction point of view, the camber alteration on the front wheels can easily be determined as a function of the wheel travel over the angle of alteration $\Delta\sigma$ of the kingpin inclination if elasticities are ignored. On double wishbone suspensions, arcs with the suspension control arm lengths *e* and *f* must be drawn around the points C and D (in other words the suspension control arm axes of rotation) and, in the normal position, the centres of the outer ball joints marked as points 1 and 2 (Fig. 3.50). A point 3 is determined on the upper arc and an arc with the path 1,2 drawn around it to give point 4. The line connecting them, 3,4, then has the alteration angle $\Delta\sigma$ to the path 1,2, if the wheel compresses by the path s_1. If it goes into negative camber (as in the example), $\Delta\sigma$ must be subtracted from the camber angle $\varepsilon_{W,0}$ in the normal position i.e.

$$\varepsilon_W = \varepsilon_{W,0} - \Delta\sigma \text{ (e.g. } -40' - 2° = -2°40'\text{)} \tag{3.4d}$$

In the case of positive camber, $\Delta\sigma$ would have to be added:

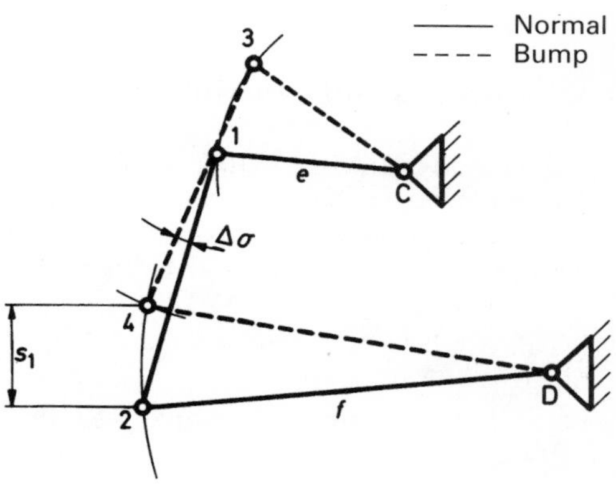

Fig. 3.50 Construction determination of the kingpin inclination alteration $\Delta\sigma$ on double wishbones which is equal to the camber alteration.

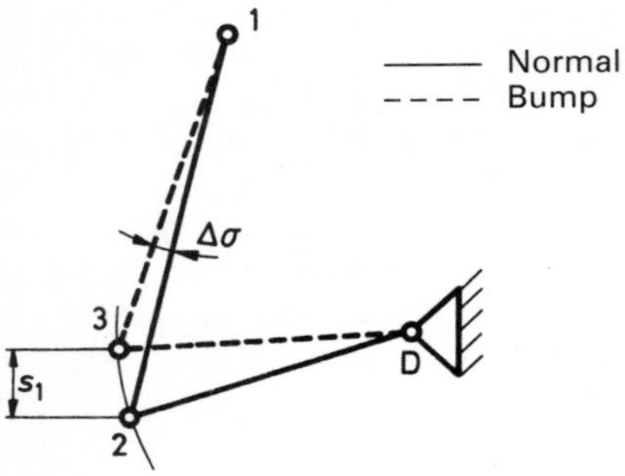

Fig. 3.51 Construction for determining the camber and kingpin inclination alteration on the McPherson strut and strut damper.

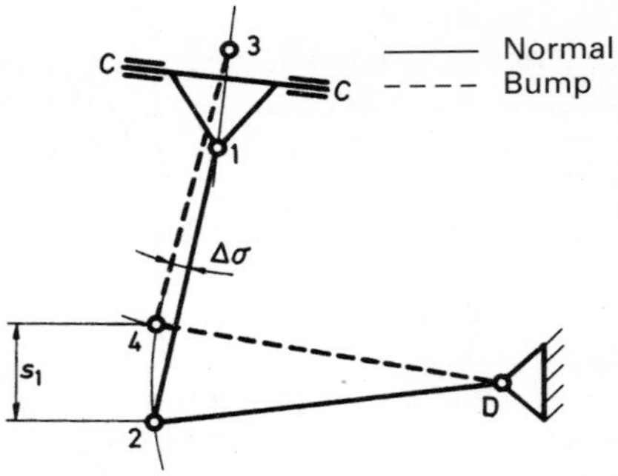

Fig. 3.52 Construction for determining the camber and kingpin inclination alteration on the longitudinal and transverse axes.

$$\varepsilon_W = \varepsilon_{W,0} + \Delta\sigma$$

On McPherson struts and strut dampers, the distance 1,2 is shortened when the wheel is in bump travel, the upper mounting point is in the wheel house and only the lower point 2 moves to 3. $\Delta\sigma$ is again the angle between the two connecting lines (Fig. 3.51).

The upper suspension control arm of the longitudinal link suspension (Fig. 3.52) requires a vertical to be created on the axes of rotation CC through the point 1 so that point 4 can be obtained using an arc around 3 and the length 1,2. If the axes CC were to deviate more from the horizontal, $\Delta\sigma$ (and therefore the camber alteration, Fig. 3.32) would improve.

An arc around vertical axis P must be drawn on the swing axle (Fig. 3.12). The tangents drawn to this one after the other give the camber alteration which must be subtracted from or added to $\varepsilon_{W,0}$. The same applies to the semi-trailing link axle where the arc needs only to be drawn around P_2 (rear view, Fig. 3.36).

3.5.4 Roll camber during cornering

When the body rolls, the camber of individually suspended wheels also changes, on the outside of a bend by the angle $\Delta\varepsilon_{W,k,o}$ and on the inside by $\Delta\varepsilon_{W,k,i}$ (Fig. 1.5). The mean value of the two $\Delta\varepsilon_{W,\varphi} = 0.5\ (\Delta\varepsilon_{W,k,o} + \Delta\varepsilon_{W,k,i})$ together with the kinematic body roll angle φ_k gives the

$$\text{roll camber coefficient } k_{\varepsilon,W,\varphi} = d\varepsilon_W/d\varphi \tag{3.5}$$

A wheel that is cambered positively to the ground on the outside of a bend by the angle $\varepsilon_{W,o} = \varepsilon_{W,0} + \Delta\varepsilon_{W,k,o}$ and one that is inclined on the inside of the bend by the angle $\varepsilon_{W,i} = \varepsilon_{W,0} - \Delta\varepsilon_{W,k,i}$ can experience an additional camber due to the vertical force elements (Fig. 3.53):

$$F_{\varepsilon,W,o} = F_{Z,W,o} \sin \varepsilon_{W,o} \text{ and } F_{\varepsilon,W,i} = F_{Z,W,i} \sin \varepsilon_{W,i} \tag{3.5a}$$

The softer the suspension control arm bearings have to be, and the shorter the path c on double wishbones (Fig. 1.5) or the distance l–o between piston and rod guide on McPherson struts and strut dampers (Fig. 1.11), the worse the roll camber becomes. The diameter of the piston rod (see Section 5.8.1) and the basic kinematics of the suspension also have an influence.

The body roll camber factor can be determined by tilting the body over to both sides and measuring the body roll angle and the camber angle. The wheel travel in compression and rebound can be plotted on the y-axis instead of the body roll angle (Figs 3.54 and 3.55), and the body roll angle can be easily calculated from this using the tread width $b_{f\ or\ r}$:

$$d\varphi = \frac{s_1 + s_2}{b_{f\ or\ r}} \text{ (rad) and } d\varphi = 57.3\ d\varphi \text{ (degree)} \tag{3.6}$$

The compound crank axle of the VW Golf has a track $b_r = 1444$ mm and where the path is $s_1 + s_2 = 80$ mm, the body roll angle is

$$d\varphi = 80/1444 = 0.00554 \text{ rad} = 3.17° = 3°10'$$

The progressive spring characteristic of this passenger car means the wheel on the outside of the bend only moves in bump a little relative to the amount by

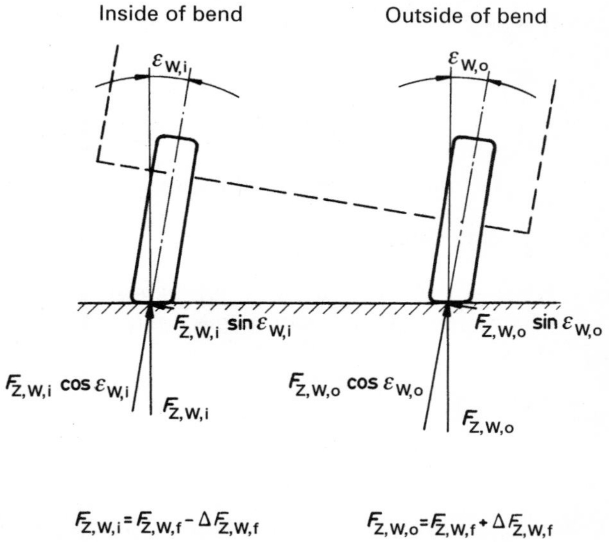

Fig. 3.53 When the body (and therefore also the wheels) incline, the vertical force element $F_{Z,W,o} \sin \varepsilon_{W,o}$ on a left-hand bend pushes the wheel on the outside of the bend (here the right-hand one) further into positive camber and the force $F_{Z,W,i} \sin \varepsilon_{W,i}$ pushes the one on the inside into an (equally unfavourable) negative camber (see also Fig. 1.6).

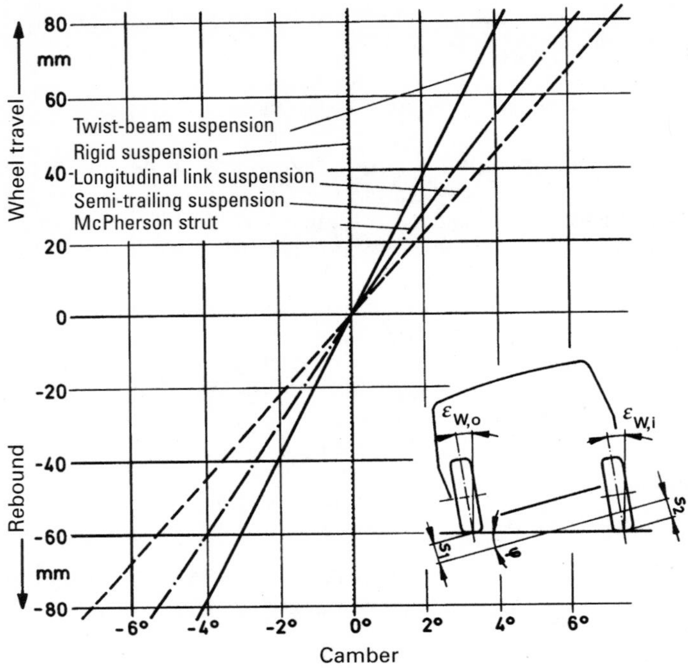

Fig. 3.54 Camber alteration relative to the ground of various rear-wheel suspensions in the case of reciprocal springing; with the exception of the rigid axle, the wheel on the outside of the bend goes into positive camber and the one on the inside into negative camber on all configurations. Wheel travel when the wheels compress and rebound is entered on the axis of the ordinate. The body roll angle φ is easy to calculate using the path differences Δs_1 and Δs_2 (see Equation 3.6).

which the opposite wheel rebounds (see Section 5.4.2). Given the permissible axle load, the following paths are assumed:

$$s_1 = 27 \text{ mm} \quad \text{and } s_2 = 53 \text{ mm}$$

The following values arise:

$$\begin{aligned} &\text{Camber} && \varepsilon_{W,o} = -0.1°; \quad \varepsilon_{W,i} = -3.55° \\ &\text{Camber alteration} && d\varepsilon_{W,k} = (\varepsilon_{W,o} - \varepsilon_{W,i})/2 \\ & && d\varepsilon_{W,k} = [-0.1 - (-3.5)]/2 = 1.7° \end{aligned} \tag{3.7}$$

and (referring to Equation 3.5) as a body roll camber factor

$$k_{\varepsilon,W,\varphi} = d\varepsilon_{W,k}/d\varphi = 1.7/3.44 = 0.49$$

The average roll camber factors for the following axles are:

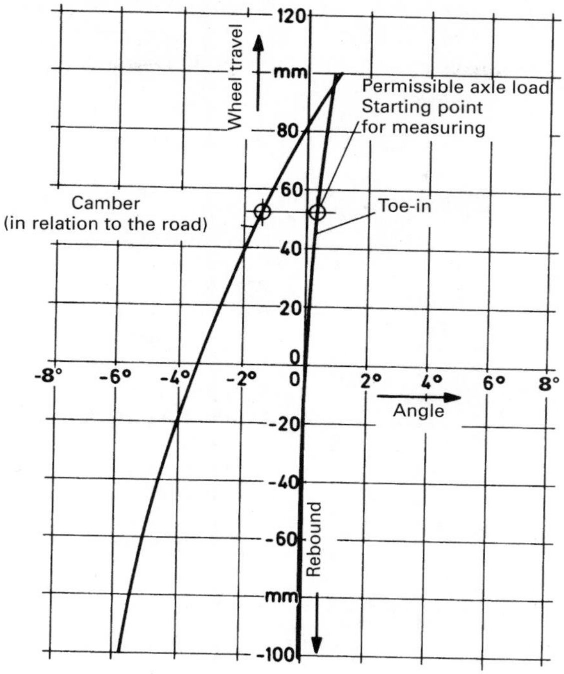

Fig. 3.55 Values for toe-in and camber angle measured by VW on the compound crank axle of a Golf with reciprocal springing, entered as a function of the wheel travel relative to the body. The bump-travel wheel on the outside of the bend goes into positive camber and the rebound-travel one on the inside of the bend goes into negative camber relative to the ground. The vehicle was measured with permissible rear axle load. Toe-in does not alter favourably. Figure 1.6 shows the body roll angle φ and Fig. 3.38 the relevant thrust centre point SM.

longitudinal link axles	1.05
McPherson struts	0.85
double wishbone suspensions	0.80
compound crank axles	0.55
rigid axles	0.0

3.5.5 Elasticity camber

In addition to the body roll camber, the camber alteration caused by the lateral forces must also be taken into consideration. In accordance with DIN 70 000, $\Delta\varepsilon_{W,k,e}$ is the proportion of the camber of a wheel that can be ascribed to the elasticity in the suspension and the steering, and is caused by forces acting between the tyre and road or by their moments.

Figure 3.56 shows the values calculated on the McPherson strut front axles of two passenger cars and Fig. 3.57 those measured on various rear axles. If there are no test results available, the following can be taken as the elasticity camber coefficient (per kilonewton):

$$d\varepsilon_{W,k}/dF \approx 22'/1\ \text{kN} \tag{3.7a}$$

For further details, see Refs [2] and [9].

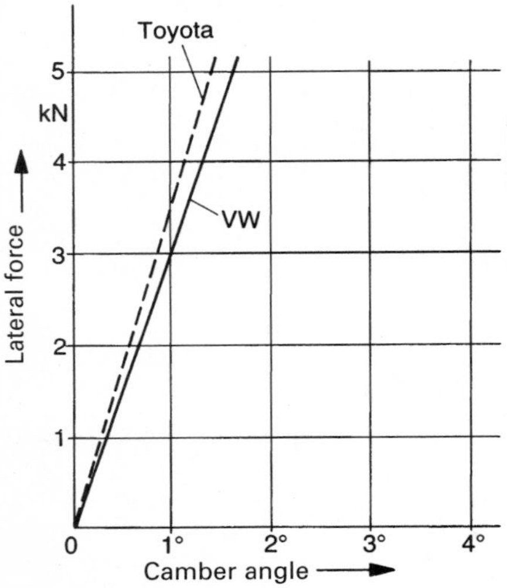

Fig. 3.56 Camber alteration measured on the driven McPherson front suspension of a lower mid-size passenger car with lateral forces directed inwards and applied statically at the centre of tyre contact. Wheel disc elasticity was eliminated on the measurements, and caster (which has no influence here), was ignored.

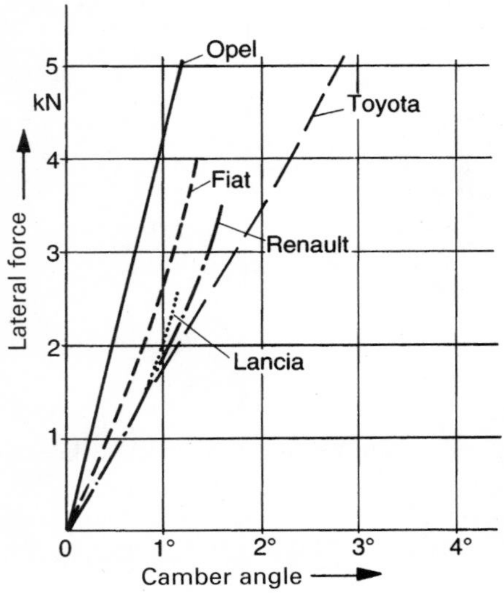

Fig. 3.57 Elastic camber change measured on various non-driven rear axles of mid-size passenger cars with lateral forces introduced statically in the middle of the centres of tyre contact. The type of axles were:

Opel:	twist beam suspension
Fiat:	twist beam suspension
Lancia:	McPherson strut
Toyota:	McPherson strut
Renault:	trailing link suspension

The low elasticity of the compound crank axles is clearly visible. Considering the caster would also give the same results.

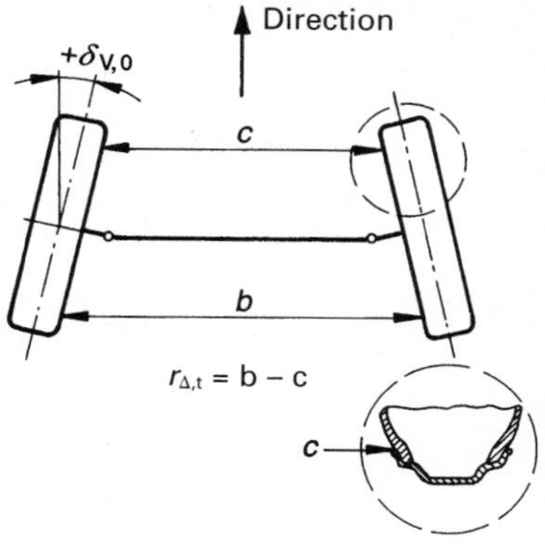

Fig. 3.58 The toe-in $r_{\Delta,t}$ of both wheels in accordance with the German standard DIN 70 020 is the difference in dimension $b - c$ in mm, measured on the rim flanges at the level of the wheel centre.

3.6 Toe-in and self-steering

3.6.1 Toe-in and crab angle, data and tolerances

In accordance with standard DIN 70 000, the static toe-in angle $\delta_{V,0,l \text{ or } rs}$ is the angle that results in a standing vehicle (reference status), between the vehicle centre plane in the longitudinal direction and the line intersecting the centre plane of one left or right wheel with the road plane. It is positive, when the front part of the wheel is turned towards the vehicle longitudinal centre plane and negative ('toe-out') when it is turned away.

The total toe-in angle $\delta_{V,0,t}$ is obtained by adding the toe-in angle of the right and left wheels. The total value is sometimes still given in millimetres (as stated in DIN 70 020, part 1). The toe-in is then the dimensional difference $r_{\Delta,t} = b - c$ (Fig. 3.58), by which the rim flanges at the back are further apart than at the front. The toe-in should be measured at the height of the wheel centre, when the vehicle is empty, with the wheels pointing straight forward; $r_{\Delta,t}$ therefore relates to both wheels of one axle. Expressed in degrees, the toe-in angle $\delta_{V,0}$ of a wheel corresponds to the tyre slip angle α_f (see Section 2.8.1); i.e. where there is toe-in, the front wheels of a vehicle are set to slip (drift), with the disadvantage of an increase in rolling resistance (Equation 2.4) of

$$\Delta F_R \approx 0.01\ F_R \text{ per } \delta_{V,0} = 10' \tag{3.7b}$$

The toe-in dimension r_Δ of just one wheel is included in determining the toe-in angle $\delta_{V,0}$ (i.e. $r_{\Delta,t}/2$):

$$\text{in radians } \delta_{V,0} = r_\Delta / D \tag{3.8}$$

$$\text{in minutes } \delta'_{V,0} = r_\Delta / D \times 57.3 \times 60 \tag{3.8a}$$

r_Δ should be taken at the rim flanges, which is why its distance D must be considered. With a given toe-in dimension, e.g. $r_\Delta = 2$ mm there is a larger angle

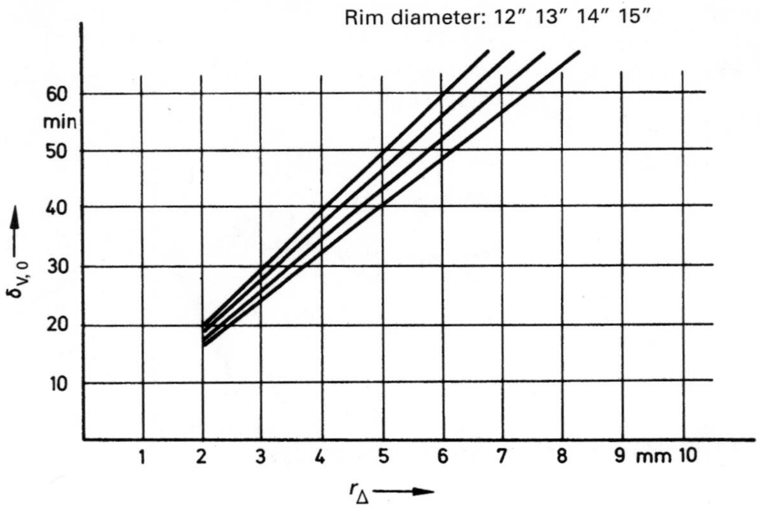

Fig. 3.59 Toe-in angle $\delta_{V,0}$ as a function of rim size and toe-in r_Δ in mm, measured on one front wheel.

on small 12″ rims than on ones with a 15″ diameter. Figure 3.59 shows the influence of the rim diameter and Fig. 2.11 the individual dimensions: $D = d + 2\,b$.

A tyre moving in a straight line has the lowest tyre wear and rolling resistance. When it rolls, a rolling resistance force F_R, directed from front to back, arises at the centre of tyre contact, which generates a moment with the lever arm r_a, which is absorbed via the tie rod to the steering (Figs 3.60 and 3.111, and Equation 2.4).

As a result of existing compliance, particularly in the suspension control arm bearings, this moment pushes the wheel backwards slightly and, in order to make it run straight when the vehicle is moving, 'slip' is set as toe-in when it is stationary. In front-wheel drive vehicles, the traction forces directed from back to front

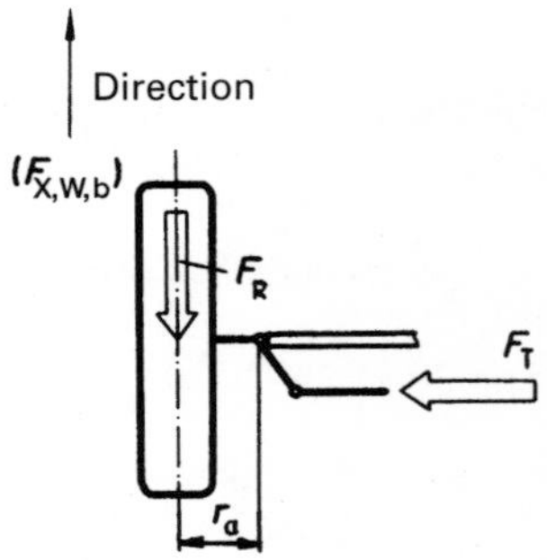

Fig. 3.60 The rolling resistance causes a longitudinal force F_R in the wheel centre, which pushes the wheel backwards into toe-out via the lever r_a; for reasons of simplification, the steering axis EG (Fig. 3.103) is assumed to be vertical in this and the next illustration. The moment $M_R = F_R r_a$ causes the force F_T to arise in the tie rod. Braking force $F_{X,W,b}$ operates in the same direction as F_R but has a different lever (Figs 3.108 and 3.109, and Section 7.1.7 in Ref. 3).

Fig. 3.61 On front-wheel drive vehicles, tractive forces $F_{X,W,a}$ attempt to push the wheels into toe-in. The tie-force F_T arises on both sides; the same applies to driven rear axles (Fig. 3.64).

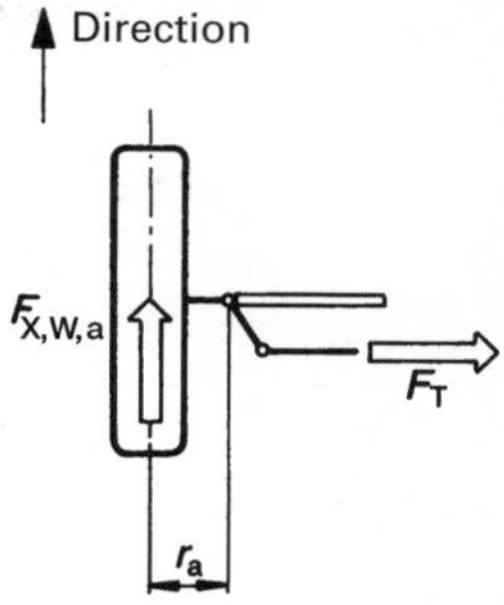

attempt to push the wheels together at the front edge (Fig. 3.61), so toe-out (i.e. negative toe-in) alignment can be beneficial. As a result of the built-in elastokinematics (Figs 3.83 and 3.86) and in order not to cause a deterioration in the driving stability in the overrun (coasting) condition (i.e. when the driver removes his foot from the accelerator), front-wheel drive vehicles may also be set with toe-in.

In addition to the absolute value of the total toe-in, tolerances must be specified for both front wheels which, because they can be adjusted by changing the tie rod length (Fig. 4.13), only need to be $\Delta\delta_{V,0,t} = 5'$ per wheel. Average values in factory information for toe-in are

$$\text{on rear-wheel drive vehicles } \Delta\delta_{V,0,t} = +15' \pm 10' \tag{3.8b}$$

$$\text{on front-wheel drive vehicles } \Delta\delta_{V,0,t} = 0° \pm 10' \tag{3.8c}$$

With semi-trailing links it is possible to alter the toe-in on the rear axle by swivelling the axis of rotation of the suspension control arms (Figs 1.15, 1.16 and 3.62) and, on 'double wishbone suspensions', by a lateral length alteration on one suspension control arm (Figs 1.1 and 1.62). Tolerances of $\Delta\delta_{V,0,t} = \pm 5'$ can be maintained where there is a setting. If this has not been provided in the design, values of $\Delta\delta_{V,0,t} = \pm 25'$ are almost inevitable, if tight component tolerances are not to render manufacturing uneconomical.

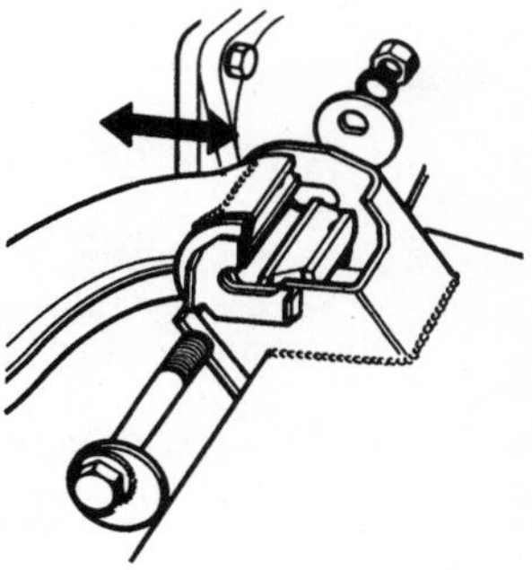

Fig. 3.62 Hexagonal bolts with eccentric discs, which come into contact with lateral collars, can be provided for setting camber and toe-in on both semi-trailing links (illustration: Ford).

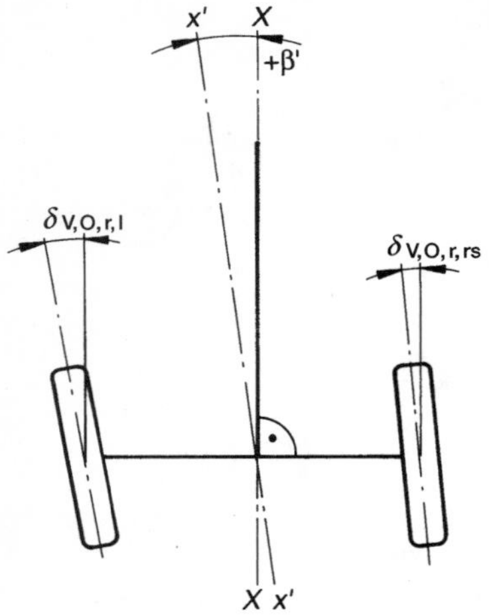

Fig. 3.63 The difference between the toe-in angle $\delta_{V,0,r,l}$ on the left and $\delta_{V,0,r,rs}$ on the right rear wheel determines the size of the axle drive (heading) angle $\pm\beta'$. It is positive if the median points forward and left (see also Fig. 3.75).

Regardless of whether the rear axle is steered or not, toe-in angles of the same size, both left and right, are required to ensure that the direction of movement x′–x′ of the vehicle corresponds to its longitudinal axis X–X (Figs 3.63 and 3.75). The German standard DIN 70 027 therefore specifies that the so-called crab angle β', must be quoted, i.e. half the total toe-in angle of the rear axle:

$$\beta' = (\delta_{V,0,r,rs} - \delta_{V,0,r,l})/2 \qquad (3.8d)$$

Where it is possible to set the toe-in, $\beta' = \pm 10'$ can be maintained; if there is no facility for setting toe-in on independent wheel suspensions or the vehicle is fitted with a twist beam suspension or a rigid axle (Fig. 3.75), up to $\beta' = \pm 25'$ must be allowed to enable economical production.

Taking as an example a passenger car with $\delta_{V,0,r,l} = -10'$ and $\delta_{V,0,r,rs} = +5'$ in accordance with Fig. 3.63:

$$\beta' = [+5' - (-10')]/2 = +7.5'$$

This means the angle is positive.

The toe-in on the rear axle of passenger cars is $\delta_{V,0,t} = 10\text{–}20'$; the drawing information for a vehicle with independent wheel suspension would then, for example, be:

toe-in 15′ ± 10′, crab angle maximum ±15′

BMW already specifies this condition for all models:

geometrical crab angle 0° ± 15′

and, on vehicles with compound crank axles, VW specifies

maximum permissible deviation from the direction of travel 25′.

3.6.2 Toe-in and steering angle alteration owing to wheel bump-travel kinematics

Even more important than a toe-in which has been correctly set on the stationary vehicle, is whether this is maintained when the vehicle is moving or whether it changes as a consequence of the wheels travelling in bump and rebound. This can be the fault of inadequate steering kinematics (see Section 4.6) or deliberately introduced to achieve certain handling properties. A change in the gradient of the toe-in characteristic as a function of wheel travel should be avoided, as handling properties then change unpredictably for the driver with variations in the load.

To avoid increased tyre wear and rolling resistance or impeding directional stability (as shown in Fig. 3.64 and curve 1 in Fig. 3.65) no toe-in change should

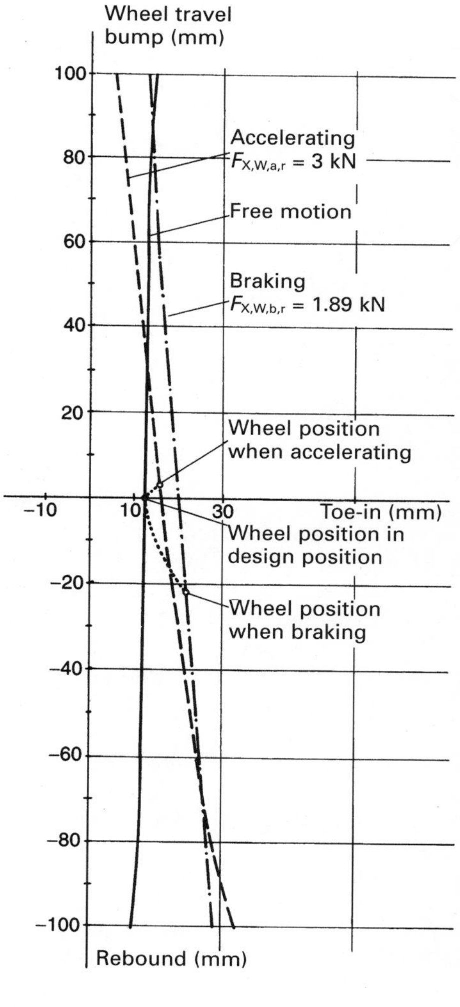

Fig. 3.64 Kinematic toe-in alteration of one wheel on the multi-link independent rear suspension of the Mercedes Benz S class with barely any deviation from the static value $\delta_{V,0,r}$ = 12′. The illustration also shows the behaviour of the wheel when subjected to a constant drive-off force $F_{X,W,a}$ = 3 kN (Fig. 3.113) introduced in the wheel centre and an opposed braking force $F_{X,W,b}$ = 1.89 kN acting at the centre of tyre contact (Fig. 3.108), all beginning in the design position (see Section 5.3.4). As the tyre and spring compresses when the vehicle moves off, it goes $+\Delta\delta_{e,r}$ = 3′ further into toe-in and, for elastokinematic reasons, further into toe-in by $+\Delta\delta_{e,r}$ = 10′ when the brakes are applied. The rear axle stabilizes the braking process (see Section 3.6.5.1).

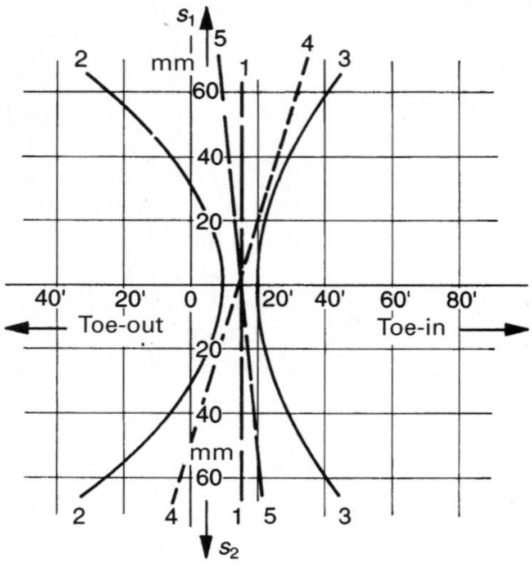

Fig. 3.65 Possible alteration of toe-in of one wheel (in minutes) as it bump and rebound travels, due to an incorrect tie rod length or position.

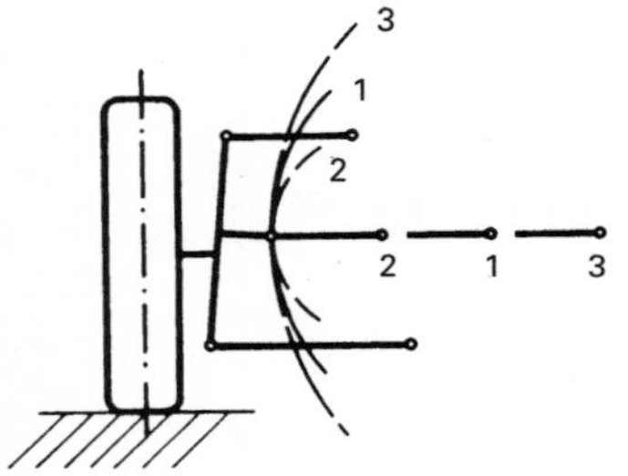

Fig. 3.66 Too short a tie rod (point 2) causes rebound both the bump and rebound travelling wheel to go into toe-out. However, too long a tie rod (point 3) causes toe-in in both directions (see Fig. 3.65).

occur when the wheels compress or rebound. The wheel travel upwards (s_1) and downwards (s_2) is plotted on the y-axis of the figures, whereas on the x-axis positive toe-in is plotted to the right for one wheel each time, and negative toe-in (i.e. toe-out) plotted to the left. The ideal curve 1 would be difficult to achieve at the design stage and certain deviations from the ideal shape have to be accepted.

A toe-in alteration can be the result of incorrect tie rod length or position. Provided that the steering arms are behind the front axle (Figs 3.60, 4.3 and 4.4), the example of a double wishbone suspension can be used to explain how different length tie rods act (Fig. 3.66). If they are too short (point 2), they pull the wheels together at the back both during bump and rebound travel, and go into toe-out as shown in curve 2 of Figs 3.65 and 3.67. Tie rods, which are too long, push the wheels apart in the direction of toe-in, curve 3; in both cases the graph displays a high curvature.

If, when the tie rods are the correct length, the inner joint 4 is too high (or the outer one too low, Fig. 3.68), when the wheel rebounds, the back of the wheel is drawn inwards and toe-out occurs; whereas, when it compresses, the wheel goes into toe-in. This results in approximate straight line running but at an angle

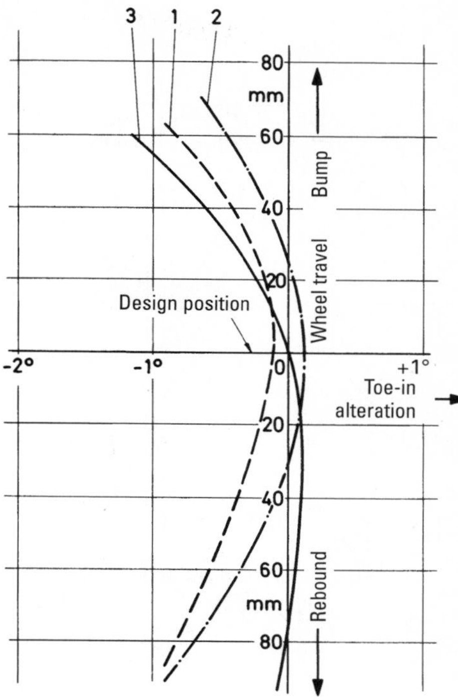

Fig. 3.67 As can be seen in Figs 3.66 (point 2) and 4.46, too short tie rods on McPherson struts cause the toe-in curve to be bent. If the steering arms are behind the axle, both bump and rebound travelling wheels go into toe-out. The figure shows values measured on the left front wheel on three front-wheel drive vehicles of the lower mid-size range. Curve 3 shows a 'roll–steer effect' on the front axle. This measure, which tends towards understeering, is achieved due to the difference in height between the inner and outer tie rod joint (which can be seen in Fig. 3.68 under point 5).

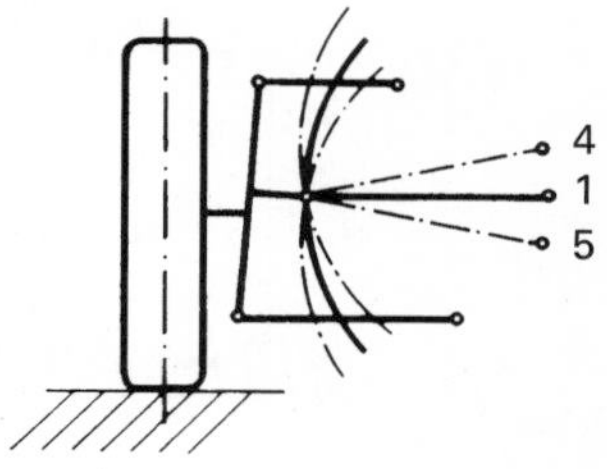

Fig. 3.68 An inner tie rod joint that is too high (point 4) produces curve 4 (Fig. 3.65), and one that is too low generates curve 5.

(curve 4 in Fig. 3.65). A tie rod joint 5, which sits too low on the inside or too high on the outside has, as the corresponding curve shows, the opposite effect – as do steering arms, which point forwards in all observed cases (Figs 4.37 and 4.39).

3.6.3 Toe-in and steering angle alteration due to roll

As a further example, Figs 3.69 and 3.70 show a deliberate steering alteration. When cornering, the outer bump-travelling wheel goes into toe-out and the inner rebounding one into toe-in. The steering angle is thus reduced slightly under the

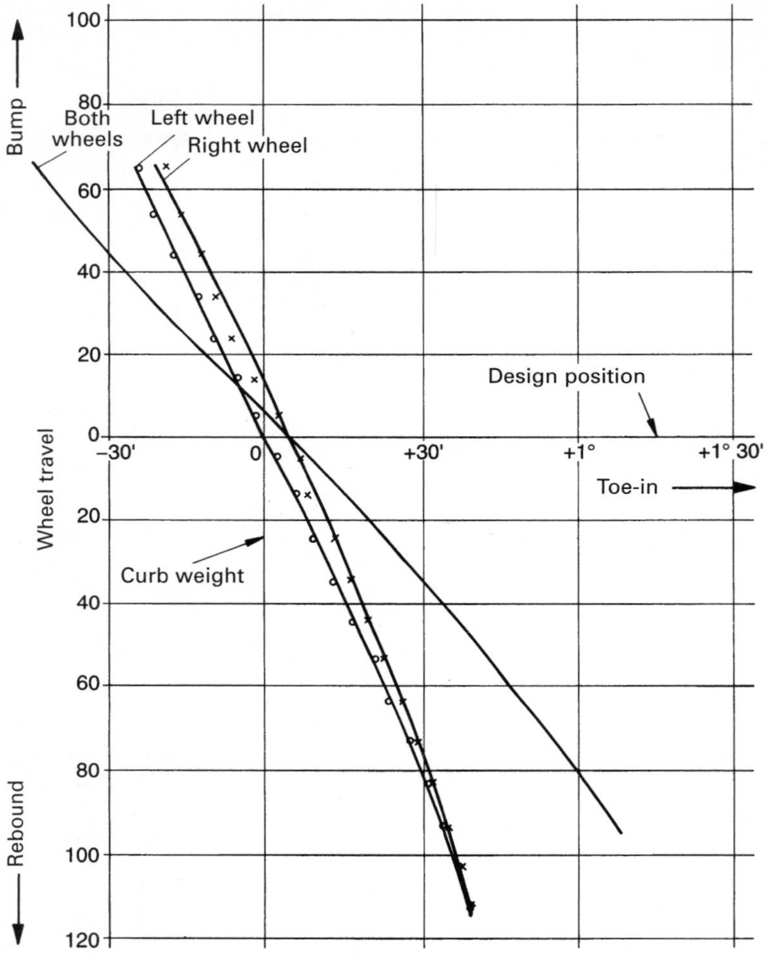

Fig. 3.69 Toe-in alteration recorded on an Opel Omega (1999) indicating body roll understeering on the front axle. The individual wheels were measured to obtain the total toe-in. The design position relates to the vehicle with three passengers each weighing 68 kg; the height of the unladen vehicle is also marked.

influence of the body inclination in order to achieve body roll understeering on the front axle or to improve handling when changing lanes (Fig. 3.71, see also curve 3 in Fig. 3.67).

As described in Section 3.6.4, rear axles can tend to lateral force oversteer – which can lead to an overswing of the vehicle's rear end (Fig. 3.72). To compensate for this and make the overall handling of the vehicle neutral, designers like to make the rear axle body roll understeer (Fig. 3.73). On individual wheel

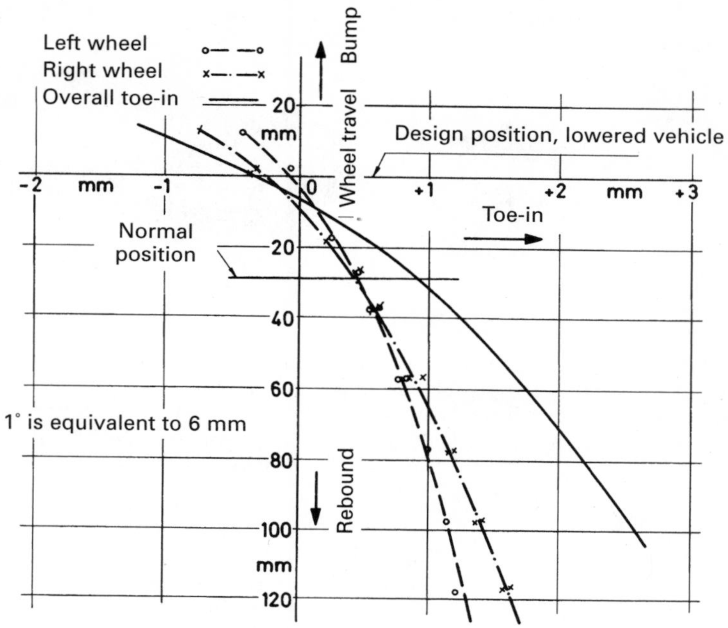

Fig. 3.70 Toe-in alteration measured on a VW Golf GTi that has been lowered by $\Delta s = 30$ mm. In the normal position (also marked as specified by the manufacturer), as the wheels bump and rebound, the alteration values (which have a negative influence on directional stability and tyre wear) are less than in the lowered condition. The (now minimal) residual small compression spring travel can be seen clearly.

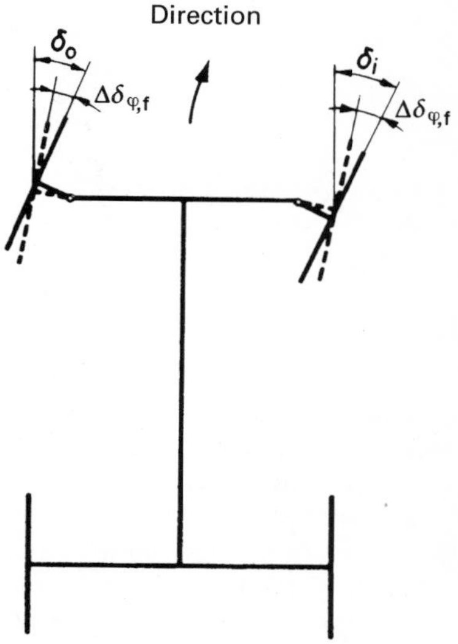

Fig. 3.71 If the bump-travelling wheel on the outside of the bend goes into toe-out and the rebounding one on the inside of the bend into toe-in under the influence of the body roll inclination (or due to lateral forces), the steering input is slightly reduced by the angle $\Delta\delta_{\varphi,f}$. The axle understeers.

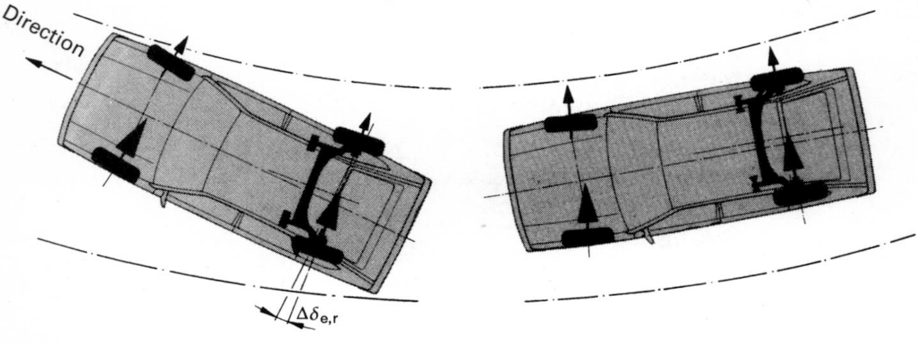

Fig. 3.72 Under the influence of a lateral force, the rear axle can take on the proportion of the steering angle $\Delta\delta_{e,r}$ – or the suspension links may be deformed accordingly – so that the vehicle oversteers to the inside edge of the bend (left and Fig. 2.42). To correct this, VW install track-correcting bearings which largely prevent oversteering (see Section 2.3.5 in Ref. [2]). Another possibility is to allow body roll understeering of the axle, (see Figs 1.30, 1.31, 3.77 and 3.78).

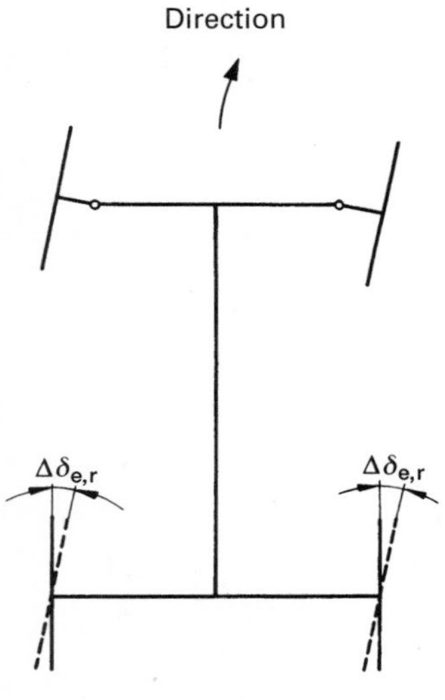

Fig. 3.73 To reduce the tendency to oversteer, the rear wheel suspension can be designed so that body roll or lateral force understeering of the axle is possible, i.e. under the influence of the body roll (or lateral forces) the compressing wheel on the outside of the bend goes into toe-in and the rebounding one on the inside of the bend into toe-out in proportion to the steering angle $\Delta\delta_{e,r}$.

suspensions the bump-motion wheel on the outside of the bend in this case must go into toe-in and the rebounding inner one into toe-out; Figs 3.20 and 3.74 show this type of alteration curve (see also Section 2.12).

As they are directly linked to one another, the wheels of rigid-axle and twist-beam suspension have no toe-in alteration where the springing is parallel. However, due to design tolerances or incorrect installation, the axle can sit at an angle in the vehicle, i.e. one wheel has toe-in and the other toe-out in respect of the longitudinal axis of the vehicle. In this case, the direction of movement x′–x′

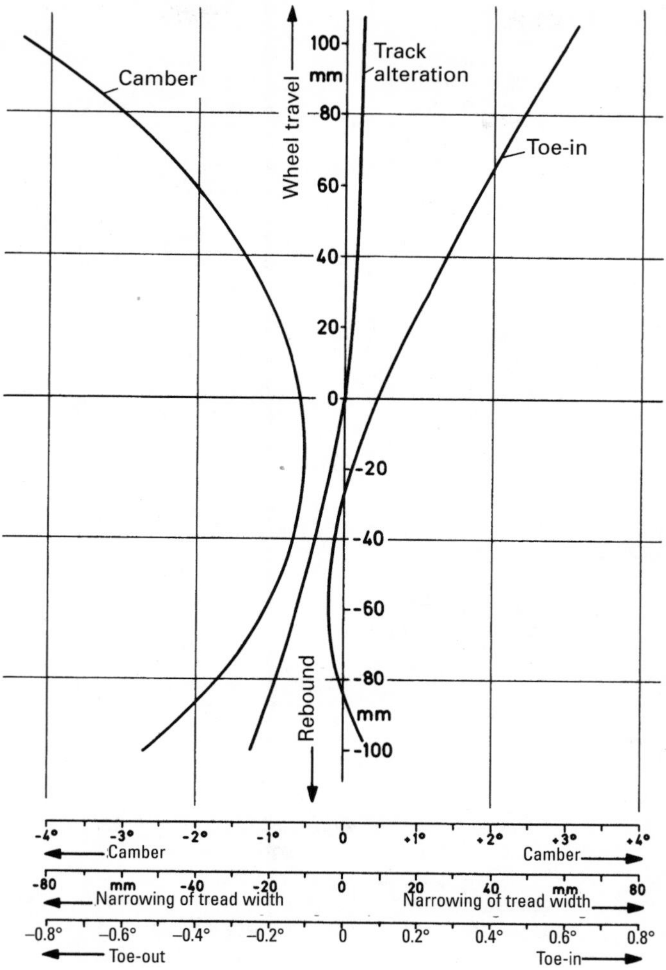

Fig. 3.74 Kinematic properties of an Audi A6 Quattro (1996) as the rear wheels compress and rebound. The relatively small tread width alteration of the two wheels, the favourable negative camber as the springs compress and the toe-in alteration (of one wheel), which points to roll understeering of the rear axle, are clearly visible.

of the vehicle and its longitudinal axis X–X deviate from one another by the crab angle (Figs 3.63 and 3.75).

On compound crank axles, the bearing points O shown in Fig. 3.37 move under the centre of the wheel when the vehicle is loaded, resulting in negative angles κ. This results in increasing body roll understeering, respectively, decreasing roll oversteering with load and therefore an improved roll-steer factor (Fig. 3.77).

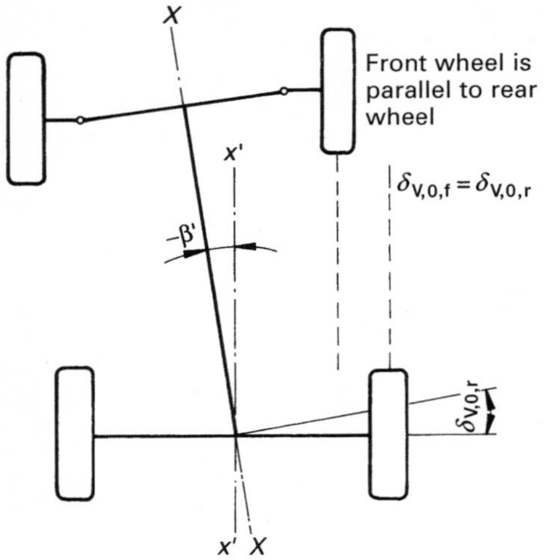

Fig. 3.75 If the rigid rear axle is not fitted at a right angle to the vehicle's longitudinal axis X–X, i.e. if the vertical on it deviates from the direction of movement x′–x′ by the crab angle β′, a slight steering input is necessary to make the vehicle move in a straight line. The figure also shows how the self-steering of the rear axle makes it necessary to turn the front wheels if the vehicle is to move in a straight line on an uneven road surface under reciprocal springing (Fig. 1.21). The axle can displace by the angle $\delta_{V,0,r} = \beta'$ (Figs 1.28 and 3.63).

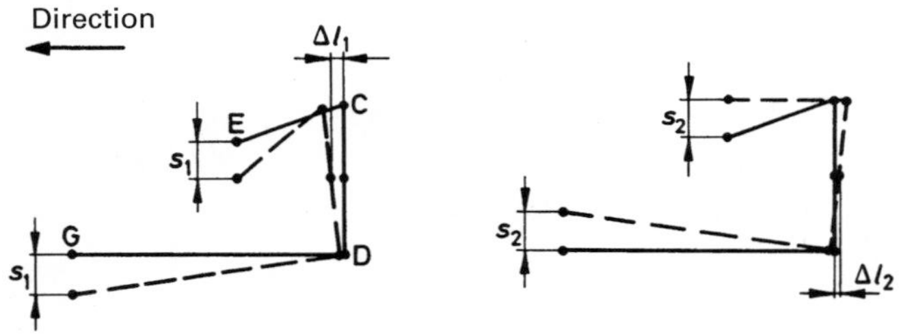

Fig. 3.76 If the body of a rigid rear axle on the outside of the bend led by two trailing link pairs has bump-travel in path s_1 – caused by trailing links at an angle to one another and of different lengths, as can be seen in Fig. 3.161 – the axle centre is drawn forwards slightly by the path Δl_1 (left) and pushed backwards by Δl_2 on the inside by s_2 rebounding. As a result of this, the rigid axle moves by an angle and roll understeers. This reduces the tendency to oversteer of standard vehicles.

Even with rigid axles, body roll understeering can be achieved by – as shown in Figs 1.28, 1.29 and 3.76 – the axle being drawn forwards on the outside of the bend when the body inclines and backwards on the inside of the bend. The alteration $\Delta\delta_{\varphi,r}$ of the steer angle in the axle as a whole, divided by the alteration $\Delta\varphi$ of the kinematic roll inclinations, is termed the 'roll steering factor' (Fig. 3.77).

A rigid rear axle which self-steers when the body inclines also self-steers when going in a straight line on an uneven road. The steering effect this causes occurs not only on reciprocal bend springing (Fig. 1.21) but also on unilateral springing. This is the reason why 'self-steering', which can only be compensated for by spontaneously turning the front wheels (see Fig. 3.74), is limited.

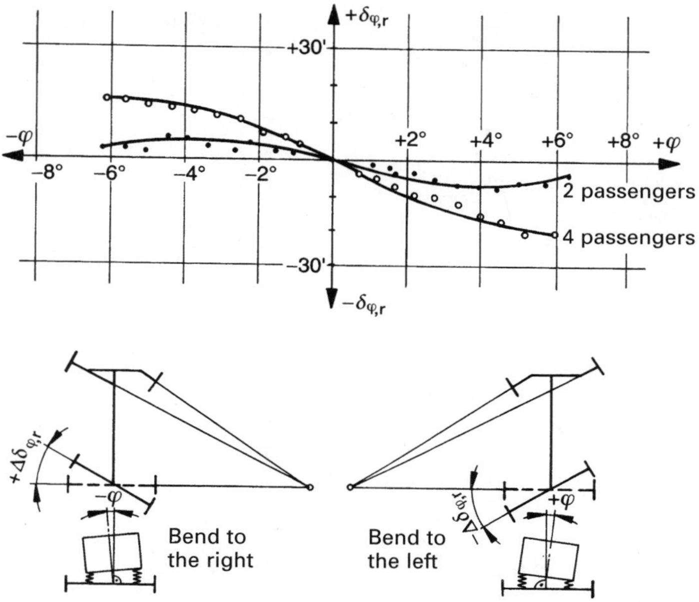

Fig. 3.77 Angled position by the angle $\Delta\delta_{\varphi,r}$ as a function of the roll angle φ measured on the driven rigid rear axle of a conventional passenger car occupied with two and four people. When there are two people in the vehicle and $\varphi = 4°$, $\Delta\delta_{\varphi,r} = 6'$. The body roll-steering factor would then be $\Delta\delta_{\varphi,r}/\Delta_{\varphi} = 0.1°/4° = 0.025$.

When there are four people in the vehicle, this increases to 0.075; the tendency of this vehicle to oversteer is therefore reduced, depending on the load.

3.6.4 Toe-in and steering angle alteration due to lateral forces

Increasing lateral forces try to push the turned-in front wheels with the lever of the kinematic caster $r_{\tau,k}$ and the caster offset $r_{\tau,T}$ (Fig. 3.120) into the straight running position. As a result of the elastic compliance in the system this reduces the steering angle and lateral force understeering takes place.

To achieve this on the rear wheels (as shown in Fig. 3.73), the wheel on the outside of the bend has to go into toe-in and the one on the inside of the bend in the direction of toe-out.

To some extent, exactly the opposite of this can be seen in Fig. 3.79. The rear wheels of the twist-beam crank axle (Opel and Fiat) on the outside of the bend are pushed into toe-out by the lateral force $F_{Y,W,o}$ and the ones on the inside of the bend into toe-in by $F_{Y,W,i}$. The result is lateral force oversteering (Fig. 3.72), which is also noticeable on the longitudinal link axle of the Renault (Fig. 1.63) and is also slightly evident on the McPherson struts of the Lancia (Fig. 1.12). Toyota moves the two transverse links 1 and 2 (Figs 3.2 and 3.80) backwards in parallel, therefore achieving the elastokinematic steering angle alteration $\Delta\delta_{e,r}$ on the outside of the bend and (as shown in Fig. 3.79) toe-out on the inside.

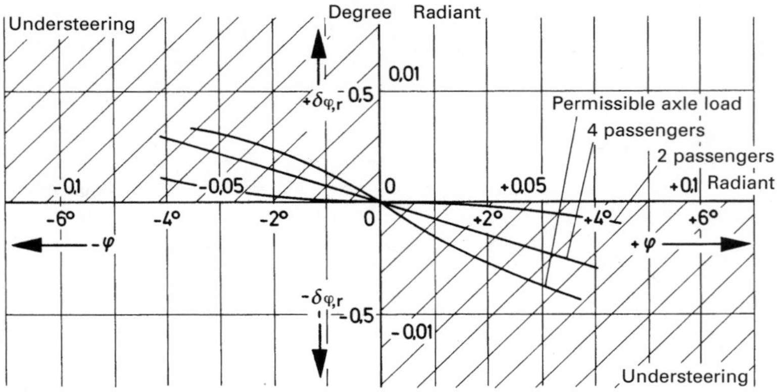

Fig. 3.78 Roll-steering measured on a VW Polo; increasing the load increases the understeering of the twist-beam suspension. At $\varphi = 4°$, the roll-steer factor is 0.025, 0.07 and 0.1 depending on the load.

For the measurement, the lateral force was applied statically in the centre of the tyre contact; shifting it backwards by the caster offset $r_{\tau,T}$ = 10 to 40 mm would cause all toe-in curves to turn counter-clockwise. The Toyota Corolla would then have a slight tendency to lateral force understeer, whereas there is an increased tendency on all other passenger cars to oversteer.

Another way of reducing this would be to give the rear wheels negative caster $-r_{\tau,k}$ (Figs 3.117 and 3.144); however, this must be greater than that of the tyre, which itself reduces as the slip angles α increase (Figs 2.50 and 3.119). This can be achieved on double wishbone suspensions (Fig. 3.145 and Section 5.3.4 in Ref. 2); the negative caster increases on the bump-travelling wheel on the outside of the bend and under load.

Even with rigid axles lateral force understeering is possible. If the panhard rod is behind the axle casing (Figs 1.61 and 3.81), the effective distance a between the lateral forces $F_{Y,W,r,o}$ and $F_{Y,W,r,i}$ on the two rear wheels and the rod force $F_{T,y}$ results in a pair of forces that generate the forces $\pm F_x$ in the trailing links and – due to the elasticity in the rubber bearings – causes the desired self-steering.

3.6.5 Toe-in and steering angle alteration due to longitudinal forces

3.6.5.1 During braking

Toe-in leads to stabilization of the vehicle braking. This means better straight-running behaviour and it can be achieved both by negative kingpin offset (Fig. 6.12) and by an elastokinematic toe-in alteration.

The front end of the vehicle moves towards bump when the brake is activated. If (as shown in Fig. 3.69) the body roll has been kinematically designed to understeer, both front wheels go into toe-out, i.e. with a positive kingpin offset,

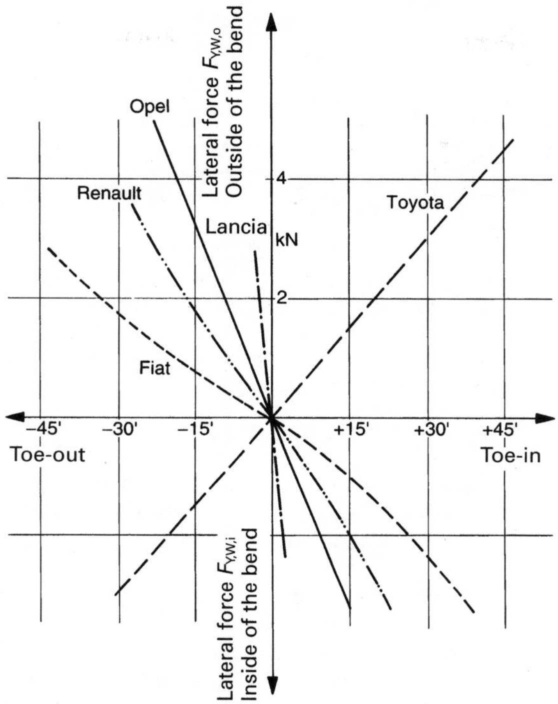

Fig. 3.79 Lateral forces introduced statically in the centre of the tyre contact of different rear axles produce, in the Toyota, a steering angle change $\Delta\delta_{e,r}$ in the direction of the toe-in on the outside of the bend, but toe-out on the other vehicles tested; these exhibit a lateral force steering tending towards oversteering. The vehicles are fitted with twist-beam suspension (Opel/Vauxhall and Fiat), McPherson strut (Lancia and Toyota) and trailing link suspension (Renault, Fig. 1.63). If the lateral force operates in the other direction (i.e. from the inside out), there is toe-in instead of toe-out on the vehicles. The toe-in alteration in minutes appears on the *x*-axis and the force in kilonewtons on the *y*-axis.

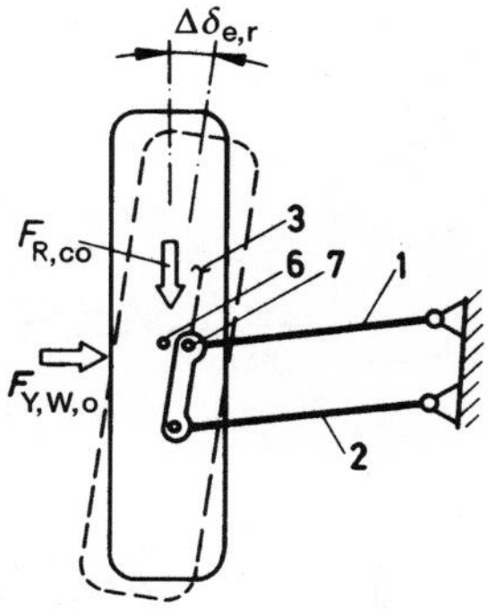

Fig. 3.80 Under the influence of the lateral force $F_{Y,W,o}$ acting on the outside of the bend behind the wheel centre by the tyre caster $r_{\tau,T}$, the mountings of the transverse link 1 flex more than the brace 2, which is offset backwards; point 6 moves to 7 and the toe-in angle $\Delta\delta_{e,r}$ occurs elastokinematically (see also Fig. 3.2).

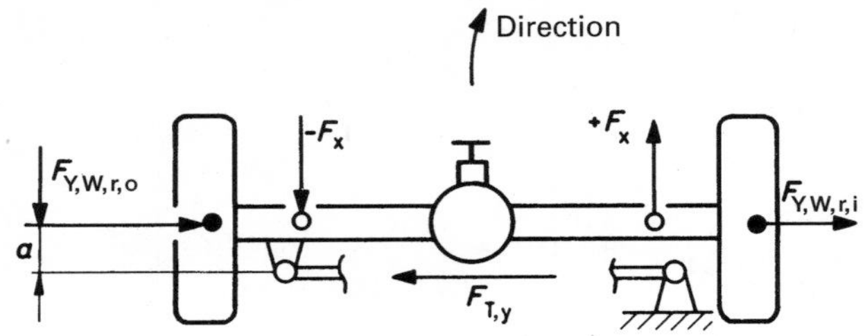

Fig. 3.81 The effective distance *a* between the lateral forces $F_{Y,W,r,o \text{ and } i}$ on the wheels of the rigid axle and the force $F_{T,y}$ on the panhard rod at the back leads to a force pair which generates the forces $\pm F_x$ in the longitudinal links and can cause lateral force understeering due to the compliance of the rubber mountings. If the rod is in front of the axle, oversteering is possible.

and they continue to travel in the same direction in which they were already being pushed by braking forces $F_{X,W,b}$ (Figs 3.60 and 6.11). To limit this effect, the necessary counter-steering in the direction of toe-in can be achieved, $r_\sigma = 0$ or there is a small positive kingpin offset at ground. The only prerequisite for this is a top view angle ξ between transverse link 1 and tie rod 7 (Fig. 3.82).

Using a Mercedes model as an example: the front of the longitudinal rod 4 is anchored at point 6 on the suspension control arm, and the back carries the supporting bearing 5. Under the influence of the braking force $F_{X,W,b}$ the defined longitudinal elasticity of part 5 yields, the lower guiding joint G moves out to 4 and the outer tie rod joint U moves to 9. As points G and U move in different arcs and the tie rod joints are also less laterally compliant than the bearing D of the transverse link 1, both front wheels are pushed into toe-in in spite of the opposing moment $M_b = F_{X,W,b}\ r_b$ seen in Fig. 3.109.

In the same way, individually suspended rear wheels can experience an elastokinematic toe-in alteration during braking (Figs 3.2 and 3.64). For further details, see Refs [2] and [6].

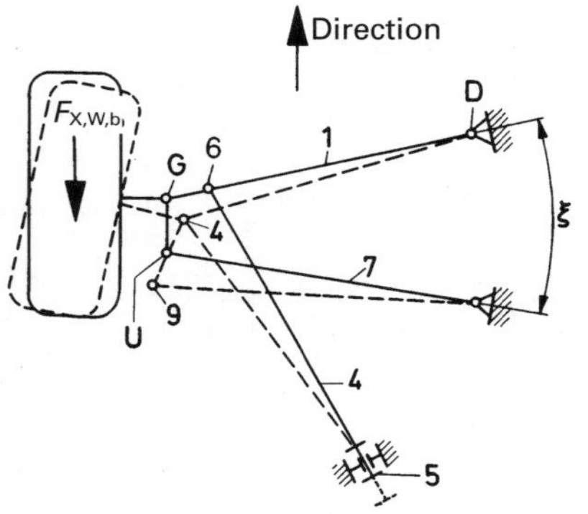

Fig. 3.82 A positive top view angle ξ between the tie rods 7 and the transverse links 1 close to them (mostly the lower ones) can cause an elastokinematic toe-in alteration during braking.

3.6.5.2 Longitudinal suspension without toe-in alteration

Nowadays, manufacturers fit only steel radials to series production vehicles. However, unlike the cross-ply tyres used in the past, these have the disadvantage of dynamic rolling stiffness (see Section 2.2.2). The very stiff belt causes longitudinal oscillations which, on independent wheel suspension, are transferred to the body via the steering knuckle and the tie rod and – particularly on cobbles, rough concrete and at speeds below 80 km h^{-1} – can cause an unpleasant droning noise inside the vehicle. The vibrations can be stopped if the steering knuckle is given a precisely defined longitudinal mobility (compliance). This is a task that is not easy to fulfil at the design stage because neither a toe-in alteration may occur, nor a lateral force at the centre of tyre contact (Fig. 3.6) under the influence of the paths of $s \geq \pm 10$ mm (Fig. 3.1), as straight rolling ability and rolling resistance would deteriorate.

On the front axle it can be solved using a transverse link, which has an outrigger pointing backwards (or forwards; Figs 3.83 and 3.84), and which is supported at the side in a rubber bearing with a highly progressive, precisely defined spring rate. The important thing is that stiff bearing elements, which only yield a little under cornering lateral forces and braking forces, sit in the pivots D and G.

If a transverse link anchored at point D controls the wheel, it can have a hole

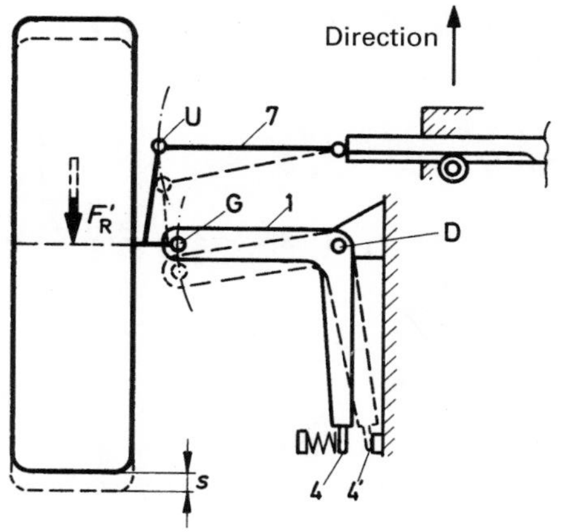

Fig. 3.83 To achieve the necessary longitudinal springing BMW fits the boomerang-shaped (bell-crank) control arm 1 (shown separately in the next figure) on the front axle of the Z3 Roadster. Under the effect of longitudinal forces, it rotates around the (only slightly compliant) ball joint D and is supported with the outrigger 4 by means of a large rubber mounting on the body. In the lateral direction this bearing has an initially soft, but then highly progressive, springing curve.

Tie-rod 7 lies at the height of the control arm and is almost parallel to the line linking the bearing points GD; points U and G therefore move on an arc of around the same radius and longitudinal wheel movements do not cause any toe-in alteration. As shown in Fig. 3.111, the rolling resistance F_R, which varies in size, must be observed in the wheel centre as F'_R.

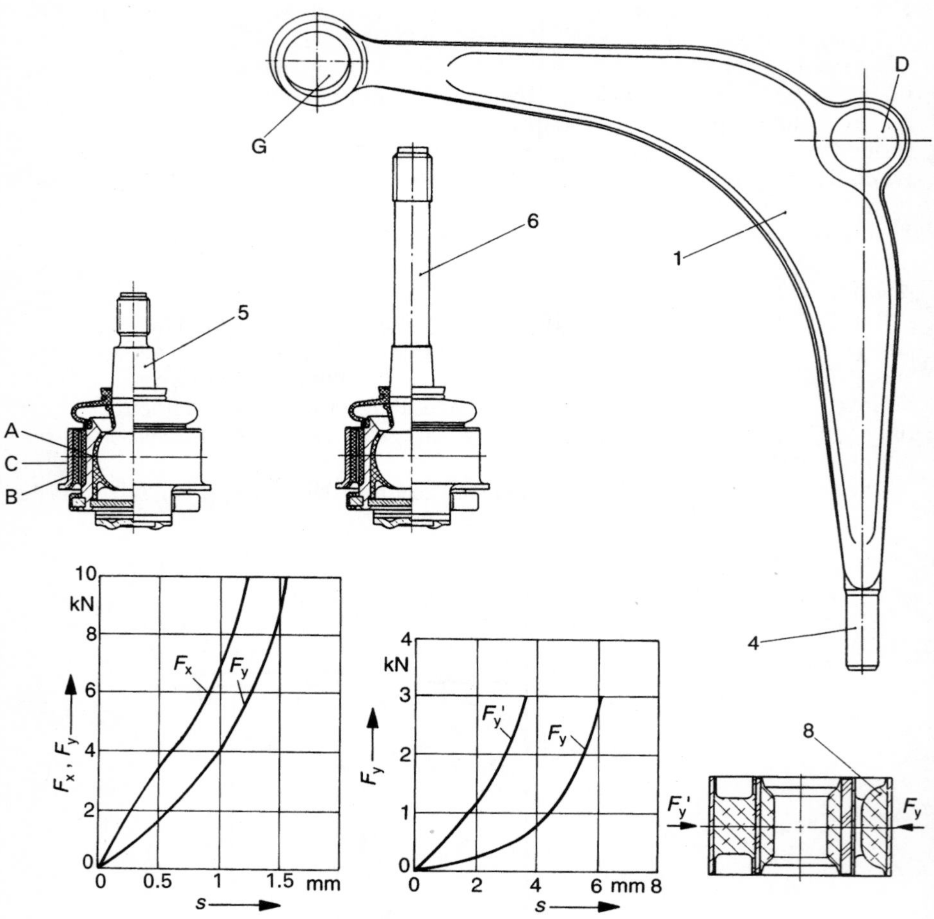

Fig. 3.84 Front boomerang-shaped suspension control arm of the BMW Z3 Roadster. The guiding joint 5 links the suspension control arm 1 with the suspension strut and is press-fitted from below into hole G; the inner joint 6 sits in hole D. The suspension control arm rotates around this part under the influence of longitudinal forces and is supported by outrigger 4 on the transversely elastic bearing 8. Its progressive compliance in the *y* direction is shown by the illustration on the right. In the vertical direction (*z*) the bearing is stiffer.

In part 5 the rubber ring C is vulcanized in between the joint housing A and the outer ring B, and is – as can be seen from the illustration on the left – laterally more compliant (F_y) than in the longitudinal (*x*) direction (illustration: Lemförder Fahrwerktechnik).

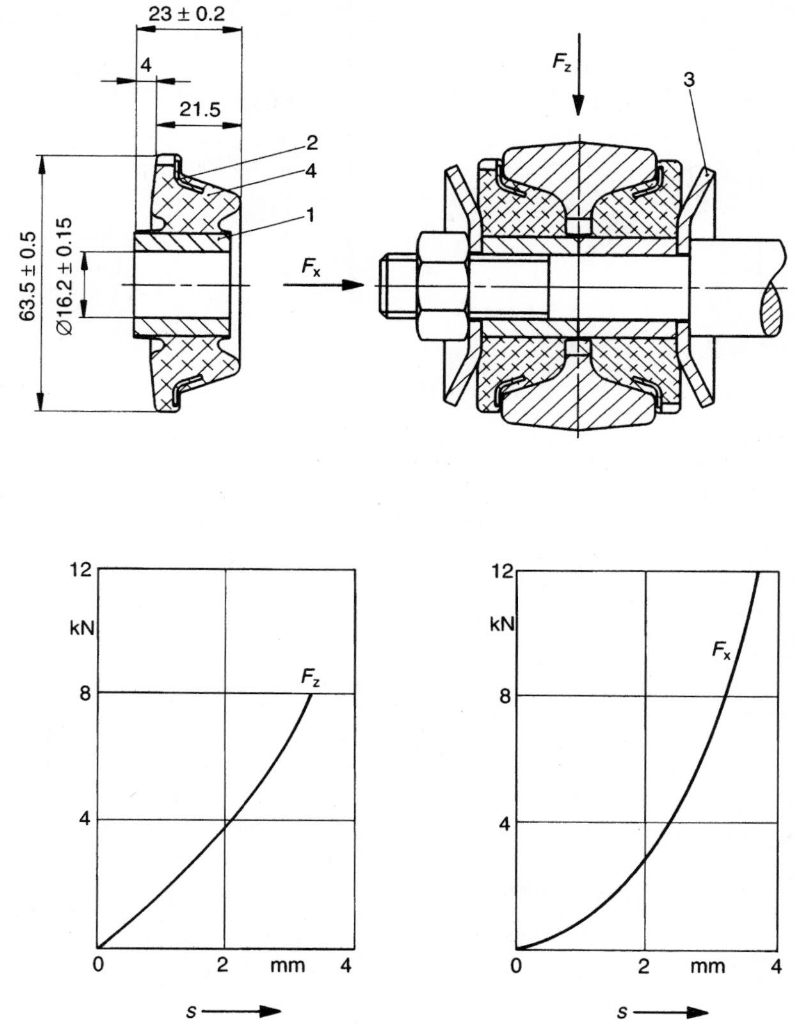

Fig. 3.85 Mounting of the anti-roll bar fitted at the front in the transverse links on the Audi 6 (built until 1996) (Fig. 1.57). The two rubber parts in the suspension control arms are vulcanized to the inner tube 1 and ring 2. Under the influence of longitudinal forces F_x one part comes into contact at the dome-shaped washer 3 and the other part relaxes. As can be seen on the left, the rubber part 4 projects beyond the sleeve 1; when fitted this achieves the necessary pre-tensioning. Ring 2 ensures that it sits firmly in the suspension control arm, so that the mounting can transmit vertical forces F_z without complying too much. The diagrams show the longitudinally progressive characteristic curve and the almost vertical linear characteristic curve of both bearings when fitted (illustration: Lemförder Fahrwerktechnik).

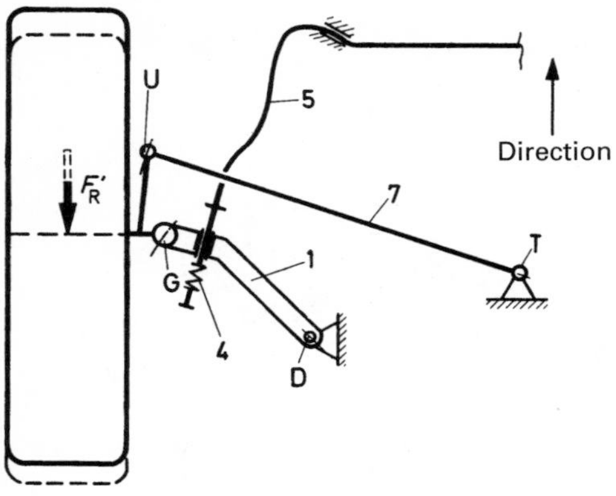

Fig. 3.86 An A arm can be replaced by two individual suspension arms: one is transverse (position 1) and carries lateral forces; the other is longitudinal (position 5) and transfers forces in this direction. A longitudinally compliant bearing (position 4) in a hole in part 1 absorbs the dynamic rolling hardness of the radial tyres. As in the Audi A6 (Fig. 1.57), and in the classic McPherson strut, part 5 can also be the arm of the anti-roll bar.

in which a longitudinally elastic rubber bearing sits (Figs 3.85, 3.86 and 1.57). The inner tube of this part is supported on the anti-roll bar 5 or a tension or compression rod strut 4, pointing either backwards or forwards.

On driven independent rear wheel suspensions it is especially important that the trailing or semi-trailing arms be controlled as well as possible to avoid elastic camber and toe-in alterations. The three or four rubber bearings, which link the suspension subframe and the differential with the body, then have to be designed so that the dynamic rolling hardness of the radial tyres is absorbed (points 2, 3 and 4 in Fig. 1.15). This task is carried out by the bearings in the longitudinal struts on rigid axles and by the rubber elements sitting in the pivot points O on compound crank axles (as shown in Figs 1.30, 1.61 and 3.87).

3.6.5.3 Toe-in alterations due to front-wheel tractive forces

As can be seen in Fig. 1.50, on a transverse engine the differential is relocated from the middle of the vehicle to the manual transmission that is sited at the side, resulting in drive shafts of different lengths. When the vehicle moves off in the lower gears, the front end rebounds and the shorter (left-hand) shaft takes on a steeper working angle α_l to the wheel axis than the longer one (right-hand, Fig. 3.88). The clockwise/anti-clockwise moments about the steering axes EG which combine to bring about toe-in result from the bending deflections due to rotation of the drive-shafts:

$$M_{Z,W,a,l \text{ or } rs} = {}^1\!/_2\, F_{X,W,A}\, r_{stat} \tan \alpha_{l \text{ or } rs} \qquad (3.8e)$$

(For $F_{X,W,A}$ see Equation 6.36 and r_{stat} in Section 2.2.5.)

As the angle α is larger on the left, a slightly higher moment can arise there than on the other side, with the risk of the vehicle pulling to the right. If the driver takes his foot off the accelerator quickly, a braking moment is generated by the engine, the front end dips and a steering effect in the other direction is inevitable. This is the main reason why (as shown in Figs 1.51 and 1.57) front-wheel drive vehicles with powerful engines necessarily have an intermediate shaft, and drive shafts of equal lengths.

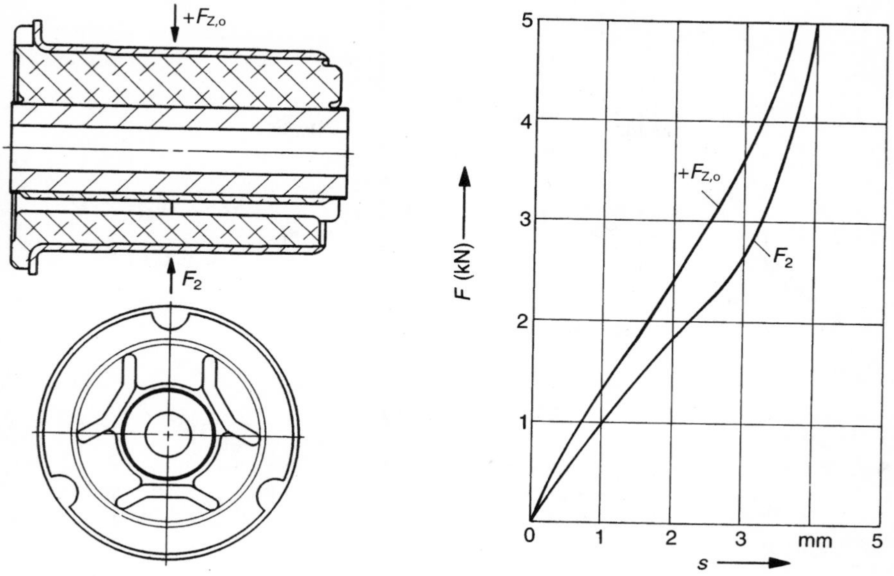

Fig. 3.87 Elastic bearing in the front eyes of the twist-beam suspension of the Audi A6 (1996). The indents in the rubber part give the necessary elasticity. The bearing must be soft enough in the longitudinal direction to absorb the dynamic rolling hardness of the tyres (when the axle shown in Fig. 1.61 is controlled precisely) and not very compliant in the vertical direction to be able to absorb safely the forces $F_{Z,o}$ which occur during braking (Fig. 3.160) (illustration: Lemförder Fahrwerktechnik).

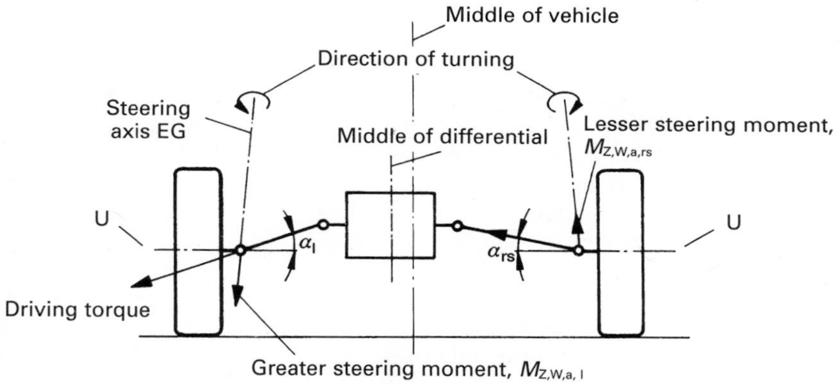

Fig. 3.88 When the engine is transverse, the differential is no longer in the centre of the vehicle and an intermediate shaft is necessary (Fig. 1.51), otherwise the drive shafts are not the same length. If they are at different angles α, different moments can occur around the steering axes, causing the steering to pull to one side. $\alpha_l = \alpha_{rs}$ can be achieved by tipping the differential by up to 2°.

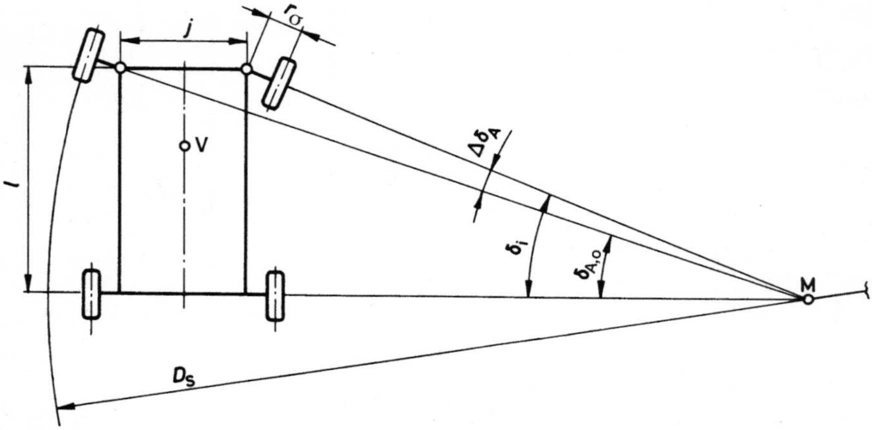

Fig. 3.89 Kinematic relationships in accordance with Ackermann between the steering angle $\delta_{A,o}$ on the wheel on the outside of the bend and δ_i on the inside of the bend. The illustration also shows the $\Delta\delta_A$ and the track circle diameter D_S (see Fig. 1.69).

3.7 Steer angle and steering ratio

References 1 and 9 deal with this area in detail. Section 4.7 covers steering kinematics.

3.7.1 Steer angle

When the vehicle is moving very slowly and 'free of lateral forces', it will only corner precisely when the verticals drawn in the middle of all four wheels meet at one point – the centre of the bend M. If the rear wheels are not steered, the verticals on the two front wheels must intersect with the extension of the rear axle centre line at M (Figs 3.89 and 1.69) whereby different steer angles δ_i and $\delta_{A,o}$ occur on the front wheels on the inside and outside of the bend. The nominal value $\delta_{A,o}$ of the outer angle – also known as the Ackerman angle – can be calculated from the larger inner angle δ_i:

$$\cot \delta_{A,o} = \cot \delta_i + j/l \tag{3.9}$$

where l is the wheelbase and j the distance between the two steer axis extensions (Figs 3.90 and 3.103), measured at the ground, i.e.

$$j = b_f - 2\, r_\sigma \tag{3.10}$$

here the kingpin offset r_σ is negative, the integer is positive (Fig. 3.113).

The differential steer angle $\Delta\delta_A$ included in Fig. 3.89 (also known as the toe

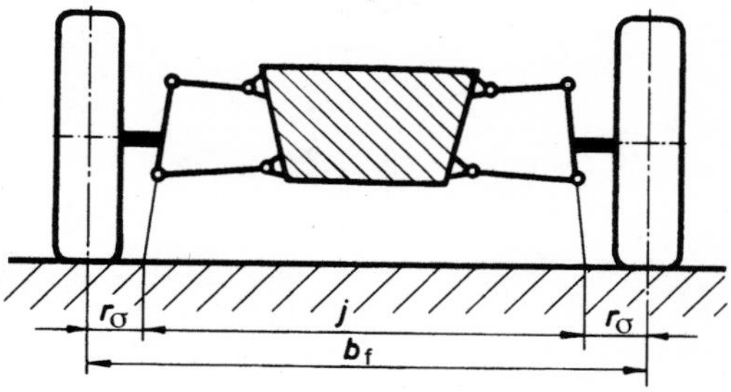

Fig. 3.90 Path designations on the front axle; b_f is the tread width on the front and r_σ the (in this case) positive kingpin offset on the ground (scrub radius).

difference angle) must always be positive for the nominal values calculated (nominal curve in Fig. 3.92).

$$\Delta\delta_A = \delta_i - \delta_{A,o} \tag{3.11}$$

The theoretical track circle diameter D_S can be calculated using the angle $\delta_{A,o}$ (Fig. 3.89), i.e. the diameter of the circle which the outer front wheel traces with the largest steering angle (see also Equation 2.10). The turning circle of a vehicle should be as small as possible to make it easy to turn and park. The formula

$$D_s = 2\left(\frac{l}{\sin \delta_{A,\sigma,max}} + r_\sigma\right) \tag{3.12}$$

derived using the illustration shows that this requirement necessitates a short wheelbase and a large steer angle $\delta_{A,o}$ on the outer wheel of the bend. This in turn requires an even greater steering angle applied to the wheel at the inside of the turning circle, though this is limited by the fact that the tyre must not come into contact with the wheel arch or any of the front-axle components. The wheel house cannot be brought too far into the sides of the front foot well as the pedals (on both left and right-hand drive vehicles) would then be at an angle to the direction in which the driver faces and foot-space would be restricted. In front-wheel drive vehicles, room also needs to be allowed for snow chains (Figs 2.8 and 3.102) and the largest working angle of the drive joints (Figs 1.3 and 1.53).

3.7.2 Track and turning circles

The inner angle δ_i is therefore limited, whereas the wheel angle on the outside (for functional reasons a smaller angle) is not. This may be the same size as the inner one. The disadvantage is that it impairs the cornering behaviour of the vehicle (Fig. 3.91), but with the advantage that the track circle becomes smaller and the lateral tyre force capacity on the outside of the bend increases. For this reason, the outer steering angle is larger on most passenger cars, i.e. the actual value δ_o (without index A) is greater than the nominal angle $\delta_{A,o}$ calculated according to Ackerman (Fig. 3.92) by the steering flaw $\Delta\delta_F$. In other words, the required steering deviation is as follows:

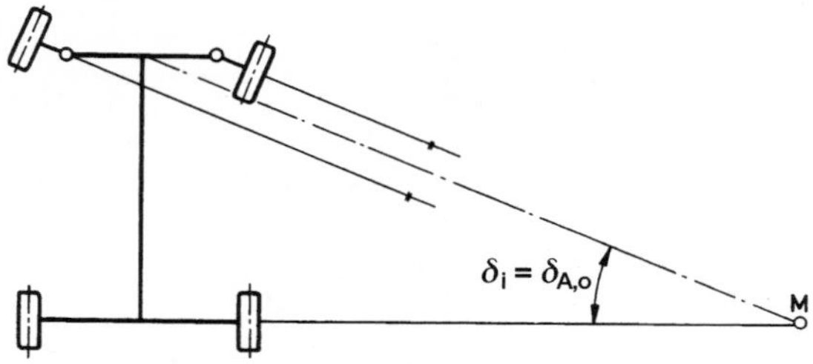

Fig. 3.91 To use the space available in the wheel house, it is an obvious idea to turn the wheel on the outside of the bend inwards by as much as the wheel on the inside of the bend; the wheels are then turned parallel and $\Delta\delta_A$ is zero. It is possible to increase the cornering force by turning the outer wheel more (compared with the wheel on the inside of the bend, Fig. 3.92).

$$\Delta\delta_F = \delta_o - \delta_{A,o} = \Delta\delta_A - \Delta\delta \tag{3.13}$$

where $\Delta\delta = \delta_i - \delta_o$ expresses the so-called differential steer angle.

The turning circle diameter D_S shown in Fig. 3.89 can be reduced by deliberately accepting a steering deviation. In addition to $\Delta\delta_F$, the angle $\delta_{A,o,max}$, in other words the largest outer nominal angle according to Ackermann calculated using Equation 3.9, must also be known. A series of test measurements has shown that a reduction by $\Delta D_S \approx 0.1$ m per 1° steering deviation can be achieved; the formula which should include all dimensions in metres would then be

$$D_S = 2\left(\frac{l}{\sin \delta_{A,o,max}} + r_\sigma\right) - 0.1\Delta\delta_F\,(\mathrm{m}) \tag{3.14}$$

A front-wheel drive vehicle with conventional steering geometry can be used as an example.

The data when the wheels are turned to the right are:

$$l = 2.677\ \mathrm{m};\ b_f = 1.47\ \mathrm{m};\ r_\sigma = -0.015\ \mathrm{m};\ \delta_{i,max} = 42°;\ \delta_{o,max} = 35°40'$$
$$j = 1.47 - [2\,(-0.015)] = 1.5\ \mathrm{m}$$
$$\cot \delta_{A,o} = \cot 42° + 1.5/2.677 = 1.671;\ \delta_{A,o} = 30°55'$$
$$\Delta\delta_F = 35°40' - 30°55' = 4°45'$$
$$D_S = 2\,[2.677/\sin 30°55' + (-0.015)] - 0.1 \times 4.75°;$$
$$D_S = 9.91\ \mathrm{m}$$

The turning circle diameter measured on the passenger car was $D_{S,t} = 9.92$ m.

The turning circle radius is basically only a theoretical value which can be calculated at the design stage; for the driver it is the swept turning circle kerb to kerb that is important, in other words the distance between two normal height kerbs (Fig. 3.93) standing parallel to one another, between which the driver can just turn the vehicle. This circle diameter $D_{tc,kb}$ can be measured but can also be

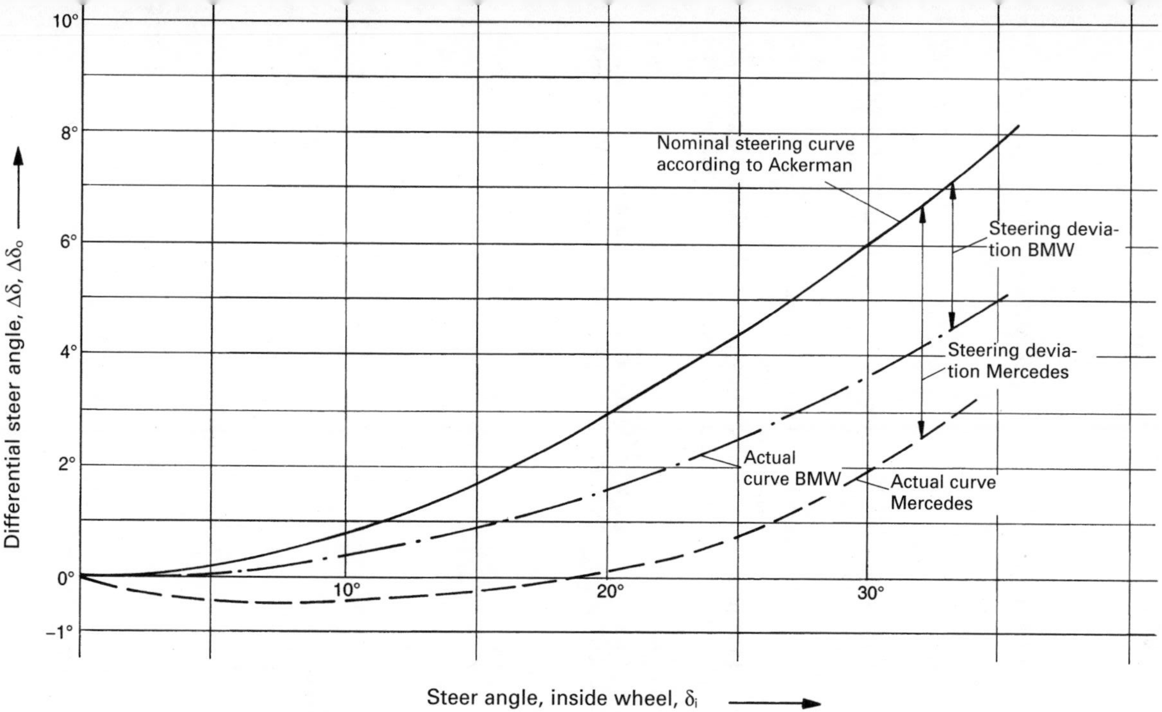

Fig. 3.92 Required, nominal steering curve for two standard passenger cars with the same wheelbase and approximately the same track calculated in accordance with Equation 3.9. The mean value of the actual curve measured when the wheels are turned to the left and right is included, and the steering deviation $\Delta\delta_F$ (also known as the steering error) is also marked. The steering angle δ_i of the wheel on the inside of the bend is entered on the *x*-axis, and the differential steer angle $\Delta\delta = \delta_i - \delta_o$ (which relates to the actual curve) and $\Delta\delta_A = \delta_i - \delta_{A,o}$ (which is valid for the nominal curve according to Ackermann) are marked on the *y*-axis.

In the workshop manuals $\Delta\delta$ must be indicated with a tolerance at $\delta_i = 20°$; on the BMW 3-series it would be $\Delta\delta = 3°$ and on the Mercedes $\Delta\delta = 10'$. The differential steer angle of the Mercedes, which is negative up to $\delta_i \approx 20°$ indicates that the wheel on the outside of the bend is turned more than the one on the inside and so the lateral force absorbed by the front axle when it corners – and with it the steering response – is increased.

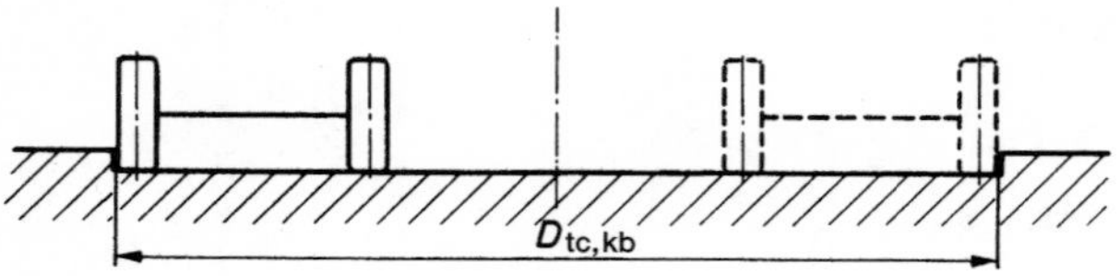

Fig. 3.93 Turning circle kerb to kerb $D_{tc,kb}$; an important dimension for the driver when turning the vehicle.

calculated easily using the turning circle diameter D_S and the actual width of the tyre (Figs 2.11 and 2.15):

$$D_{tc,kb} = D_S + B \quad (m) \tag{3.15}$$

However, the swept turning circle, the diameter of which D_{tc} is greater than that of the circle by the front overhang length $L_{ex,f}$ (see the caption to Fig. 1.67), is probably a more important dimension.

According to DIN 70 020, D_{tc} is the diameter of the smallest cylindrical envelope in which the vehicle can turn a circle with the largest steering input angle (Fig. 3.94). The smallest turning circle can be calculated at the design stage, but is easier to measure and appears as manufacturer's information in the specifications or as a measurement value in test reports.

The radius $R_{r,o}$ of the turning circle, which the rear wheel on the outside of the bend traces, or $R_{r,i}$ – that traced by the wheel on the inside of the bend – can be calculated from the known turning circle diameter D_S (see also Fig. 1.69). These are:

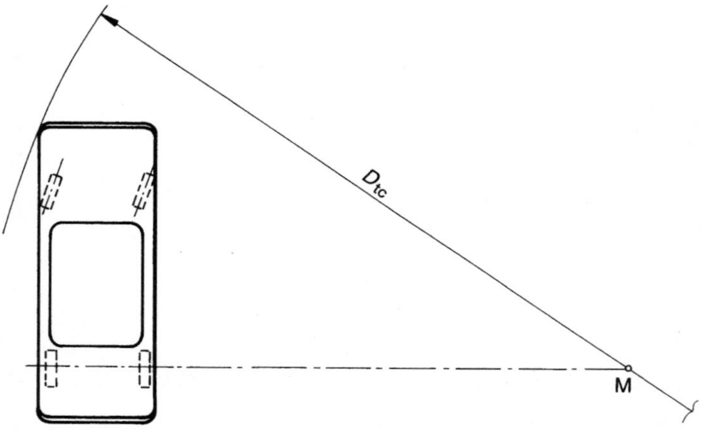

Fig. 3.94 The swept turning circle D_{tc} is the arc described by the parts of the vehicle protruding furthest outwards when the wheels are turned in at the largest steering angle.

$$R_{r,o} = \left((D_S/2 - r_\sigma)^2 - l^2\right)^{\frac{1}{2}} + \frac{b_r - j}{2} \tag{3.16}$$

$$R_{r,i} = R_{r,o} - b_r \tag{3.16a}$$

The formulae indicate that the longer the wheelbase l, with the radius of the track circle D unchanged, the vehicle requires more width ($R_{r,o}$ and $R_{r,i}$ become smaller).

3.7.3 Kinematic steering ratio

The kinematic steering ratio i_S is the ratio of the alteration δ_H of the steering wheel angle to the minimal alteration δ_m of the mean steer angle, of a pair of steered wheels, where steering is operated free of moments and begins from the on-centre (straight ahead) position. Initially, the steering compliance and the alteration of the ratio during steering are ignored:

mean steering angle $\delta_m = (\delta_o + \delta_i)/2$ (3.17)

kinematic ratio $i_S = \delta_H/\delta_m$ (3.18)

The equations are only valid when there is a greater input range (e.g. $\delta_m = 20°$) or a ratio which remains approximately constant over the whole steering range (Fig. 3.95). However, if this changes (Fig. 3.96), the steering wheel angle

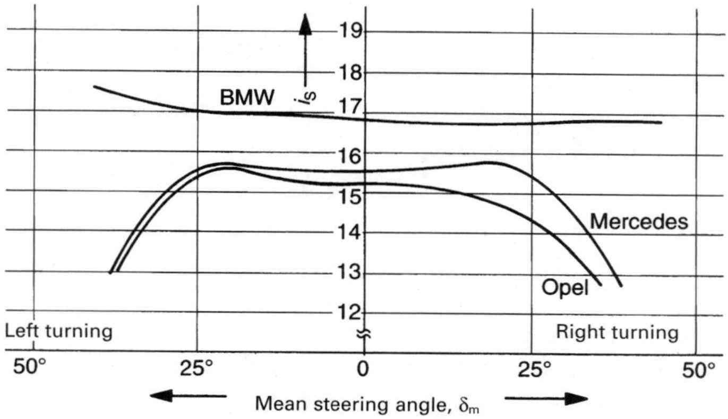

Fig. 3.95 Overall steering ratio i_S (see Section 4.3), measured on three conventional passenger cars with power-assisted recirculating ball steering. Although the BMW has a ratio which remains almost constant throughout the turning range it reduces on both sides from around δ_m = 20° on the Vauxhall/Opel and the Mercedes, so the driver needs fewer turns of the steering wheel to park. These two model groups have an opposed steering square positioned behind the axle (Figs 1.41, 4.12 and 4.30), whereas the BMW uses a synchronous one which also sits behind the axle (Fig. 4.3).

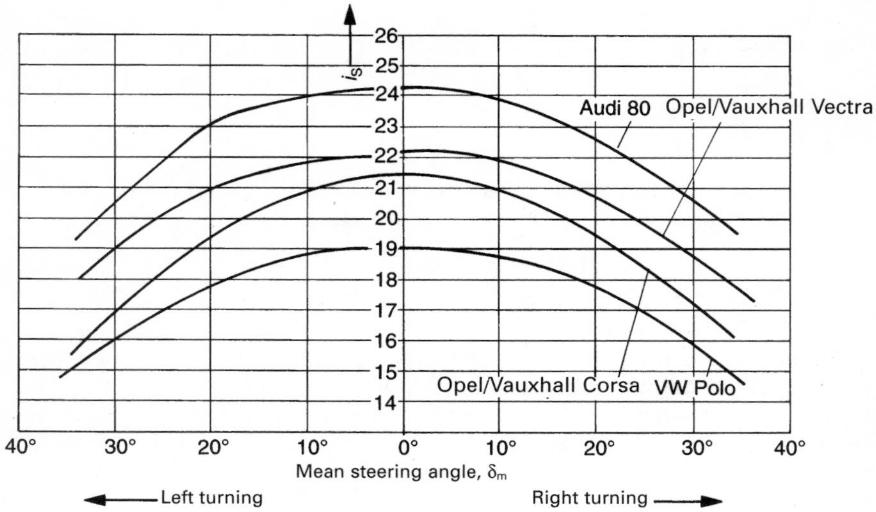

Fig. 3.96 Total steering ratio i_S (Equation 3.19) measured on four front-wheel drive passenger cars with manual (non-power-assisted) rack and pinion steering, superimposed on the mean steering angle δ_m of the wheels (Equation 3.17). It is important to note the relatively severe drop in ratio as the wheels are turned more (due to the steering kinematics, see Section 4.2). To limit the forces on the steering wheel when the vehicle is being parked, heavy vehicles, such as the Audi 80 and the Opel/Vauxhall Vectra, have the larger ratio $i_{S,0}$ = 24.2 or 22.2 in the straight running position. All vehicles have a constant steering gear ratio i'_S, i.e. not the varying split ratio seen in Fig. 3.97.

proportions $\Delta\delta_H$ must be assumed, as well as the resulting minimal mean steering wheel proportion $\Delta\delta_{m,min}$ relating to both wheels:

$$i_S = \Delta\delta_H/\Delta\delta_{m,min} \tag{3.19}$$

If the overall steering ratio relates to the on-centre position a zero should be given as the index $i_{S,0}$.

As shown in Figs 4.3 and 4.36 to 4.38, steering gears with a rotational movement need a steering square arrangement of the linkage, in which the length and position of the tie rods and steering arms allow almost every type of steering ratio as a function of the input angle. However, the entire steering system has more component parts and is more expensive (see Section 4.3).

The more economical design is rack and pinion steering, although this has the disadvantage that – as can be seen in Fig. 3.96 – for kinematic reasons the ratio reduces as steering angles increase. On power-assisted steering systems, the reduction in ratio has a favourable effect on the handling properties. In the straight running position, a more generous ratio is desirable on passenger cars at high speeds in order not to make the steering too sensitive, whereas reducing the ratio could be better for cornering and making parking and manoeuvring possible with fewer turns of the steering wheel.

The hydraulics (or electrics) (see Ref. [1]) support the increasing activation forces at greater steering angles, however this is not the case on vehicles without power-assisted steering. Here, forces can become disproportionately high because the fall in ratio cannot be reduced, especially on front-wheel drive vehicles. The reasons for this are:

- the steering gear is located in the narrow space available between the dash-panel and engine;
- the fixing points have to be laterally stiff;
- toe-in alteration (Fig. 3.67) must be avoided;
- the need to produce the actual steering curve (Fig. 3.92).

The design position of the tie rods in the top view is also a consideration. It makes a difference whether these – as shown in Figs 4.4 and 4.39 to 4.41 – are situated in front of or behind the centre of the axle (or intersect with it) and whether the inner joints are screwed into the sides of the steering rack (outer take-off) or must be fixed in the centre (centre take-off). The influence of the kingpin inclination angle and caster offset angle and the size of the steering arm angle, (Fig. 4.32) also have to be taken into account.

Series measurements have shown that, on front-wheel drive vehicles, the reduction in ratio from the 'on-centre position' to full lock is 17–30%.

Standard passenger cars have space under the engine–gearbox–block; this is the reason for the significantly lower reduction of only 5–15%.

Rear-engine vehicles offer even more space under the front-end boot. Of these, passenger cars were found with rack and pinion steering systems in which the ratio does not change throughout the entire input range.

The curve shown in Fig. 3.96 of the steering ratio of the Vauxhall Cavalier exhibits $i_{S,O} = 22.2$, with the wheels in the straight-ahead position and at a mean steering angle of $\delta_m = 35°$ the value $i_{S,min} = 17.7$; $i_{S,min}/i_{S,O} = 0.80$, i.e. the reduction is 20%.

The steering gear manufacturer ZF has developed a system to counteract the disadvantage of the reduction in ratio on non-assisted steering systems. For this purpose the steering rack varies its pitch from t_1 to t_2 (Fig. 3.97). This causes the rolling circle diameter of the pinion gear to reduce on both sides from d_1 to d_2 when the wheels are turned to the off-centre range position. The path s_2 shortens as the wheels are turned more and therefore the ratio i_S in the steering gear itself increases. The consequence is more turns of the steering wheel from stop to stop, but a reducing steering wheel moment (Fig. 3.98).

3.7.4 Dynamic steering ratio

The true steering ratio, as experienced by the driver, would be the dynamic ratio i_{dyn}; this comprises the proportions resulting from steering $\Delta\delta_H$ and the elastic

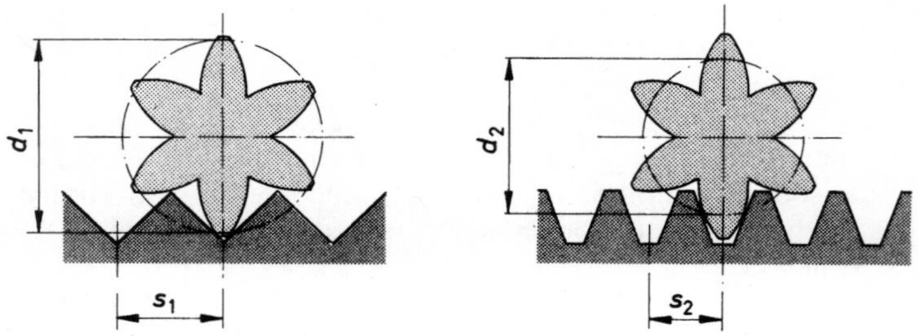

Fig. 3.97 If the steering rack is designed in such a way that the pinion gear is given a larger pitch circle (d_1, left) in the middle than on the outside (d_2, right), the rack travel reduces from s_1 to s_2 as the wheels are turned more: the ratio becomes larger, the rack moment smaller (illustration: ZF).

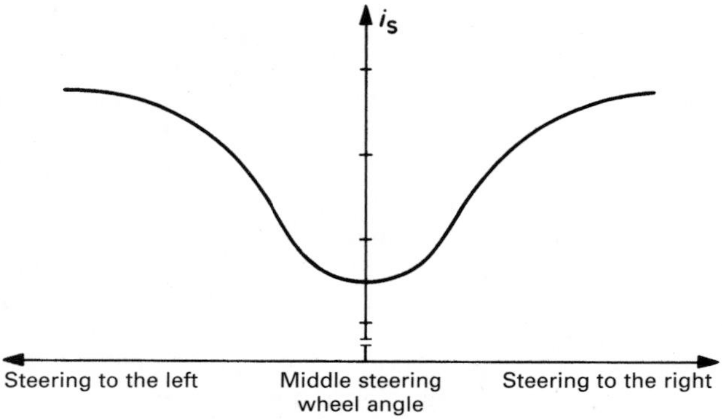

Fig. 3.98 Ratio i_S, generated in the steering gear itself when (as shown in Fig. 3.97) the steering rack has a varying split (illustration: ZF).

$\Delta\delta_{H,e}$ (Fig. 3.99) portion. To calculate the group of curves a given steering angle range $\Delta\delta_H$ must be assumed on both wheels (e.g. 0° to 5°, 0° to 10°, 0° to 15°, etc.) and the respective mean value determined in each case (here δ_m = 2.5°, 5°, 7.5° etc.) to be able to take the kinematic steering ratio i_S at these points on the respective curves. The dynamic ratio depends on the height of the steering wheel moment M_H, so that only one point of a given curve can be considered in each instance. The equation is

$$i_{dyn} = i_S + (\Delta\delta_{H,e}/\Delta\delta_H) \tag{3.20}$$

Figure 3.100 shows the dynamic steering ratio measured on a standard passenger vehicle. As an example i_{dyn} at M_H = 5 N m in the range $\Delta\delta_H$ = 0°–5° can be calculated. Taken from the lower curve (for i_S) the overall steering ratio is

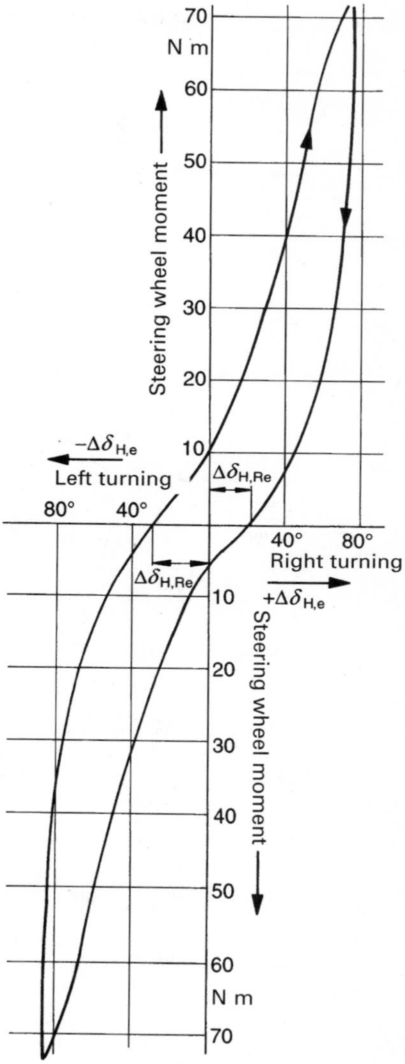

Fig. 3.99 Characteristic result of a steering compliance measurement on a passenger car with rack and pinion steering that records the steering wheel angle as a result of elasticities. It shows the compliance $\Delta\delta_{H,e}$ when the wheels are turned to the left and right and the steering wheel moments M_H increase; the wheels were locked during the measuring process. If the curve is steep, there is a high $C_H = M_H/\Delta\delta_{H,e}$ value, i.e. low steering elasticity. The greatest moment $M_H = \pm 70$ N m corresponds to a force of $F_H = 184$ N per hand with a steering wheel diameter of 380 mm. This should be enough to permit conclusions about the elasticity behaviour during driving. The hysteresis also shows the residual angles $\Delta\delta_{H,Re}$ that remain when the wheels are turned and the vehicle is stationary.

$i_S = 21$. In accordance with Fig. 3.99, the mean value of the steering wheel proportions as a result of elasticities is $\Delta\delta_{H,e} = 17°$. This gives:

$$i_{dyn} = 21 + 17/5 = 24.4$$

This value should then be entered at $\delta_m = 2.5°$. The smaller the steering angle range, and the greater M_H becomes, the more the dynamic ratio increases; if, for example, M_H is 15 N m, i_{dyn} already has a value of 31.

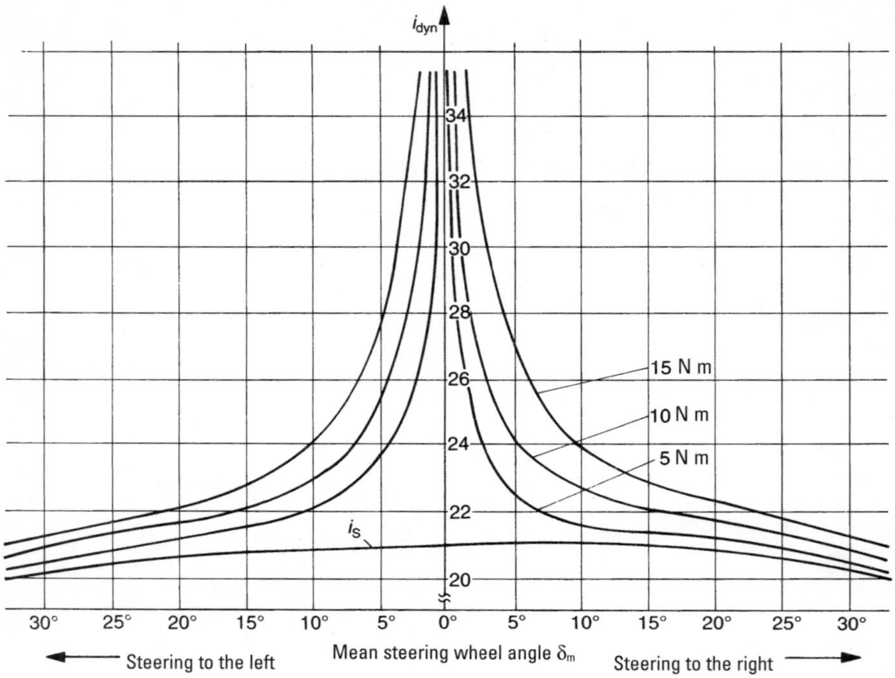

Fig. 3.100 Typical curve of the dynamic steering ratio i_{dyn} of a vehicle with rack and pinion steering entered as a function of the mean steering angle δ_m and the steering wheel moments M_H = 5, 10 and 15 N m. The kinematic total ratio i_S measured on the same vehicle was entered for comparison; this falls from $i_{S,0}$ = 21 (in the centre position) to $i_{S,min}$ = 19.7 (where δ_m = ±35°), in other words by only 6°.

3.8 Steering self-centring – general

If there were no self-centring torque on the steering axle (Z) on the front wheels of a vehicle, straight-ahead driving would be impaired and only a small force would be needed to turn it into the bend; no feedback on steering torque – the most important source of information about lateral force conditions – would be forthcoming. When the bend had been negotiated, the steering wheel would have to be turned back and would not go back of its own accord to the straight-ahead position. The driver would have no feel for cornering speed and handling with the risk that they would not be able to return the steering to the normal position fast enough when coming out of a bend. Sections 1.4.1, 1.5 and 1.6.2 refer to the correlations with the various types of drive and Fig. 1.35 shows the differences.

There are several ways of self-centring the steering at the end of a bend with, in each instance, one of the three forces acting on the centre of tyre contact (vertical force $F_{Z,W}$, lateral force $F_{Y,W}$ or longitudinal force $F_{X,W}$) having a lever to generate moments. They have been given indices to differentiate them and

Fig. 3.101 The forces occurring between tyres and road tyre contact point *W* are transferred from the suspension W to the body. This is shown on the left front wheel for the vertical force $+F_{Z,W}$, the rolling resistance or braking force $-F_{X,W,b}$ and the lateral force $+F_{Y,W}$ (see also Fig. 3.3) acting from the inside out which increases the moment.

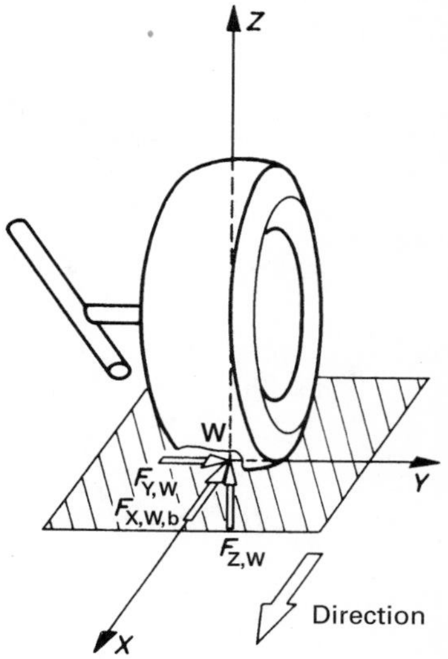

which indicate their point of action the direction of the righting force (Fig. 3.101) or other associated aspects whose length is contingent on the type of tyre (see index T):

- $M_{Z,W,Z}$ moment from vertical force $F_{Z,W}$, kingpin offset r_σ and kingpin inclination σ (Figs 3.105 and 3.107).
- $M_{Z,T,Y}$ moment from lateral force $F_{Y,W}$ and lateral force lever $n_{\tau,k}$ (Figs 3.119 and 2.49).
- $M_{Z,T,X}$ moment from rolling resistance force F_R, and lateral force lever $n_{\tau,k}$ (Fig. 3.123).
- $M_{Z,W,Y}$ moment from lateral force $F_{Y,W}$ and caster $n_{\tau,k}$ (Figs 3.121 and 3.127).

In addition, there can be self-centring torques on front-wheel drive vehicles caused by the tractive forces ($M_{Z,W,a,l \text{ and } rs}$; Fig. 3.129), by the body roll when drive shafts lie diagonally (Figs 1.6 and 3.88) and drive joints, whose centres lie outside the steering axis ($M_{Z,W,A,f}$; Fig. 3.102). Braking forces would also right the wheel on the outside of the bend while turning the (lesser loaded) wheel on the inside of the bend further (see $M_{Z,W,b}$; Equation 3.26a).

In accordance with the German standard DIN 70 000, the steering moment M_S is the sum of all moments around the steering axis of the steered wheels. These 'self-induced' moments are introduced by the driver, whereas self-centring after the vehicle has negotiated a bend is a question of the driving condition and the coefficients of friction. This difference is merely being pointed out. The vertical force $F_{Z,W}$, which influences all righting moments and is also called wheel load, is half the weighed front axle force $F_{Z,V,f}$ determined

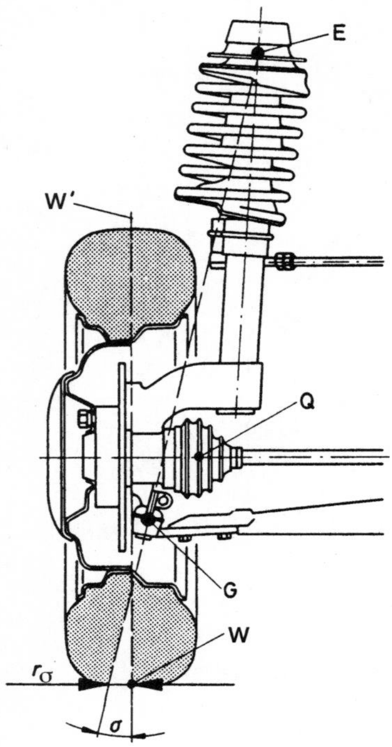

Fig. 3.102 Left front axle of an Audi with negative kingpin offset r_σ = –18 mm and an almost vertical damper unit; the spring was angled to reduce the friction between the piston rod and rod guide. For reasons of space, the CV-joint centre Q had to be shifted inwards; the space allowed for snow chains can be seen here (see Fig. 2.8).

in the design position (see Section 5.3.4), i.e. when there are three people each weighing 68 kg in the vehicle:

$$F_{Z,W} = F_{Z,V,f}/2 \text{ and } F_{Z,V,f} = m_{V,f}g \text{ (kN)} \quad (3.21)$$

As can be seen, the level of the front axle load $m_{V,f}$ is also a consideration here and so we sometimes speak of 'weight self-centring'.

Using $F_{Z,W}$ we can obtain:

the lateral force	$F_{Y,W} = \mu_{Y,W} F_{Z,W}$
the rolling resistance force	$F_R = k_R F_{Z,W}$, sometimes also
the tractive force	$F_{X,W,a} = \mu_{X,W} F_{Z,W}$ (see Equations 6.36 and 6.37a)
the braking force	$F_{X,W,b} = \mu_{X,W} F_{Z,W}$

The values for $\mu_{Y,W}$, k_R and $\mu_{X,W}$ are given in Sections 2.8.3, 2.6.1 and 2.7, and Section 3.10.3 contains a summary of all righting moments.

The opinion still sometimes expressed that the steering is centred by the vehicle front end lifting when the wheels are turned would only apply at zero caster. As Fig. 3.165 shows, at $\tau = 0°$ the body lifts on both wheels ($-\Delta H$), but if there is caster, the wheel on the outside of the bend moves upwards, the most highly loaded side of the body sinks and instead of self-centring it, the weight would turn the steering further. However, the less loaded side on the inside of the bend lifts.

3.9 Kingpin inclination and kingpin offset at ground

3.9.1 Relationship between kingpin inclination and kingpin offset at ground (scrub radius)

According to ISO 8855, the kingpin inclination is the angle σ which arises between the steering axis EG and a vertical to the road (Figs 3.103 and 3.107). The kingpin offset is the horizontal distance r_σ from the steering axis to the intersecting point of line N′N in the wheel centre plane with the road. Values on present passenger cars are:

$\sigma = 11°$ to $15°30'$ and
$r_\sigma = -18$ mm to $+20$ mm

As shown in Fig. 2.8, r_σ can also depend on the tyre width.

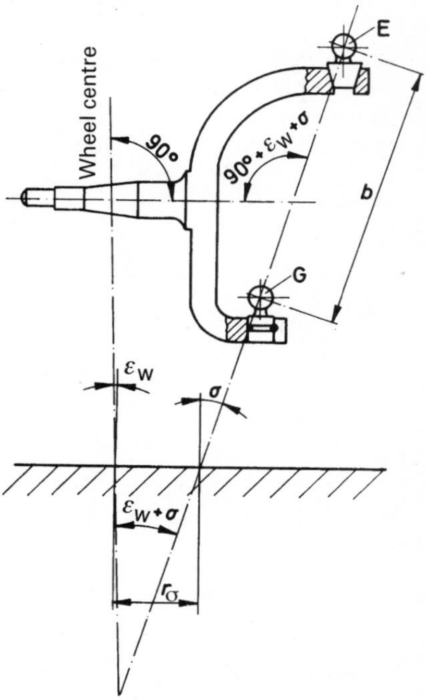

Fig. 3.103 The precise position of the steer axis – also known as kingpin inclination axis – can only be determined if the centre points E and G of the two ball joints are known. The total angle of kingpin inclination and camber ($\sigma + \varepsilon_W$) must also be included when dimensioning the steering knuckle as an individual part.

Larger kingpin inclination angles are necessary to give the vehicle a small or negative kingpin offset. In commercial vehicles, tractors and building-site lorries, the inclination of the kingpin is often equivalent to the angle σ (Fig. 1.3), whereas the wheels are controlled by ball joints on the front axles of passenger cars. On double wishbone suspensions, the steering axis therefore goes through the centres of the ball sockets E and G indicated (Figs 1.38, 3.120 and 3.103); the engineering detail drawing must show the total angle of camber and kingpin inclination.

The McPherson strut and strut damper have a greater effective distance between the lower ball joint G and the upper mounting point E in the wheel house (Figs 1.8 and 3.102); however, the upper axle parts are next to the wheel, so attention should be paid to creating enough clearance for the rotating tyre (possibly for snow chains). As a result, a higher inclination of the steering axis and a higher angle σ have to be accepted. In addition, as can be seen in the illustrations, point G has been shifted to the wheel to obtain a negative kingpin offset. The steering axis then no longer matches the centre line of the suspension strut (Figs 1.8, 3.30 and 3.104).

Due to the relationship between camber and kingpin inclination shown in Fig. 3.103, the angle σ does not need to be toleranced on double wishbone suspensions. The permissible deviations on the overall angle $\varepsilon_W + \sigma$ are given in the detailed drawing of the steering knuckle. If the camber has been set correctly on this type of suspension, the kingpin inclination angle will also be correct. However, the important thing is (as specified in the camber tolerance) that the deviation between left and right does not exceed 30′, otherwise the steering could pull to one side if the caster angle τ on the left- and right-hand sides differ (see Section 3.10.7).

On McPherson struts and strut dampers, the steering knuckle is usually bolted to the damping unit (Figs 1.56 and 5.54). In this case there may be play between the bolts and holes or the position may even be used for setting the camber (Fig. 3.104). In this case it is sensible to tolerance the kingpin inclination angle

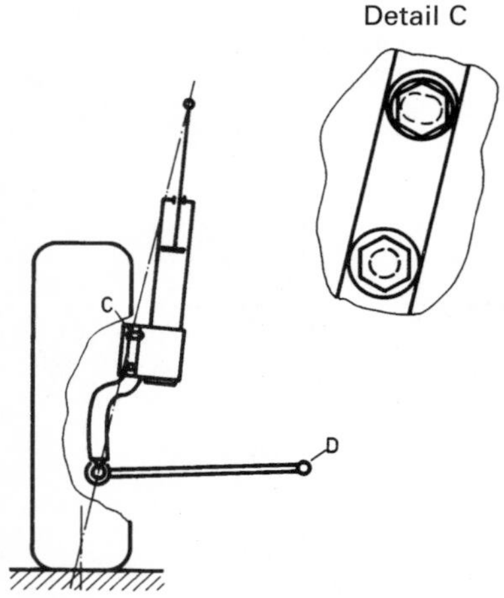

Fig. 3.104 Camber can be set at the bracket between steering knuckle and suspension strut using an excenter on the upper bolt C; the lower screw is then used as a pivot. The kingpin inclination, which is important for driving behaviour, cannot be corrected in such cases. The steering axis entered here does not lie in the damper centre line.

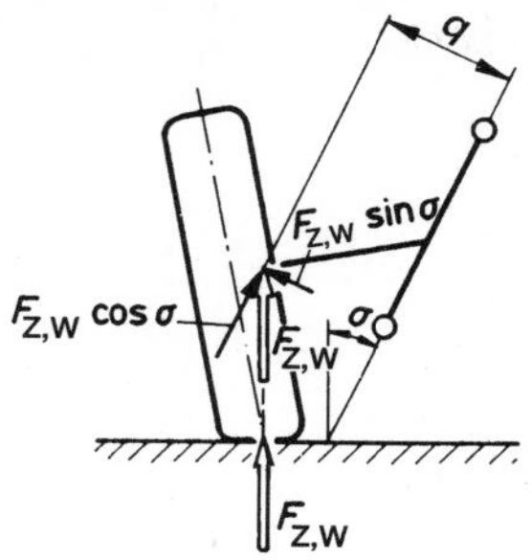

Fig. 3.105 For static observation, the vertical force $F_{Z,W}$ must be shifted to the wheel axis and resolved into its components. The distance to the steering axis is equivalent to the vertical force lever q, the size of which depends on the kingpin offset r_σ and the angle σ.

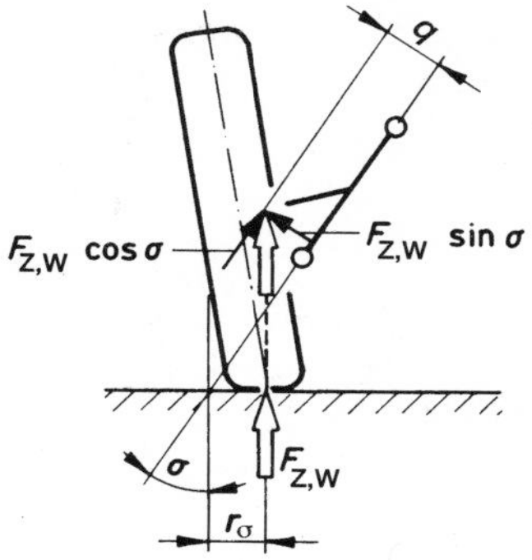

Fig. 3.106 The negative kingpin offset reduces the vertical force lever q. However, its length is one of the determining factors in the self-aligning torque $M_{Z,W,Z}$. To maintain its level, the kingpin inclination angle σ would have to be increased.

because, provided the camber is correct, the kingpin inclination does not have to be.

There is also a direct correlation between the alteration to camber and kingpin inclination when the wheels bump and rebound. As described in Section 3.5.2, the aim is to make the compressing wheel go into negative camber, as this leads to small changes in camber at body roll, but an increase in kingpin inclination by the same angle. Strictly speaking, the calculation by drawing of the camber alteration, shown and described in Figs 3.50 to 3.52, relates to the kingpin inclination, and for this reason the angle alteration $\Delta\sigma$ is also entered.

To obtain the self-aligning torque $M_{Z,W,Z}$ which is important for righting, the vertical force $F_{Z,W}$, which is always present on the centre of tyre contact, must, for static consideration, be shifted up to the wheel axis and resolved there in the direction of the steering axis:

$F_{Z,W} \cos\sigma$ and, vertical to it, $F_{Z,W} \sin\sigma$ (Figs 3.105 and 3.106)

The steering lever or vertical force lever q at the resolution point is:

$$q = (r_\sigma + r_{dyn} \tan\sigma) \cos\sigma \tag{3.21a}$$

The equation will apply provided that $\cos\varepsilon_W = 1$, a condition that applies to normal camber angles. If the vehicle has caster, the force components $F_{Z,W} \sin\sigma$ must be further resolved by the angle τ (see Equation 3.3). The parameter r_{dyn} can be calculated using Equation 2.2.

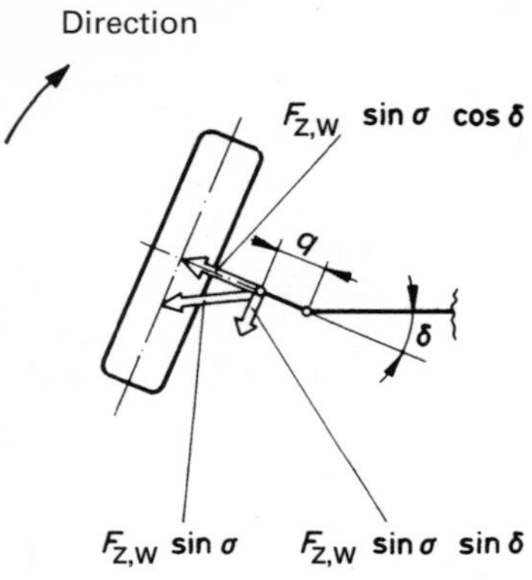

Fig. 3.107 When the wheels are turned by the angle δ, the vertical force component $F_{Z,W}$ sin σ gives the self-aligning torque $M_{Z,W,Z}$; the extent of this weight-related self-alignment depends on the kingpin inclination angle σ, the lever q, the front axle load $m_{V,f}$ and the caster (Fig. 3.147).

When the wheels are turned, the force $F_{Z,W}$ sin σ is at the angle δ to the wheel axis (Fig. 3.107) and the component $F_{Z,W}$ sin σ sin δ will, with smaller steering angles, give the approximate righting moment based on the whole axle:

$$M_{Z,W,Z} = F_{Z,V,f} \ \sin\sigma \ \sin\delta_m q \tag{3.22}$$

(For $F_{Z,V,f}$ see Equation 3.21 and δ_m in Equation 3.17.)

The exact solution has to take the changing kingpin inclination angle (due to lateral forces when the wheels are turned and due to the body roll) into account as well as the positive and negative caster that occurs (Figs 3.48, 3.53, 3.56 and 3.132). The influence of the paths $r_{\tau,T}$ and r_T in the tyre contact area (Figs 3.119 and 3.120) also has to be considered. Both can have a significant influence on the size of $M_{Z,W,Z}$ during cornering. On the outside of the bend, the lateral force $F_{Y,W,O}$ reduces the kingpin offset by $-r_T$ (or causes it to become more negative) whereas on the inside of the bend, it increases or becomes less negative (Fig. 3.127).

There is also a load alteration during cornering, whereby $F_{Z,W,f,o} > F_{Z,W,f,i}$ and also δ_i and δ_o are not always of the same size, so that different moments always occur on individual wheels. The kingpin offset r_σ, which appears in Equation 3.21a, influences the level of the self-aligning torque $M_{Z,W,Z}$; if this offset is large, the righting increases, if r_σ decreases or even becomes negative (owing to the shorter lever q), the moment reduces (Fig. 3.106).

The more $M_{Z,W,Z}$ increases, the more the front axle becomes longitudinally sensitive. There is, therefore, a clear tendency towards a small positive or negative kingpin offset.

If $M_{Z,W,Z}$ is to remain at the same level, the kingpin inclination angle has to be enlarged with the disadvantage that, when the wheels are turned, the wheel on the outside of the bend goes in the more positive camber direction, which makes more space necessary because the brake disc has to be shifted into the disc wheel (Figs 3.102 and 2.23). With a given path $r_{\sigma,1}$, the necessary angle σ_1 can be calculated from the existing values r_σ (in mm) and σ_0:

$$\tan \sigma_1 = \frac{r_{\sigma,1}}{2B} + \sqrt{\left(\frac{r_{\sigma,1}}{2B}\right) + A/B} \tag{3.23}$$

where

$$A = (r_\sigma + r_{dyn} \tan \sigma_O) \sin \sigma_O \cos \sigma_O$$
$$B = r_{dyn} - A$$

The dynamic tyre radius r_{dyn} can be determined from the rolling circumference C_R (or $C_{R,dyn}$, see Section 2.2.8):

$$r_{dyn} = C_R/(2\pi) \tag{3.24}$$

Taking as an example a standard passenger car with the tyre size 185 R 14 90 S, which has a rolling circumference of 1965 mm, the axle settings were $\sigma_O = 5°54'$ and $r_\sigma = 73$ mm.

The aim is to find the kingpin inclination angle σ_1 with a negative kingpin offset $r_{\sigma 1} = -18$ mm:

$$r_{dyn} = 1965/2\pi = 313 \text{ mm}$$
$$A = (+73 + 313 \tan 5°54') \sin 5°54' \cos 5°54'$$
$$A = 11 \text{ mm: } B = 311 - 11 = 302 \text{ mm}$$
$$\tan \sigma_1 = \left(\frac{-18}{2 \times 302}\right) + \sqrt{\left[\frac{-18}{2 \times 302}\right]^2 + \frac{11}{302}} = 0.0298 + 0.1912$$
$$\tan \sigma_1 = 0.211;\ \sigma_1 = 12.46° = 12°28'$$

The following would then appear on the drawing and in the workshop manual:

kingpin inclination 12°30′

a normal value for a negative kingpin offset. The parameter $r_{\sigma,1}$ can be more easily calculated as a function of the amended kingpin inclination angle σ_1:

$$r_{\sigma,1} = \frac{A}{\sin \sigma_1 \cos \sigma_1} r_{dyn} \tan \sigma_1 \tag{3.25}$$

3.9.2 Braking moment-arm

During a braking process carried out with the brake mounted on the steering knuckle or wheel carrier, the braking force $F_{X,W,b}$ tries to turn the wheel with the brake acting on the moment-arm (Fig. 3.108):

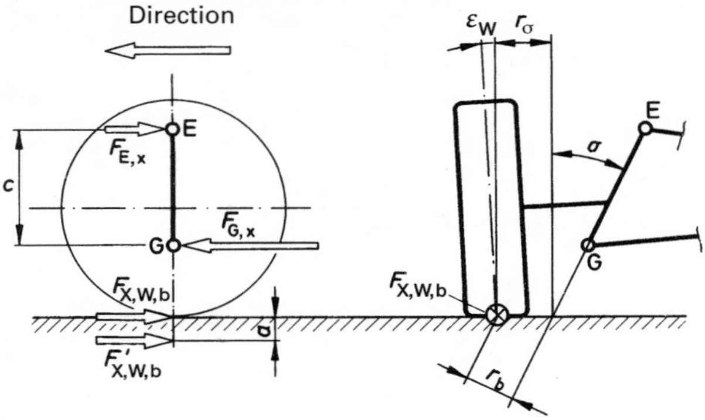

Fig. 3.108 The braking force $F_{X,W,b}$ has the lever $r_b = r_\sigma \cos \sigma$ to the steering axis EG; shifted vertically on this axis, $F_{X,W,b}$ acts by the amount a below ground and causes the greatest force in point G: $F_{G,x} = F_{X,W,b} + F_{E,x}$ (see also Fig. 3.155). When there is caster, $F_{X,W,b}$ must be resolved at the centre of tyre contact around the angle τ (Fig. 3.115).

$$r_b = r_\sigma \cos \sigma \tag{3.26}$$

around the steering axis, i.e. the moment

$$M_{Z,W,b} = F_{X,W,b} \cos \tau \; r_b \tag{3.26a}$$

is generated, which, as Fig. 3.109 shows, results in the tie rod force F_T and, where r_σ is positive, pushes the wheel into toe-out (for caster angle τ see Fig. 3.115).

The longer the path r_σ, the more the moment $M_{Z,W,b}$ increases and the larger the influence of uneven front brake forces on the steering – which is the reason for keeping r_σ as low as possible or even making it negative (Figs 3.102 and 3.106). Thus, as shown in Fig. 6.12, brakes that do not respond equally cause a counter-steering effect, which can reduce or eliminate the yaw response of the

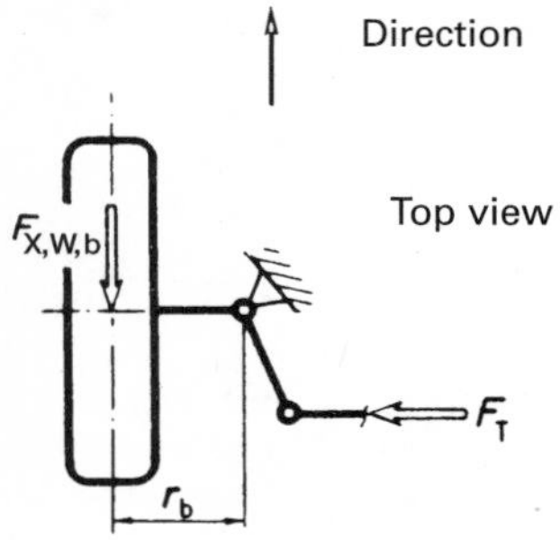

Fig. 3.109 If the brake is in the wheel, the braking force $F_{X,W,b}$ causes the moment $M_{Z,W,b} = F_{X,W,b}\, r_b$, which tries to push the wheel into toe-out and causes the tie rod force F_T. The steering axis is assumed to be vertical to simplify the calculation.

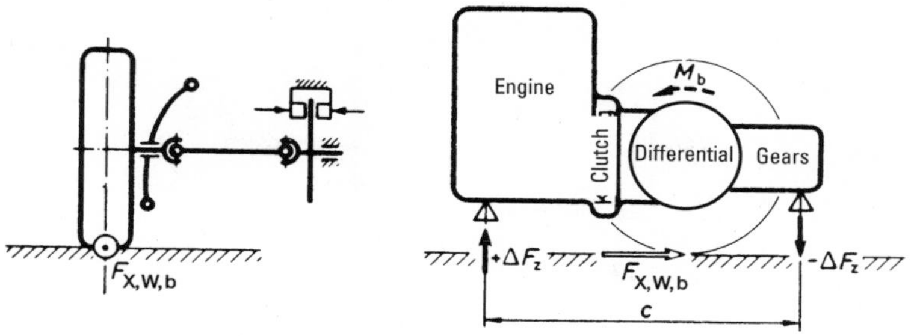

Fig. 3.110 If a front-wheel drive vehicle has an inside brake, the engine mounting must absorb not only the drive-off moment, but also the braking moment; the reaction forces $\pm\Delta F_z$, the size of which depends on the effective distance *C*, occur in the rubber buffers.

body, which is also true for an elastokinematic toe-in alteration (Figs 3.2 and 3.82). The longitudinal force $F_{X,W,b}$ arising on the ground produces the reaction forces $F_{E,X}$ and $F_{G,X}$ in the pivot points of the steering knuckle. In order to be able to determine their size, $F_{X,W,b}$ must be shifted towards the braking force lever on the extension of the steering axis EG. Therefore, with positive kingpin offset $F_{X,W,b}$ lies below the ground by the amount *a* and is shown in the side view of Fig. 3.108 as $F'_{X,W,b}$:

$$a = r_b \sin\sigma = +r_\sigma \cos\sigma \sin\sigma \tag{3.27}$$

If r_σ is negative, $F'_{X,W,b}$ moves above ground (Fig. 3.156) and $F_{G,X}$ becomes smaller.

If the brake is on the inside on the differential, the braking moment is transmitted via the universal joints to the engine and causes the bearing reaction forces ΔF_z in the engine mounting (Fig. 3.110):

$$\pm\Delta F_z = (F_{X,W,b}\, r_{dyn})/c$$

The smaller the wheel radius (r_{dyn}) and the larger the effective distance *c*, the lower the forces and therefore also the compliance in the rubber mountings. The braking force $F_{X,W,b}$ which occurs at the centre of tyre contact must, in such cases, be shifted to the centre of the wheel (like the rolling resistance force F'_R in Fig. 3.111), because a shaft bearing can only transfer forces, and not moments, in its effective direction. Just like F'_R, $F'_{X,W,b}$ acts on the longitudinal force lever r_a, also known as the disturbance and traction force moment-arm:

$$r_a = r_\sigma \cos\sigma + r_{dyn} \sin(\sigma + \varepsilon_W) \tag{3.28}$$

With it, $F_{X,W,b}$ causes the moment:

$$M_{Z,W,b} = F_{X,W,b} \cos\tau\, r_a \tag{3.28a}$$

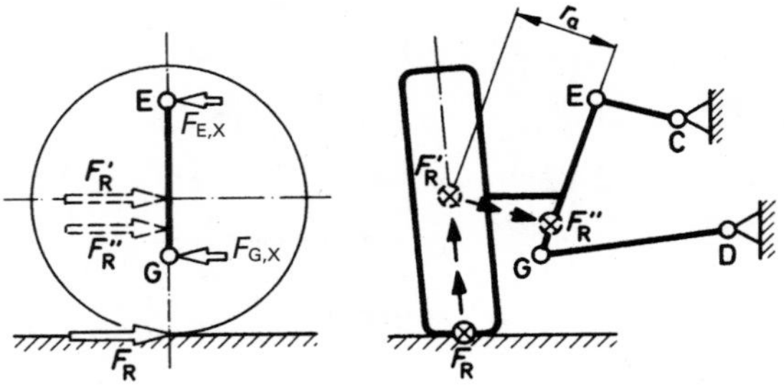

Fig. 3.111 When the wheel is rolling in a straight line, the rolling resistance force F_R must be observed as F_R' in the wheel centre; its distance to the steering axis is r_a. This so-called longitudinal force lever depends on the kingpin offset r_σ and the smaller this can be, the further up F_R acts as F_R'' on the steering axis and the more evenly points E and G are stressed in the longitudinal direction. The same static conditions apply to the braking force if the brake is located on the inside on the differential (see also Figs 3.113 and 3.154).

which also occurs when $r_\sigma = 0$. In the equations, note must be taken of the plus or minus signs; in the case of a negative kingpin offset, the first element ($-r_\sigma \cos \sigma$) must be subtracted from the second. Figure 3.115 contains $\cos \sigma$.

Because $r_a > r_b$, there is a higher moment when the brake is on the inside at the differential, which has a more pronounced influence on the steering. The reaction force $F_{G,X}$ in the lower ball joint, however, becomes much smaller. To determine the forces $F_{E,X}$ and $F_{G,X}$ $F'_{X,W,b}$ has to be shifted vertical to the steering axis and, in the side view, comes to lie below the wheel centre as $F''_{X,W,b}$ (Fig. 3.154) by

$$a = r_a \sin \sigma \tag{3.28b}$$

3.9.3 Longitudinal force moment-arm

Figure 3.111 shows the rolling resistance force F_R, which always occurs when the vehicle is moving. It generates the same moment left and right:

$$M_{Z,T,X} = F_R \cos \tau \; r_a \tag{3.28c}$$

which is absorbed at the tie rods (Fig. 3.112); any caster angle τ must be considered. If the moments are of the same size, the vehicle moves in a straight line, but if they are different it can pull to one side. Tyres that have a different rolling circumference (Fig. 2.15) or front axles where the angles $\sigma + \varepsilon_W$ differ to the left and the right can be the reason for this. The factor $r_{dyn} \sin (\sigma + \varepsilon_W)$ primarily determines the length of the moment-arm r_a (see Equation 3.28). On a bend, the outer wheel experiences a force increase ($F_{Z,W} + \Delta F_{Z,W}$, Fig. 1.6) and the inner one a reduction equivalent to $F_{R,o} > F_{R,i}$. Where there is no caster, the wheel on

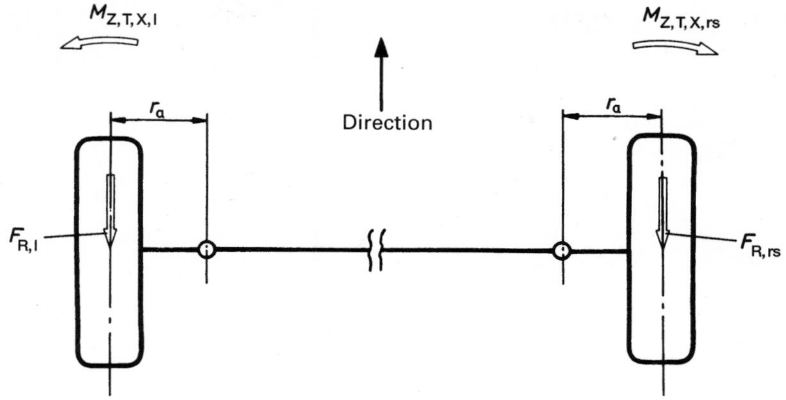

Fig. 3.112 The rolling resistance force F_R pushes the wheels backwards via the longitudinal force moment-arm r_a, i.e. into toe-out $-r_{\Delta,f}$. A moment arises on both sides, which is absorbed and cancelled out at the tie rods. In the case of caster the angle τ must also be observed (see Fig. 3.115).

the outside of the bend rights itself more than the one on the inside is trying to turn into the bend.

The tractive forces $F_{X,W,a}$, which occur at the contact points of the front wheels on front-wheel drive vehicles, cause moments acting in the opposite direction, but also have to be taken into consideration in the centre of the wheel (Fig. 3.113), i.e. in vehicles of this design, a smaller longitudinal force moment-arm r_a will be particularly important. Citroën has achieved this by shifting the ball joints E and G in to the wheel centre plane (Fig. 3.114). This means that

$$\varepsilon_W + \sigma = 0, \;\; r_\sigma = 0 \;\text{ and, therefore, }\; r_a = 0$$

The longitudinal force moment-arm should be as short as possible. Comparison with the formula for the vertical force lever q (Equation 3.21a) shows the difficulties:

Fig. 3.113 The negative kingpin offset on the ground favourably shortens the longitudinal force moment-arm r_a. The tractive force $F_{X,W,a}$ relates to one wheel and, in case of caster, must be resolved around the angle τ in the wheel centre $F_{X,W,a}$.

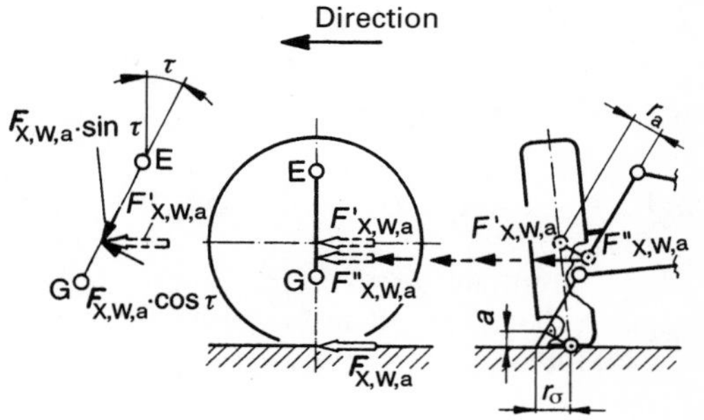

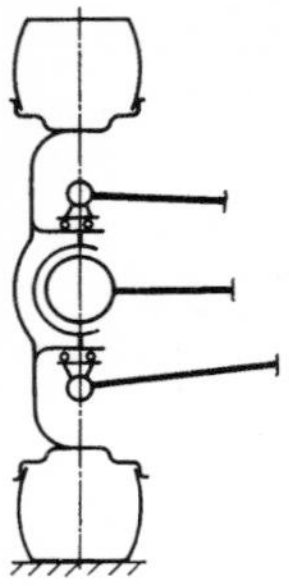

Fig. 3.114 Section through the centre axle steering of the model GSA which Citroën no longer builds; swivel and supporting joints are in the wheel centre plane and r_σ is zero.

$$q = r_\sigma \cos\sigma + r_{dyn} \sin\sigma \text{ and}$$
$$r_a = r_\sigma \cos\sigma + r_{dyn} \sin(\sigma + \varepsilon_W)$$

If, for example, the camber is $\varepsilon_W = 0°$, $q = r_a$ if there is only a small or no caster angle τ and the vehicle moves unimpaired in a straight line.

However, during cornering, q changes significantly while r_a remains virtually unchanged (Fig. 1.6).

3.9.4 Alteration to the kingpin offset

To improve cornering behaviour, disc wheels with lower wheel offsets e are sometimes used (Fig. 2.23) or (in the past) spacer rings were laid between the wheel and brake disc to give the advantage of a slightly wider track (around 2–4%), but with the disadvantage of up to 100% greater kingpin offset at ground. The result is a more noticeable disturbance effect on the steering when the road is uneven and particularly when the front brakes do not pull evenly.

If, as on almost all passenger cars with negative kingpin offset, the two brake circuits are designed to be diagonal, these measures cannot be implemented. The negative kingpin offset would either change from negative to positive or become too positive when there is an elastokinematic toe-in alteration on the front axle (Fig. 3.82), and toe-out would occur during braking instead of toe-in.

Figure 2.8 shows the alteration of r_σ depending on the tyre width, using the example of the VW Golf III.

3.10 Caster

3.10.1 Caster trail and angle

We differentiate between the kinematic caster trail $r_{\tau,k}$ of the wheel, the caster angle τ, the caster offset $n_{\tau,}$, the tyre caster $r_{\tau,T}$, the lateral force lever $n_{\tau,k}$ and the elastokinematic caster $r_{\tau,e}$. Dynamic measurements are contained in Section 5.2.3 of Ref. [9].

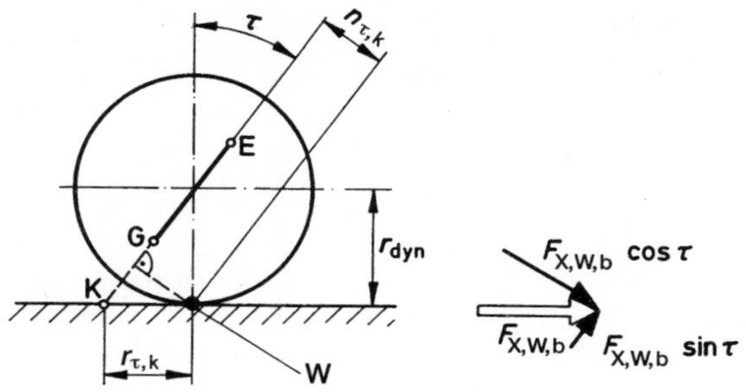

Fig. 3.115 If the extension of the steering axis goes through the ground at point K in front of the wheel centre, the distance arising is the kinematic caster trail $r_{\tau,k}$ (Case 1). A vertical to EG, drawn through the centre of tyre contact W, when projected onto the xz plane, gives the lateral force lever $n_{\tau,k}$ (Equation 3.30).

Longitudinal forces which arise, such as the braking force $F_{X,W,b}$ (or the rolling resistance F_R), must be resolved at the centre of tyre contact (or as F_R' in the wheel centre, Fig. 3.111) by the angle τ.

In accordance with DIN 70 020 (and also DIN 70 000), τ is the angle between the steering axis EG projected onto an xz plane and a vertical, drawn through the wheel centre (Case 1, Fig. 3.115), and $r_{\tau,k}$ the distance between the points K and W on the ground. The castering of the wheel centre of contact W behind the intersection K can also be achieved by shifting the axes of rotation in front of the wheel centre: $+n_\tau$ (Case 2, Fig. 3.116). On some front-wheel drive vehicles, owing to the increased self-centring moment caused by tractive forces, negative caster was designed, which can be achieved with a reversed angled steering axis (Case 3, Fig. 3.117) or by positioning an axis EG behind the wheel centre and inclining it by the angle τ, which leads to negative caster offset $-n_\tau$ (as can be seen in Figs 3.118 and 3.139). For the following reasons, linking a positive caster angle and $-n_\tau$ is popular from the point of view of construction:

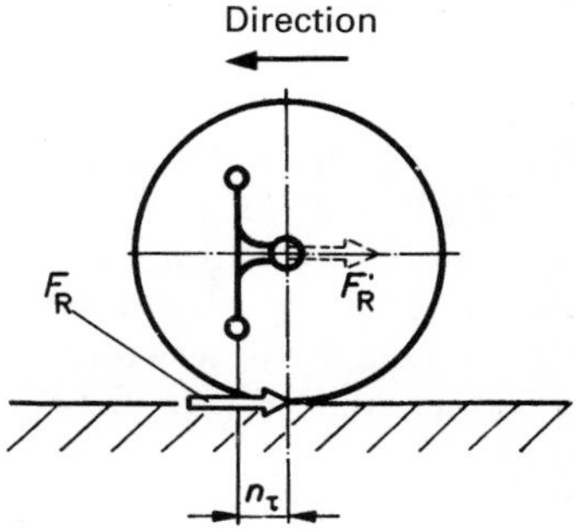

Fig. 3.116 Caster can also be achieved by shifting the wheel centre behind the steering axis (Case 2); if this is vertical, as shown, the (here) positive caster offset is equal to the moment arm: $n_\tau = r_{\tau,k} = +n_{\tau,k}$. Rolling resistance forces F_R acting at the centre of tyre contact must be observed as F_R' in the wheel centre.

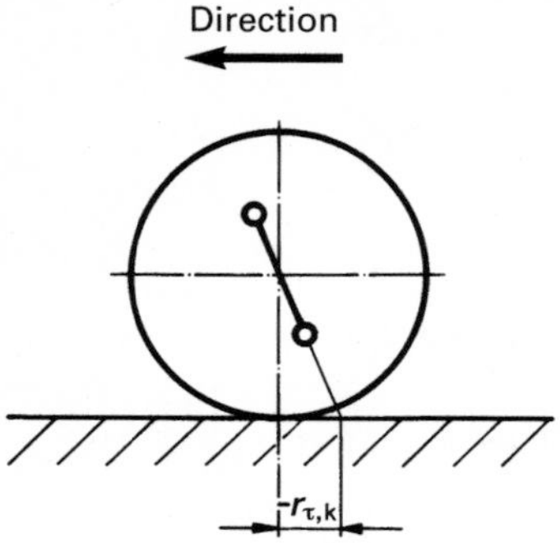

Fig. 3.117 Caster (Case 3): a steering axis, which is inclined opposite by the angle $-\tau$, results in negative caster $-r_{\tau,k}$, and the associated disadvantage of a more positive camber on the outside of the bend when the wheels are turned. However, where the angles $-\tau$ are small, the tyre caster $r_{\tau,T}$ balances out the negative caster trail (Fig. 3.121). On the independent rear wheel suspensions the steering knuckle (here not the steering axis) can be given negative caster to achieve lateral force understeering (see Figs 3.144 and 3.145).

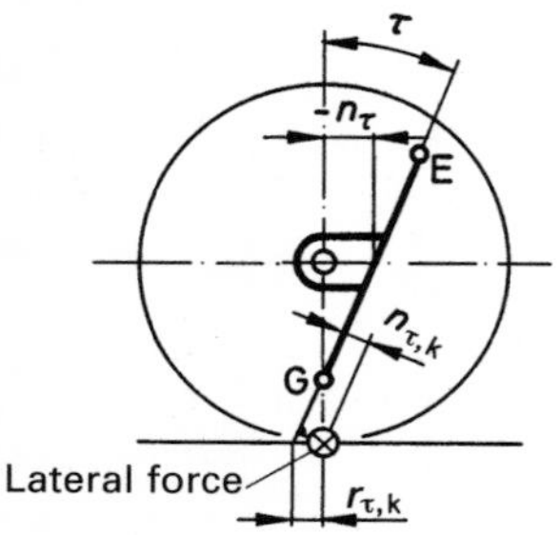

Fig. 3.118 Front axle properties can be improved by a negative caster offset n_τ; the caster trail $r_{\tau,k}$ on the ground shortens by this amount and the camber alteration when the wheels are turned becomes more favourable.

- the kinematic caster trail $r_{\tau,k}$ is smaller, i.e. the influence on the steering resulting from uneven road surfaces reduces;
- the camber alteration is increased when the wheels are turned (Fig. 3.132).

The trail $r_{\tau,k}$ and the moment-arm n_τ of the lateral force (i.e. the path projected onto the vertical plane xz) both with and without negative offset $-n_\tau$ can be easily determined using the dynamic rolling radius r_{dyn} (see also Section 3.10.7):

$$r_{\tau,k} = r_{dyn} \tan \tau$$
$$r_{\tau,k} = r_{dyn} \tan \tau - n_\tau \tag{3.29}$$

$$n_{\tau,k} = r_{dyn} \sin \tau$$
$$n_{\tau,k} = r_{dyn} \sin \tau - n_\tau \cos \tau \tag{3.30}$$

During a bend, the area of tyre contact deforms due to the slip angle α (Fig. 3.119). The lateral force $F_{Y,W}$ therefore is offset by the amount $r_{\tau,T}$ – known as tyre caster – behind the wheel centre (Figs 3.120 and 2.50). The tyre caster of practically $r_{\tau,T}$ = 10 mm to 40 mm must therefore be included in all static and elastokinematic observations. Without and with caster offset the overall path $r_{\tau,t}$ is then as follows (Fig. 3.121):

$$r_{\tau,t} = r_{\tau,k} + r_{\tau,T} \tag{3.31}$$

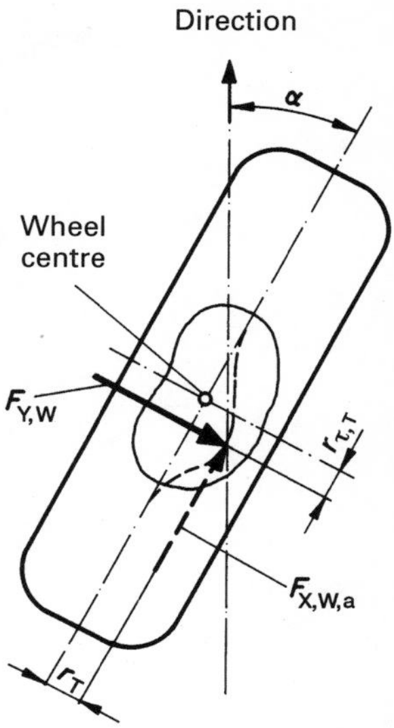

Fig. 3.119 The tyre contact patch (also known as the 'tyre print', Fig. 2.9) of a tyre rolling at an angle under the influence of lateral forces deforms in the shape of a kidney; this means the point of application of the vertical force $F_{Z,W}$ and the lateral force $F_{Y,W}$ moves by the trail $r_{\tau,T}$ – the tyre caster – behind the wheel centre and the tyre self-aligning torque $M_{Z,T,Y} = F_{Y,W} \times r_{\tau,T}$ occurs. If the vehicle has front-wheel drive, $F_{X,W,a}$ acts at a point in the tyre contact area, offset by r_T from the wheel centre plane, as does the rolling resistance force $F_{R,co}$ on a bend. Tyre caster is between $r_{\tau,T}$ = 10 mm and 40 mm; lateral offset is $r_T \approx 3$ mm per $\mu_{Y,W} = 0.1$ (see Section 2.10.2 and Figs 3.127 and 3.128).

If the slip angle α is specified instead of the coefficient of friction $\mu_{Y,W}$, Equation 2.4c will apply. Camber has a similar effect; negative camber diminishes r_T and positive increases it:

$$\pm r_T = 6 \text{ mm per } \Delta\varepsilon_{W,k} = \pm 1°$$

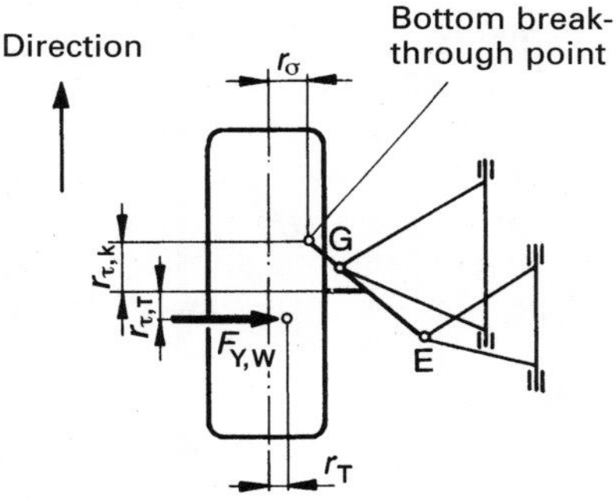

Fig. 3.120 The extension of the steering axis EG, which is three-dimensionally at an angle due to kingpin inclination and caster, penetrates the ground in front of the wheel centre and gives (in the example) the positive kingpin offset r_σ and the kinematic caster trail $r_{\tau,k}$. On a bend, the lateral force acts offset to the tyre caster $r_{\tau,T}$ in the tyre contact area. The total trail (index τ, t) is therefore $r_{\tau,t} = r_{\tau,k} + r_{\tau,T}$ and – in accordance with Fig. 3.119 – the kingpin offset (overall on the outside of the bend) $r_{\sigma,t} = r_\sigma - r_T$.

$$n_{\tau,t} = n_{\tau,k} + r_{\tau,T} \cos \tau \tag{3.32}$$

$$n_{\tau,t} = n_{\tau,k} + (r_{\tau,T} - r_{\tau,W}) \cos \tau \tag{3.32a}$$

If precise calculations are required, the elastokinematic caster $r_{\tau,e}$ must be used instead of $r_{\tau,k}$, although this can only be determined by experiment on the vehicle (see Section 5.2.3 in Ref. [9]).

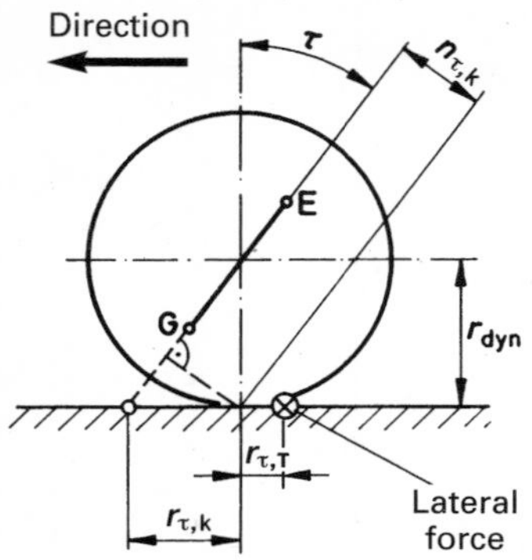

Fig. 3.121 Due to the tyre caster $r_{\tau,T}$, which is always present during cornering, the lateral force lever is extended and becomes (Fig. 2.50)

$$n_{\sigma,t} = n_{\tau,k} + r_{\tau,T} \cos \tau$$

In order to make them clearer, the path $r_{\tau,T}$ is not shown in some of the following figures.

3.10.2 Caster and straight running

Caster can be compared with the tea trolley effect, where the pulled wheel takes on the direction of pull and the wheel centre adopts a position behind the axis of rotation 1 (Fig. 3.122). The tensile force and the opposed force F_R generated by the rolling resistance are on an effective line, in other words in a stable ratio to one another because the guiding and wheel axis lie behind one another. The same effect also exists (despite kingpin offset and kingpin inclination) on the wheels of a vehicle if these can be rotated around axes. The wheels are set to caster on both sides and are linked by tie rods.

If unevenness in the road surface or a steering input pushes the wheels out from the straight-ahead direction, by the angle δ, the rolling resistance components $F_R \sin \delta$ (as shown in Fig. 3.123) move both wheels back via the force lever $n_{\tau,k}$ (or the overall lateral force lever $n_{\tau,t}$) until they roll in a straight line again. The components $F_R \cos \delta$ (left and right) compensate, and only subject the tie rods to pressure. Negative caster on the wheels could have the opposite effect and the vehicle would become unstable.

On a vehicle moving in a straight line, caster would not only have advantages but also disadvantages. Uneven road surfaces cause alternating lateral forces on the centres of tyre contact (see Section 4.2 in Ref. [3]) and these, together with

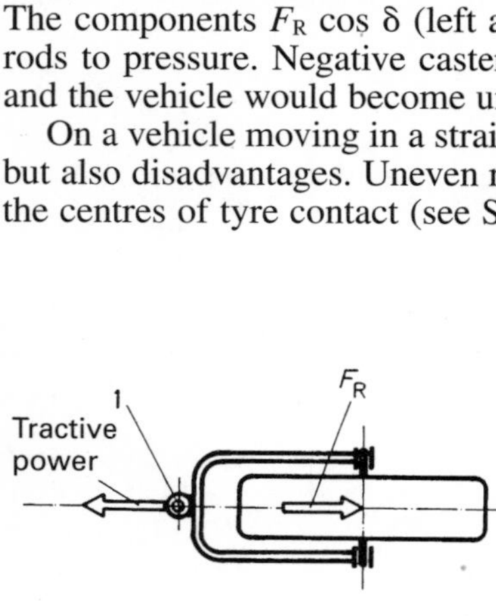

Fig. 3.122 If the rolling resistance force F_R acts behind the steering axis 1, the wheel follows in a stable manner in the direction in which it is pulled.

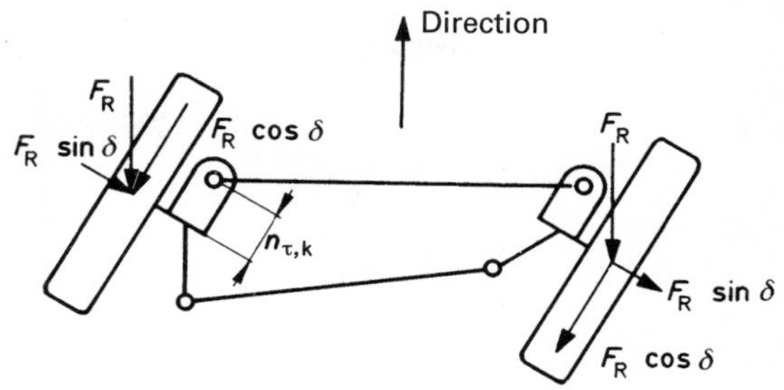

Fig. 3.123 When the vehicle is travelling in a straight line, caster has a stabilizing effect. Fig. 3.147 shows the necessary further resolution of the force components by the angle τ.

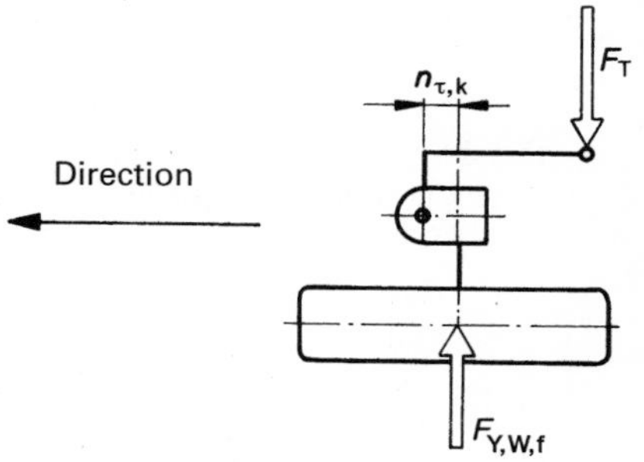

Fig. 3.124 Lateral forces $F_{Y,W,f}$ caused by uneven ground, in conjunction with the caster moment-arm $n_{\tau,k}$ cause the forces F_T in the tie rods.

the lever in $n_{\tau,k}$ (or $n_{\tau,t}$, Equations 3.30 to 3.32a) cause moments around the steering axis (Fig. 3.124), which are supported on the tie rods and can cause steering disturbances and vibrations. Furthermore, there is increased wind sensitivity due to the fact that a wind force acting on the body (Fig. 3.125) causes lateral forces $F_{Y,W}$ in the opposite direction, on the centres of tyre contact. In addition, the front forces $F_{Y,W,f}$ together with the caster moment-arm $n_{\tau,k}$ (or $n_{\tau,t}$) result in moments that turn the vehicle in the direction of the wind, i.e. further in the direction in which the body is already being pushed by the wind. This also applies to driving on (diagonally) inclined roads and leads to increased steering moment.

3.10.3 Righting moments during cornering

The alteration to the caster and kingpin inclination (or camber) angle, which is influenced by the body roll inclination (Figs 3.53 and 3.143) and is caused by the steering angle of the wheels (Figs 3.132 and 3.135), results in an alteration to all levers on which vertical, lateral and longitudinal forces are acting. Observation of each individual wheel would mean looking at these very complicated kinematic relationships, and errors would almost be inevitable because of the additional elastokinematic movements.

Determination of the righting moments based on the whole axle and on the position of the vehicle parallel to the ground is – particularly in the case of small steering angles and low cornering speeds – sufficiently precise. The caster $r_{\tau,T}$ = 10 mm to 40 mm (Fig. 3.120) (not present when the vehicle is moving in a

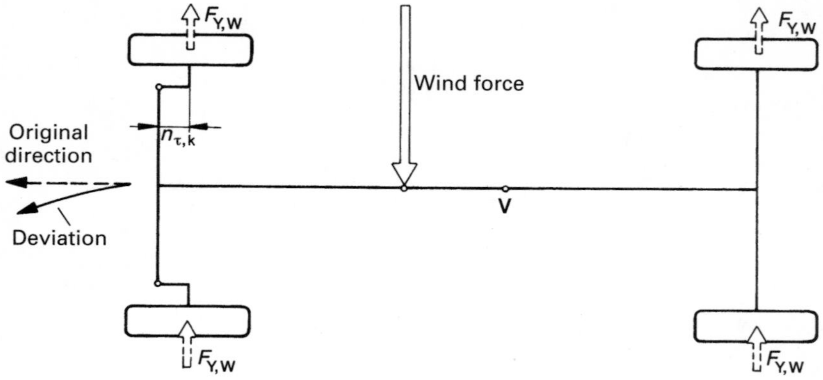

Fig. 3.125 Caster can increase the wind sensitivity of a vehicle. The point at which the wind acts is usually in front of the centre of gravity V; a moment arises which seeks to turn the vehicle and causes the wheels to steer in the same direction.

straight line) must, nevertheless, be included in the equation. Righting moments should be indicated in newtons, whereby 1 kN mm = 1 N m. The front axle caster angle will affect the Equation 3.22 righting moment, due to the vertical forces generated:

$$M_{Z,W,Z} = F_{Z,V,f} \sin \sigma \cos \tau \sin \delta_m q \tag{3.33}$$

The caster angle in the lever $n_{\tau,k}$ is considered in the case of the righting moment due to lateral force. Here, the kingpin inclination must also be included in the calculation (Fig. 3.126):

$$F_{Y,W,f,o} + F_{Y,W,f,i} = \mu_{Y,W} F_{Z,V} = F_{Y,W,f,t} \tag{3.33a}$$

$F_{Y,W,f,t}$ acts around the lateral force lever $n_{\tau,t}$ offset behind the wheel centres (Figs 3.120 and 3.127, and Equations 3.32 and 3.32a):

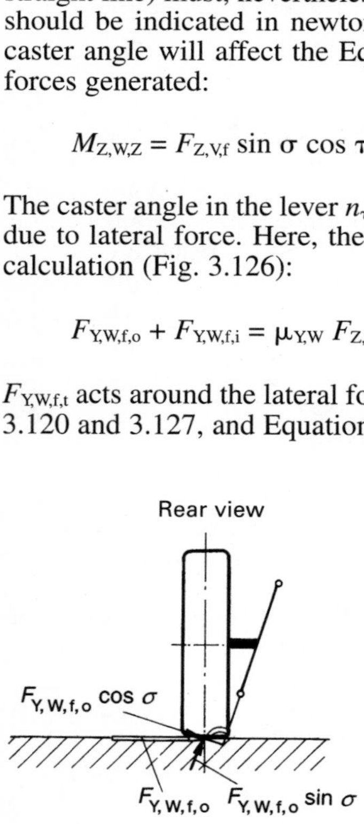

Fig. 3.126 The lateral forces acting on the centres of tyre contact of the front wheels must be resolved in the direction of the steering axis and vertical to it, shown here for the wheel on the outside of the bend. $F_{Y,W,f,o} \cos \sigma$ then has a righting effect and $F_{Y,W,f,o} \sin \sigma$ strengthens the vertical force $F_{Z,W,f,o}$.

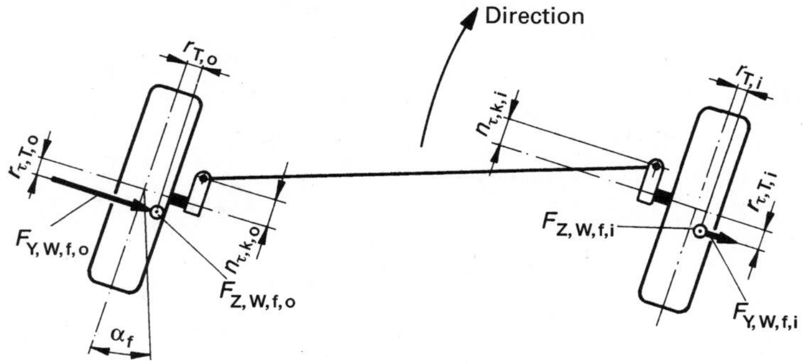

Fig. 3.127 On wheels that roll at an angle of α_f, the lateral cornering forces $F_{Y,W,f}$ act behind the wheel centres, offset by the tyre caster $n_{\tau,T}$ and push the centres of tyre contact (and therefore also the vertical forces $F_{Z,W,f}$, Fig. 3.119) to the bend centre by the trail r_T. The marked forces and paths are of different sizes on the outside (o) and inside (i) of the bend:

$F_{Z,W,f,o} = F_{Z,W} + \Delta F_{Z,W,f,i}$ and $F_{Z,W,f,i} = F_{Z,W} - \Delta F_{Z,W}$
$F_{Y,W,f,o} = \mu_{Y,W} F_{Z,W,f,o}$ and $F_{Y,W,f,i} = \mu_{Y,W} F_{Z,W,f,i}$

The steering axis is shown – simplified – standing vertically.

$$M_{Z,W,Y} = F_{Y,W,f,t} \cos \sigma \; n_{\tau,t} \tag{3.34}$$

Section 2.8.4 describes the coefficient of friction $\mu_{Y,W}$.

If the axis of rotation is vertical in the side view, Case 2 ($\tau = 0$, Fig. 3.116), the formula remain unchanged; only $n_\tau + r_{\tau,T}$ appears for $n_{\tau,t}$, i.e. the path around which the wheel centre is located behind the steering axis, together with the tyre caster.

If the vehicle has negative caster (Case 3, Fig. 3.117), the lateral force could cause the wheels to steer into the bend if this were not counterbalanced by caster $r_{\tau,T}$ and by the tractive force $F_{X,W,a}$ which also has a righting effect on front-wheel drive vehicles (Figs 3.119 and 3.129). In the case of negative caster, only $r_{\tau,T} - r_{\tau,k}$ needs to be inserted into the equation.

The increased rolling resistance force $F_{R,co,f}$ during cornering on the outside and the inside is

$F_{R,co,o} = k_{R,co} F_{Z,W,f,o}$ and
$F_{R,co,i} = k_{R,co} F_{Z,W,f,i}$

and seen together this must be resolved into two components:

$F_{R,co} \cos \alpha_f$ and
$F_{R,co} \sin \alpha_f$

In the case of positive caster, Case 1, the last component has a righting effect on both wheels (Fig. 3.128 and Equations 3.32 and 3.32a):

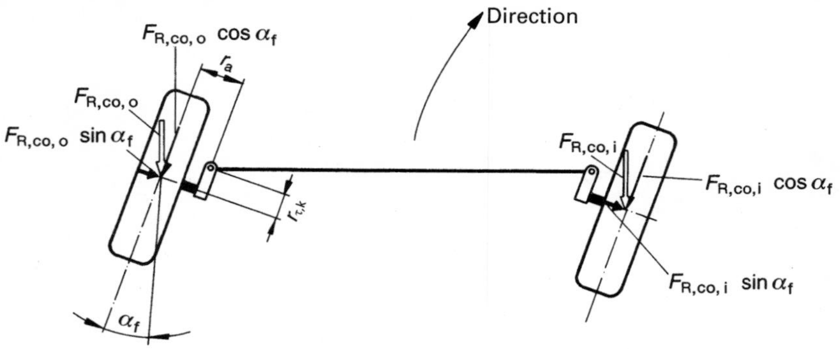

Fig. 3.128 The rolling resistance forces $F_{R,co,o}$ and $F_{R,co,i}$, which have increased on a bend due to the tyre slip, must be resolved by the angle α_f; the component $F_{R,co}$ cos α_f then appears in the wheel centre with the lever r_a. The greater α_f, and the longer the caster trail $r_{\tau,k}$, the stronger the self-righting due to $F_{R,co}$ sin α_f.
For reasons of clarity, the tyre caster $r_{\tau,T}$ and lateral offset r_T (Fig. 3.119) have been ignored here and the steering axis has been shown vertically.

$$M_{Z,T,X,1} = k_{R,co} F_{Z,V,f} \sin\sigma \sin\alpha_f \, n_{\tau,t} \; [\text{or } (n_\tau + r_{\tau,T})] \tag{3.35}$$

The rolling resistance force is (in accordance with the transfer of wheel forces $\Delta F_{Z,W,f}$ during cornering, Fig. 1.6) greater on the outside of the bend than on the inside, so that the difference in vertical force $\Delta F_{Z,W,f}$, together with cos α_f, can be a factor:

$$F_{Z,W,o} - F_{Z,W,i} = 2\,\Delta F_{Z,W,f} \tag{3.35a}$$

$$M_{Z,T,X,2} = k_{R,co}\, 2\,\Delta F_{Z,W,f} \cos\alpha_f \cos\tau\, (r_a - 2\, r_T) \tag{3.35b}$$

This deals with the longitudinal forces that act at the wheel centres – shifted to the middle of the bend (Figs 3.111 and 3.119); it is possible to calculate the coefficient of rolling resistance $k_{R,co}$ required using Equations 2.4a to 2.4c.

The previous figures also refer to the tractive forces $F_{X,W,a}$ on front-wheel drive vehicles (related to one wheel). These must be resolved first in the wheel centre by the angle τ (Fig. 3.113) and considered offset by r_T, in the rolling direction of the wheel. Provided that the differential divider the moment equally to each front wheel when the wheels are turned as the wheel load changes $\pm\Delta F_{Z,W}$, the tractive force component $F_{X,W,A} = 2F_{X,W,a}$ (relating to the entire axle) would cause the following moments (Fig. 3.129):

$$\begin{aligned} M_{Z,W,A} &= -F_{X,W,a} \cos\tau(r_a - r_T) + F_{X,W,a} \cos\tau\,(r_a - r_T) \\ &= F_{X,W,a} \cos\tau\, 2r_T \\ &= F_{X,W,A} \cos\tau\, r_T \end{aligned} \tag{3.36}$$

The size of the force $F_{X,W,A}$ depends either on the coefficient of friction $\mu_{X,W}$ ($F_{X,W,A} = \mu_{X,W}\, F_{Z,V,f}$, see Equation 2.5) or on the drive torque (see Equation 6.36);

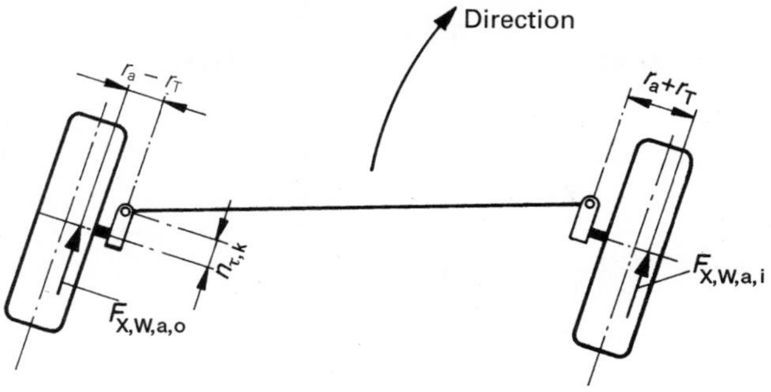

Fig. 3.129 At $r_a + r_T$, the tractive force $F_{X,W,a,i}$ on the inside of the bend has a larger moment-arm than that on the outside of the bend $F_{X,W,a,o}$ at $r_a - r_T$; the steering axis is shown vertically, for simplification.

the lateral offset length r_T is contained in the caption to Fig. 3.119 (see also Section 2.10.3.4).

3.10.4 Kingpin inclination, camber and caster alteration as a consequence of steering

Due to the spatial movement of the steering axis (onto which the vertical force $F_{Z,W}$ must be shifted, see Figs 3.105 and 3.107) the righting moment for one wheel can only be calculated precisely if the kingpin inclination, when the wheels are turned, is taken into account. If, in the zero position, the steering axis is inclined exclusively by the angle σ_0, i.e. there is either no caster or this has been achieved by shifting the wheel centre (Fig. 3.116), then it is easy to determine the kingpin inclination angle $\sigma_{o \text{ or } i}$ which becomes smaller in both input directions:

$$\text{outside of the bend: } \tan \sigma_o = \tan \sigma_0 \cos \delta_o \tag{3.37}$$

$$\text{inside of the bend: } \tan \sigma_i = \tan \sigma_0 \cos \delta_i \tag{3.37a}$$

As shown in Fig. 3.103, kingpin inclination and camber are directly related, i.e. if either one changes the other one must change too. This means that the camber values $\varepsilon_{W,o \text{ or } i}$, adopted by the wheel on the outside and the one on the inside of the bend when the wheels are turned, can be determined simultaneously:

$$\varepsilon_{W,o} = (\sigma_0 + \varepsilon_{W,0}) - \sigma_o \quad \text{and} \quad \varepsilon_{W,i} = (\sigma_0 + \varepsilon_{W,0}) - \sigma_i \tag{3.38}$$

σ_0 and $\varepsilon_{W,0}$ are the angles prevailing when the wheels point straight ahead in the design load or the particular load position (this applies equally to τ_0). If the steering axis is also inclined by the positive caster angle τ_0, the two auxiliary angles σ' and δ' must first be calculated using σ_0 and τ_0:

$$\tan \delta' = \frac{\tan \tau_0}{\tan \sigma_0} \text{ and } \tan \sigma' = \frac{\tan \tau_0}{\sin \delta'} \quad (3.39)$$

They can then be used to determine directly the angles $\delta_{o\,or\,i}$ on the wheels on the outside and inside of the bend:

$$\text{outside of the bend: } \tan \sigma_o = \tan \sigma' \cos (\delta' - \delta_o) \quad (3.39a)$$

$$\text{inside of the bend: } \tan \sigma_i = \tan \sigma' \cos (\delta' + \delta_i) \quad (3.39b)$$

Equation 3.38 again applies to the camber $\varepsilon_{W,o\,or\,i}$. Using a passenger car with the following axle settings as an example:

$$\varepsilon_{W,0} = 15', \quad \sigma_0 = 9°53' \quad \text{and} \quad \tau_0 = 10°4'$$

This gives $\delta' = 45.54°$ and $\sigma' = 13.97°$ and, where δ_o and $\delta_i = 20°$, the values are as follows:

$$\sigma_o = 12°39', \quad \varepsilon_{W,o} = -2°31', \quad \sigma_i = 5°53' \quad \text{and} \quad \varepsilon_{W,i} = +4°15'$$

The wheel on the outside of the bend therefore goes into negative camber on this vehicle, while the inner one goes into positive camber. This is not the same as the case of a front-wheel drive vehicle with the following axle settings (Fig. 3.130):

$$\varepsilon_{W,0} = +40', \quad \sigma_0 = 12°25' \quad \text{and} \quad \tau_0 = +36'$$

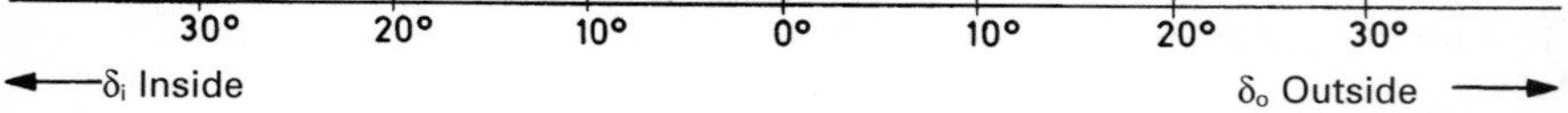

Fig. 3.130 Camber alteration measured and calculated as a function of the steering angle on a front-wheel drive vehicle. Due to the large kingpin inclination $\sigma_0 = 12°25'$, the wheels on both the inside and outside of the bend go into positive camber.

The values measured are higher than those calculated because the camber of the vehicle tested was in the plus tolerance. The calculation was made on the basis of the manufacturer's information ($\varepsilon_W = 20'$ and $\tau = 0°$) and this accounts for the slightly different inclination of the curves.

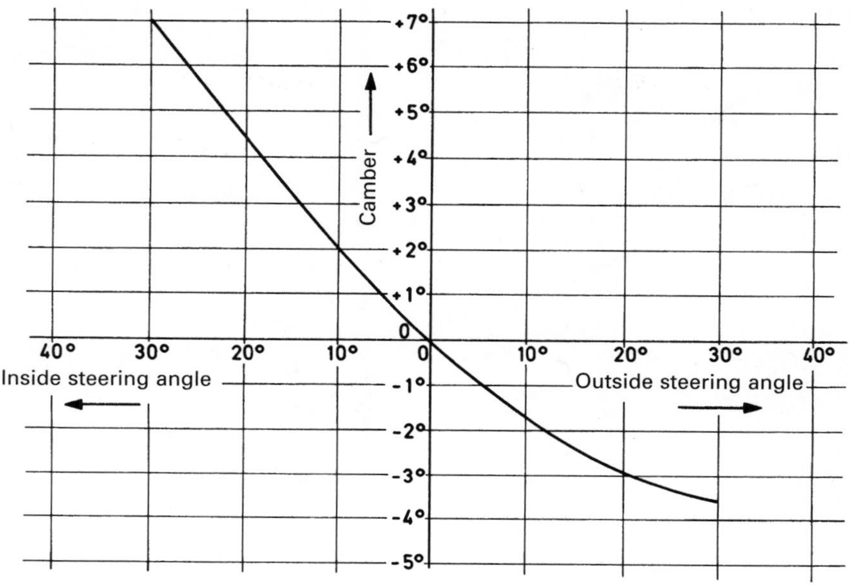

Fig. 3.131 Camber alteration measured on a Mercedes as a function of the steering angle. The axle settings in the design position were $\varepsilon_W = 0°$, $\sigma = 14°40'$, $\tau = 10°10'$, $r_\sigma = -14$ mm and the negative caster offset $n_\tau = -28$ mm.

Owing to the front-wheel drive and the manual non-power-assisted steering, the vehicle has a minor caster and therefore positive camber on the turned front wheel on the outside of the bend.

Mercedes Benz designed their passenger car with a non-driven strut damper front suspension to have negative caster offset $-n_\tau$ (Fig. 3.118) and a large angle τ. Figure 3.131 shows the success of this design: the wheel on the outside of the bend goes into severe negative camber and the one on the inside of the bend goes favourably into positive camber.

For demonstration purposes, the camber alteration based on $\varepsilon_{W,0} = 0°$, $\sigma_0 = 6°$ and various caster angles were calculated (Fig. 3.132); a larger kingpin inclination would only have resulted in a higher curvature for all curves. It can be clearly seen how an increase in the angle τ_0 improves the lateral grip properties of the entire front axle as the wheel on the outside of the bend goes into more negative camber and the one on the inside into positive camber.

Trail and caster angle alter in exactly the same way as kingpin inclination and camber alteration when the wheels are turned. In a passenger vehicle with rear-wheel drive for example, $r_{\tau,k}$ is 6.5 mm when the wheels are pointing straight ahead. The trail increases on the inside of the bend when the wheel is turned, with a decrease on the outer wheel (Fig. 3.133). Negative caster occurs as $\delta_a \approx 8°$, and at $\delta_o = 30°$ it amounts to $r_{\tau,k} \approx -30$ mm, which would lead to the outer wheel turning into the bend under lateral force if there were no caster.

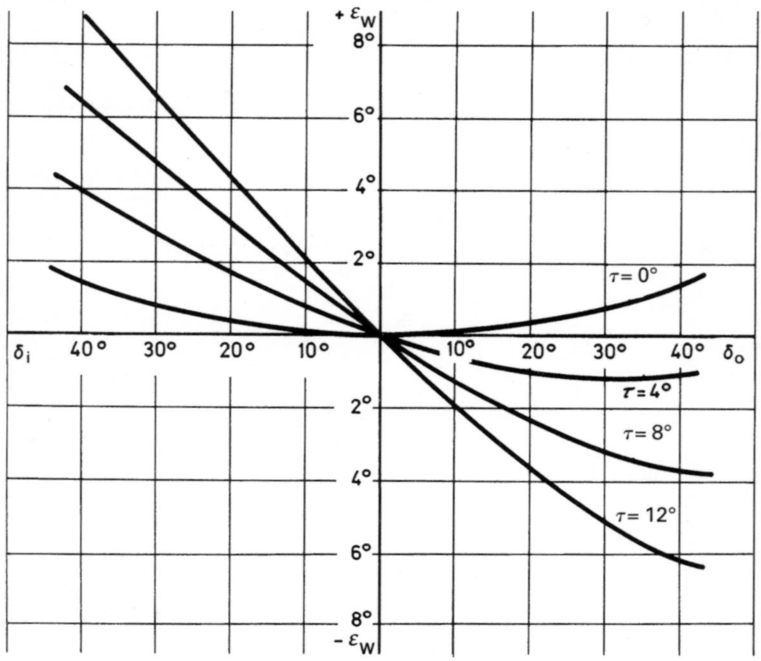

Fig. 3.132 Camber angles $\varepsilon_{W,o}$ and $\varepsilon_{W,i}$, as a function of the steering angle δ_o (outside of bend) and δ_i (inside of bend). The influence of the various caster angles τ can be clearly seen. Values given: $\sigma = 6°$ and $\varepsilon_W = 0°$.

The caster angle alteration can be calculated just as simply as that of the king-pin inclination:

$$\text{outside of the bend: } \tan \tau_o = \tan \sigma' \sin (\delta' - \delta_o) \tag{3.40}$$

$$\text{inside of the bend: } \tan \tau_i = \tan \sigma' \sin (\delta' + \delta_i) \tag{3.40a}$$

If the vehicle has $\tau_0 \approx 0°$, only the kingpin inclination angle σ_0 plays a role thereby simplifying the formulas as follows:

$$\text{outside of the bend: } \tan \tau_o = -\tan \sigma_0 \sin \delta_o \tag{3.41}$$

$$\text{inside of the bend: } \tan \tau_i = +\tan \sigma_0 \sin \delta_i \tag{3.41a}$$

The equation shows that negative caster can occur on the wheel on the outside of the bend even with a small steering input; this is clearly demonstrated in Figs 3.134 to 3.136, which show a comparison of curves calculated with various τ_0 and σ_0 angles and also one measured curve.

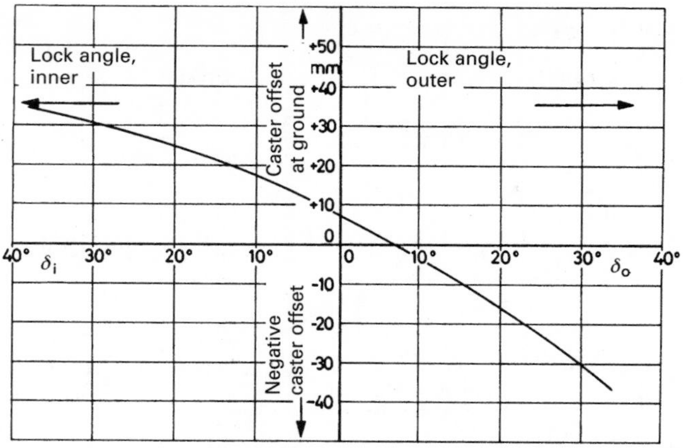

Fig. 3.133 The length of a caster trail $r_{\tau,k}$ at the ground alters depending on the steering input, shown using the example of a standard passenger car and the axle settings:

$\varepsilon_W = +20'$, $\sigma = 11°5'$, $\tau = 8°20'$
$n_\tau = -32.5$ mm and $r_\sigma = +56$ mm

The large kingpin inclination angle results in the deviation of the curve from the horizontal.

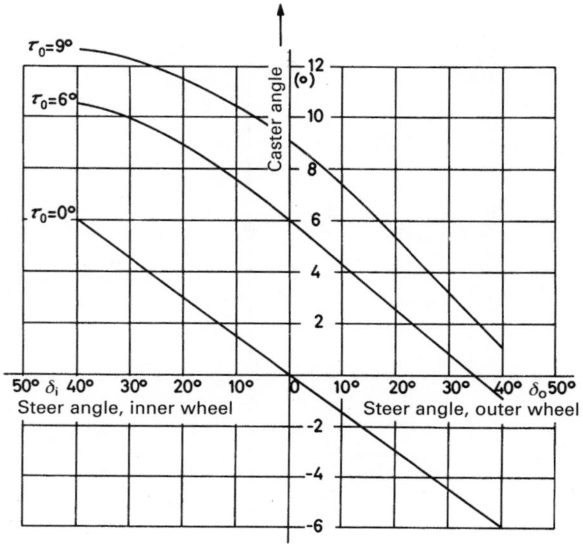

Fig. 3.134 Caster angles calculated as a function of $\sigma = 9°$ and $\tau_0 = 0°$, 6° and 9°. The smaller the τ_0 value in the normal position, the faster negative caster occurs on the wheel on the outside of the bend.

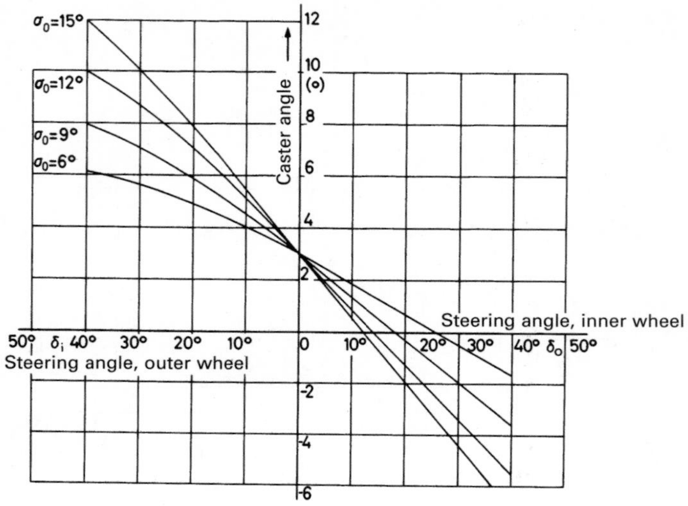

Fig. 3.135 Caster angles calculated as a function of the steering input with τ = 3° and σ_0 = 6°, 9°, 12° and 15°. The larger the kingpin inclination, the sooner the wheel on the outside of the bend goes into negative caster ($-\tau$).

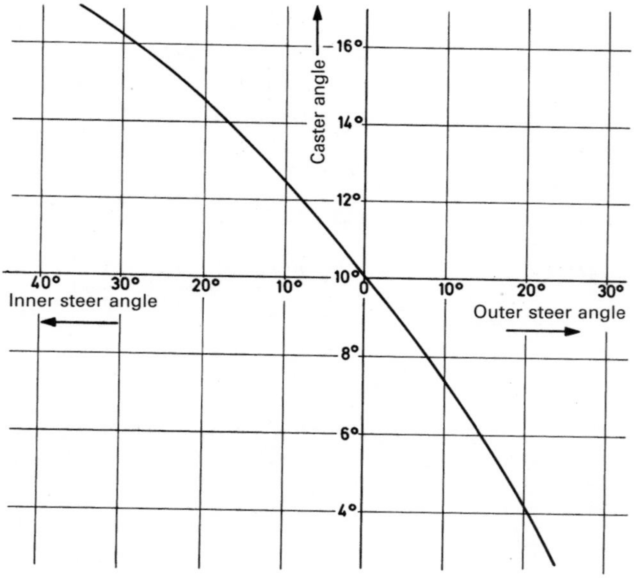

Fig. 3.136 Caster alteration measured as a function of the steering angle on the wheels of a Mercedes. The kingpin inclination angle of σ = 14°40′ is the determining factor for the severe curvature of the curve and the caster angle τ = 10°10′ for the angle position.

3.10.5 Kinematic caster alteration on front-wheel travel

If there are two people seated in the front of a vehicle, the body moves into bump travel almost parallel and the caster hardly changes. However, if two or three people are seated in the back, or the boot at the back of the vehicle is loaded, it is a very different story. The rear axle springing complies more strongly than that of the front axle and the body's position, which was almost parallel to the ground, alters by $\Delta\theta_{Bo,t} = 1°$ to $2\frac{1}{2}°$ (Fig. 3.137). The caster angle increases by the same amount $\Delta\tau$ – something which designers should bear in mind when specifying axle settings.

The increase in the caster angle under load is likely to be the main reason why the steering is heavier on a fully laden vehicle even though this sometimes causes the front axle load to be reduced. An alteration in caster has its disadvantages, as this in turn causes the self-righting torque to alter, but it is unavoidable if the brake dive on the front axle is to be kept within limits by means of vehicle pitch poles (see Section 6.3.2).

On double wishbone suspensions, the axes of rotation 1 and 2 of the two suspension control arms are usually parallel to one another (Fig. 3.138); in the standard configuration of the McPherson strut and strut damper there is a right angle between the centre line of the damping part and suspension control arm (Fig. 3.139). In such cases – regardless of the position of the compressed or rebounded wheel – the caster is retained. This is not the case where there are different angles between the suspension control arm axes of rotation (Fig. 3.140), or the damper centre and suspension control arm (Fig. 3.141).

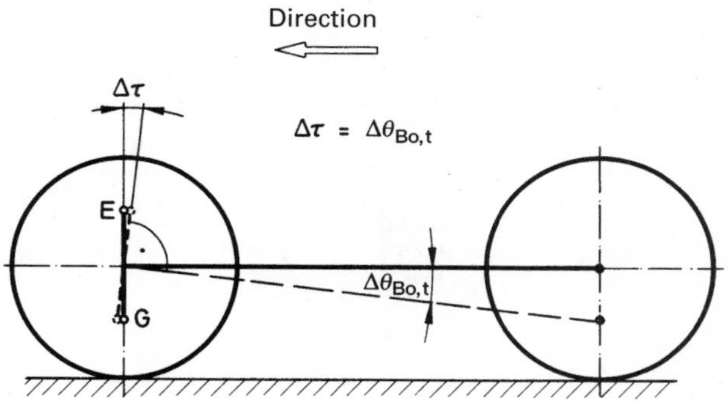

Fig. 3.137 When loaded, the body tail sinks further than the front; the caster angle τ increases by its angle alteration $\Delta\theta_{Bo,t}$ (see also Fig. 6.15).

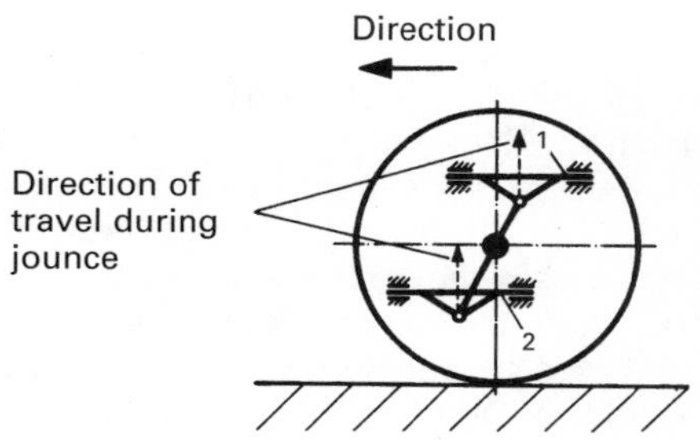

Fig. 3.138 On most double wishbone suspensions, the axes of rotation 1 and 2 are parallel to one another; in such cases, caster does not change when the wheels compress and rebound.

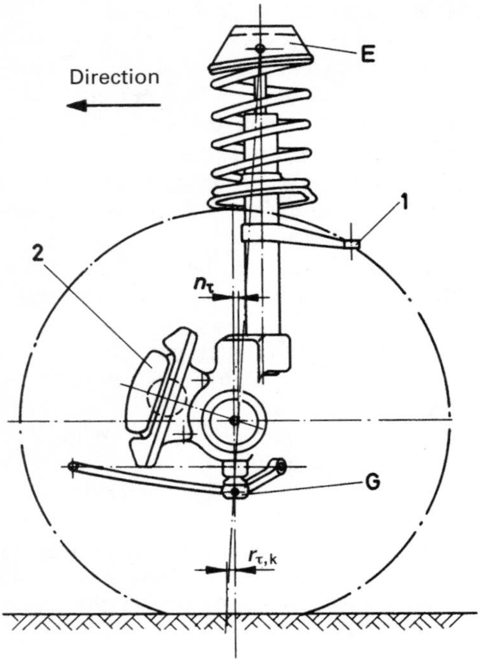

Fig. 3.139 If the line EG and control arm axis form a right angle on the McPherson strut and strut damper there is no caster alteration. Point G moves vertical to the suspension control arm axis when the wheels bottom out, i.e. parallel to the line EG. The axle shown has a negative caster offset $-n_\tau$ and the lower link G shifted forwards. The line EG gives a small caster angle and the trail $r_{\tau,k}$.

The steering arm 1 is positioned high up and inclined backwards; the disc brake calliper 2 is at the front, giving the disadvantage of a higher wheel bearing load during braking. (See Section 7.4 in Ref. [6].)

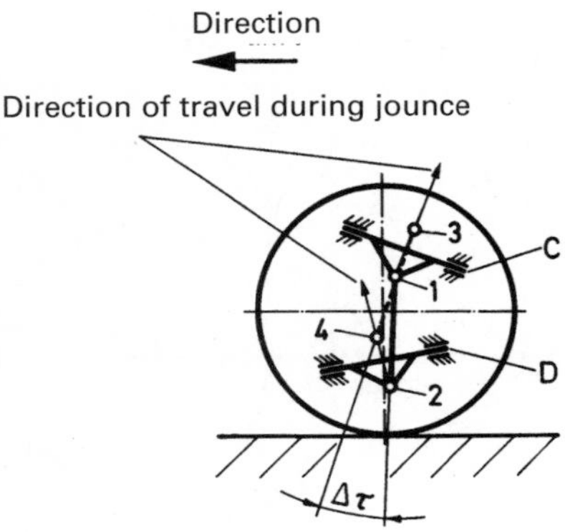

Fig. 3.140 To create a virtual centre of rotation pole on the front axle (see Fig. 3.155) on double wishbone suspensions, the axes of rotation C and D must be inclined against one another. The disadvantage of this is that when the wheels compress, point 1 moves to point 3 and point 2 to 4, increasing the caster angle by $\Delta\tau$, equivalent to twisting the steering knuckle by this angle.

When the front wheel compresses, the upper ball joint 1 of the steering knuckle moves backwards and the lower one forwards, resulting in an increase in caster. Rebounding has the opposite effect – the caster (if in the normal position) decreases and may even become negative. In the case of McPherson struts and strut dampers, point 2 moves to 4, parallel to the axes of rotation, and compresses the damping element, which is fixed in point 1. This shortens and there is rotation by the angle $\Delta\tau$.

As can be seen in Figs 3.155 and 3.156, a virtual centre of rotation O_f, which lies behind the front axle, would be more readily achieved with a double wishbone

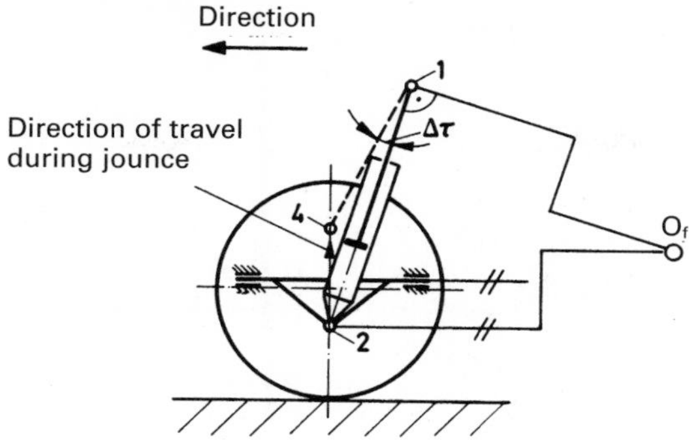

Fig. 3.141 When the McPherson strut or strut damper compresses, point 2 moves to 4 and the caster angle increases by $\Delta\tau$. The intersection of a parallel to the suspension control arm axis of rotation (drawn through point 2) and a vertical on the damper centre line in point 1 gives the vehicle the pitch pole O_f. The steering knuckle fixed to the damping part also rotates by this angle.

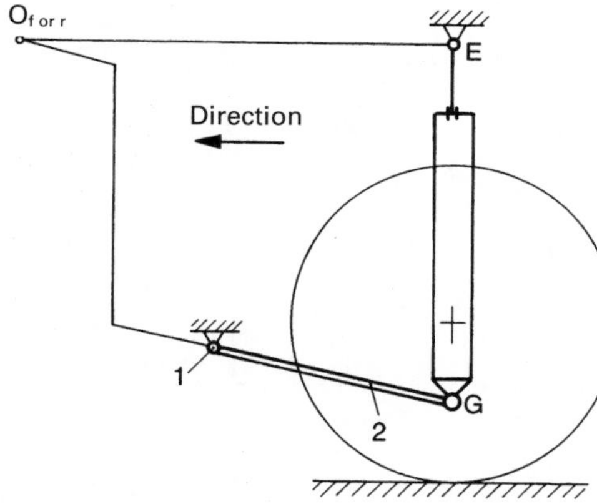

Fig. 3.142 So as not to reduce the ground clearance of the front axle and the front overhang (Fig. 1.67), the back 1 of the anti-roll bar must be raised. The arms 2, which support the lower transverse links in the longitudinal direction, therefore drop backwards. The result is a vehicle virtual centre of rotation O_f in front of the axle, which causes the front end to be pulled further down during braking and adverse lift of the front end when a front-wheel drive vehicle moves off (Figs 4.1 and 3.143).

In the case of a rear wheel suspension, this position of the vehicle pitch pole O_r would be favourable.

suspension than with McPherson struts and strut dampers (Fig. 3.141). If the anti-roll bar is located in front of the axle and used for the longitudinal wheel control, its rear-end must be raised to provide enough ground clearance (Fig. 3.142). This will cause the centre of rotation to lie in front of the axle and will also draw the front end down when the brakes are applied. Figure 3.143 shows the kinematic caster

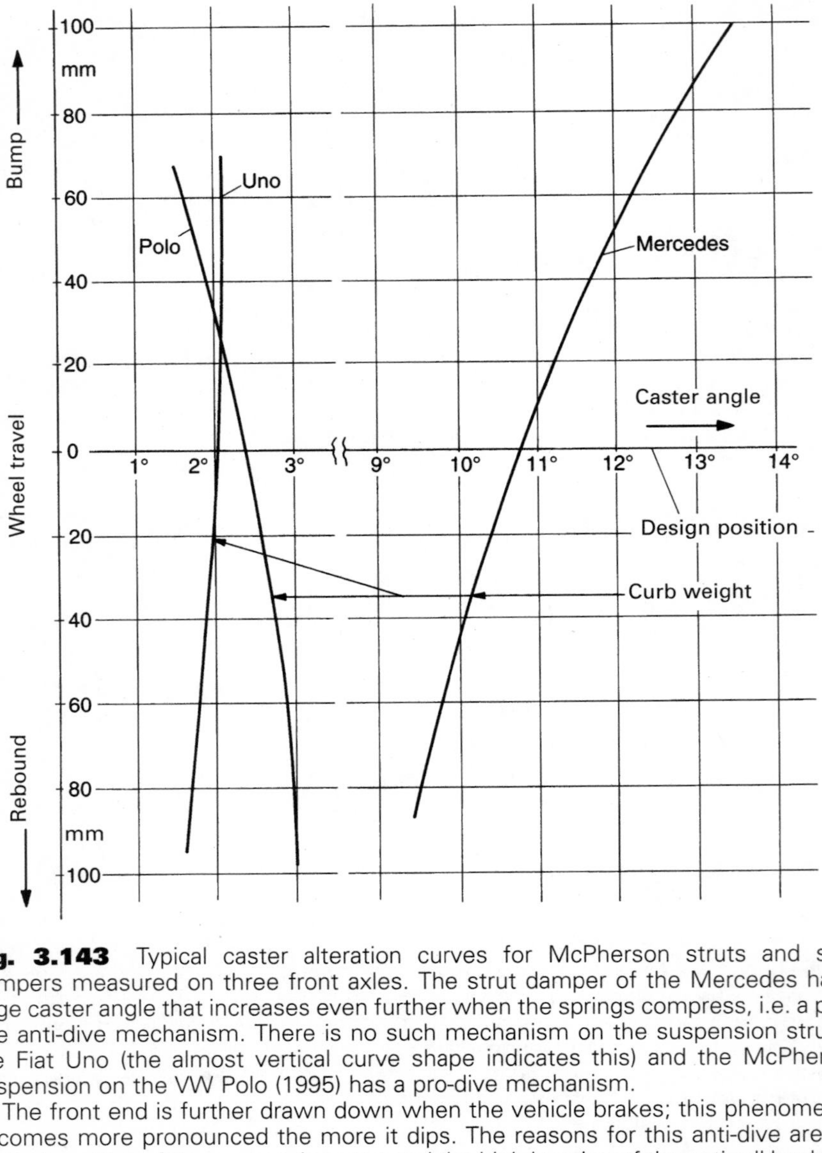

Fig. 3.143 Typical caster alteration curves for McPherson struts and strut dampers measured on three front axles. The strut damper of the Mercedes has a large caster angle that increases even further when the springs compress, i.e. a positive anti-dive mechanism. There is no such mechanism on the suspension strut of the Fiat Uno (the almost vertical curve shape indicates this) and the McPherson suspension on the VW Polo (1995) has a pro-dive mechanism.

The front end is further drawn down when the vehicle brakes; this phenomenon becomes more pronounced the more it dips. The reasons for this anti-dive are the vertical position of the suspension strut and the high location of the anti-roll bar back; the virtual centre of rotation is therefore far in front of the axle. Figures 3.139, 3.143 and 4.1 give details. On the Mercedes, the lateral force lever $n_{\tau,k}$ on the compressing wheel on the outside of the bend increases; this means, therefore, that speed-dependent lateral force understeering occurs.

alteration measured on three passenger cars with spring dampers or McPherson axles. The curve shape clearly shows whether there is an 'anti-dive' or a 'pro-dive' mechanism.

From a design point of view, the alteration angle $\Delta\tau = f(s)$ can easily be determined by drawing verticals to the suspension control arm axes of rotation C and D through the centres 1 and 2 of the wheel joints, as shown in Fig. 3.140. Fixed paths must be marked off on one of the two verticals and, using a compass with the path 1–2 the corresponding point on the other determined. The angle $\Delta\tau$ of the connecting line 3–4 to the initial position 1–2 is the caster alteration. In the case of McPherson struts and strut dampers (Fig. 3.141) the upper point 1 is fixed in the wheel house, so that the distance 1–2 shortens when the spring compresses (path 1–4) and lengthens when it rebounds.

Figure 3.140 shows that when the vehicle is designed with a virtual centre of rotation, the steering knuckle rotates by the angle $\Delta\tau$ – clockwise at the front

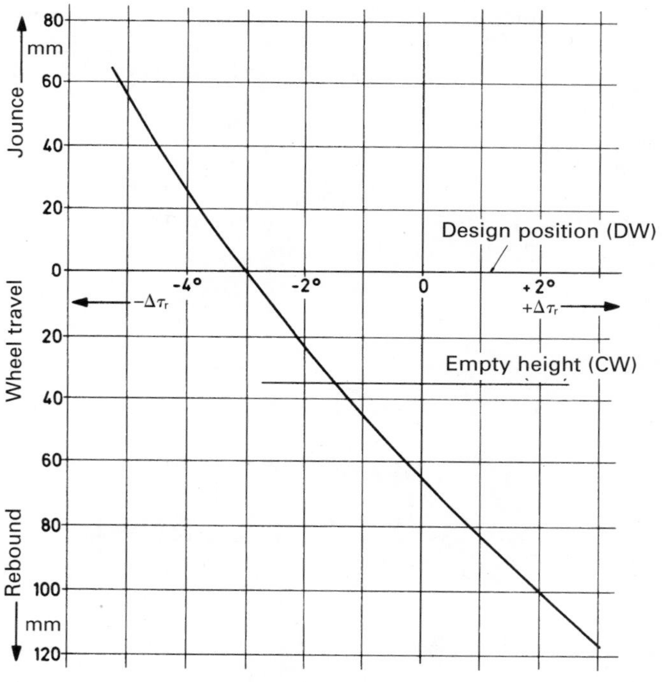

Fig. 3.144 Alteration $\Delta\tau_r$ in the theoretical negative caster angle, measured as a function of the compression and rebound travel on the rear axle of a Mercedes. The company specifies the trail as $r_{\tau,k} = -15$ mm; in the design position this would correspond to an angle of $\tau_r \sim -3°$. This increases as the springs compress, and it decreases or goes into positive caster as they rebound. The inclined position of the curve indicates high virtual centre of rotation that move further upwards when the wheels rebound and therefore progressively reduce the brake dive. Furthermore, the negative caster trail increases on the compressing outer wheel during cornering, resulting in favourable lateral force understeering, increasing with speed.

(Figs 3.141 and 3.143) and anticlockwise at the rear axle (Fig. 3.144); this is demonstrated by the shape of the curves in the above figures. This wheel travel-dependent rotation causes a changing relative speed between stator and rotor in the wheel sensors of all wheel slip control systems, which adversely affects the response speed of the ABS and ASR traction control. Comprehensive information is given in Chapter 3 of Ref. [7].

3.10.6 Wheel travel-dependent rotation of the rear steering knuckle

On the multi-link independent rear suspensions fitted in Mercedes models, five rods are used to control the steering knuckle with four of them providing lateral force reaction support (see Section 5.3.4 in Ref. [2]). The extensions of the two upper rods intersect in pole E and those of the lower rods in G. The lines connecting the two poles give the theoretical steering axis EG. A negative caster was set (Figs 3.117 and 3.145) to obtain lateral force understeering (Fig. 3.73).

The tyre caster $r_{\tau,T}$, which reduces this, must also be considered. The lateral force lever is then (Fig. 3.120, see also Equation 3.32):

$$n_{\tau,t} = r_{\tau,T} \cos \tau_r - n_{\tau,k} \tag{3.41b}$$

If anti-dive behaviour is desired, the centre of rotation O_r must lie in front of the rear axle, as shown in Figs 3.142 and 3.153.

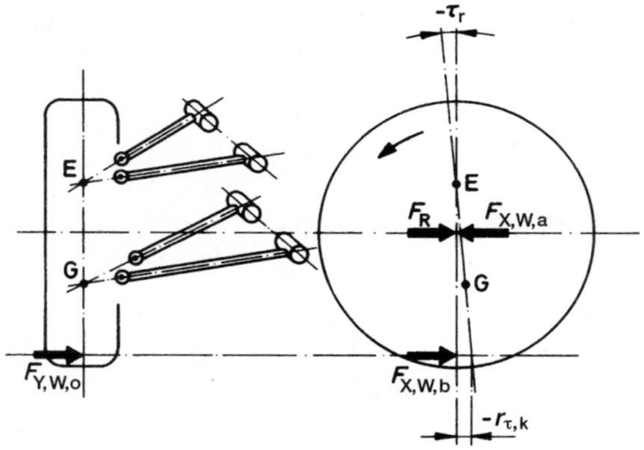

Fig. 3.145 If, on a multi-link rear suspension, there are four bars supporting the lateral forces, when viewed from the rear, their extensions meet in the points E and G. When they are connected in the side view, the result can be the theoretically negative caster angle $-\tau_r$ and the caster trail $-r_{\tau,k}$ on the ground.

Where the brake is on the outside, the braking force $F_{X,W,b}$ should be regarded as acting at the centre of tyre contact. The rolling resistance force F_R and the tractive force $F_{X,W,a}$ have to be shifted into the wheel centre.

A parallel development is the multi-link rear axle, which is becoming more and more popular. This contains a trailing link (that forms one piece with the steering knuckle), with a pivot in front of the axle centre which, simultaneously, represents the centre of rotation O_r (Figs 1.1, 1.18, 1.62 and 1.77; see also Section 5.3 in Ref. [2]). The kinematic movement of the wheel carrier corresponds to that of the longitudinal link suspension (Figs 3.159 and 6.17).

3.10.7 Resolution of the vertical wheel force on caster

If the steering axis EG on a double wishbone suspension is angled by the caster angle τ, the lower ball joint lies in front of the wheel centre and the upper one behind it. If the spring is supported on the lower suspension control arm, its force $F_{G,z}$ may be the same size as the vertical wheel force less the weight of the axle side (Fig. 3.146, Equation 5.3), but the moment $M_Z = F_{G,z}\,(f - e)$ occurs, causing the forces $F_{E,x}$ and $F_{G,x}$. The compliance present causes the caster angle to reduce. If the spring were on the top, it would increase.

Where there is caster (Case 1), the vertical force component $F_{Z,W} \cos \sigma$, shown in Fig. 3.105, would be further resolved by the angle τ, i.e. in $F_{Z,W} \cos \sigma \cos \tau$ and $F_{Z,W} \cos \sigma \sin \tau$ (Fig. 3.147). The last component tensions the wheels via the lever q at the front (Fig. 3.148). If the caster angles τ on the left and right wheels are different, the same will apply to the tensioning forces, i.e. the vehicle could deviate from the direction of travel if the steering wheel were let go, and would pull to one side when held (Fig. 3.149 and Equation 3.41c). A 2° difference means that there is a 30–40 N higher tie rod force on the side with the greater angle τ.

If the caster is achieved by relocating the wheel centre to the back (Case 2), the component $F_{Z,W} \sin \sigma$ pushes the wheels together at the front via the force lever n_τ (Figs 3.150 and 3.151), i.e. even here the parallel position of the left and right steering axes to one another plays a role.

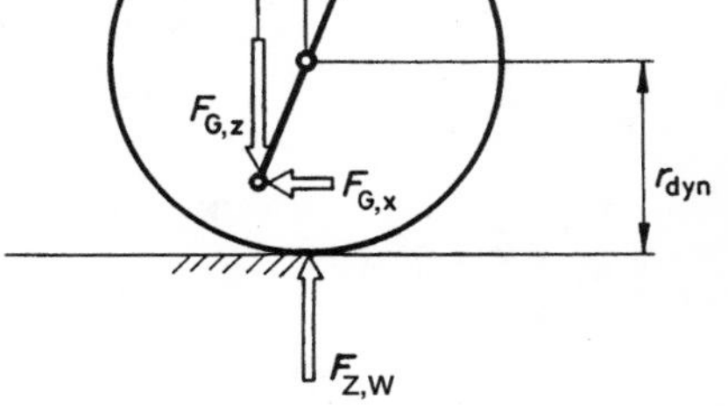

Fig. 3.146 If the spring is supported on the lower suspension control arm and if the front axle has caster, the supporting ball joint will be in front of the wheel centre. Forces $F_{Z,W}$ and $F_{G,z}$ form a moment, which generates the reaction forces $-F_{E,x}$ and $+F_{G,x}$ in the direction of the suspension control arm axes of rotation. In the example these are assumed to be parallel to the ground.

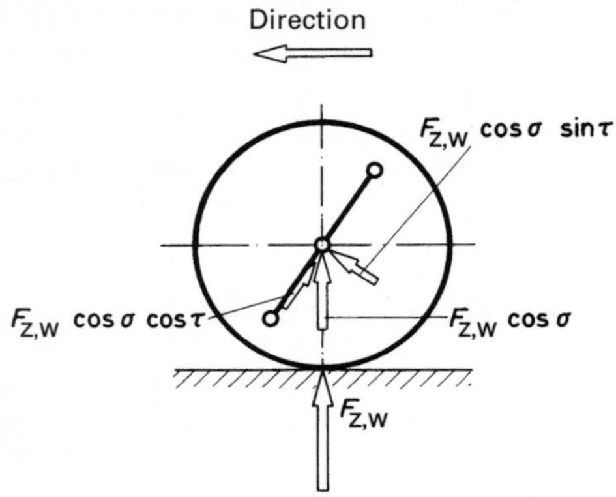

Fig. 3.147 If the steering axis is at the caster angle τ in the side view, the vertical force component $F_{Z,W} \cos \sigma$ calculated in the rear view in Fig. 3.105 must be further resolved.

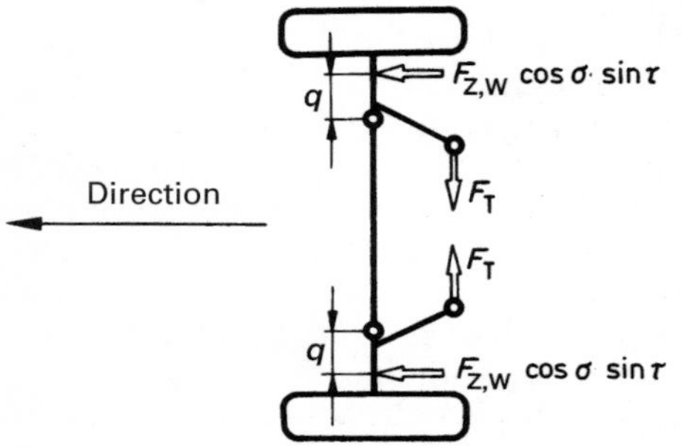

Fig. 3.148 The forces $F_{Z,W} \cos \sigma \sin \tau$ push the front wheels together at the front via the levers q (i.e. into toe-in) both when the vehicle is in a stationary position and moving in a straight line, and generate the forces F_T in the tie rods. The caster angles τ left and right may therefore only deviate slightly from one another (see Equation 3.42a).

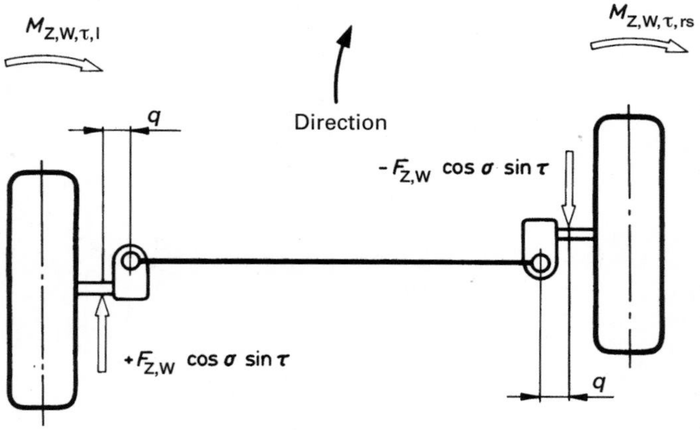

Fig. 3.149 Caster on the left and negative caster on the right front wheel (or caster angles τ of different sizes) cause the vehicle to pull to the right when travelling in a straight line. This is caused by opposed moments:

$$M_{Z,W,\tau} = \pm F_{Z,W} \cos \sigma \sin \tau q \tag{3.41c}$$

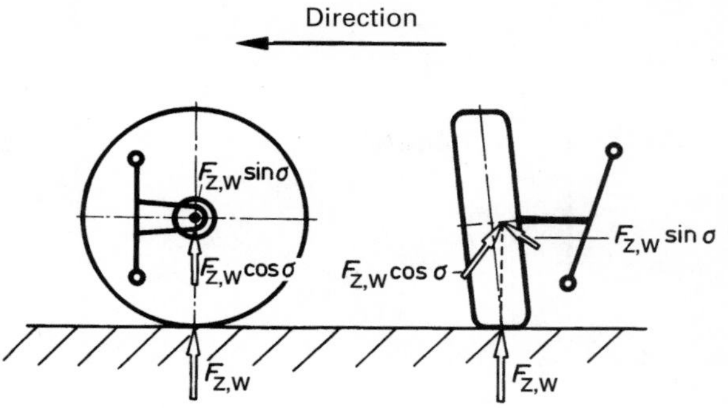

Fig. 3.150 With caster (Case 2, Fig. 3.116) achieved by setting the wheel centre back, the vertical force component $F_{Z,W}\sin\sigma$ comes to be located behind the steering axis.

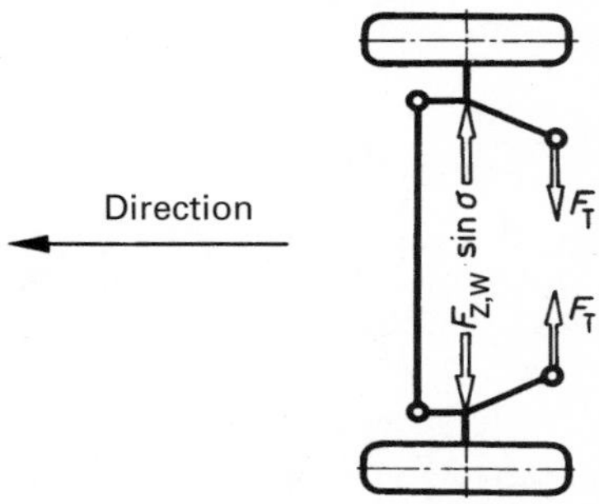

Fig. 3.151 Left and right vertical force components $F_{Z,W}\sin\sigma$ push the front wheels into toe-in when the vehicle is stationary and when it is moving in a straight line, and put the tie rods under stress (forces F_T).

The camber angle ε_W (and therefore also the kingpin inclination angle σ) should be largely the same left and right (see Equations 3.4b and Fig. 3.103).

In addition, equal kingpin inclination angles are required on both wheels, and because these are generally directly related to the camber (see Section 3.9), only a small camber deviation between left and right front wheels is permissible (see Equation 3.4c).

Where the σ angles are different, the length of the vertical force lever $q = (r_\sigma + r_{dyn}\tan\sigma)\cos\sigma$ (Equations 3.46 and 3.21a), which appears in all formular, changes and, with caster as in Case 2 above, the vertical force $F_{Z,W}\sin\sigma$ is no longer the same on the right and the left. Both instances cause the steering wheel to pull to one side.

The negative offset n_τ shown in Fig. 3.118 – together with the angle τ – requires a more in-depth look at the correlations (Fig. 3.152). The vertical force $F_{Z,W}$ on the wheel axes resolved in the direction of the kingpin inclination, gives $F_{Z,W}\cos\sigma$ and $F_{Z,W}\sin\sigma$. The first component must be further divided up in the side view into $F_{Z,W}\cos\sigma\sin\tau$ and $F_{Z,W}\cos\sigma\cos\tau$. As can be seen in the top view, when the vehicle is moving in a straight line, there are two opposing moments on each wheel (which can cancel each other out):

$$M_{Z,W,\tau,t} = F_{Z,W}\,(\cos\sigma\sin\tau\, q - \sin\sigma\cos\tau\, n_\tau) \tag{3.42}$$

Further details are given in Section 7.2 of Ref. [3].

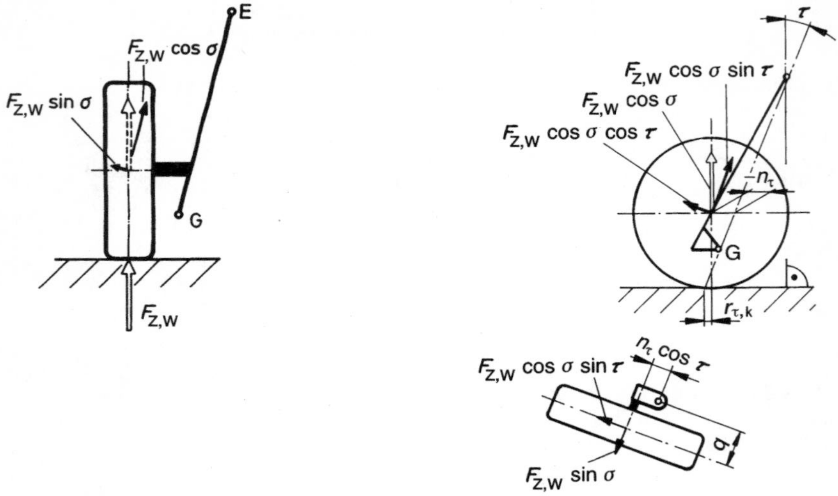

Fig. 3.152 Force ratios on front axles with negative caster offset $-n_\tau$. The opposed moments $F_{Z,W} \sin \sigma n_\tau \cos \tau$ and $F_{Z,W} \cos \sigma \sin \tau q$ an cancel one another out.

3.10.8 Settings and tolerances

The caster value of the empty vehicle should appear on the drawing and in workshop manuals. Optical measurement is also carried out in this load condition, as specified in DIN 70 020.

Where there is no caster offset, in order to ensure favourable steering self-centring, passenger cars of a standard design have caster angles of around 4° to 8°. However, in the case of a designed offset $-n_\tau$, the values can rise to $\tau = 8°$ to 11°. The type of steering system is also a factor here. If it is power assisted (see Sections 4.2.5 and 4.3.3), the steering moment must also right the parts in the hydraulics. In such cases a greater caster angle is preferable. If no power steering is available, lower angles have generally to be designed to limit the steering effort, especially during parking manoeuvres.

Front-wheel drive vehicles are set to $\tau = 1°$ to 4°. The righting moment, which is strengthened by the tractive forces $M_{Z,T,Y}$ (Fig. 3.119), means that caster values are not absolutely necessary.

In addition to the absolute value, a tolerance is required, which is usually around ±30′ but can be as much as ±1°30′ to make manufacturing more cost-effective. The additional requirement (as in the case of camber, see Equation 3.4c) that there should be no greater difference than 30′ between left and right wheels is necessary to prevent the vehicle pulling to one side (Fig. 3.149). The details given on the drawing would then be:

$$\tau = 4° \pm 30' \quad \text{maximum difference between left and right } 30' \qquad (3.42a)$$

3.11 Anti-dive and anti-squat mechanisms

3.11.1 Concept description

The anti-dive mechanism reduces the amount by which the front end of the vehicle dips or the tail rises when the brakes are applied. It can – in the case of brakes which are outside in the wheels – only be achieved if there are pitch poles O_f and O_r between the axles at the front, at the rear, or on both axles (Fig. 3.153).

The anti-squat mechanism reduces the amount by which the tail drops on rear-wheel drive vehicles or the front end lifts (on front-wheel drive vehicles). It acts during acceleration and only on the driven axle. On independent wheel suspensions it is important for the pole to be higher than the wheel centre of the driven axle (as can be seen in Figs 3.156 and 3.160) or, on a rigid axle, the differential is located in the axle housing (Figs 1.22 and 1.43). For further details, see Sections 6.3.2 and 6.4.1 in Ref. 2.

The anti-dive and anti-squat angle is also a consideration here; ε and κ are entered in Fig. 3.160; the greater these can be the better is the pitch equalization.

3.11.2 Vehicle pitch axis front

Left and right suspensions are generally identical so the pivot axes determined by the momentary position of the suspension control arms are in the same position on both sides, which leads to the so-called pitch axes. If these are at infinity (i.e. for practical purposes they do not exist, Fig. 3.138) the longitudinal forces are concentrated in the wheel centre, which applies if the brake is located on the inside (on the differential). Here the brake dive can be countered by the two double wishbones being set at an angle in the same direction (Fig. 3.154).

As can be seen from the illustration, the brake force operating as $F''_{X,W,b}$, shifted from the wheel centre vertical to the steering axis (shown in Fig. 3.111 for the rolling resistance), causes the reaction forces $F_{E,x}$ and $F_{G,x}$ in the suspension

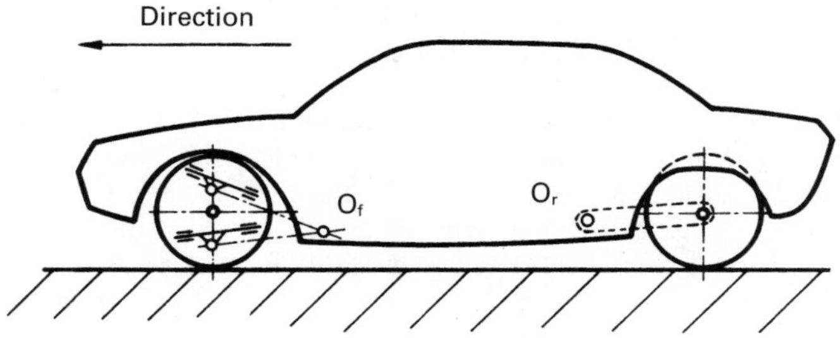

Fig. 3.153 The pitch axis is obtained by linking virtual centres of rotation front and rear. If these are available with O_f (at the front) and O_r (at the rear), the body is supported at this point in the longitudinal direction when the brakes are applied, assuming the brakes are on the outside of the wheels.

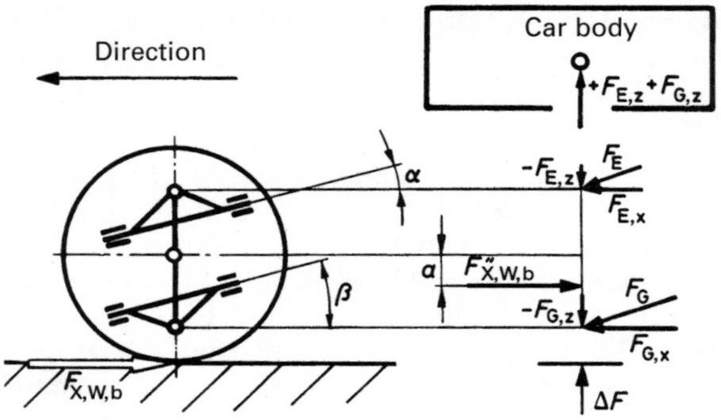

Fig. 3.154 If the front brake is on the inside on the differential, brake dive can be compensated by disposing the suspension control arms in the same direction, but at an angle. The braking force must be regarded as being under the wheel centre by $a = r_a \sin \sigma$ (see Equation 3.28b). When it compresses, the wheel moves forward. The diagonal springing angle is $\kappa = (\alpha + \beta)/2$.

control arms, which (due to the angled position) cause the vertical component $-F_{E,z} = F_{E,x} \tan \alpha$ and $-F_{G,z} = F_{G,x} \tan \beta$. The sum of forces in one effective direction must be zero, i.e. $+F_{E,z}$ and $+F_{G,z}$ work against the vehicle front-end bump travel. Two suspension control arms, which are placed at an angle in this manner, have the advantage of no caster change but the disadvantage that they move forward during jounce (in other words in the direction of the obstacle force). The Citroën GSA had this type of suspension control arm configuration and therefore an almost 100% anti-dive system (see Section 5.2.4 in Ref. [2]).

Where the brake is on the outside (as shown in Figs 3.153 and 6.16), it is also necessary for the suspension control arms to be at an angle to achieve a pitch axis and, therefore, reaction forces in the vertical direction. However, the two suspension control arms must be inclined against one another. The right-hand side of Fig. 3.155 shows the statics with the (compared with Fig. 3.154) significantly increased components $F_{G,z}$, caused by the higher force $F_{G,x} = F_{X,W,b} + F_{E,x}$ in the case of outside brakes (in the case of inside brakes $F_{G,x} = F_{X,W,b} - F_{E,x}$). All front-wheel drive vehicles built in Germany have a negative kingpin offset. The prerequisite for the counter-steering effect, which can be achieved in this way (Fig. 6.12), is a brake inside the wheel. By angling the lower suspension control arm, on a double wishbone suspension it is possible to reduce both the brake dive and the squat. The brake force $F''_{X,W,b}$ acting now on the steering axis by the amount a above ground causes the component $F_{G,z}$ supporting the body (Fig. 3.156) and – as shown on the right – the drive-off force $F''_{X,W,a}$ acting below the wheel centre causing the force $-F_{G,z}$ pulling downwards. The upper suspension control arm is horizontal. Its job can also be done by a vertically positioned McPherson strut or strut damper. In this type of suspension there is an anti-dive and anti-squat mechanism.

The pitch axis on double wishbones can be shown on a drawing using parallels to the suspension control arm axes of rotation C and D, drawn through the

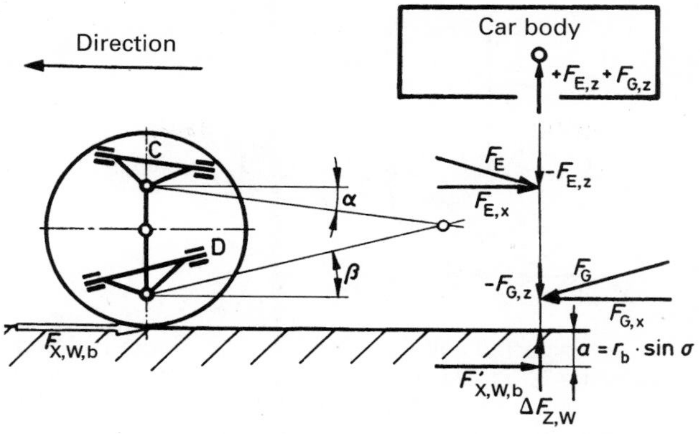

Fig. 3.155 To reduce brake dive when the brakes are on the outside, the suspension control arms must be inclined against one another. The forces $F_{E,x}$ and $F_{G,x}$ must be calculated assuming the braking force $F'_{X,W,b}$ below ground by $a = r_b \sin \sigma$ (or in the case of negative kingpin offset, above ground by the same amount, Equation 3.27). The components acting against front-end dip are then $+F_{E,z}$ and $+F_{G,z}$; in the case of $-r_\sigma$, all forces are smaller. In the case of caster $F'_{X,W,b} = F_{X,W,b} \cos \tau$ (Fig. 3.115).

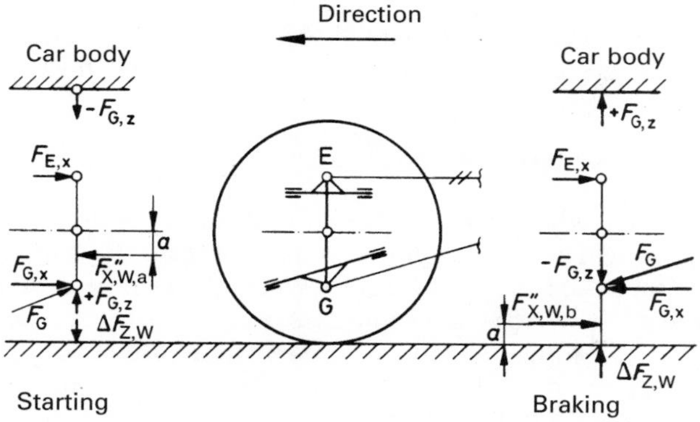

Fig. 3.156 In front-wheel drive vehicles, both the lifting of the vehicle as it moves off and front-end brake dive can be reduced by disposing the lower suspension links only at an angle, if (as is usually the case) the brake is in the wheel.

In the case of a negative kingpin offset, $F'_{X,W,b}$ acts on the steering axis by the amount a above ground (see Equation 3.27).

centres of the ball joints E and G (Fig. 3.155). The McPherson strut and strut damper require a vertical to be set up on the direction of movement of the damper in point E, the intersection of which with the suspension control arm parallel passing through G, gives the point O_f (Fig. 3.141), or an extension of the tension rods or anti-roll bars absorbing the longitudinal forces (Fig. 3.142).

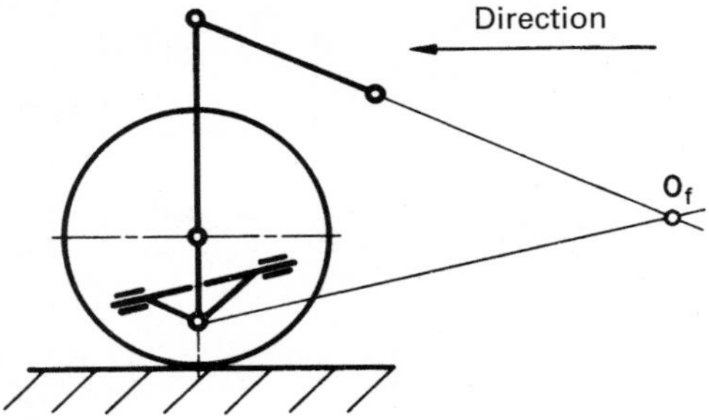

Fig. 3.157 To determine the vehicle virtual centres of rotation O_f on the longitudinal transverse axle the upper suspension control arm must be lengthened and a parallel must be drawn to the suspension control arm axis of rotation through the ball joint centre. When the front end moves towards bump, O_f moves towards the wheel, this being the equivalent of a progressive anti-dive mechanism.

On the trailing link axle, in order to get the axis O_f, the upper control arm must be extended and again a parallel to the axis of rotation must be drawn through the lower wheel joint (Fig. 3.157). When the front end of the vehicle jounces, the upper suspension control arm moves to a greater angle of inclination and the pole O moves closer to the wheel. This means a progressively increasing anti-dive system, which also applies if, in the case of a double wishbone suspension, the braking forces are absorbed upwards through a trailing or semi-trailing link (Figs 1.39 and 3.32) or by the legs of the anti-roll bar.

3.11.3 Pitch axes rear

The requirement for a reduction in the brake dive demands a pitch axis that is close to the wheel and as high as possible; however, both of these result in a severe caster change on the front axle. Here – particularly where the vehicle is fitted with ABS (see the end of Section 3.10.5) – a compromise must be struck between both criteria. On the rear axle, the picture is different. The virtual centres of rotation O_r can be positioned close in front of the axle whereby the length of the suspension control arms and the ABS performance objectives represent the limits.

Too short a suspension control arm gives unfavourably large rotation angles $\pm\kappa$ (Fig. 3.158) to achieve the desired spring travel s_1 and s_2. The wheel base change associated with the pitch axis and Δl should not affect the handling properties. As proof, we can look at the earlier Renault models on which there were different wheel bases left and right.

The trailing link suspension and the compound crank axle (Figs 1.2, 1.13 and 1.31) have the best position of the pitch pole among the wheel suspensions commonly used as rear axles. These lie in the centre of the suspension control arm axis of rotation and the force $-F_{O,z}$, which draws the rear end down during braking, is in accordance with Fig. 3.159:

$$F_{O,z} = F_{X,W,b}\, g/d \qquad (3.44)$$

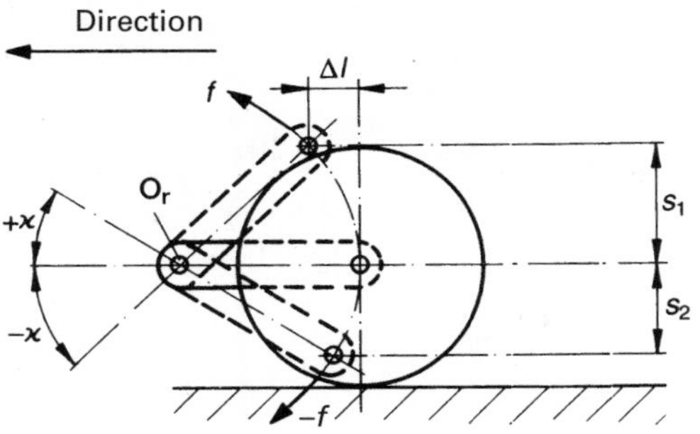

Fig. 3.158 Longitudinal links on a rear axle have the advantage of the favourable virtual centre of rotation O_r. The suspension control arm should be as short as possible, however, with the required spring travel s_1 and s_2, the angular deflections $\pm\kappa$, which may arise must not be too large. These would lead to significant diagonal springing *f*; the driver will hardly be aware of the change in wheel base associated with this.

Fig. 3.159 On trailing link and multi-link suspensions with axes of rotation parallel to the ground, the mounting point on the body is also the pitch axis. The higher the axis lies (path *g*) and the closer it is to the wheel (path *d*), the more the force $-F_{O,z}$ pulls the tail end down during braking.

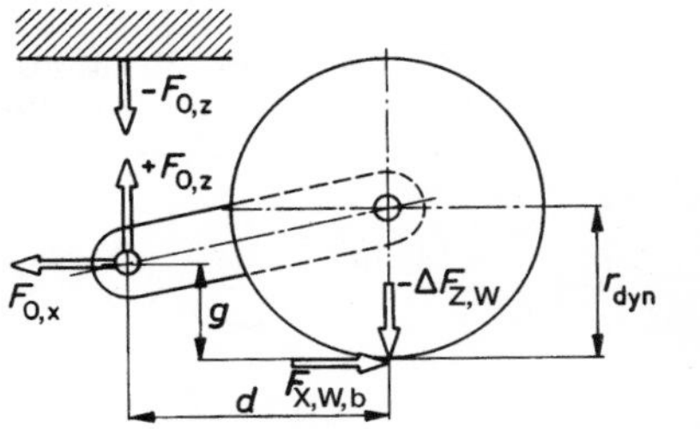

i.e. the higher the height *g* and the shorter the distance *d* can be, the stronger the effect.

Figure 3.159 also applies to all 'multi-link axles' as well as rigid axles carried on two trailing links:

- the twist-beam axle (Fig. 1.61)
- the drawbar or A-bracket axle (Fig. 1.60)
- the off-road vehicle axle shown in Fig. 1.43.
- the multi-link axles in Figs 1.1, 1.62 and 1.77.

For the semi-trailing link axle, the top view must be drawn first to ascertain the virtual centre of rotation (Figs 3.160 and 3.36). Using the angle α, the distance *d* (pitch pole O to wheel centre) is determined and then, in the rear view, the height *g* of point O is also determined. The side view then shows the actual position.

If a watt linkage or a pair of control arms per side is used to link the rigid rear axle (Fig. 3.41), the centre lines of the suspension arms must be extended and made to intersect to obtain O_r (Fig. 3.161 and Section 3.4.5). The shorter upper suspension control arms ensure that the pitch pole moves favourably towards the axle when the vehicle is loaded (in other words the tail sinks at points E and G) and therefore the anti-dive mechanism is reinforced.

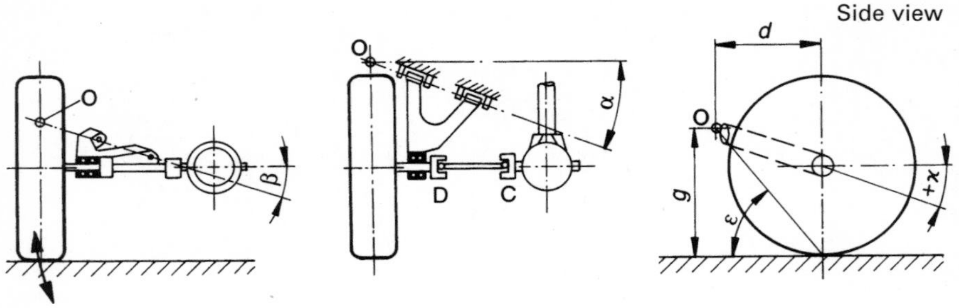

Fig. 3.160 On the semi-trailing link suspension the point at which the extension of the axis of rotation goes through the plane of the wheel centre gives the pitch axis O. The brake reaction support angle ε can be calculated from the existing paths:

$$\tan \varepsilon = g/d \tag{3.45}$$

The same applies to the anti-squat (or diagonal springing) angle κ (see Section 5.4.4 and Fig. 3.30):

$$\tan \kappa = (g - r_{dyn})/d \tag{3.46}$$

except that here the sign of the integer is important. In the case of $+\kappa$ (as shown) when the vehicle accelerates, the squatting tail is pushed upwards and in the case of $-\kappa$, it is pulled further down.

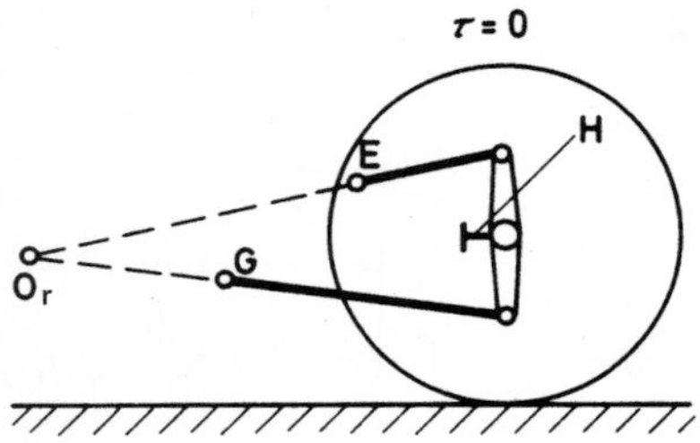

Fig. 3.161 If a rigid rear axle is controlled by two trailing link pairs, its extensions give the pitch axis O_r. When the vehicle is laden, points E and G on the body side move down, i.e. O_r moves towards the wheel in a favourable manner.

If, as can be seen in Fig. 1.43, the differential is contained in the axle housing, the moments coming from the engine and driving the wheels are vertical to one another (Fig. 1.22); because the forces on the axle housing are jointly supported, the anti-squat mechanism is, at the same time, supported at the pitch axis.

3.12 Chassis alignment

3.12.1 Devices for measuring and checking chassis alignment

The handling properties of a vehicle, both in the steady-state as well as the transient region, are determined by the kinematics – the change in position of the

Fig. 3.162 Computer-aided wheel kinematics measuring device of the Chassis/-Simulation Technology Laboratory of the University of Applied Science, Cologne. Changes in the position of the wheels in the course of body lifting and lowering movements can also be measured by means of actuators which are integrated into the test stand.

wheel during travel or roll movements of the body – and the elastokinematics – the movements of the wheel with longitudinal or lateral forces in the tyre contact area – of the wheel suspension. As even slight changes in the position of the wheels can have a big effect on handling properties and the wear and tear of tyres, particular importance is attached to the accuracy of the measuring technology used and a precise knowledge of the way in which control of the wheels depends on forces and movements.

The wheel kinematics are measured by mechanical, optical or computer-aided measuring devices (Fig. 3.162). The vehicle, loaded in accordance with the construction or other specifications of the manufacturer, is placed on four sliding plates which are made level with each other. These permit the forceless horizontal movement of the wheels. The castor, toe-in, camber, kingpin and crab angle as well as the track change are measured by sensors attached to the spin axes.

In order to measure the elastokinematic properties of wheel suspensions, test stands are used that introduce both longitudinal and lateral forces into the tyre contact area. Different wheel travel positions and contact area angles can also be represented to simulate the roll movement of the vehicle body during cornering (Fig. 3.163). Forces are introduced into the vehicle either by fixing the vehicle body on the test stand and applying forces and movements to the four wheels, or by applying forces and movements to the vehicle body, with the tyre contact areas representing a fixed plane. Particular attention is paid to the attachment of the body to the test stand regardless of the kind of forces which are introduced, as unwanted flexibility leads to false results. Where possible, the body should be held in the

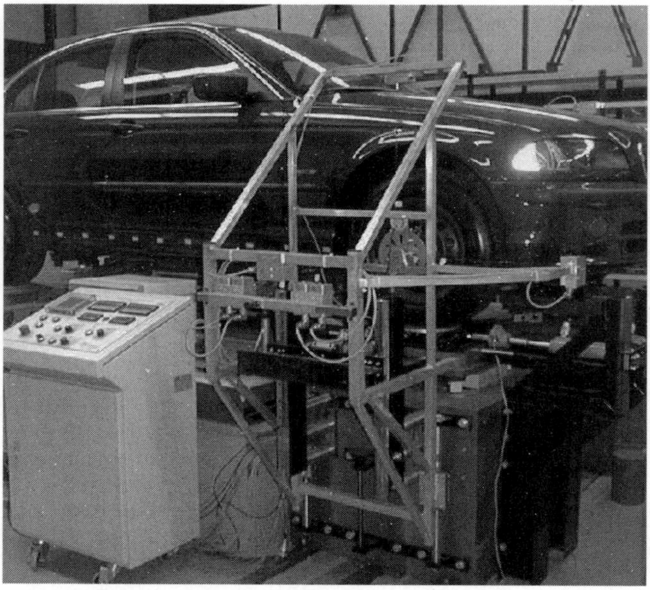

Fig. 3.163 Elastokinematics test stand of the Chassis/Simulation Technology Lab-oratory of the University of Applied Science, Cologne. The movements of the wheels are measured by a combined system of filament potentiometers and inclinometers constructed on the basis of a system of error and fault minimization; forces are measured by means of force transducers. A computer system is used for measurement, stipulation of the required values and evaluation. Forces can be applied with the normal tyres left in place or via wheel replacement carriers; air-sprung sliding plates are used to minimize friction.

immediate vicinity of the wheel suspensions, for example on or in the longitudinal frame side rails, the suspension strut domes or the auxiliary frame connection points. In order to prevent unwanted elastic deformation of the tyres, wheel replacement carriers are used which reproduce the exact force application conditions (such as tyre and diameter, offset, effective lever arm with traction or braking forces).

The standards governing the accuracy of measurement of the forces, displacements and angles to be ascertained are very high; Fig. 3.164 gives reference values for these.

3.12.2 Measuring the caster, kingpin inclination, camber and toe-in alteration

3.12.2.1 Measurement conditions

In repair workshop manuals, the axle settings (with a few exceptions) relate to the vehicle when empty, and when checking only the manufacturer's specified values, this condition should be assumed. To eliminate the influencing friction

	Measurement area	Measurement precision	Unit of measurement
Wheel load	20 000	40	N
Wheel vertical travel	±150	0.5	mm
Longitudinal force	±10 000	20	N
Longitudinal (fore-and-aft) travel	±75	0.2	mm
Lateral force	±10 000	20	N
Lateral displacement	±75	0.2	mm
Camber angle	±10	0.01	degrees
Steering wheel angle	±5	0.025	degrees
Steering angle	±45	0.2	degrees
Caster angle	±10	0.02	degrees
Steering wheel moment	±20	0.2	Nm
Steering wheel angle	±900	1	degrees

Fig. 3.164 Necessary measuring ranges and accuracy of elastokinematic characteristics

in the suspension parts, the vehicle should be briefly settled by hand on both axles before measurement begins.

The initial basis for all alterations resulting from the wheels compressing and rebounding is the design position. The vehicle carries a load based on three people with a weight of 68 kg each. Preferably, if this is done, use dummies which can be filled with water, as these exactly reflect the masses and mass distributions of the occupants to be represented. The static settings are determined in this condition. Even load distribution (apart from the 'swinging' or dynamic load transfer) is important because otherwise the body can tilt and therefore take on different camber on the left and the right. It is therefore essential that the third person should sit in the middle of the rear seat. The bump travel between the empty condition and the design position should be taken from the wheel house arches so that the vehicle can later be drawn down as far as possible against a fixed resistance for the static measurements.

3.12.2.2 Measuring the camber angle

A spirit level or electronic measuring device can be used to measure the static camber angle precisely if the zero position of the device corresponds to the wheel centre plane. If the wheel is turned slowly, the device holder can be aligned.

3.12.2.3 Measuring the caster angle

Determining the static angle can (regardless of the measurement condition) require a steering angle input of $\delta = \pm 20°$. The greater the angle τ the more the body sinks down over the wheel on the outside of the bend and is accordingly pushed up over the wheel on the inside of the bend (Fig. 3.165). The body tilts slightly, and the positive camber on the side of the vehicle on the outside of the

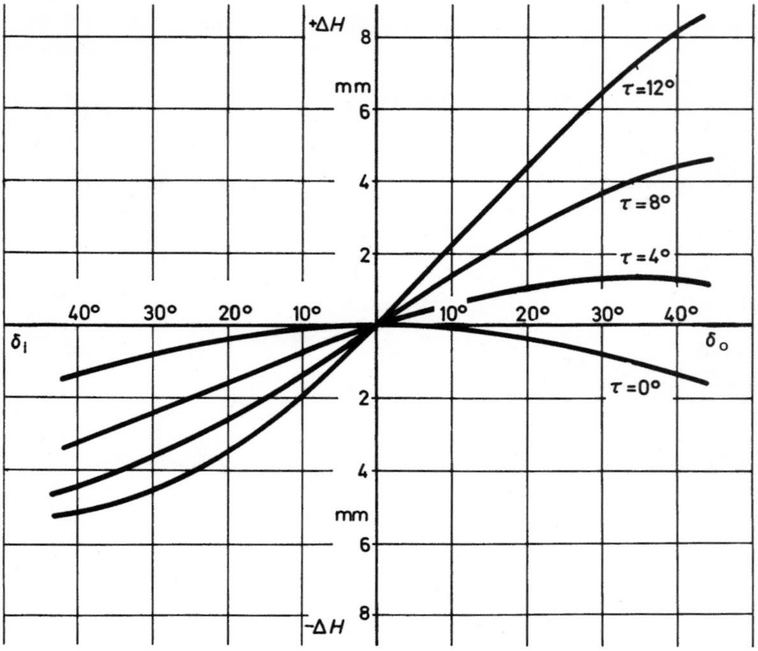

Fig. 3.165 Lift heights ΔH calculated for the wheel on the outside and the one on the inside of the bend as a function of the steering angle with the settings σ_0 = 6°, r_σ = +25 mm and various caster angles. Where τ = 0°, the centre of tyre contact of both wheels moves below ground (ΔH becomes negative), which is equivalent to lifting the body. The larger is τ, the more the body is raised on the inside of the bend ($-\Delta H$ at δ_i), but drops on the outside of the bend. When kingpin inclination and caster are measured, these relationships must be taken into account. In the case of $r_\sigma = 0$, straight lines rather than sets of curves are produced and when the kingpin offset on the ground is negative, the curves bend in the other direction.

bend consequently increases, and that on the inside of the bend reduces. The associated decrease in the kingpin inclination on the outside (and increase on the inside) of the bend can lead to a measurement error, if the body is not braced against a fixed resistance to obtain the necessary horizontal position during the measurement process.

3.12.2.4 Measuring the caster alteration

To avoid a pitch angle distorting the measurement (Fig. 3.137), the body should be drawn down parallel (or pushed up parallel). Only the alteration values $\pm\Delta\tau$, are ascertained and these must be deducted from or added to the initial data in the design position. The simplest way of doing this is to determine the rotation of the wheel with a measuring device. It is important that the brakes be locked. The floating plates (on which the wheels stand) flex longitudinally and laterally.

No other measurement method can be used on the rear axle; Figs 3.143 and 3.144 show curves recorded in this way.

3.12.2.5 Measuring the kingpin angle

Once the static caster angle has been determined in the empty condition, a mean value between left and right should be calculated to eliminate the angle τ_0 calculated in this manner, by raising the tail end (or lowering it in the case of negative caster) and thereby obtaining steering axes that are vertical from the side view. To measure the angle σ, steering inputs (where possible up to $\delta \pm 20°$) are necessary, and the kingpin inclination angle is determined via the three-dimensional movement of the wheel centre plane. The modification values should be the same for left and right inputs. Figure 3.132 indicates the correlations clearly. If the vehicle has the wheelbase l, the necessary lift height Δh_τ in the middle of the rear axle would be:

$$\Delta h_\tau = l \sin \tau_0 \tag{3.43}$$

In the raised position, the caster must then be zero, but it is always worthwhile checking.

3.12.2.6 Checking kingpin inclination and camber

As shown in Fig. 3.103 the sum of camber and kingpin inclination ($\varepsilon_W + \sigma$) left and right should be the same. If the deviation exceeds 30′, this may be a measurement error, the result of an accident, or an assembly inaccuracy on McPherson struts and strut dampers (Fig. 3.104).

3.12.2.7 Measuring kingpin inclination and camber alteration

The two are identical and pure alteration values can easily be determined. ($\pm\Delta\varepsilon_{W,k} = \pm\Delta\sigma$, see Figs 3.50 and 3.51). Only spirit levels or electronic measuring devices need to be fixed to the wheels and corrected by the caster angle, which changes as the vehicle is drawn up or down parallel with the brakes locked. The values should then be added to or subtracted from the data determined in the design position. Figures 3.48 and 3.49 indicate curves measured in this way.

3.12.2.8 Measuring the toe-in alteration and drive axle angle

The static toe-in angle $\Delta_{V,0,f \text{ or } r}$ (at the front or back, see Equation 3.8) can be determined nowadays with opto-electronic measuring devices. The alteration values for the front and rear axle should then be recorded as a function of the wheel travel s_1 and s_2 – separately from the basic values – for the left and right wheel and added to the basic values. The wheel position is measured relative to the body, so it is sensible to work with optical devices and to fix the scale (or the mirror) to the body itself. Lateral movements of the vehicle, when it is raised or pulled down, could otherwise lead to errors when reading off the figures. Drive axle angle β' indicated in Fig. 3.63 can be determined with the aid of the stationary toe-in angle.

4

Steering

This chapter gives only the essential aspects of the subject: details are given in Refs [1] and [2] and the connections relating to four-wheel drive passenger cars are described in Ref. [9], Section 5.2.

The steering system is type-approved on all new passenger cars and vans coming on to the market; it is governed by the following EC directives.

70/311/EWG	91/662/EWG
74/297/EWG	92/62/EWG

Figures 4.1, 1.46, 1.57 and 1.72 show the complete steering system of a front-wheel drive passenger vehicle with left-hand steering.

4.1 Steering system

4.1.1 Requirements

On passenger cars, the driver must select the steering wheel angle to keep deviation from the desired course low. However, there is no definite functional relationship between the turning angle of the steering wheel made by the driver and the change in driving direction, because the correlation of the following is not linear (Fig. 4.2):

- turns of the steering wheel;
- alteration of steer angle at the front wheels;
- development of lateral tyre forces;
- alteration of driving direction.

This results from elastic compliance in the components of the chassis. To move a vehicle, the driver must continually adjust the relationship between turning the steering wheel and the alteration in the direction of travel. To do so, the driver

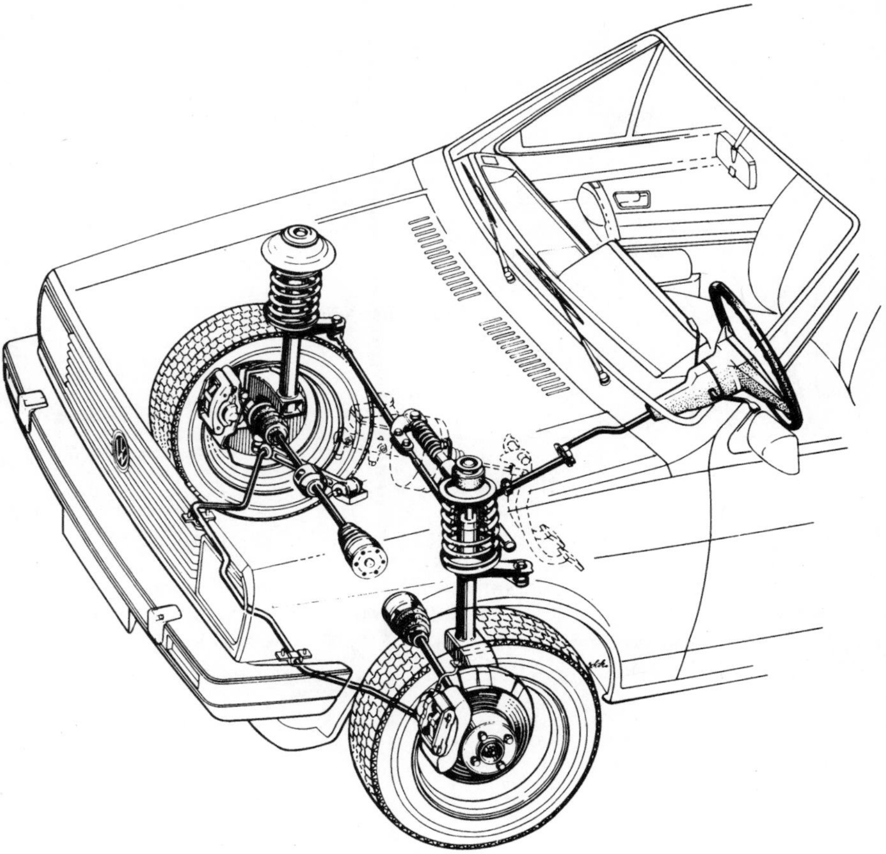

Fig. 4.1 Damper strut front axle of a VW Polo (up to 1994) with 'steering gear', long tie rods and a 'sliding clutch' on the steering tube; the end of the tube is stuck onto the pinion gear and fixed with a clamp. The steering arms, which consist of two half shells and point backwards, are welded to the damper strut outer tube. An 'additional weight' (harmonic damper) sits on the longer right drive shaft to damp vibrations. The anti-roll bar carries the lower control arm. To give acceptable ground clearance, the back of it was designed to be higher than the fixing points on the control arms. The virtual pitch axis is therefore in front of the axle and the vehicle's front end is drawn downwards when the brakes are applied (Figs 3.142 and 3.143).

will monitor a wealth of information, going far beyond the visual perceptive faculty (visible deviation from desired direction). These factors would include for example, the roll inclination of the body, the feeling of being held steady in the seat (transverse acceleration) and the self-centring torque the driver will feel through the steering wheel. The most important information the driver receives comes via the steering moment or torque which provides him with feedback on the forces acting on the wheels.

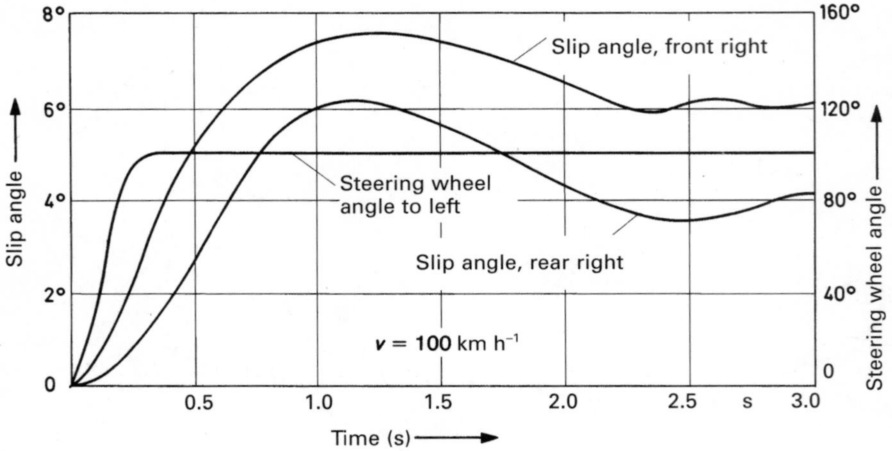

Fig. 4.2 Delayed, easily manageable response of the right front wheel when the steering wheel is turned by 100° in 0.2 s, known as step steering input. A slip angle of $\alpha_f \approx 7°$ on both front tyres is generated in this test. The smaller angle α_r on the rear axle, which later increases, is also entered. Throughout the measurement period it is smaller than α_f (*x*-axis), i.e. the model studied by Mercedes Benz understeers and is therefore easy to handle.

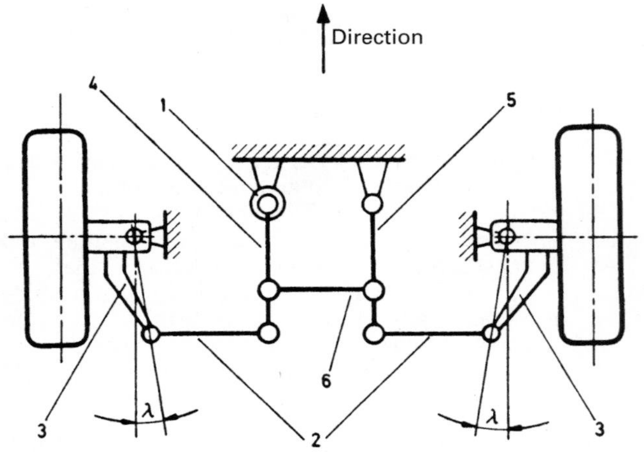

Fig. 4.3 Synchronous steering A-bar on the front suspension of a left-hand drive passenger car or light van; on the right-hand drive vehicle, the steering gear is on the other side. The steering arm (3) and the pitman arm (4) rotate in the same direction. The tie rods (2) are fixed to these arms.

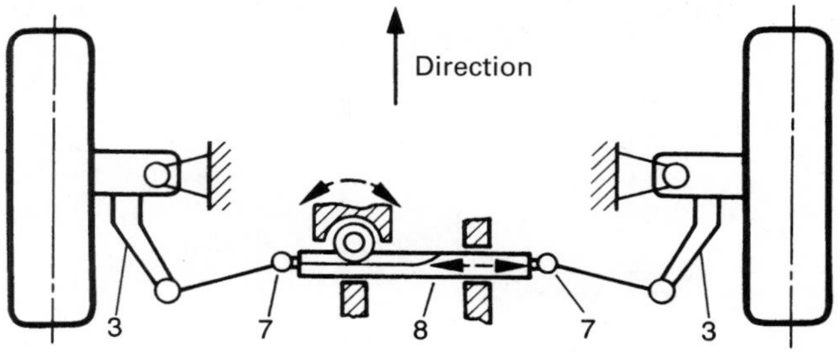

Fig. 4.4 Rack and pinion steering with the steering linkage 'triangle' behind the front axle. The spigots of the inner tie rod joints 7 are fixed to the ends of the steering rack 8 and the outside ones to the steering arms 3 (see also Figs 1.40 and 1.54).

It is therefore the job of the steering system to convert the steering wheel angle into as clear a relationship as possible to the steering angle of the wheels and to convey feedback about the vehicle's state of movement back to the steering wheel. This passes on the actuating moment applied by the driver, via the steering column to the steering gear 1 (Fig. 4.3) which converts it into pulling forces on one side and pushing forces on the other, these being transferred to the steering arms 3 via the tie rods 2. These are fixed on both sides to the steering knuckles and cause a turning movement until the required steering angle has been reached. Rotation is around the steering axis EG (Fig. 3.103), also called kingpin inclination, pivot or steering rotation axis (Fig. 1.3).

4.1.2 Steering system on independent wheel suspensions

If the steering gear is of a type employing a rotational movement, i.e. the axes of the meshing parts (screw shaft 4 and nut 5, Fig. 4.15) are at an angle of 90° to one another, on independent wheel suspensions, the insides of the tie rods are connected on one side to the pitman arm 4 of the gear and the other to the idler arm 5 (Fig. 4.3). As shown in Figs 4.12 and 4.36 to 4.38, parts 4 and 5 are connected by the intermediate rod 6. In the case of steering gears, which operate using a shift movement (rack and pinion steering), it is most economical to fix the inner tie rod joints 7 to the ends of the steering rack 8 (Fig. 4.4).

4.1.3 Steering system on rigid axles

Rack and pinion steering systems are not suitable for steering the wheels on rigid front axles, as the axles move in a longitudinal direction during wheel travel as a result of the sliding-block guide. The resulting undesirable relative movement between wheels and steering gear cause unintended steering movements. Therefore only steering gears with a rotational movement are used. The intermediate lever 5 sits on the steering knuckle (Fig. 4.5). The intermediate rod 6

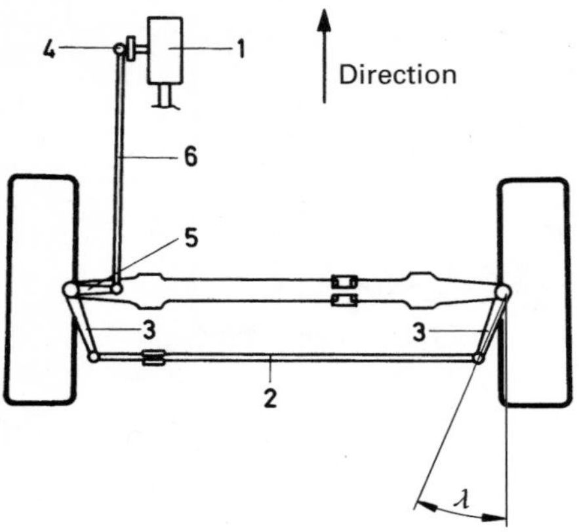

Fig. 4.5 On rigid axles, apart from the two steering arms 3, only the tie rod 2, the idler arm 5 and the drag link 6 are needed to steer the wheels. If leaf springs are used to carry the axle, they must be aligned precisely in the longitudinal direction, and lie vertical to the lever 5 when the vehicle is moving in a straight line. Steering arm angle λ is an essential factor in the relationship between the outer and the inner curve steering angles.

links the steering knuckle and the pitman arm 4. When the wheels are turned to the left, the rod is subject to tension and turns both wheels simultaneously, whereas when they are turned to the right, part 6 is subject to compression. A single tie rod connects the wheels via the steering arm.

However, on front axles with leaf springs, the pitman arm joint 4, which sits on the steering gear 1, must be disposed in such a manner that when the axle is at full suspension travel, the lower joint 8 describes the same arc 9 as the centre of the front axle housing (Figs 4.6 and 1.37). The arc 9 must be similar to the curved path 7, otherwise there is a danger of the wheels experiencing a parallel

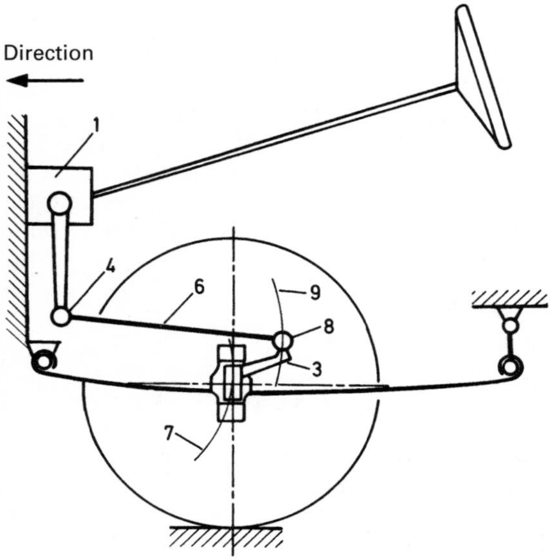

Fig. 4.6 Side view of a rigid front axle showing the movement directions 9 and 7 of the drag link and axle housing during bump and rebound-travel. The path of point 7 is determined by the front half of the leaf spring and can be calculated on a spring-balance by measuring the change in length when a load is added to and removed from the spring.

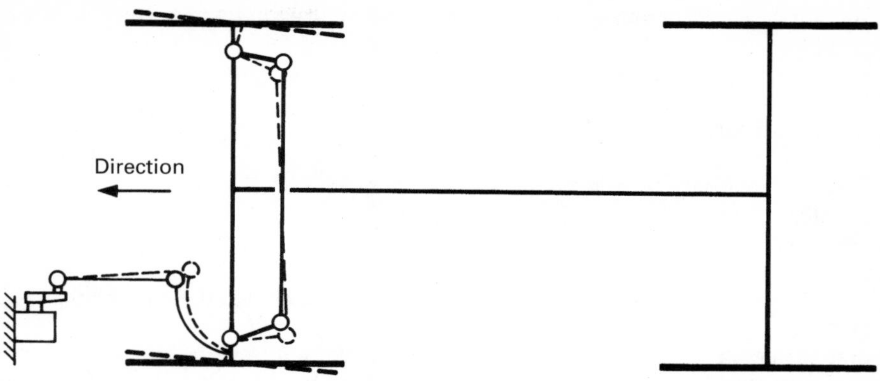

Fig. 4.7 If the movement curve 7 of the axle housing and curve 9 of the rear steering rod joint do not match when the body bottoms out, the wheels can turn and therefore an unwanted self-steering effect can occur.

toe-in alteration when the suspension reaches full travel, i.e. both being turned in the same direction (Fig. 4.7). If a rigid axle is laterally controlled by a panhard rod, the steering rod must be parallel to it.

Its construction is similar to that of the intermediate rod of the steering linkage shown in Fig. 4.13; length adjustment and ball joints on both sides are necessary.

4.2 Rack and pinion steering

4.2.1 Advantages and disadvantages

This steering gear with a shift movement is used not only on small and medium-sized passenger cars, but also on heavier and faster vehicles, such as the Audi A8 and Mercedes E and S Class, plus almost all new light van designs with independent front wheel suspension. The advantages over manual recirculating ball steering systems are (see also Section 4.3.1):

- simple construction;
- economical and uncomplicated to manufacture;
- easy to operate due to good degree of efficiency;
- contact between steering rack and pinion is free of play and even internal damping is maintained (Fig. 4.10);
- tie rods can be joined directly to the steering rack;
- minimal steering elasticity compliance (Fig. 3.99);
- compact (the reason why this type of steering is fitted in all European and Japanese front-wheel drive vehicles);

- the idler arm (including bearing) and the intermediate rod are no longer needed;
- easy to limit steering rack travel and therefore the steering angle.

The main disadvantages are:

- greater sensitivity to impacts;
- greater stress in the case of tie rod angular forces;
- disturbance of the steering wheel is easier to feel (particularly in front-wheel drivers);
- tie rod length sometimes too short where it is connected at the ends of the rack (side take-off design, Fig. 3.67);
- size of the steering angle dependent on steering rack travel;
- this sometimes requires short steering arms 3 (Fig. 4.4) resulting in higher forces in the entire steering system;
- decrease in steering ratio over the steer angle (Fig. 3.96) associated with heavy steering during parking if the vehicle does not have power-assisted steering;
- cannot be used on rigid axles.

4.2.2 Configurations

There are four different configurations of this type of steering gear (Fig. 4.8):

Type 1 Pinion gear located outside the vehicle centre (on the left on left-hand drive and on the right on right-hand drive) and tie rod joints screwed into the sides of the steering rack (side take-off).

Type 2 Pinion gear in vehicle centre and tie rods taken off at the sides.

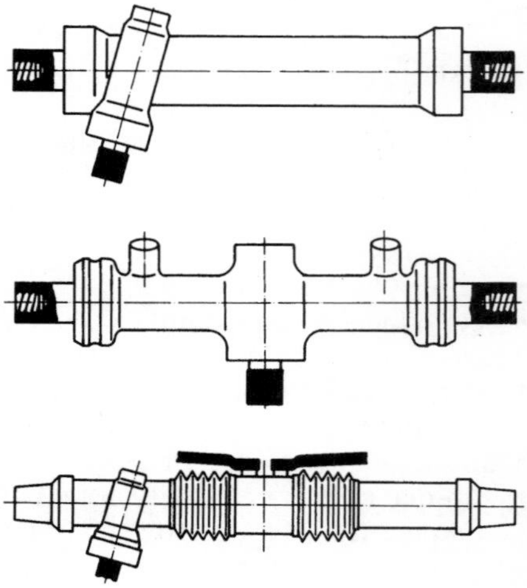

Fig. 4.8 The three most common types of rack and pinion steering on left-hand drive passenger cars; right-hand drive vehicles have the pinion gear on the other side on the top and bottom configurations (shown in Fig. 4.39). The pinion gear can also be positioned in the centre to obtain longer steering rod travel.

Type 3 Pinion gear to the side and centre take-off, i.e. the tie rods are fixed in the vehicle centre to the steering rack.

Type 4 'Short steering' with off-centre pinion gear and both tie rods fixed to one side of the steering rack (Fig. 4.1).

Types 1 and 3 are the solutions generally used, whereas Type 2 was found in some Porsche vehicles, and Type 4 used to be preferred by Audi and VW. For design details, see Section 3.2.4 in Ref. 1.

4.2.3 Steering gear, manual with side tie rod take-off

Type 1 (Fig. 4.8) is the simplest solution, requiring least space; the tie rod joints are fixed to the sides of the steering rack (Fig. 4.9), and neither when the wheels are turned, nor when they bottom out does a moment occur that seeks to turn the steering rack around its centre line. It is also possible to align the pinion shaft pointing to the steering tube (Figs 1.57, 4.24 and 4.29) making it easy to connect the two parts together. Using an intermediate shaft with two joints (Figs 1.49 and 4.26) enables the steering column to bend at this point in an accident. In this event the entire steering gear is turned when viewed from the side (i.e. around the y-axis).

Figure 4.10 is a section showing how, on all rack and pinion steering systems, not only can the play between the steering rack and the pinion gear be easily eliminated, but it also adjusts automatically to give the desired damping. The pinion gear 21 is carried by the grooved ball bearing 20; this also absorbs any axial forces. Ingress of dirt and dust are prevented by the seal 31 in a threaded ring 43 and the rubber cap 45. The lower end of the pinion gear is supported in the needle bearing 23.

In a left-hand drive passenger car or light van, the steering rack 3 is carried on the right by a plastic bearing shell and on the right by guide 15, which presses the steering rack against the pinion gear. On a right-hand drive vehicle this arrangement is reversed. The half-round outline of the guide 15 does not allow radial movement of the steering rack. To stop it from moving off from the pinion gear, when subject to high steering wheel moments (which would lead to reduced tooth contact), the underside of the guide-bearing 15 is designed as a buffer; when it has moved a distance of $s \leq 0.12$ mm it comes into contact with the screw plug 16.

Depending on the size of the steering system, coil spring 14 has an initial tension force of 0.6 kN to 1.0 kN, which is necessary to ensure continuous contact between steering rack and pinion gear and to compensate for any machining imprecision, which might occur when the toothing is being manufactured or the steering rack broached or the pinion gear milled or rolled. The surface of the two parts should have a Rockwell hardness of at least 55 HRC; the parts are not generally post-ground due to the existence of a balance for the play. Induction-hardenable and annealed steels such as Cf 53, 41 Cr 4 and others are suitable materials for the steering rack, case-hardened steels such as 20 MnCr 5, 20 MoCr 4, for example, are suitable for the pinion gear. In order to ensure a good

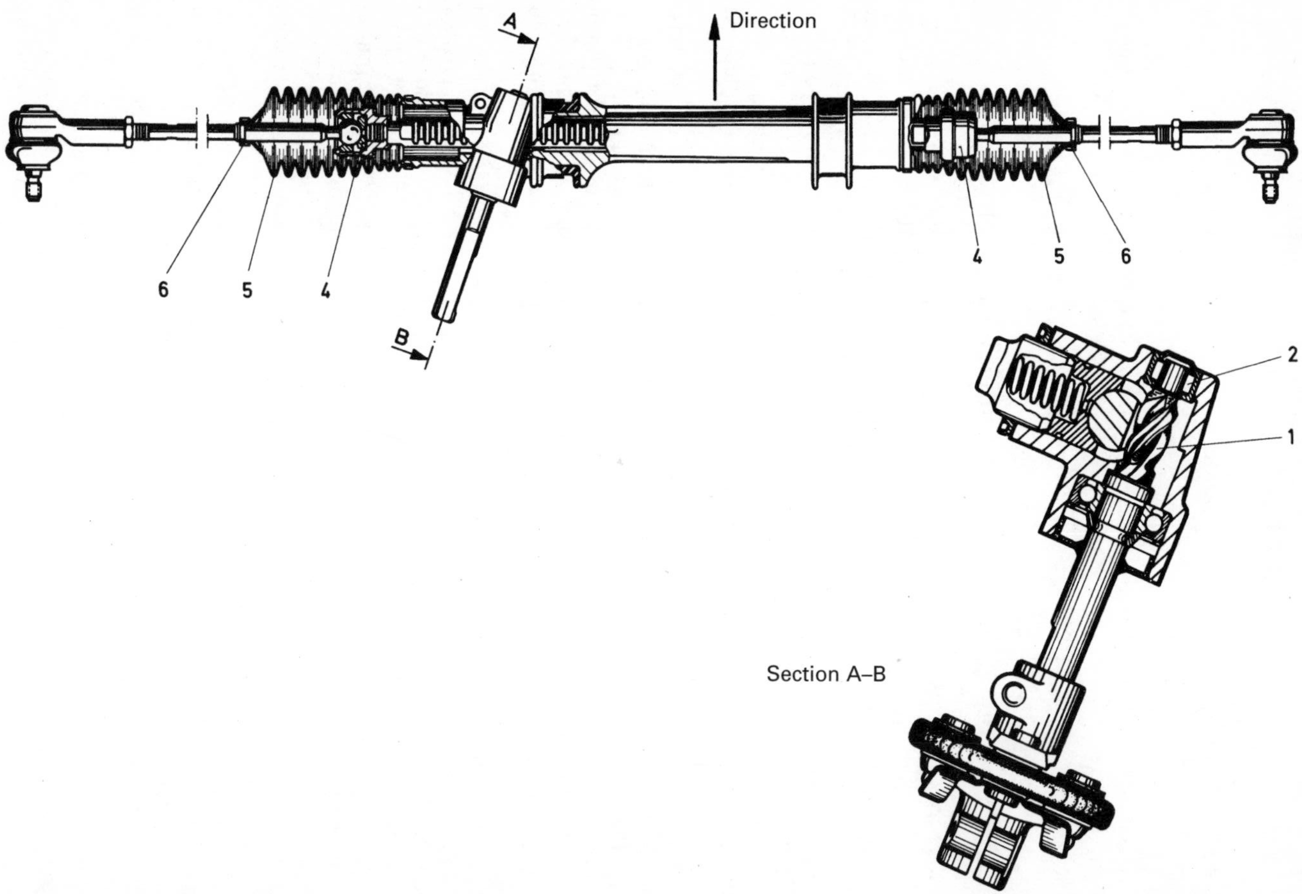
Direction
A
B
6
5
4
4
5
6
2
1
Section A–B

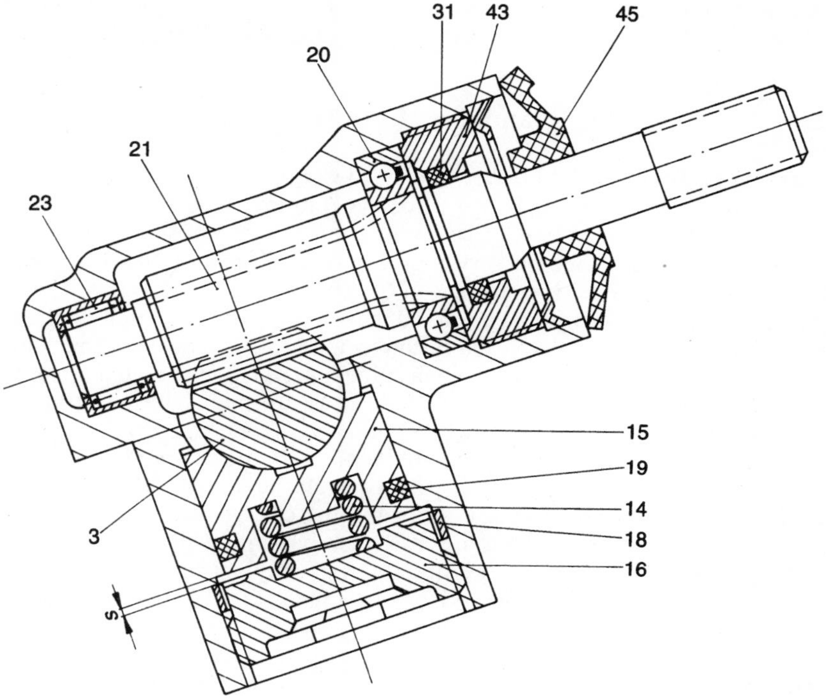

Fig. 4.10 Rack- and- pinion steering by ZF; section through pinion gear, bearing and rod guide. The distance ring 18 is used for setting the play *s*, and the closing screw 16 is tightened against it. The O-ring 19 provides the damping function and prevents rattling noises.

response and feedback of the steering, the frictional forces between guide-bearing 15 and gear rack 3 must be kept as small as possible.

Sealing the steering rack by means of gaiters to the side (Fig. 4.9) makes it possible to lubricate them with grease permanently, and lubrication must be provided through a temperature range of −40°C to +80°C. It is important to note that if one of the gaiters is damaged, the lubricant can escape, leading to the steering becoming heavier and, in the worst case, even locking. Gaiters should

---

Fig. 4.9 Rack and pinion steering on the Vauxhall Corsa (1997). The tie rod axial joints 4 bolted to the side of the steering rack and the sealing gaiters 5 can be seen clearly. To stop them from being carried along when the toe-in is set (which is done by rotating the middle part of the rod) it is necessary to loosen the clamps 6.

The pinion 1 has been given a 'helical cut', due to the high ratio, and is carried from below by the needle bearing 2. The bearing housing has been given a cover plate to facilitate assembly and prevent dirt ingress.

therefore be checked at every service inspection. They are also checked at the German TÜV (Technischer Überwachungs Verein) annual vehicle inspection.

4.2.4 Steering gear, manual with centre tie rod take-off

As shown in Figs 1.57 and 4.1, and as described in Section 4.7.3.2, with McPherson struts and strut dampers the tie rods must be taken off from the centre if the steering gear has to be located fairly high up. This is because the steering tie rods must thus be very long in order to prevent unwanted steering movements during wheel travel (Fig. 4.46).

In such cases the inner joints are fixed in the centre of the vehicle to the steering rack itself, or to an isolator that is connected to it. The designer must ensure that the steering rack cannot twist when subject to the moments that arise. When the wheels rebound and compress, the tie rods are moved to be at an angle, something which also happens when the wheels are steered. The effective distance *a* between the eye-type joints of the tie rods and the steering rack centre line, shown in Fig. 4.11, gives a lever, via which the steering could be twisted. Two guide pieces which slide in a groove in the casing stop this from happening. However, the need to match the fit for the bearing of the steering rack and the guide groove can lead to other problems. If they are too tight, the steering will be heavy, whereas if they are too loose, there is a risk of rattling noises when the vehicle is in motion.

As the steering forces are introduced at a relatively large distance from the bearing points of the steering axle (suspension strut support bearings at the top, ball-and-socket joint on the transverse link at the bottom), elastic (flexural) deformations occur on the suspension strut and shock-absorber strut. As a result, steering precision and response characteristics worsen.

Fig. 4.11 Top view of the rack and pinion steering of the front-wheel drive Opel (Vauxhall) Astra (up to 1997) and Vectra (up to 1996); the steering arms on the McPherson strut point backwards and the steering gear is located relatively high. For this reason the tie rods have to be jointed in the middle and (in order not to come into contact with the gear housing when the wheels are turned) have to be bent. The guide-bearing in the groove of the housing prevents the steering rack from twisting. On the inside, both tie rods have the eye-type joint shown in Fig. 5.45; the distance *a* to the steering rack centre, which causes a bending moment, and a torque (when the wheels bump and rebound) is also shown. The two bolts 6 gripping into the steering rack are secured.

Once the screws 3 and 4 have been loosened, toe-in to the left and right can be set by turning the connecting part 5.

The steering gear has two fixing points on the dashpanel, which are a long way apart and which absorb lateral force moments with minimal flexing.

As also shown in Fig. 4.10, the pinion is carried by a ball and a needle bearing (positions 20 and 23) and is also pressed onto the steering rack by a helical spring. The illustration shows the possible path *s* of the rack guide. Figures 4.46 to 4.48 show the reason for the length of the tie rods on McPherson struts and strut dampers.

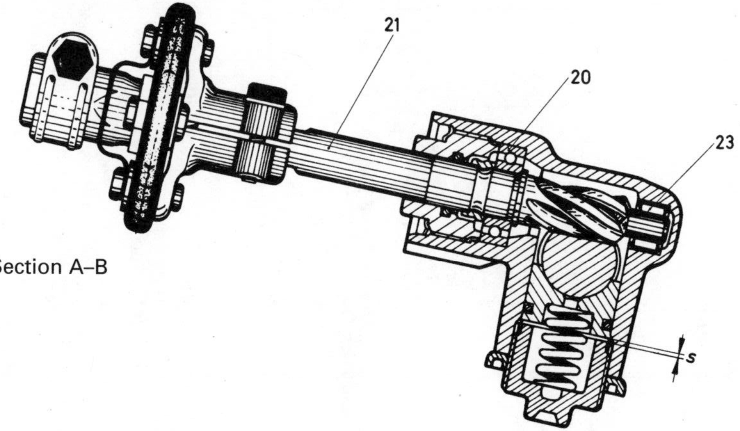

Section A–B

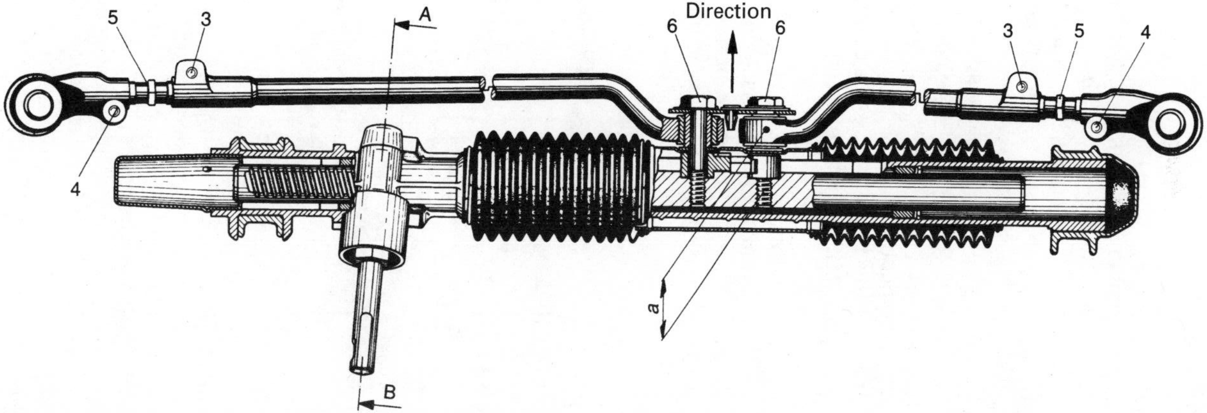

4.3 Recirculating ball steering

4.3.1 Advantages and disadvantages

Steering gears with a rotating movement are difficult to house in front-wheel drive passenger cars and, in a standard design vehicle with independent wheel suspension, also require the idler arm 5 (see Fig. 4.3) and a further intermediate rod, position 6, to connect them to the pitman arm 4; the tie rods are adjustable and have pre-lubricated ball joints on both sides (Figs 4.13 and 4.14).

This type of steering system is more complicated on the whole in passenger cars with independently suspended front wheels and is therefore more expensive than rack and pinion steering systems; however, it sometimes has greater steering elasticity, which reduces the responsiveness and steering feel in the on-centre range (see Section 3.7.4).

Comparing the two types of configuration (without power-assisted steering) indicates a series of advantages:

- Can be used on rigid axles (Figs 4.5 and 1.37).
- Ability to transfer high forces.
- A large wheel input angle possible – the steering gear shaft has a rotation range up to ±45°, which can be further increased by the steering ratio.

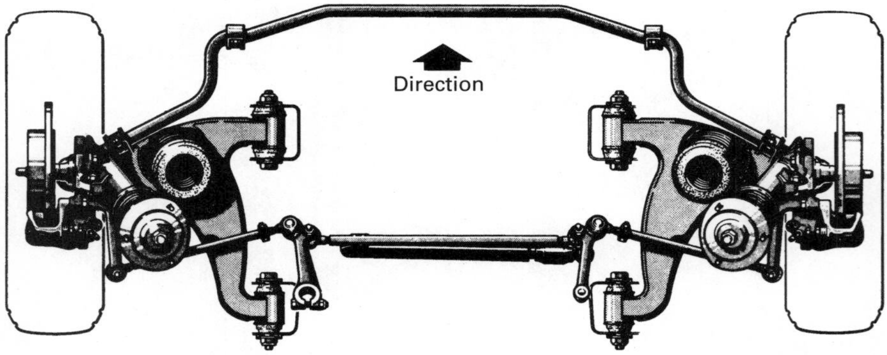

Fig. 4.12 Top view of the strut damper front axle on a Mercedes vehicle. The intermediate rod and the tie rods are fixed side by side on the pitman and idler arms and one grips from the top and the other from the bottom into the two levers; the steering square is opposed. The steering damper is supported on the one side at the intermediate rod and on the other side on the suspension subframe.

The anti-roll bar is linked to the lower wishbone type control arms whose inner bearings take large rubber bushings. The defined springing stiffness of these bearings, together with the inclined position of the tie rods (when viewed from the top) means that when the vehicle corners, there is a reduction in the steering input, i.e. elastic compliance in the steering, tending towards understeering (Fig. 3.82). The strut dampers are screwed to the steering knuckles; the negative kingpin offset is $r_\sigma = -14$ mm.

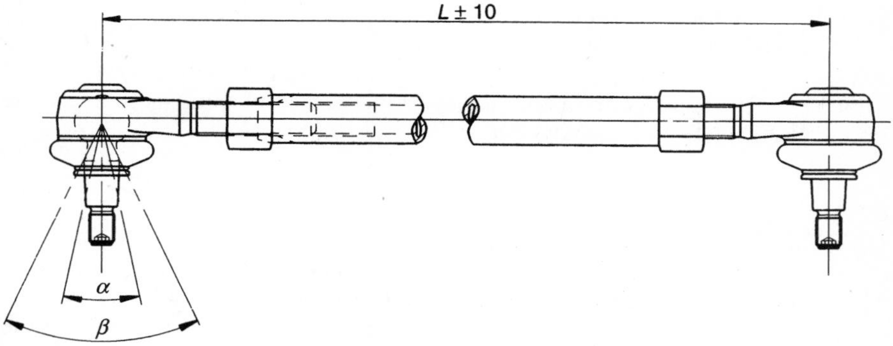

Fig. 4.13 Configuration of an adjustable tie rod with pre-lubricated joints and buckling-resistant central tube, the interior of which has a right-hand thread on one side and a left-hand thread on the other. It can usually be continuously adjusted by ±10 mm. When toe-in has been set, the length on the right and left tie-rod may differ, resulting in unequal steering inputs and different size turning circles; for this reason, the central tube should be turned the same amount on the left and right wheel.

The configuration shown in the illustration is used on rigid front axles and as a drag link (illustration: Lemförder Fahwerktechnik).

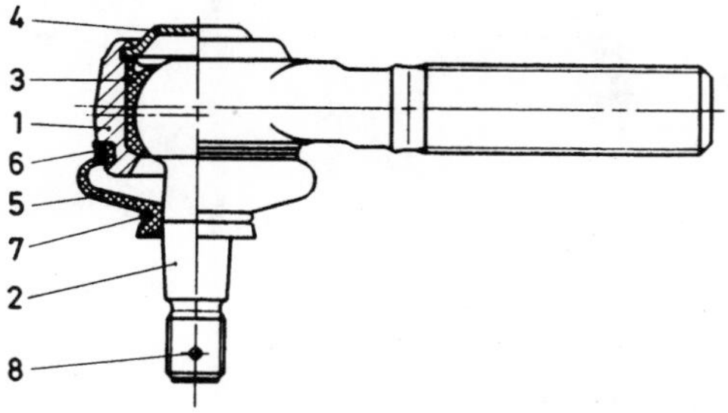

Fig. 4.14 Lemförder Fahrwerktechnik pre-lubricated tie rod joint, used on passenger cars and light vans. The joint housing 1 has a fine thread on the shaft (M14 × 1.5 to M22 × 1.5) and is made of annealed steel C35V; surface-hardenable steel 41Cr4V is used for the ball pivot 2.

The actual bearing element – the one-part snap-on shell 3 made from polyacetal (e.g. DELRIN, made by Dupont) – surrounds the ball; the rolled-in panel cover 4 ensures a dirt- and waterproof seal. The polyurethane or rubber sealing gaiter 5 is held against the housing by the tension ring 6. The gaiter has a bead at the bottom (which the second tension ring 7 presses against the spigot) and a sealing lip, which comes into contact with the steering arm.

The ball pivot 2 has the normal 1:10 taper and a split pin hole (position 8). If there is a slit or a hexagonal socket (with which the spigot can be held to stop it twisting), a self-locking nut can be used instead of a slotted castle nut and split pin.

- It is therefore possible to use long steering arms.
- This results in only low load to the pitman and intermediate arms in the event of tie rod diagonal forces occurring.
- It is also possible to design tie rods of any length desired, and to have steering kinematics that allow an increase in the overall steering ratio i_S with increasing steering angles. The operating forces necessary to park the vehicle are reduced in such cases (see Section 3.7.3).

4.3.2 Steering gear

The input screw shaft 4 (Fig. 4.15) has a round thread in which ball bearings run, which carry the steering nut 5 with them when the steering wheel is rotated. The balls which come out of the thread at the top or the bottom (depending on the

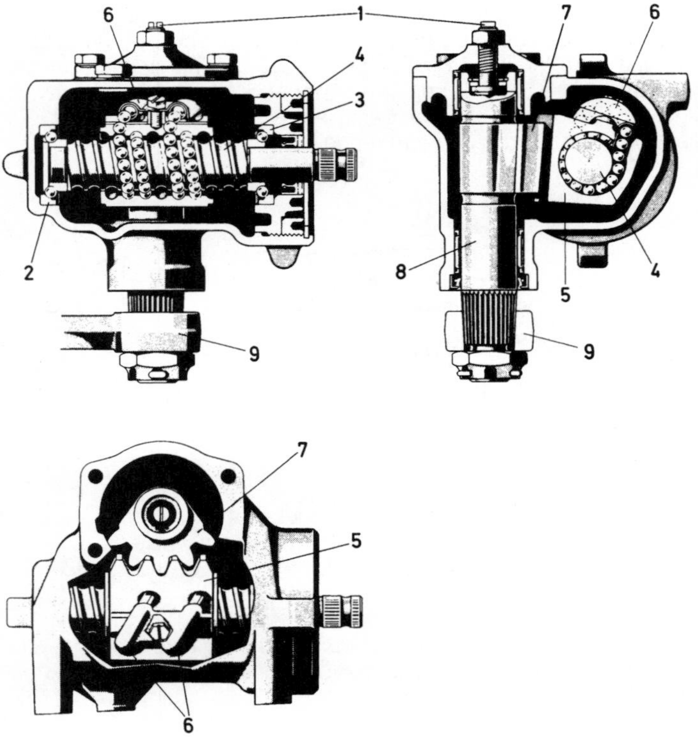

Fig. 4.15 Mercedes Benz recirculating ball steering suitable for passenger cars and light vans; today, apart from in a few exceptional cases, this is only fitted as a hydraulic power-assisted version. Pitman arm 9 is mounted onto the tapered toothed profile with a slotted castle nut 11 (Fig. 4.24).

direction of rotation) are returned through the tube 6. The nut has teeth on one side which mesh with the toothed segment 7 and therefore with the steering output shaft 8. When viewed from the side, the slightly angular arrangement of the gearing can be seen top right. This is necessary for alignment bolt 1 to overcome the play of the wheels when pointing straight ahead, by axial adjustment. If play occurs in the angular ball bearings 2 and 3, the lock-nut must be loosened and the sealing housing cover re-tightened.

Only a few standard design larger saloons can be found on the road with manual recirculating ball steering. For reasons of comfort, newer passenger cars of this type have hydraulic power-assisted steering. The same applies to commercial vehicles; only a few light vans are still fitted with manual configurations as standard and even these are available with power-assisted steering as an option.

4.4 Power steering systems

Power steering systems have become more and more widely used in the last few years, due to the increasing front axle loads of vehicles on the one hand and the trend towards vehicles with more agile steering properties and hence direct transmission steering systems on the other. With the exception of some members of the 'sub-compact' class, power steering systems are optionally or automatically included as one of the standard features.

Manual steering systems are used as a basis for power steering systems, with the advantage that the mechanical connection between the steering wheel and the wheel and all the components continues to be maintained with or without the help of the auxiliary power. The steering-wheel torque applied by the driver is detected by a measurement system located in the region of the input shaft of the steering gear or in the steering tube, and additional forces or moments are introduced into the system. This follows a characteristic curve (valve characteristic) or group of curves depending on the height of the steering-wheel torque, if another quantity, e.g. driving speed, is entered as a signal. The steering boost is thereby reduced, with the aim of achieving better road contact at higher speeds. An exact functional description of such systems can be found in Chapter 10 in Ref. [1].

4.4.1 Hydraulic power steering systems

Hydraulic power steering systems are still the most widely used. The method of using oil under pressure to boost the servo is sophisticated and advantageous in terms of cost, space and weight. Sensitivity to movements caused by the road surface and hence the effect of torsional impacts and torsional vibrations passing into the steering wheel is also noticeably reduced, particularly with rack and pinion steering. This can be attributed to the hydraulic self-damping. It might also be the reason why it is possible to dispense with an additional steering shock absorber in most vehicles with hydraulical rack and pinion steering, whereas it is required for the same vehicles with manual steering (see Section 4.6).

The oil pump is directly driven by the engine and constantly generates hydraulic power. As hydraulic power steering systems have to be designed in such a way that a sufficient supply volume is available for fast steering movements even at a low engine speed, supply flow limiting valves are required. These limit the supply flow to about 8 l per minute in order to prevent the hydraulic losses which would otherwise occur at higher engine speeds. Depending on the driving assembly and pump design, the additional consumption of fuel can lie between 0.2 and 0.7 l per 100 km.

Assemblies which are added to provide auxiliary power are shown in Fig. 4.16, taking the example of the rack and pinion steering used by Opel in the Vectra (1997). The pressure oil required for steering boost is supplied direct to the steering valve 6 located in the pinion housing from vane pump 1 via the high-

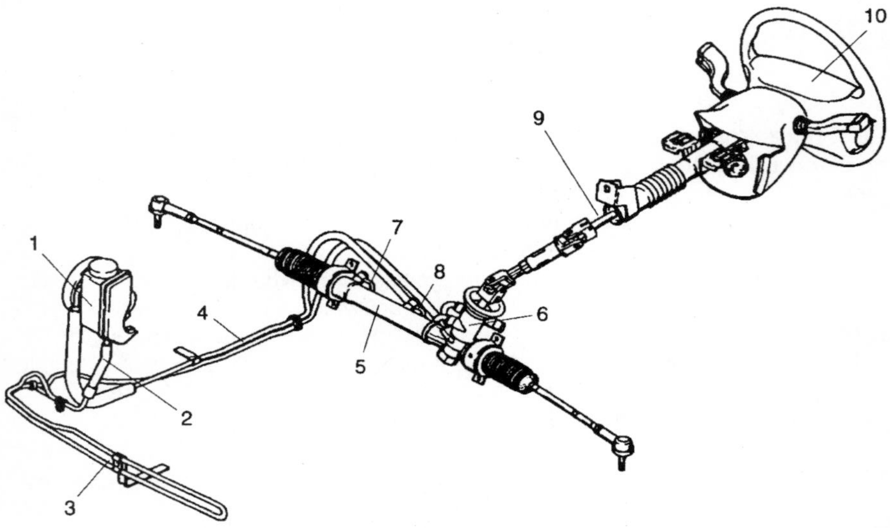

Fig. 4.16 Hydraulic power steering system of the Opel Vectra (1997). The individual components are:

1 vane pump, driven by V-belts
2 high-pressure line
3 cooling circuit
4 return line, from the steering valve to the pump
5 steering gear with external drive, attached to the auxiliary frame
6 steering valve
7/8 pressure lines to the working cylinder
9 steering column with intermediate shaft
10 steering wheel with integrated airbag.

pressure line 2 and the cooling circuit 3. From here, depending on the direction of rotation of the steering wheel and the corresponding counterforce on the wheels, distribution to the right or left cylinder line takes place (items 7 and 8). Both lead to the working cylinder which is integrated in the steering-gear housing 5. A disc located on the gear rack divides the pressure chamber. Differences in pressure generate the required additional axial force in the gear rack F_{Pi} via the active areas of the disc:

$$F_{Pi} = (p_{hyd,2} - p_{hyd,1})\,A_{Pi} \tag{4.1}$$

where A_{Pi} is the effective piston surface, here the difference between the disc and gear rack surfaces, and $p_{hyd,1 \text{ or } 2}$ are the pressures acting on the working piston. In a situation where there is no torque, for example during straight running, the oil flows direct from the steering valve 6 back to the pump 1 via the return line 4.

The method of operation of the steering valve is shown in Fig. 4.17, using the example of recirculating-ball steering. In a similar way to rack and pinion steering, it is integrated into the input shaft of the steering gear. As is the case with most hydraulic power steering systems, the measurement of the steering-wheel torque is undertaken with the use of a torsion bar 18. The torsion bar connects the valve housing 5 (part of the steering screw) to the valve pistons 9/10 in a torsionally elastic way. Steering-wheel torque generates torsion of the torsion bar. These valve pistons then move and open radial groove 13 or 14, depending on the direction of rotation. This leads to a difference in pressure between pressure chambers D1 and D2. The resultant axial force on the working piston 2 is calculated using Equation 4.2. Because $p_{hyd,2}$ also operates in the interior space of the piston behind the steering screw 5, the surface areas are the same on both sides:

$$F_{Pi} = p_{hyd,1 \text{ or } 2}\,A_{Pi} = p_{hyd,1 \text{ or } 2}\,\frac{\pi D_{Pi}^2}{4} \tag{4.2}$$

The exact description is contained in Section 5.2 in Ref. [1].

4.4.2 Electro-hydraulic power steering systems

With electro-hydraulic power steering systems, the power-steering pump driven by the engine of the vehicle via V-belts is replaced by an electrically operated pump.

Figure 4.18 shows the arrangement of the system in an Opel Astra (1997). The electrically operated power pack supplies the hydraulic, torsion-bar controlled steering valve with oil. The pump is electronically controlled – when servo boost is not required, the oil supply is reduced.

The supply of energy by electricity cable allows greater flexibility with regard to the position of the power pack. In the example shown, it is located in the immediate vicinity of the steering gear. Compared with the purely hydraulic

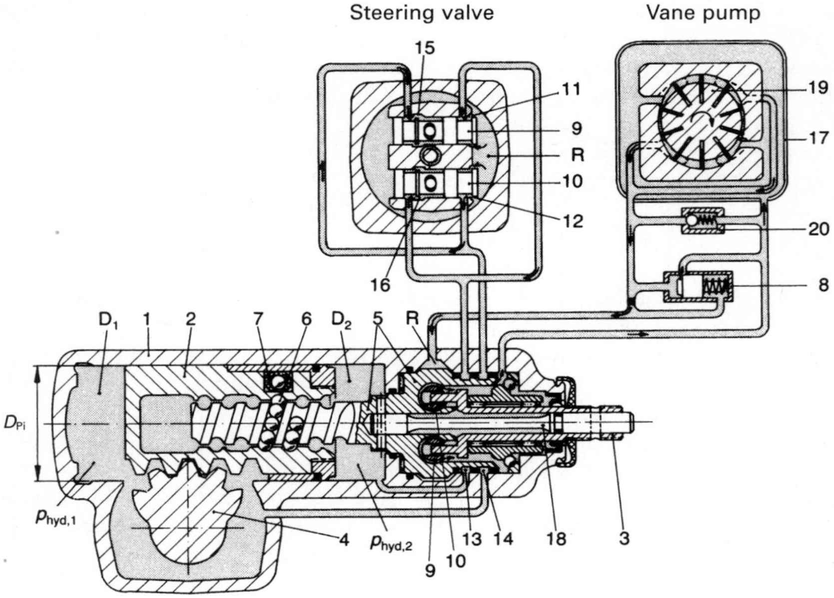

Fig. 4.17 Illustration of the principles of the ZF recirculating ball steering in the neutral position (vehicle travelling in a straight line). The steering valve, the working piston and the mechanical gear sit in a common housing. The two valve pistons of the steering valve have been turned out of their operating plane to make the diagram easier to see. The individual parts are:

1	gear housing	9/10	valve piston
2	piston with steering nut	11/12	inlet groove
3	steering spindle connection	13/14	radial groove
4	steering shaft with toothed segment	15/16	return groove
5	steering worm roller with valve body	17	fluid reservoir
6	balls	18	torsion bar
7	recirculation tube	19	hydraulic pump
8	fluid flow limitation valve	20	pressure-limiting valve

system, the lines can be made considerably shorter and there is no cooling circuit. The steering gear, power pack and lines are installed as a ready-assembled and tested unit.

To sum up, electro-hydraulic power steering systems offer the following advantages:

- The pressure supply unit (Fig. 4.19) can be accommodated in an appropriate location (in relation to space and crash safety considerations).
- Servo boost is also guaranteed by the electrical pressure supply even when the engine is not running.
- Pressure-controlled systems generate only the amount of oil required for a

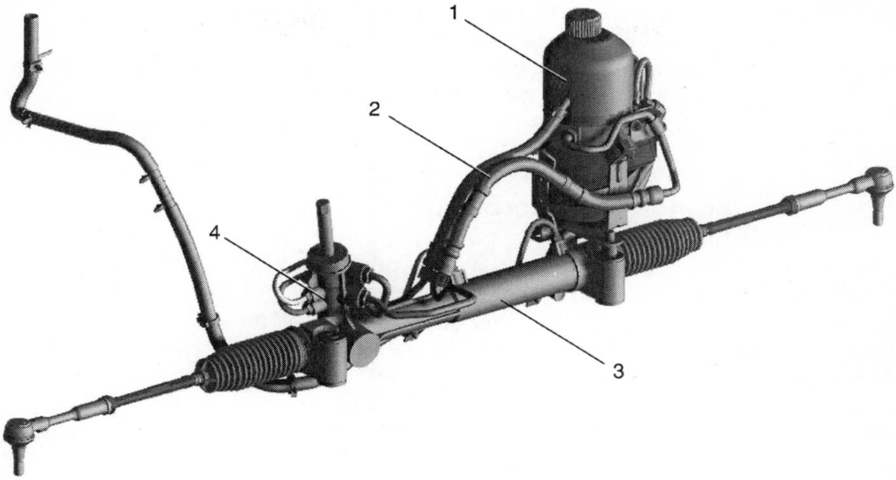

Fig. 4.18 Electro-hydraulic power steering system of the Opel Astra (1997). The individual components are:

1 electrically operated power-steering pump with integrated reserve tank ('power pack')
2 pump–steering valve hydraulic lines
3 rack and pinion steering gear with external drive, attached to auxiliary frame
4 steering valve.

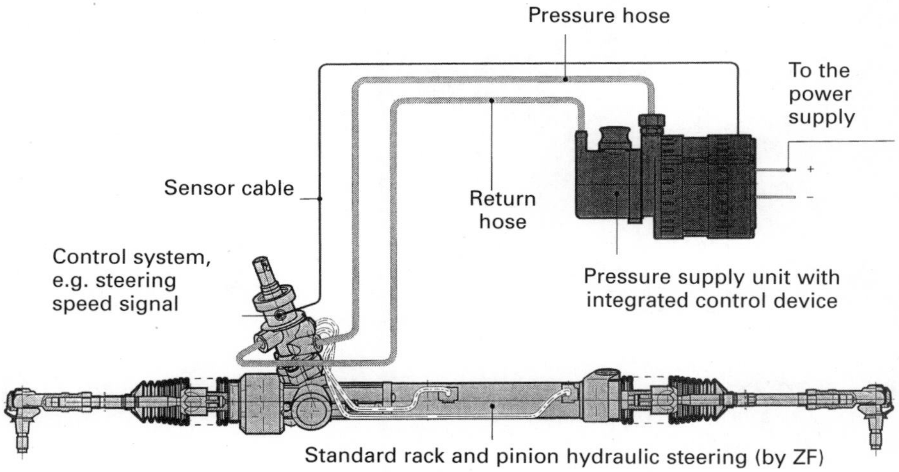

Fig. 4.19 Open-centre control system from ZF. The pressure supply unit designed as a modular unit can be fitted with different electric motors (DC motor with or without brushes) and pump fuel feed volumes (1.25–1.75 cm^3 per rpm) depending on its particular function. Oil tanks for horizontal or vertical installation are also available. Operating pressure is up to 120 bar, with a maximum power consumption of 80 A.

particular driving situation. Compared with standard power steering systems, energy consumption is reduced to as little as 20%.

- The steering characteristics (nature and amount of steering boost, sensitivity, speed dependency) can be adjusted by the control electronics individually for the particular vehicle.

4.4.3 Electrical power steering systems

The bypass of the hydraulic circuit and direct steering boost with the aid of an electric motor has additional advantages in terms of weight and engine bay space compared with electro-hydraulic steering, because of the omission of all the hydraulic components. Other advantages are obtained through more variations of the steering boost because of the purely electrical signal processing.

The electrical servo unit can be installed on the steering column (Fig. 4.20), pinion (Fig. 4.21) or gear rack (Fig. 4.22). The steering axle loads and maximum gear rack forces are, depending on the particular arrangement, about 650 kg and 6000 N, 850 kg and 8000 N or 1300 kg and 10 000 N.

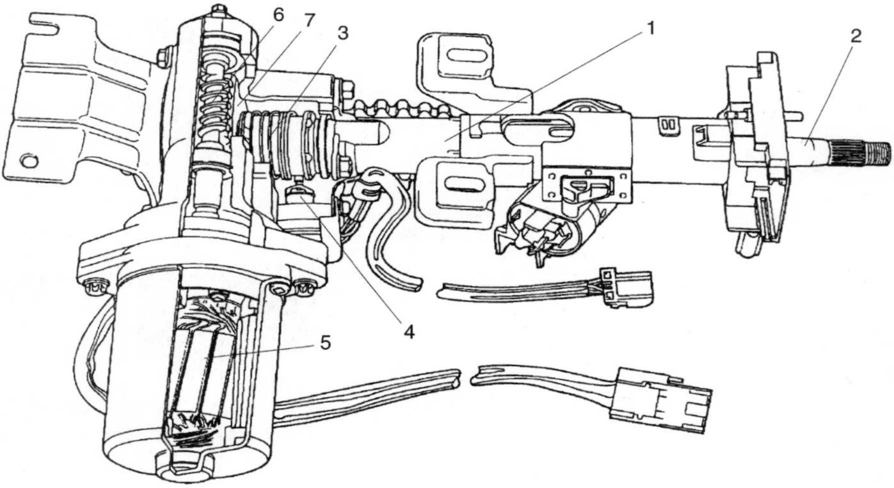

Fig. 4.20 Steering column with power-steering assembly of the Opel Corsa (1997). The individual components are:

1 column tube
2 steering tube
3 sliding sleeve with groove
4 rotary potentiometer with tap
5 servomotor
6 drive worm
7 worm gear.

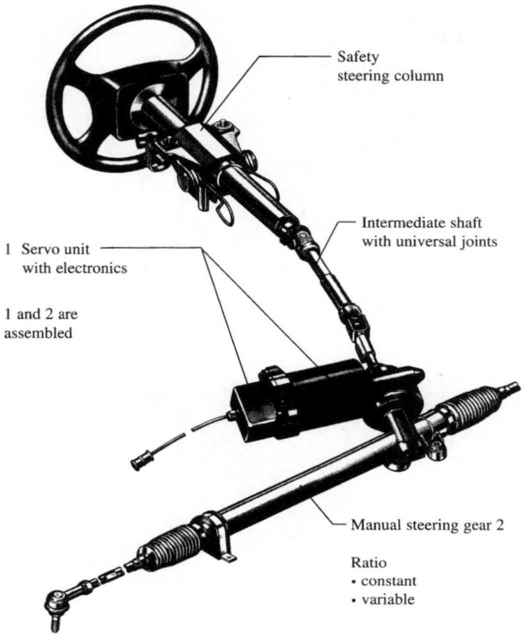

Fig. 4.21 Electrical power steering system by ZF. The servo unit acts directly upon the pinion of the rack and pinion steering. Consequently, the amount of stress to which the pinion is subjected increases by the amount of steering boost, compared with a mechanical or hydraulic power steering system.

The systems only have limited power because the current is limited by an operating voltage of 12 V. They are of interest though for smaller vehicles. In this class of vehicle in particular, electric power steering systems show their advantages, not least because of the small amount of energy required. The introduction of the increased voltage of 42 V will make the use of electrical power steering systems and wheel brakes much easier.

Figure 4.23 shows the steering system of the Opel Corsa (1997) with electric power steering. It is a system with steering-tube transmission, i.e. the intermediate spindle transmits the whole of the torque resulting from the steering wheel force and servo boost. Due to the more direct steering transmission, this torque is clearly higher than in a comparable manual steering system, something which must be taken into consideration when deciding on the size of the components which control performance.

In Fig. 4.20, the method of operation of the servo assembly (EPAS system by NSK) becomes clear: a plastic worm gear 7 is applied to the steering tube 2. This is engaged by the worm 6, which in its turn is connected to the shaft of the servomotor 5. Steering-wheel torque generates a torsional movement of the torsion bar (concealed by the sliding sleeve 3). The steering tube area is axially grooved above the torsion bar and spindle-shaped below. As the spindle rises, the sliding

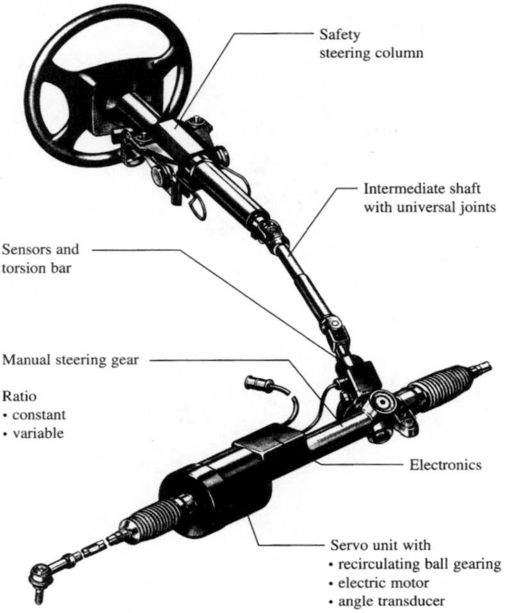

Fig. 4.22 Electrical power steering system by ZF. The servo unit acts on the gear rack itself. This system is suitable for high axle loads and steering forces. The maximum current strength is 105 A with a 12 V electric system; with a 42 V system, it is only 35 A.

sleeve makes an axial movement on the steering tube proportional to the torsion of the torsion bar. This axial movement is transmitted to the rotary potentiometer 4 via a tap. Corresponding to a group of curves, the servo boost is determined from the steering-wheel torque and driving speed signals and the servomotor 5 controlled accordingly.

More detailed functional descriptions, also of other systems, are contained in Chapter 8 in Ref. [1].

4.5 Steering column

In accordance with the German standard DIN 70 023 'nomenclature of vehicle components', the steering column consists of the jacket tube (also known as the outer tube or protective sleeve), which is fixed to the body, and the steering shaft, also called the steering tube. This is only mounted in bearings at the top (or top and bottom, positions 9 and 10 in Fig. 4.26) and transfers the steering-wheel moment M_H to the steering gear.

A compliant cardan joint (part 10 in Fig. 4.24) can be used to compensate for small angular deviations. This also keeps impacts away from the steering wheel

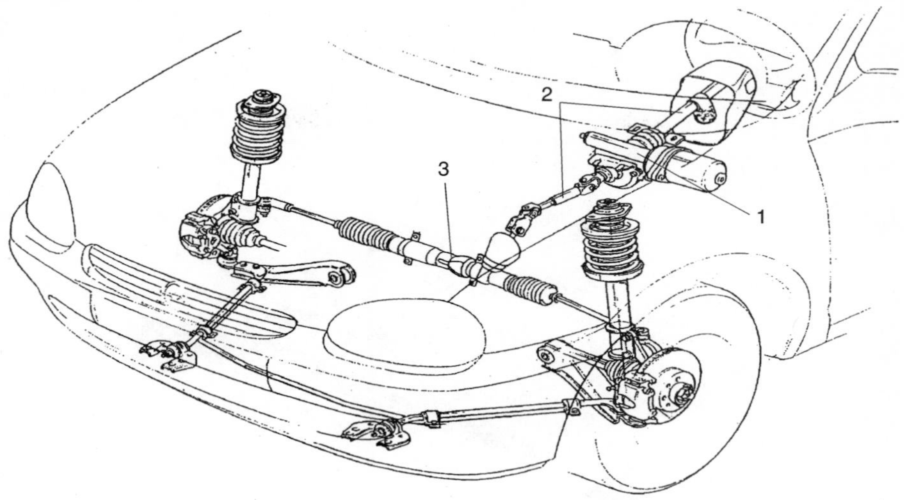

Fig. 4.23 Electric power steering system of the Opel Corsa (1997). The individual components are:

1 steering-column assembly
2 steering column with intermediate spindle
3 rack and pinion steering with external drive.

and, at the same time, performs a noise insulation function on hydraulic power-assisted steering. If the steering column does not align with the extension of the pinion gear axis (or the steering screw), an intermediate shaft with two universal joints is necessary (part 6 in Fig. 4.26). When universal joints are used, attention should be paid to their transmission properties, which are dependent on their angle of inflexion for steering wheel angle and moment, because a non-linear steering moment above the steering angle, noticeable for the driver, can occur.

The steering tube should be torsionally stiff to keep the steering elasticity low. On the other hand, it should show, together with the jacket tube, a deformation behaviour which is defined in a longitudinal direction, as steering wheel intrusion in case of a head-on crash is to be avoided while the absorption of force necessary for the functionality of the airbag (Fig. 4.25) must be safeguarded. As there is a requirement in some US states that the airbag should cushion a driver who is not wearing his seatbelt in a crash, despite the fact that seat belts are mandatory, the steering column must be designed for this borderline contingency.

Three types of steering tube configuration meet these requirements with vehicle-specific deformation paths on passenger cars:

- steering tubes with flexible corrugated tube portion (Fig. 4.24);
- collapsible (telescopic) steering tubes (Figs 4.27 and 4.28);
- detachable steering tubes (Figs 4.1 and 4.29).

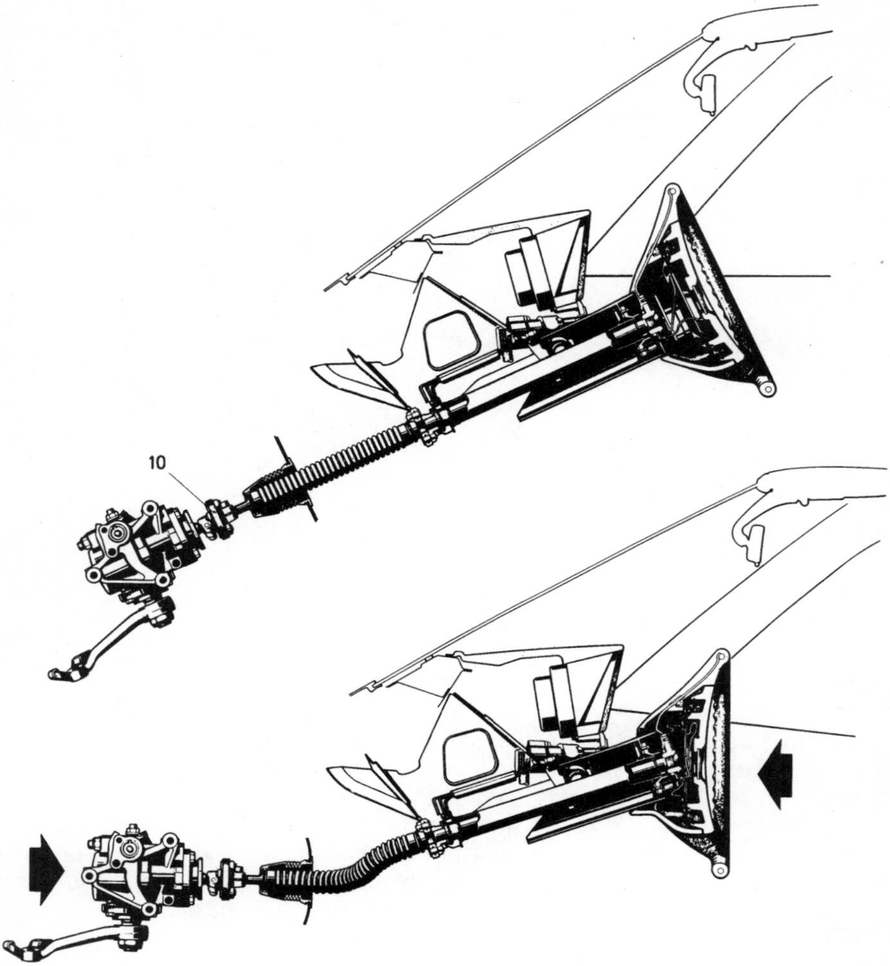

Fig. 4.24 Mercedes Benz safety steering tube and dished steering wheel; it is fixed to the recirculating ball steering gear with a compliant 'joint'. The bottom illustration shows the corrugated tube bending out in a head-on crash. The illustration also shows the energy-absorbing deformation of the steering wheel and the flexibility of the steering gear mounting.

To increase ride and seating comfort, most automobile manufacturers offer an adjustable steering column, either as standard or as an option. The position of the steering wheel can then be altered backwards and forwards as well as up and down (positions 1 and 2 in Fig. 4.30). As can be seen in the illustrations, electrical adjustment is also possible.

On light vans, which have a steering gear in front of the front axle, the steering

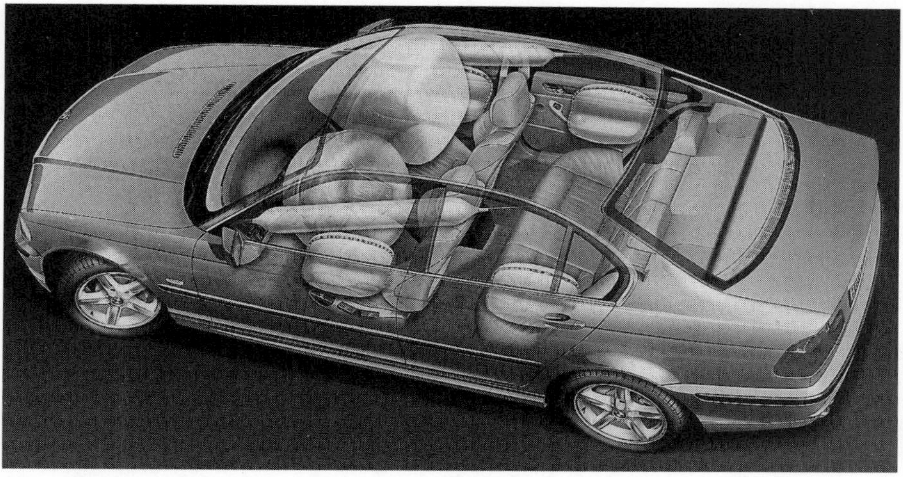

Fig. 4.25 BMW passenger car with air bags for the front, sides and head (front and back).

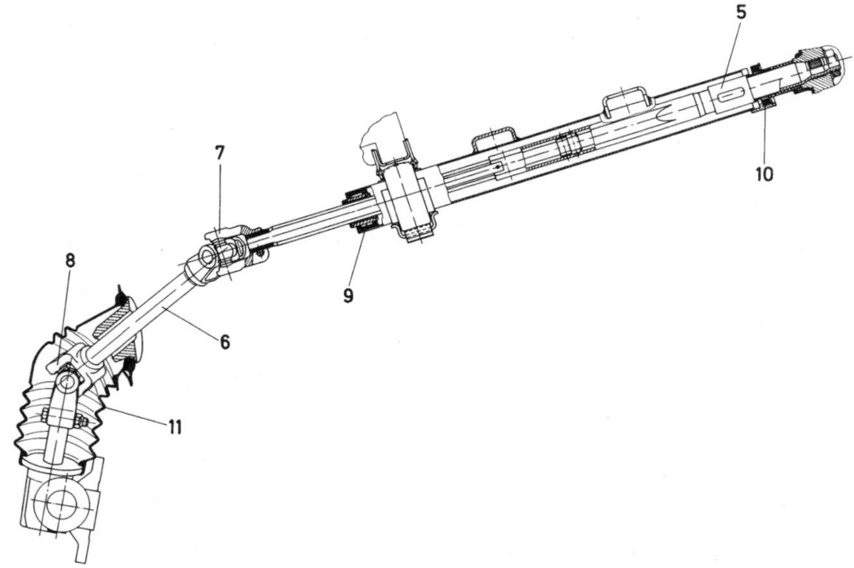

Fig. 4.26 Steering column of the VW Golf III and Vento (1996). The collapsible steering tube (Fig. 4.27) is carried from the bottom by the needle bearing 9 and through the top by the ball bearing 10 in the jacket tube; the spigot of the steering lock grips into part 5. The almost vertical pinion gear of the rack and pinion steering is linked to the inclined steering tube via the intermediate shaft 6 with the universal joints 7 and 8. The dashpanel is sealed by the gaiter 11 between this and the steering gear (illustration: Lemförder Fahrwerktechnik).

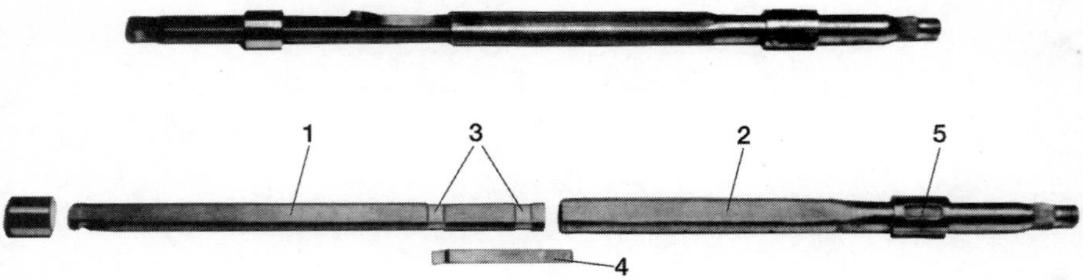

Fig. 4.27 Telescopic collapsible steering tubes consist of a lower part 1, which is flattened on the outside, and a hollow part 2, which is flattened on the inside. The two will be fitted together; the two plastic bushes 3 ensure that the assembly does not rattle and that the required shear-off force in the longitudinal direction is met. The tab 4 fixed to part 1 ensures the passage of electric current when the horn is operated. The spigot of the steering wheel lock engages with the welded-on half shells 5 (illustration: Lemförder Fahrwerktechnik).

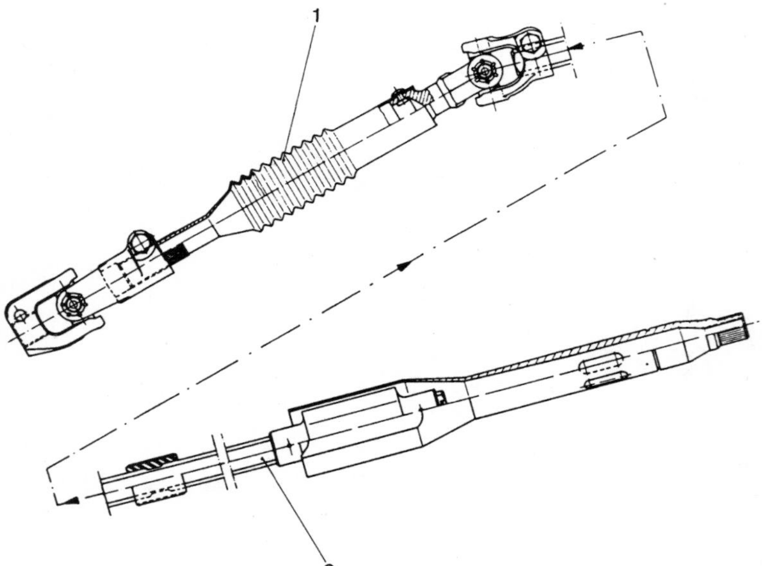

Fig. 4.28 Volvo steering column. Both the corrugated tube 1 in the intermediate shaft and the collapsible steering tube 2 meet the safety requirements. To save weight, the universal joints are made of aluminium alloy Al Mg Si 1 F31 (illustration: Lemförder Fahrwerktechnik).

Fig. 4.29 'Release clutch' used by VW on steering columns. A half-round plate sits on the short shaft that is linked to the steering pinion gear, and carries the two pins 1 which point downwards. They grip into the two holes of the clutch 2 sitting on the steering tube from the top. The jacket tube is connected to the dashboard via a deformable bracket. As shown in a head-on crash, this part 3 flexes and the pins 1 slide out of part 2.

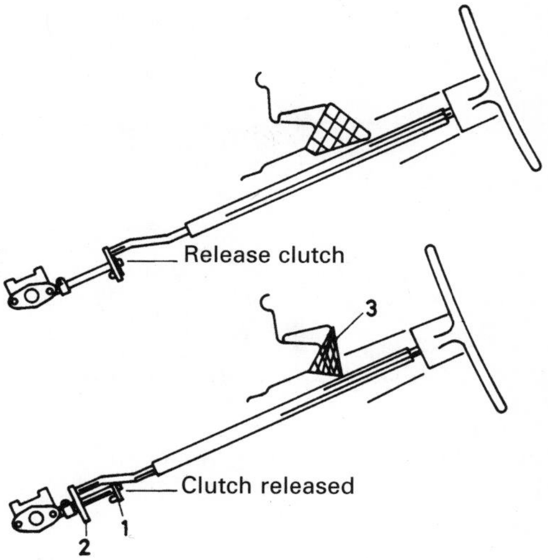

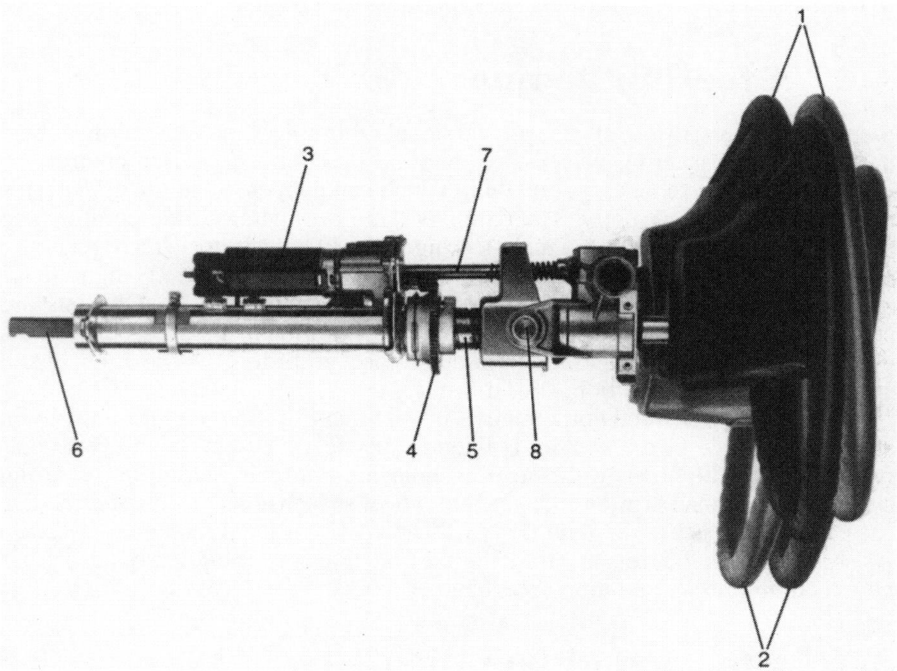

Fig. 4.30 Electrically adjustable steering column manufactured by Lemförder Fahrwerktechnik. The electric motor 3 turns a ball nut via the gears 4 and this engages with the grooves 5 of the steering tube and shifts it (position 6) in the longitudinal direction (position 1). To change the height of the steering wheel (position 2), the same unit tips around the pivot 8 by means of the rod 7.

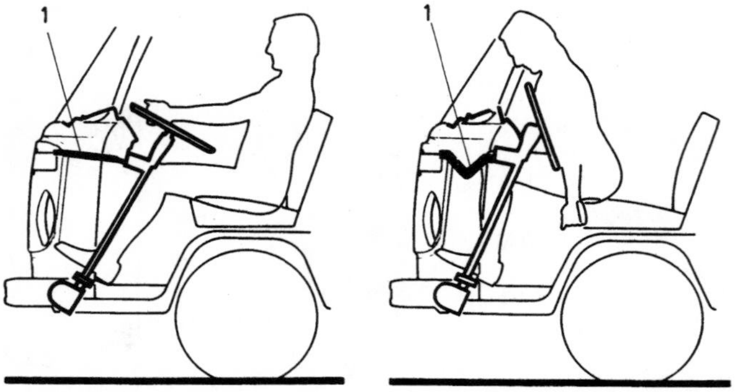

Fig. 4.31 The VW Bus Type II has an almost vertical steering column. In a head-on crash, first the steering wheel rim gives and then the retaining strut 1, which is designed so that a given force is needed to make it bend inwards.

column is almost vertical (Figs 1.7 and 1.37). In a head-on crash the outer tube bracket 1 and the steering wheel skeleton must flex (Fig. 4.31).

4.6 Steering damper

Steering dampers absorb shocks and torsional vibrations from the steering wheel and prevent the steering wheel over-shooting (also known as free control) on front-wheel drive vehicles – something which can happen when the driver pulls the steering wheel abruptly. The dampers therefore increase ride comfort and driving safety, mainly on manual steering gears. The setting, which generally operates evenly across the whole stroke range, allows sufficiently light steer-ability but stops uncontrollable wheel vibrations where the front wheels are subjected to uneven lateral and longitudinal vibrational disturbances; in this event the damper generates appropriate forces according to the high piston speeds involved (see Section 11.4 in Ref. [5]).

The dampers are fitted horizontally. As shown in Fig. 1.57, on rack and pinion steering, one side of the damper is fixed to the steering rack via an eye or pin-type joint and the other to the steering housing. On recirculating ball steering systems, the pitman arm on independent wheel suspensions or the intermediate rod can be used as a pivot point (Figs 1.39 and 4.12) or the tie rod on rigid axles. As shown in Fig. 4.5, this is parallel to the axle housing. Section 5.6.5 describes how the non-pressurized monotube damper works.

4.7 Steering kinematics

4.7.1 Influence of type and position of the steering gear

Calculating the true tie rod length u_0 (Fig. 4.32) and the steering arm angle λ (Fig. 4.3) creates some difficulties in the case of independent wheel suspensions.

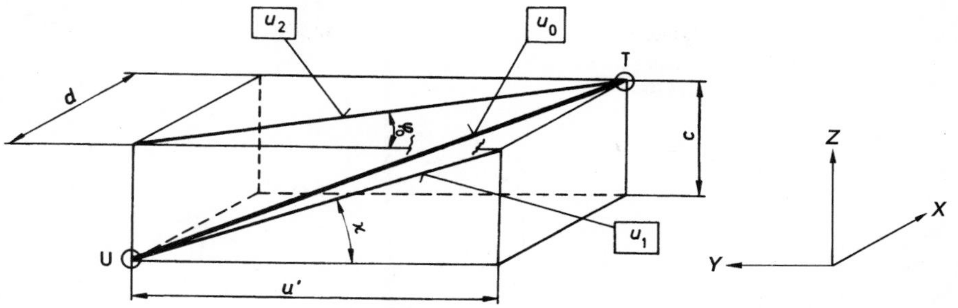

Fig. 4.32 On independent wheel suspensions, the tie rod UT is spatially inclined. The path u' (i.e. the lateral distance of points U and T from one another) or the angle κ must be determined when viewed from the rear. From the top view, the distance d or the angle ω_0 is more important; the projected lengths which appear in both views are u_1 and u_2. The true tie rod length is then:

$$u_0 = (u'^2 + c^2 + d^2)^{1/2}$$

The position of the steering column influences the position of the steering gear by the type of rotational movement. If this deviates from the horizontal by the angle ω (Fig. 4.33), a steering gear shaft, which is also inclined by the angle ω, becomes necessary. The inner tie rod joint T which sits on the pitman arm, is carried through a three-dimensional arc, influenced by this angle ω when the wheels are turned. However, the outer joint U on the steering knuckle whose steering axis is inclined inwards (Fig. 4.34) by the kingpin inclination angle σ and is often also inclined backwards by the caster angle τ (shown in Fig. 4.33). This joint therefore moves on a completely different three-dimensional arc (Figs 3.7, 3.9 and 3.11).

The construction designer's job is to calculate the steering arm angle λ (and possibly also the angle o of the pitman arm, Fig. 4.37) in such a manner that when the wheels are turned, the specified desired curve produced comes as close as possible. The achievement of the necessary balance is made more difficult still by the movements of the wheel carrier during driving: for example, wheel travel, longitudinal flexibility and vertical springing.

Figure 3.92 shows two curves that are desirable on passenger cars with an initially almost horizontal shape ($\Delta\delta \approx +30'$) and a subsequent rise in the curve to nearly half the nominal value when the wheels are fully turned. The more highly loaded wheel on the outside of the bend can even be turned further in than the inner wheel (and not just parallel to it, $\Delta\delta \approx -30'$); due to the higher slip angle that then has been forced upon it, the tyre is able to transfer higher lateral forces. When the wheels are fully turned, the actual curve should, nevertheless, remain below the nominal curve to achieve a smaller turning circle (see Equation 3.14).

The steering angle δ_o of the wheel on the outside of the bend depends on the angle of the one on the inside of the bend δ_i via the steering difference angle $\Delta\delta$:

$$\Delta\delta = \delta_i - \delta_o \text{ (axis of the ordinate, Fig. 3.92)}$$

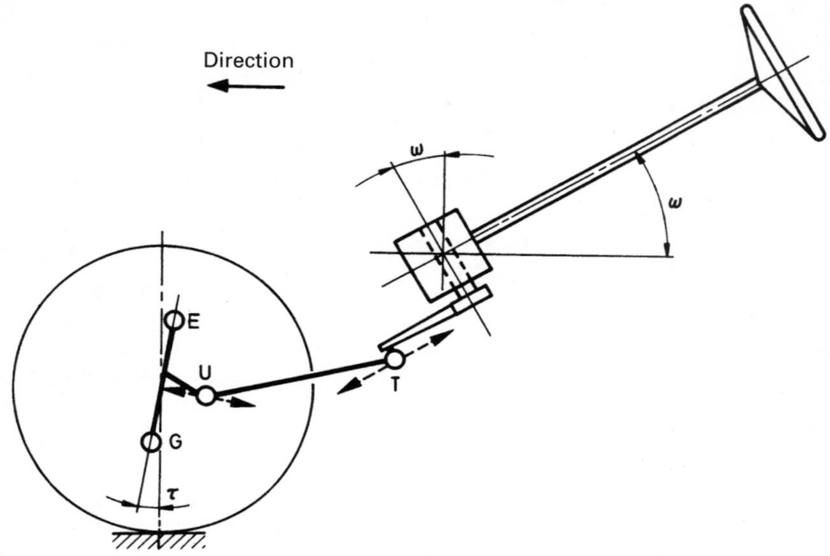

Fig. 4.33 The central points of the tie rod joints (T on the inside and U on the outside) change their position relative to one another, based on the wheel travel (vertical and horizontal) on independent wheel suspensions. The reasons for this are the different directions of movement of pitman arm and steering arm. The former depends on the inclined position of the steering gear (angle ω) and that of the point U from the inclination of the steering axis EG, i.e. the kingpin inclination σ and the caster angle τ.

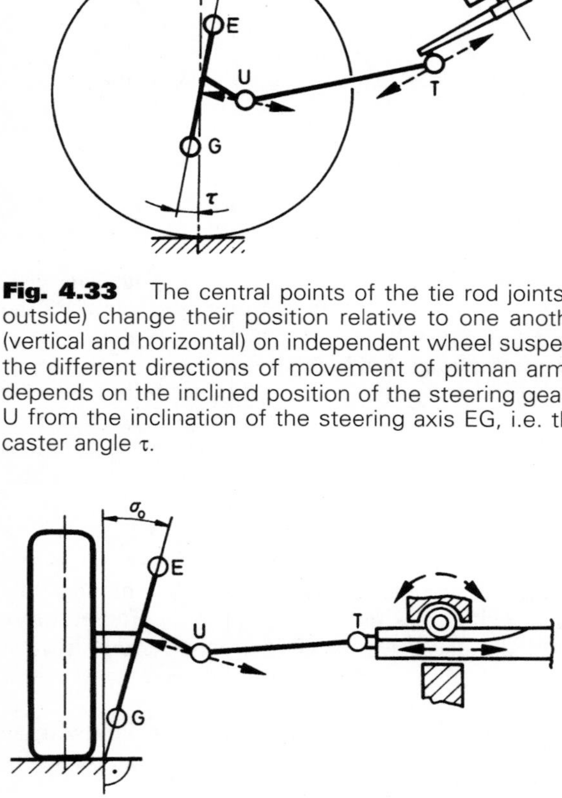

Fig. 4.34 When viewed from the rear, the inner tie rod joint T on rack and pinion steering moves parallel to the ground, whereas the outer tie rod joint U moves on an arc running vertical to the steering axis EG. Any caster angle τ must also be considered.

4.7.2 Steering linkage configuration

The main influences on Δδ are the steering arm angle λ, the inclined position of the tie rod when viewed from the top (angle φ_o, Fig. 4.32) and the angle *o* of the pitman and idler arms on steering gears with a rotational movement. The tie rod position is determined by where the steering gear can be packaged. The amount of space available is prescribed and limited and the designer is unlikely to be able to change it by more than a little. The task consists of determining the angles λ and *o* by drawing or calculation. Both also depend on the bearing elasticities, which are not always known precisely.

The configuration of the steering kinematics on rack and pinion steering is comparatively simple; here, it is only necessary to transfer a straight-line lateral shift movement into the three-dimensional movement of the steering knuckle (Fig. 4.34). However, the extension of the tie rod UT must point to virtual centre of rotation P (Fig. 4.35); this is necessary on all individual wheel suspensions for determining the body roll centre Ro and is therefore known (see Sections 3.4.3 and 4.6.3).

On steering gears with a rotational movement, the 4-bar linkage can be either in front or behind the axle and can be opposed or synchronous; Figs 4.3 and 4.36 to 4.38 show four different configurations.

From a kinematic point of view, rack and pinion steering systems have a triangular linkage that can either be in front of or behind the axle or even across it. Figures 4.4 and 4.39 to 4.41 show the individual options for left- and

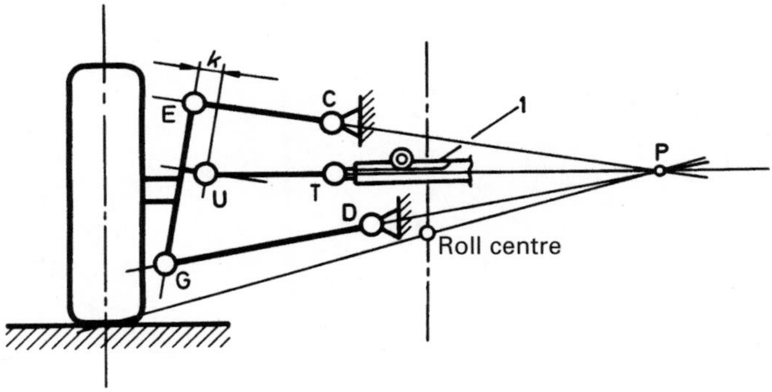

Fig. 4.35 Path and movement points necessary for determining the tie rod length and position. The position of the tie rods is given by the connecting line UP (to the pole). The illustration also shows the roll centre Ro.

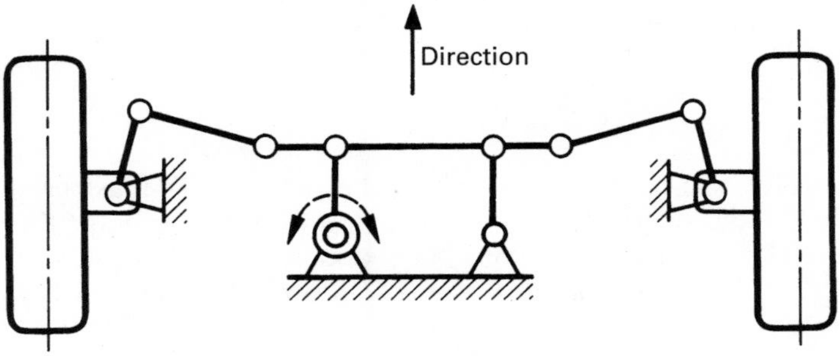

Fig. 4.36 'Synchronous' 4-bar linkage with steering arms pointing forwards. The inner joints are fixed to the sides of the intermediate rod.

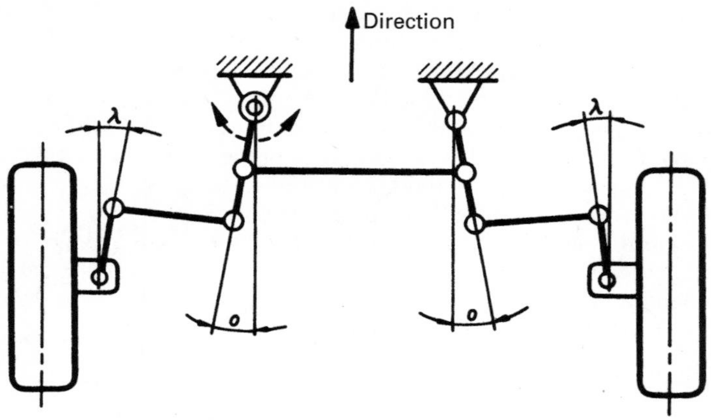

Fig. 4.37 'Opposed' 4-bar linkage located in front of the wheel centre. Steering arm and pitman arm rotate in opposite directions towards one another, similar to meshing gears. The tie rods are fixed directly to pitman and idler arms. For kinematic reasons, these can have the pre-angle *o* (see also Fig. 1.7).

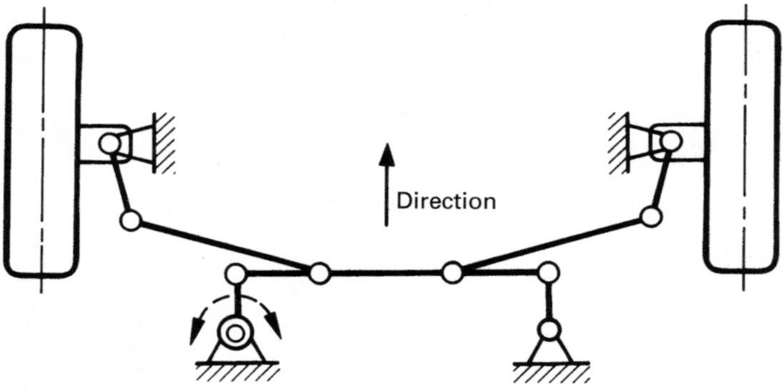

Fig. 4.38 'Opposed' 4-bar linkage located behind the wheel centre. The inner tie rod joints can be fixed to the middle part of the intermediate rod or directly to the pitman and idler arm (see Fig. 4.12).

right-hand drive vehicles and also where the pinion gear must be located – above or below the steering rack – to make the wheels turn in the direction in which the steering wheel is turned. The steering arms (negative angles λ) which point outwards, shown in Fig. 4.41, allow longer tie rods; something which is useful when the inner joints are pivoted on the ends of the steering rack (Fig. 3.67).

The significantly simpler steering kinematics on rigid axles are shown in Figs 4.5 to 4.7 and are described in Chapter 2 of Ref. [1] and Chapter 5 of Ref. [10].

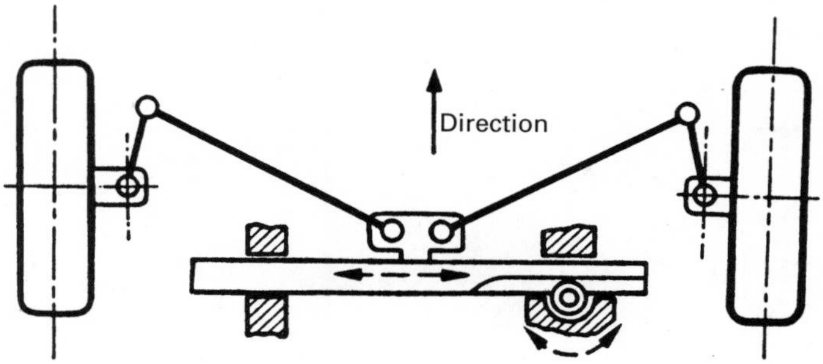

Fig. 4.39 The rack-and-pinion steering is behind and above the wheel centre and the steering arms point forward (shown for a right-hand drive vehicle). For kinematic reasons, the inner tie rod joints are fixed to a central outrigger – known as a central take-off. This type of solution (also shown in Fig. 1.57) is necessary on McPherson and strut damper front axles with a high-location steering system as the tie rods have to be very long to avoid unwanted steering angles during jounce.

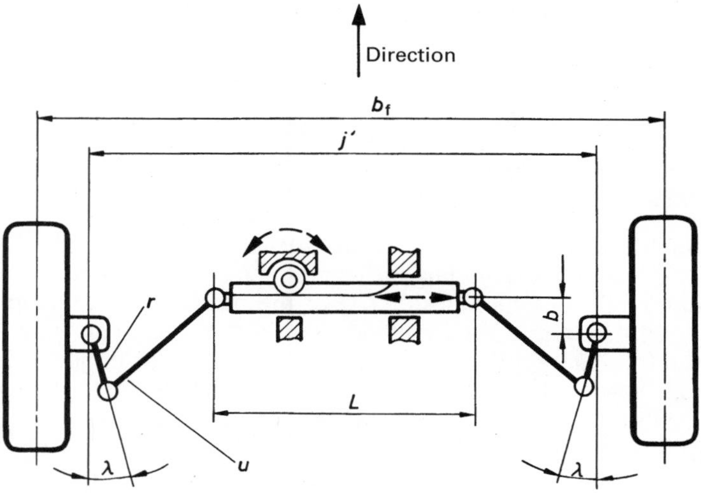

Fig. 4.40 The steering is in front of the wheel centre and the triangular linkage behind it, with the inner joints fixed to the ends of the steering rack.

4.7.3 Tie rod length and position

When the wheels compress and rebound as well as in longitudinal movement, there should not be any, or only a very specific, toe-in alteration; both depend primarily on the tie rods being the correct length and on their position. Various illustrations in Section 3.6 show the results of incorrect toe-in and the possibility of achieving a roll–steer effect on the front wheels and steer-fight during

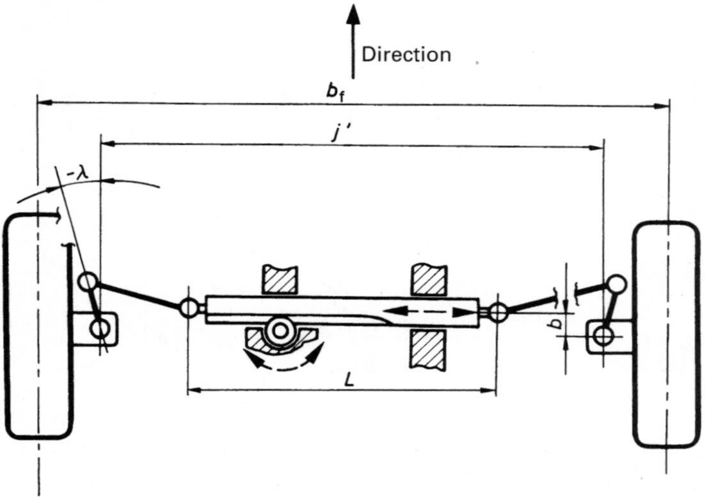

Fig. 4.41 Where rack and pinion steering and the steering triangle are shifted in front of the wheel centre, for kinematic reasons the steering arms must point outwards, making longer tie rods possible (see also Fig. 1.40).

braking. The elasticity in the steering system (Figs 3.99 and 3.100) or that in the bearings of the steering control arms, is also a contributory factor. Chapter 7 of Ref. [3], gives the calculation of the forces required for such elasticity.

4.7.3.1 Double wishbone and multi-link suspensions

There are two ways of determining the central point T of the inner tie rod joint as a function of the assumed position U of the outer joint, the template and 'virtual centre' procedure. Both methods consider one side of the front axle when viewed from the rear (here the left side, Fig. 4.42). The projected length u' of the tie rod shown in Fig. 4.32 and the angle κ, which determines its position, must be calculated. This must match the line connecting the outer joint U with pole P, which is also needed for calculating the roll centre (see Section 3.4.3).

Initially, the position of the outer tie rod joint U is unknown when viewed from the rear; to obtain an approximation of this point, the height of the steering gear must be specified (Fig. 4.35). The angle λ is assumed so that, together with the known steering arm length r, the path required for configuring it

$$k = r \sin \lambda \tag{4.3}$$

can be calculated (for r and λ see Fig. 4.40).

The templates that are used for finding point T by drawing have already been described in Section 3.3 and can be seen in Figs 3.7 to 3.11. All figures contain point U and the curve of its movement. It only remains to find point T on the connecting line UP. T would be the centre point of the arc which best covers the path of point U.

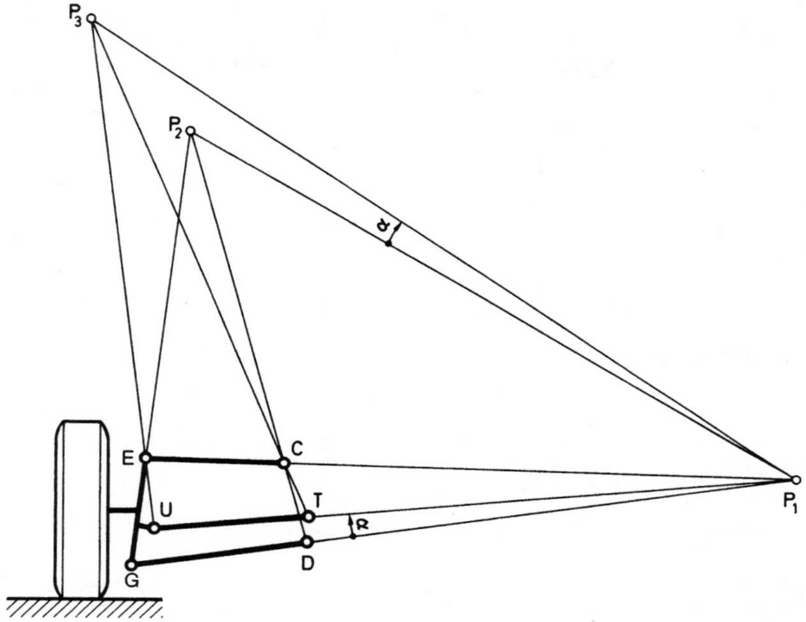

Fig. 4.42 Double wishbone suspension with steering arm pointing inwards. The tie rod is above the lower control arm.

It is likely to be simpler and more precise to determine the point T graphically, using virtual centres. First, as shown in Figs 4.42 and 3.24 to 3.28, the virtual-centre at P (marked here as P_1) must be calculated so that it can be connected to U. The extension of the paths EG and DC gives P_2, which is also required and from which a line to P_1 must be drawn. If the path UP_1 is above GD, the angle α enclosed by the two must be moved up to P_1P_2; if UP_1 were to lie below it, the line would have to be moved down. A line drawn from P_1 at the angle α must be made to intersect with the extension of the connecting path UE to give the tie rod virtual-centre P_3. To calculate the desired point T – i.e. the centre of the inner joint – P_3 is connected to C and extended.

The path k (i.e. the distance of point U from the steering axis EG, Fig. 4.35 and Equation 4.3) is the determining factor for the position of virtual-centre P_3 in the lateral direction. Figure 4.43 shows the case of point U, which lies left of the path EG. This is something that is only possible where the steering gear is located in front of the axle (Fig. 4.41). P_3 moves to the right, resulting in an inner link T moving further away from the centre of the vehicle. This is beneficial if it is to be fixed to the end of the steering rod.

A tie rod that is located above the upper suspension control arm (Fig. 4.44) causes a large angle α and P_3 that is shifted a long way to the right. Where the control arms are parallel to one another (Figs 4.45 and 3.25), P_1 is at ∞. In such cases, a line parallel to the path GD must be drawn through U and, at the same

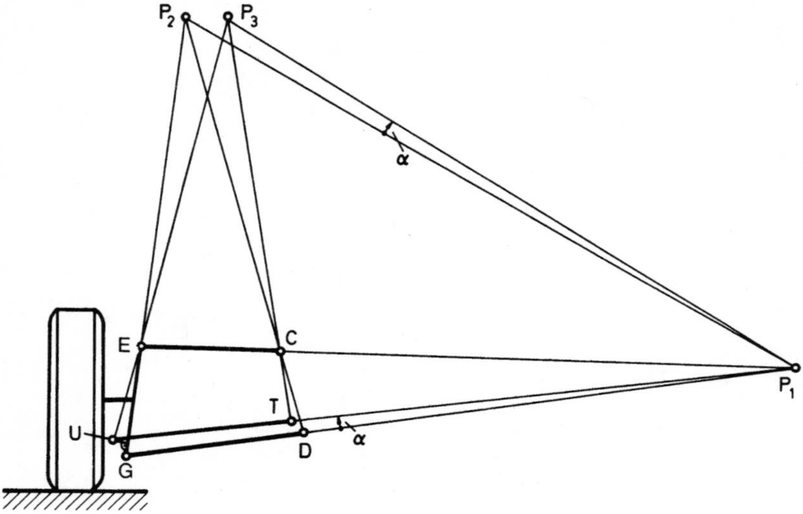

Fig. 4.43 In the case of a steering gear located in front of the wheel centre, the centre of the tie rod joint U lies outside the steering axis EG.

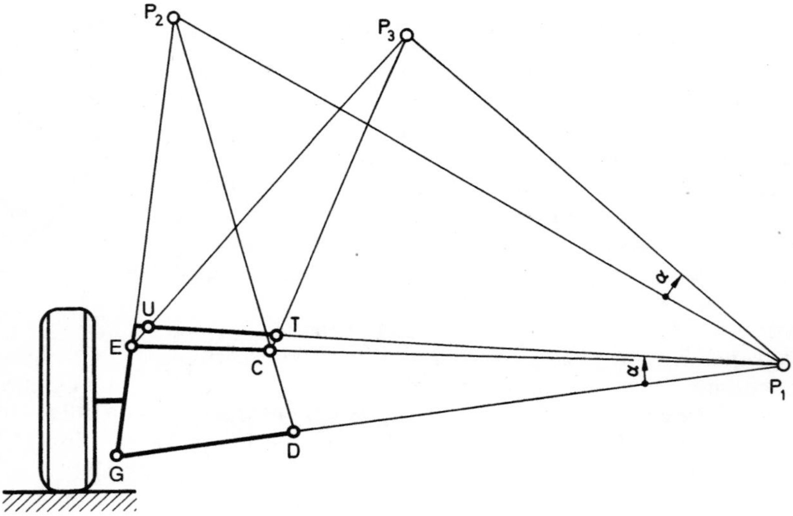

Fig. 4.44 A high-location steering gear can involve a tie rod above the upper control arm. The steering arm points backwards and towards the inside in the example.

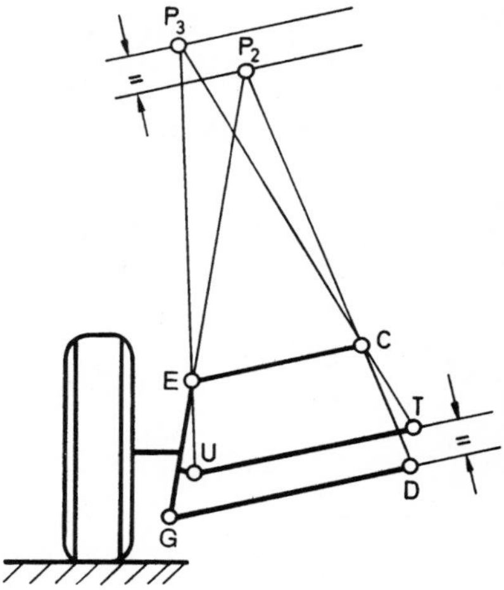

Fig. 4.45 Suspension control arms, which are parallel to one another in the design position of the vehicle, have to have a tie rod in the same position.

distance, a further one drawn through the virtual centre P_2. The intersection of this second parallel with the extension of the path UE gives P_3, which must be linked to C to obtain T.

4.7.3.2 McPherson struts and strut dampers

When the vehicle is fitted with McPherson struts or strut dampers – due to the alteration in distance between E and G when the wheels compress and rebound – point T is determined by a different method. To obtain pole P_1, a vertical to the centre line of the shock absorber is drawn in the upper mounting point E and made to intersect with the extension of the suspension control arm GD (Figs 3.29 and 4.46); P_1 linked with U gives the position of the tie rod. A line parallel to EP_1 must be drawn through G; the intersection with the extension of ED then gives the second virtual-centre P_2. The angle α, included by the paths EP_1 and UP_1, must be entered downwards to the connection P_1P_2 to obtain P_3 as the intersection of this line with the extension of the path UG. The extension of the connecting line P_3D then gives the central point T of the inner tie rod joint on UP_1.

If, in the case of $\lambda = 0°$, point U is on the steering axis EG which dominates the rotation movement (Figs 3.30 and 4.47), P_3 is on the extension of this path. The determining factor for the position of P_1 is the direction of the shift in the damping part of the McPherson strut; for this reason, the vertical in point E must be created on its centre-line (not on the steering axis EG). The important thing in this calculation is the position of point U, i.e. the extension of the connecting line UG downwards. U is shown on the steering axis EG simply for reasons of presentation.

A low mounted tie rod causes the virtual-centre P_3 to move to the right (Fig. 4.48) and this then causes a shorter rod. This situation is favourable if the inner joint needs to sit on the ends of the steering rack. The figures clearly show that the higher U, which constitutes the connection between steering arm and tie-rod,

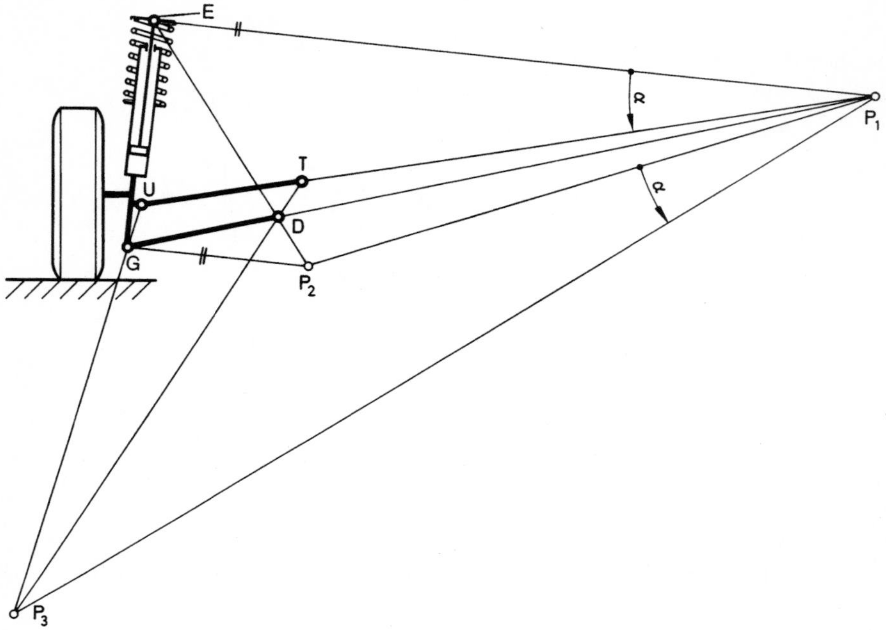

Fig. 4.46 On the McPherson strut or strut damper, the tie rod is above the lower control arm; the steering arms point inwards with the result that the outer joint U lies more to the vehicle centre.

is situated, the longer the tie rods must be, i.e. a centre take-off becomes necessary on a high-mounted rack and pinion steering (Figs 1.57, 4.11 and 4.39).

4.7.3.3 Longitudinal transverse axles

On longitudinal wishbone axles the upper point E moves in a straight line vertical to the steering axis CF and the lower point G on an arc around D (Figs 3.32 and 4.49). To obtain P_1, a parallel to CF must therefore be drawn through E and made to intersect with the control arm extension GD. A parallel to EP_1 laid through point D gives the virtual centre P_2 on the connecting line EG. The angle α enclosed by the paths EP_1 and UP_1 must be drawn downwards to the connecting line P_1P_2 to obtain the virtual centre P_3 as the intersection with the extension of the path UG. P_3 linked with D then gives the centre T of the inner tie rod joint.

4.7.3.4 Reaction on the steering arm angle λ

Figures 4.40 to 4.49 indicate that shifting the outer joint U to the side results in a slight alteration in the distance UT. However, this shift is necessary if the angle λ has to be reduced or increased with a given steering arm length r. The projected length u' of the tie rod, and therefore also its overall length u_0 (Fig.

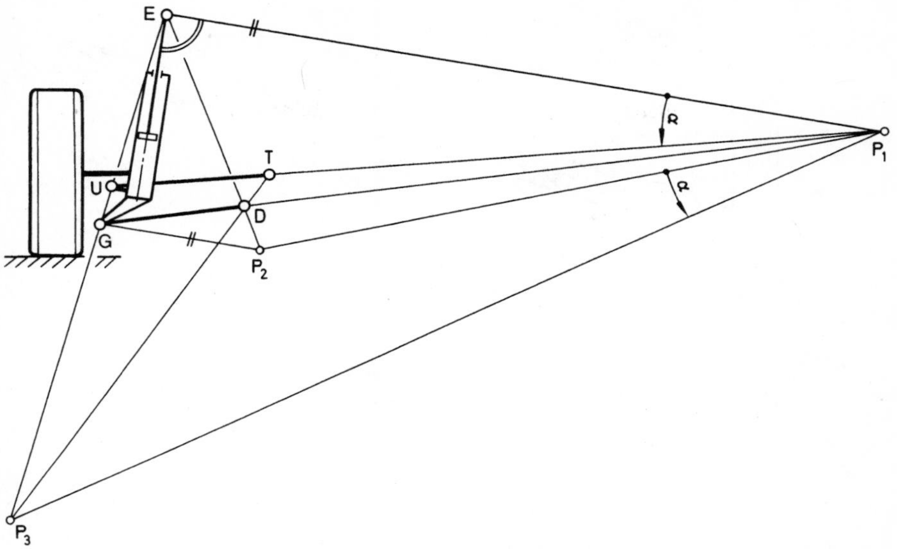

Fig. 4.47 On a McPherson strut with the joint G shifted to the wheel, the outer one, U of the tie rod, can lie in the plane of the steering axis (i.e. on the connecting line EG) when viewed from the rear. Extending the path UG is crucial for determining the virtual centre P_3, whereas the direction of movement of the damper, i.e. a vertical on the piston rod in point E, must be the starting point for calculating P_1.

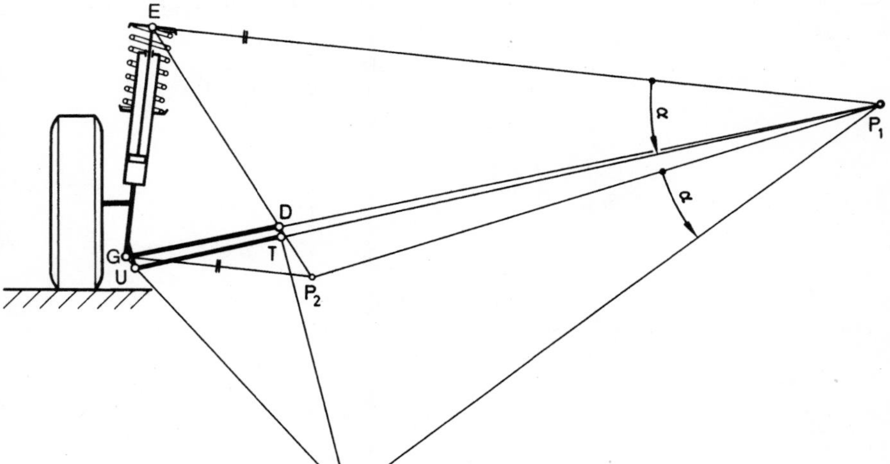

Fig. 4.48 The tie rod can also lie under the control arm when the steering arm points inwards.

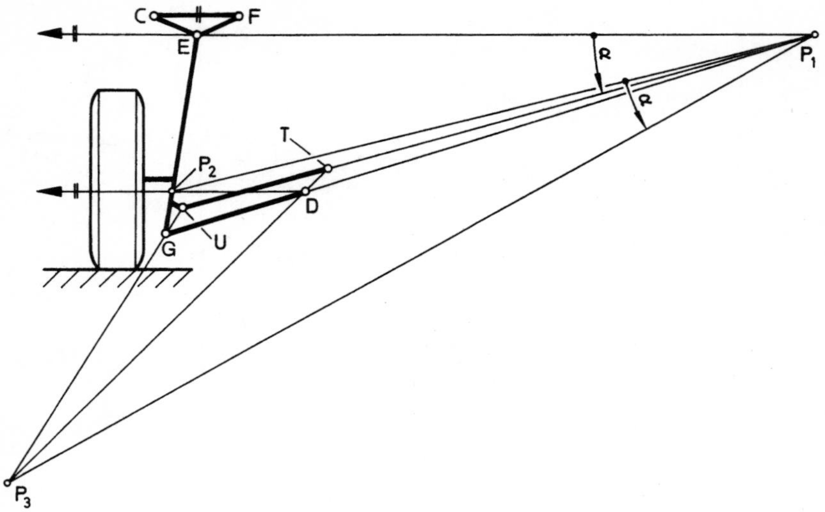

Fig. 4.49 Longitudinal transverse axle with the tie rod located above the lower control arm and the steering arm pointing inwards.

4.32), changes when viewed from the rear. However, the latter is one of the determining factors for the aspects relating to the steering angles δ_i (inside) and δ_o (outside), i.e. for the actual steering curve (Fig. 3.92). It is, therefore, likely to be essential to check the desired position of point T with the tie rod, which has become longer or shorter.

5

Springing

5.1 Comfort requirements

Springing and damping on a vehicle are mainly responsible for:

- ride comfort and dynamic wheel load.

They also play an important part in:

- handling (Fig. 5.2)
- the tendency of the body to roll and pitch.

Other important influences on the handling are the kinematic changes, and the elastokinematics, of the wheels as they bump and jounce. Details are given in Chapter 3, and in Refs 2 and 9.

The ride comfort experienced by the vehicle occupants depends on their (Stainz) fitting position (Fig. 5.1) – also in relation to the controls such as steering wheels and pedals – as well as on the acceleration and mechanical vibration acting upon them. The critical frequency range is 1–80 Hz. It is sensible to subdivide this into two ranges to which different comfort terms are allocated:

- springing or ride comfort, which lies below $n = 240\ \text{min}^{-1}$, i.e. 4 Hz;
- wheel comfort or road harshness ($f > 4$ Hz).

The split is sensible because the two frequency ranges are experienced differently by the human body and become important if individual parts of the suspension, such as springs, shock absorbers, suspension link bearings etc., are to be evaluated for their influence on comfort. The springing balance (which expresses how well the front and rear axles are matched to one another) also needs to be taken into consideration. If a vehicle does not pitch when it goes over bumps in the ground, but instead moves up and down in parallel translation, it has a good springing balance. To provide an objective evaluation of

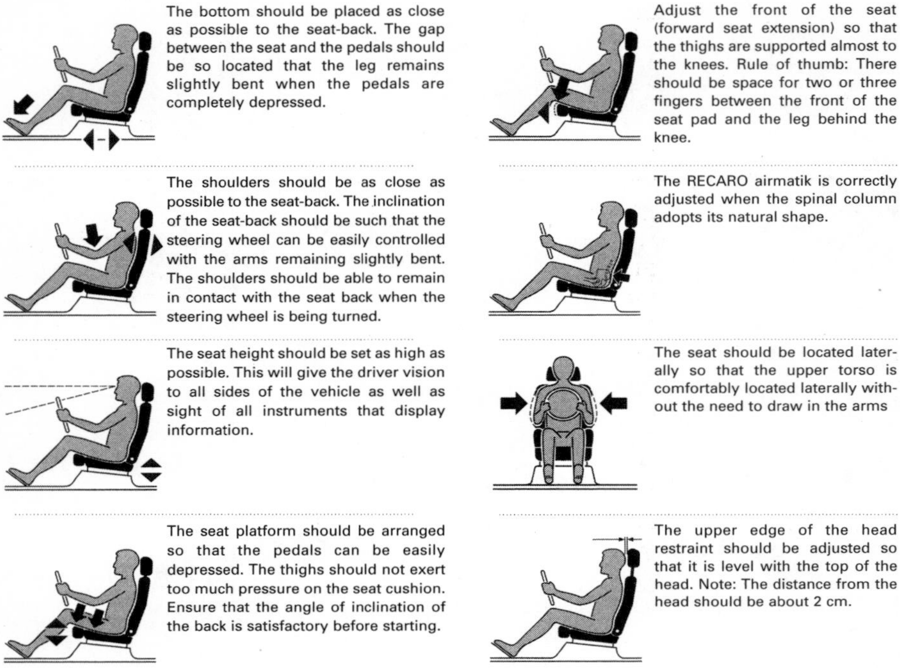

Fig. 5.1 In addition to the body suspension and damping, the seat is of crucial importance for driver comfort. The position of the seat must ensure safe, comfortable and tireless operation of the vehicle. Apart from the static properties of the seat (general seat position, possibility of adjustment), the quasi-static characteristics (possibility of slight body movements to achieve muscle relaxation), temperature and air-conditioning properties and the transmission of vibration are also important. The latter depends on the seat design (suspension and damping behaviour of the seat) and the mass of the driver; the latter in particular determines the excitation of the driver and hence the impression of comfort, specifically in the region of vertical vibration above 5 Hz.

To increase the safety of the driving environment, active ventilation of the seats by electric fans in the seat cushion and backrest cushion can be used, as well as dynamically operating pneumatic adjustment of the backrest in the shoulder and lumbar region for muscle, pelvic and spinal column relaxation can be used. By means of two hydraulic chambers attached to the surface of the seat, BMW produces a movement of the spinal column specifically intended for the prevention of scoliosis.

comfort, measurement devices are used which evaluate, based on the VDI (Verein Deutscher Ingenieure) directive 2057, the vibration that occurs (vibration stroke, speed and acceleration) dependent on frequency, in accordance with existing knowledge of how the human body experiences it. The measurement result is then available as a numerical value, the so-called K-value. Low values indicate good ride comfort, whereas high values indicate poor ride comfort (see also Chapter 7 in Ref. 9).

5.1.1 Springing comfort

This comfort range is mainly influenced by the acceleration acting on the upper part of the human body and lies in a frequency range of 1 to approximately 4 Hz. With a given vehicle body mass m_{Bo} (see Equation 6.5), the critical variables are the configuration of the rate of the body springing and rising from the body's resonant frequency. In accordance with the simplified model of the sprung-mounted car body shown in Fig. 5.7 (single mass vibrator) the mass portion $m_{Bo,\,f\text{ or }r}$ exhibits free undamped vibration at the natural frequency in accordance with Equation 5.4, and the corresponding vibration rate in accordance with Equation 5.4a.

The softer the springing, i.e. the lower the springing rate $c_{f\text{ or }r}$ of the body (front or rear), the lower the natural frequency for a specified body mass and, accordingly, the greater the ride comfort. Unfortunately, at the same time the roll increases on bends (it must be reduced by anti-roll bars, see Section 5.5.4 and Fig. 5.2), as does the tendency to pitch when the brakes are applied or when starting out (see Sections 3.11 and 6.3). Vibration values of $n_{f\text{ or }r} = 60\ \text{min}^{-1}$ (where $f = 1$ Hz) are desirable, but cannot necessarily be easily achieved (see Section 5.2).

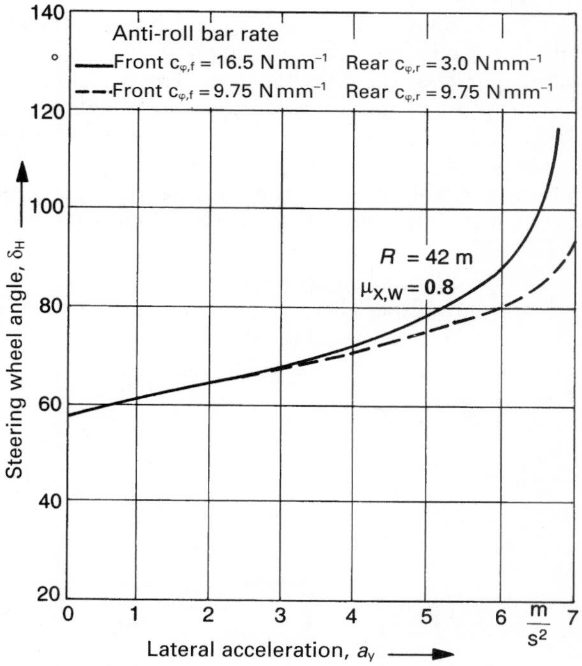

Fig. 5.2 Influence of the anti-roll bar rate on the steering angle, measured while the vehicle is steady-state driving on a circular path (R = 42 m) on a standard design passenger car with $m_{V,t}$ = 1544 kg. The understeering can be reinforced, or an incipient tendency to oversteer reduced by increasing the rate of the front anti-roll bar and/or reducing the rate of the rear one. On front-wheel drive vehicles a more highly stabilized rear suspension is usually necessary.

In the case of low lateral acceleration, and therefore also on wet or slippery roads, the anti-roll bar rate has no effect on the self-steering behaviour.

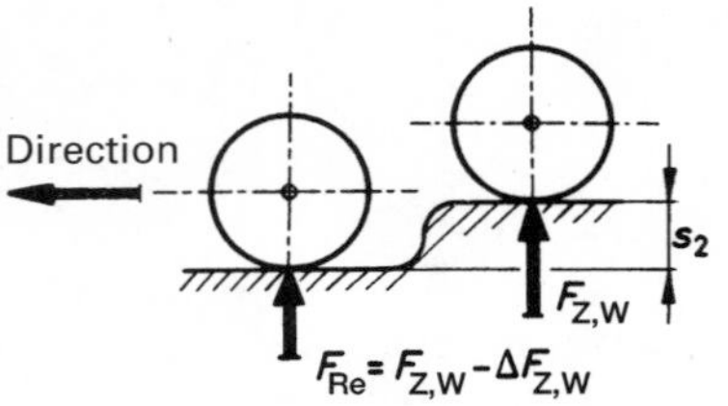

Fig. 5.3 When a wheel rebounds by the path s_2, the wheel load reduces by the amount $\Delta F_{Z,W}$. The level of the residual force which still ensures wheel grip $F_{Re} = F_{Z,W} - \Delta F_{Z,W}$ depends mainly on the springing stiffness, defined by the rate $c_{f\ or\ r}$.

Another advantage of soft springing would be the improvement in the absorbency of bumps and the wheel grip. If, for example, a front wheel loaded at $F_{Z,W}$ = 3000 N drops into a 60 mm deep pothole (Fig. 5.3), with soft linear springing at the rate c_f = 15 N mm^{-1} the residual force at the bottom of the pothole is

$$F_{Re,f} = F_{Z,W,f} - c_f\, s_2 = 3000 - 15 \times 60 = 2100\ \text{N} \tag{5.0}$$

With 'sporty' hard springing at c_f = 30 N mm^{-1}, it is only $F'_{Re,f}$ = 1200 N. The greater residual force equates with better road holding. The same can be said of a vehicle travelling over a 40 mm high bump (Fig. 5.4). With hard springing, the force transferred from the axle to the body as an impact, ignoring the damping and time influence, would be $\Delta F_{Z,W}$ = 1200 N; soft springing only transfers 600 N and therefore generates lower wheel load fluctuation.

The disadvantage (as already mentioned) is the greater body roll on bends and the concomitant lower ability of the wheels to transfer lateral forces (see Section 5.4.3 and Equation 2.16). As shown in Fig. 1.6, the wheels incline with the body on independent wheel suspensions. The wheel on the outside of the bend, which absorbs most of the lateral forces, loses negative camber (or goes into positive camber), resulting in the need for a larger tyre slip angle (see Section 2.8.5.5).

The springing comfort, and associated with it also the handling, depends not only on the weight of the vehicle and the body springing rate, but also on other variables and the interaction of the individual components:

- the load distribution (see Section 5.3.6)
- the design of the wheel suspension
- the type of mounting and design of the springs (see Section 5.3)

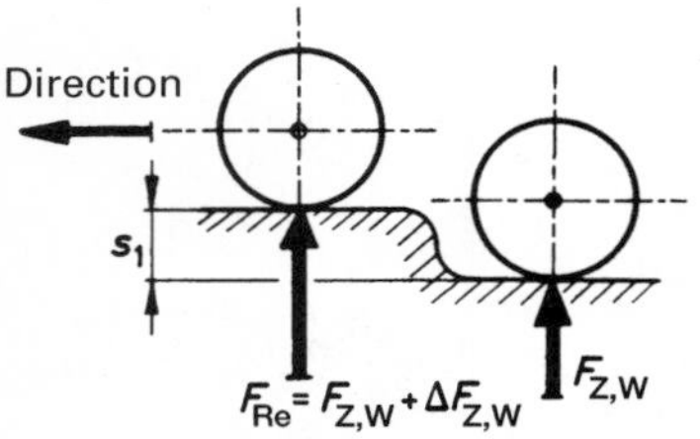

Fig. 5.4 When the wheel displaces by distance s_1 the wheel load increases by $\Delta F_{Z,W}$. The size of the increase in force in the body depends mainly on the springing hardness, i.e. the rate $c_{f\ or\ r}$.

- the anti-roll bars (Fig. 5.2 and Section 5.5.4)
- the torsional rate of the rubber bushings (Figs 3.18, 3.84 to 3.87, and 5.5)
- the shock absorbers and their mountings (see Section 5.6.7)
- the weight of the axles (see Section 6.1.3)
- the type of engine and gearbox mounting (see Chapter 10 in Ref. [5])
- the wheelbase (see Section 3.2)
- the tread width (see Section 3.3) and
- the tyres in general (see Section 2.4).

5.1.2 Running wheel comfort

Even smooth-looking road surfaces have almost invisible slight irregularities and bumps which are transferred to the body as high frequency acceleration and jolts (4–80 Hz). The vehicle occupants feel them in the underbody of the vehicle, in the seat cushion, and the driver also feels them in the steering wheel and the pedals. They determine the wheel comfort and the concomitant road harshness.

The cause of this is the often limited vibration insulation between the suspension parts and the body, i.e. the suspension links, suspension subframe and McPherson strut mount, plus the friction in the suspension control arm bearings (Fig. 5.5), the wheel joints (Fig. 1.38) and in the shock absorbers or spring dampers (Fig. 5.51; see also section 4.2 in Ref. [2] and Ref. [5]).

On McPherson struts and strut dampers the friction in the piston rod guide generated by transverse forces can be the cause (see Figs 1.8, 1.11 and 3.30; also Section 6.43 in Ref. 5). The springing does not respond as well and today's ever-wider (and therefore harder) tyres no longer absorb the bump loads – these are transferred directly to the body.

These relationships can easily be explained using the hysteresis of a springing curve (Fig. 5.6). The friction force is 200 N per wheel in the central range, i.e. starting from the centre line:

$$F_{fr} = \pm 100 \text{ N}$$

The rate of body springing at the front should be $c_f = 15$ N mm^{-1} and the height of a bump $s_1 = 6$ mm results in a spring force of

$$\Delta F_f = c_f\, s_1 = 15 \times 6 = 90 \text{ N} \tag{5.0a}$$

As $F_{fr} > \Delta F_f$, in this instance the soft springing would not absorb the bump and the suspension would pass on the force to the body (see also Section 2.5).

However, if the spring rate is $c_f = 30$ N mm^{-1}, the force would be absorbed by the spring. The problem here is reversed, as shown in Figs 5.3 and 5.4.

Soft springing creates greater difficulties in achieving the desired running wheel comfort in terms of road harshness than harder springing, particularly on front-wheel drive vehicles.

There is also the longitudinal vibration caused by the steel belts of the radial tyre, particularly on rough cobbles. Section 2.2.2 contains details and Section

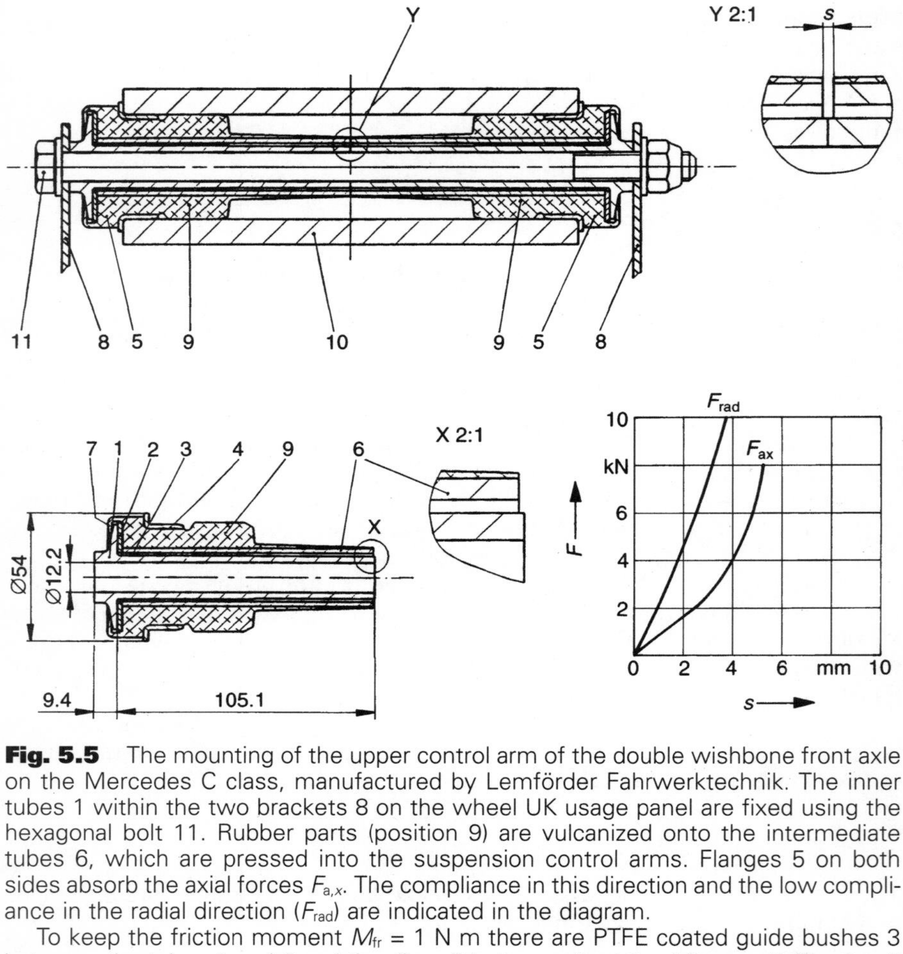

Fig. 5.5 The mounting of the upper control arm of the double wishbone front axle on the Mercedes C class, manufactured by Lemförder Fahrwerktechnik. The inner tubes 1 within the two brackets 8 on the wheel UK usage panel are fixed using the hexagonal bolt 11. Rubber parts (position 9) are vulcanized onto the intermediate tubes 6, which are pressed into the suspension control arms. Flanges 5 on both sides absorb the axial forces $F_{a,x}$. The compliance in this direction and the low compliance in the radial direction (F_{rad}) are indicated in the diagram.

To keep the friction moment M_{fr} = 1 N m there are PTFE coated guide bushes 3 between the tubes 1 and 6 and the discs 2 between the lateral flanges 5. The lips 7 provide the seal to the maintenance-free mountings. The smaller the moment M_{fr} can be, the more favourable the ride comfort and the absorbency of the springs become. The outer tubes 6 are slightly shorter than the inner ones (position 1), between them is the clearance *s,* which evens out installation tolerances and provides the longitudinal mobility to take the radial tyre rolling hardness (see Section 3.6.5.2). In the case of high (axial) braking forces, depending on the compliance of the rubber flange 5, the outer tubes 6 butt up against one another and ensure the necessary longitudinal stiffness.

For further details, see Section 2.3 in Ref. [2].

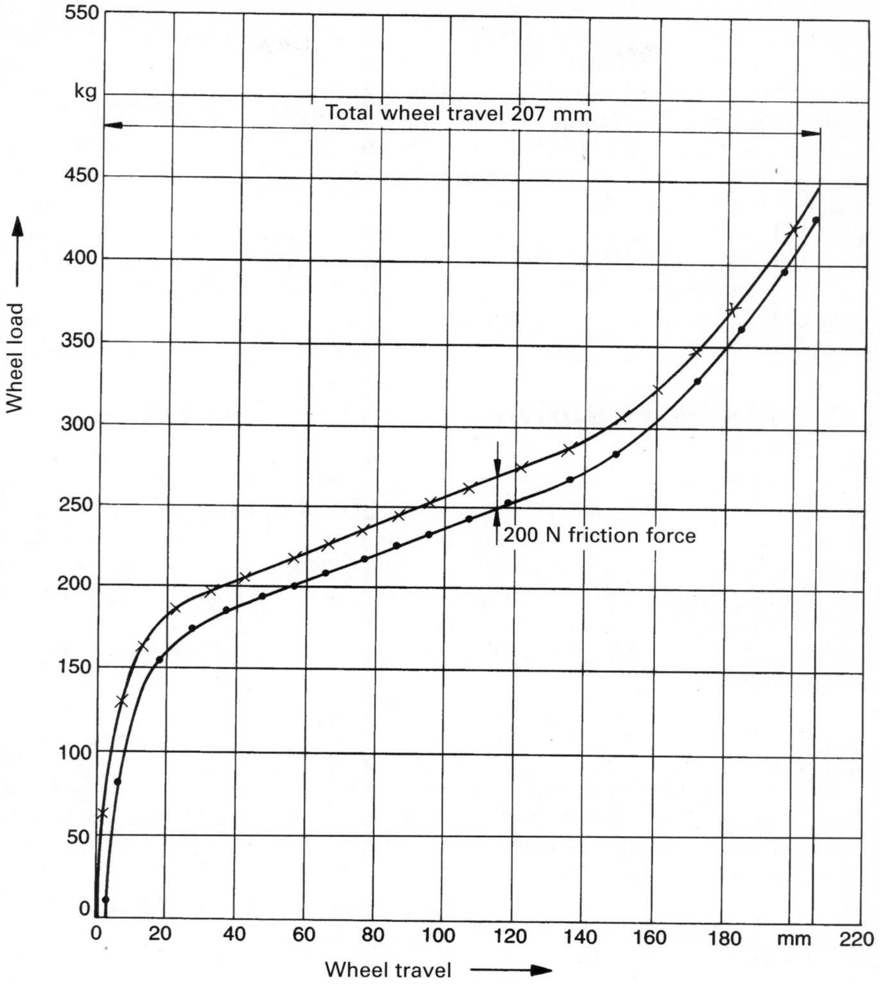

Fig. 5.6 Hysteresis of the curve of front wheel springing shown in Fig. 5.9; the line distance indicates the friction force in the suspension parts, i.e. the self-damping. This is 200 N in total, i.e. $F_{fr} = \pm 100$ N (taking the mean value as nominal).

3.6.5.2 explains how this vibration can be kept away from the body. The design complexity is likely to be greater on driven wheels than on non-driven ones.

5.1.3 Preventing 'front-end shake'

'Front-end shake' is a term used to describe short, hard jolts (in the vertical direction) in the body floor and the front end of the vehicle which, particularly on front-wheel drive vehicles, are triggered by movement of the engine on the rubber parts of the engine mountings and are in a frequency range of around 8–12 Hz. This vibration does not occur continuously, but whenever the engine mounting, the frequency of which is often very close to that of the suspension, begins to resonate. The softer these bearings can be, the less engine noise and vibration will be transmitted to the vehicle interior, but the higher will be the tendency to front-end shake. Conversely, hard mountings reduce front-end shake, but more engine noise is transferred to the vehicle interior. To solve this conflict of aims, hydraulically dampened engine mountings, so-called hydro-mounts, are used and these have a lower static spring rate and, in the event of resonance, generate far higher damping than is possible with normal elastomer mountings. See Section 10.4 in Ref. [5] for details.

5.2 Masses, vibration and spring rates

For determining the vibration rates $n_{\mathrm{f\ or\ r}}$ (front or rear) of the body and the spring rate $c_{\mathrm{f\ or\ r}}$ the front axle load $m_{\mathrm{V,\,f,\,pl}}$ (or $m_{\mathrm{V,\,f,\,max}}$) and the rear axle load $m_{\mathrm{V,\,r,\,pl}}$ (or $m_{\mathrm{V,\,r,\,max}}$) must be known in the design (normal ride height) position (see Section 5.3.4, index pl = partly loaded) and for a permissible gross vehicle weight (index max). With maximum payload the permissible rear axle load $m_{\mathrm{V,\,r,\,max}}$ is mostly fully utilized; in this instance, the resulting front axle load $m_{\mathrm{V,\,f,\,lo}}$ (index lo = loaded) needs to be calculated from the maximum gross vehicle weight $m_{\mathrm{V,\,t,\,max}}$ (see Equation 5.9):

$$m_{\mathrm{V,\,f,\,lo}} = m_{\mathrm{V,t,max}} - m_{\mathrm{V,\,r,\,max}}\ (\mathrm{kg}) \tag{5.1}$$

The mass proportions $m_{1,\mathrm{Bo,f}}$ and $m_{1,\mathrm{Bo,r}}$ of the body (which at front and rear load one axle side respectively) can be calculated using the axle load and the masses $m_{\mathrm{U,f}}$ and $m_{\mathrm{U,r}}$ of the front and rear axles (unsprung masses, based on both wheel stations of the suspension system; see Section 6.1.3).

$$m_{1,\,\mathrm{Bo,\,f}} = \frac{m_{\mathrm{V,f}} - m_{\mathrm{U,f}}}{2} \tag{5.2}$$

$$m_{1,\,\mathrm{Bo,\,r}} = \frac{m_{\mathrm{V,r}} - m_{\mathrm{U,r}}}{2} \tag{5.3}$$

The suspension masses comprise the mass of the wheels and wheel carriers. The latter can be the steering knuckles or, in the case of rigid axles, the axle housing including the differential. There is also the proportional (sometimes half) mass of the suspension parts, which flexibly connect the actual axle with the body or frame. This includes:

- suspension control arms
- tie rods

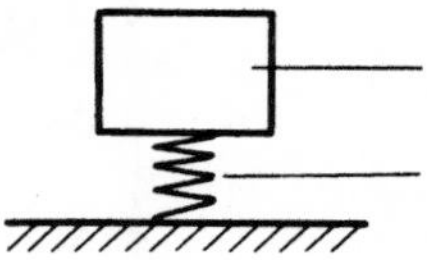

Fig. 5.7 On the simple vibration system, the level of the body frequency $n_{f\ or\ r}$ (front or rear) depends only on the weight or mass proportion $m_{1,Bo,f\ or\ r}$ of the body over a front or rear axle side and the spring rate $c_{f\ or\ r}$, which on linear springing is a quotient of force and travel: $c_{f\ or\ r} = F/s$. On progressive springing, the change in force ΔF over a minimum travel range Δs plays a part $c_{f\ or\ r} = \Delta F/\Delta s$ (see Fig. 5.12).

- axle shafts
- leaf or coil springs
- shock absorbers
- anti-roll bar arms
- panhard rod etc.

The other half of the mass is accounted for by the body. Torsion bars are in the underbody, so their mass forms part of the sprung mass.

Section 6.1.3 contains all details and Equation 6.4c contained therein makes it possible to determine the approximate weight of an axle based on its design. (see also Section 5.2 in Ref. [3]).

The spring rate $c_{f\ or\ r}$ (Fig. 5.9) is required for calculating the spring itself and the configuration of the suspension. This should appear in N mm^{-1} on drawings and as a measurement value, whereas in all calculations the unit is N m^{-1}. If this stipulation is not complied with, there is a risk of calculation errors, unless these are recognized when a dimension equation is done. With the international units the equation for the angular frequency ω is as follows (Fig. 5.7):

$$\omega = \left(\frac{c}{m}\right)^{\frac{1}{2}} \left(\frac{N}{m\ kg}\right)^{\frac{1}{2}}$$

The conversion $1\ N = 1\ \dfrac{kg\ m}{s^{-2}}$ results in:

$$\left(\frac{kg\ m}{s^2\ m\ kg} = s^{-1}\right)^{\frac{1}{2}}$$

To obtain the vibration rates $n_{f\ or\ r}$ (per minute) used in the springing layout, the angular frequency needs to be multiplied by

$$60/2\pi = 9.55\ \left(s\ min^{-1}\right)$$

Related to the body, if the damping, the influence of the mountings and the tyre were ignored, the equation (with indexes) would then be:

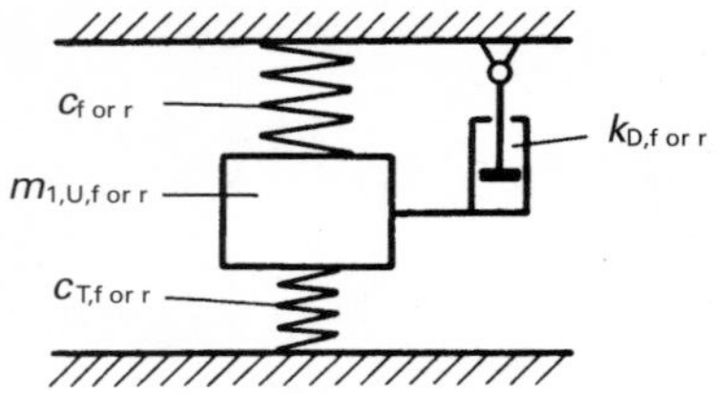

Fig. 5.8 The level of the wheel vibration rate $n_{U,f\ or\ r}$ is a function of the axle mass $m_{1,U,f\ or\ r}$, of the body springing rate $c_{f\ or\ r}$, the tyre springing rate $c_{T,f\ or\ r}$ and the damping $k_{D,\ f\ or\ r}$. The driving speed also has an influence (shown in Fig. 2.28).

$$n_{f\ or\ r} = 9.55 \left(\frac{c_{f\ or\ r}}{m_{1,\ Bo,\ f\ or\ r}} \right)^{\frac{1}{2}} \ (\mathrm{min}^{-1}) \tag{5.4}$$

The calculation of the vibration rate $n_{U,f\ or\ r}$ of one axle side (front or rear) includes half the axle mass

$$m_{1,\ U,\ f\ or\ r} = m_{U,\ f\ or\ r}/2 \tag{5.5}$$

in kilograms and the tyre spring rate $c_{T,\ f\ or\ r}$ in N m^{-1}. Figures 2.27 and 2.28 show statically measured values which increase during driving (see also Section 2.2.8). The factor k_T includes the springing hardening of around 1% per 30 km h^{-1} (see also Section 2.2.8):

$$k_T \approx 1.04 \text{ at } 120 \text{ km h}^{-1} \tag{5.5a}$$

The equation for the axle vibration rate is then (Fig. 5.8):

$$n_{U,f\ or\ r} = 9.55 \left(\frac{k_T \cdot c_{T,\ f\ or\ r} + c_{f\ or\ r}}{m_{1,\ U,\ f\ or\ r}} \right)^{\frac{1}{2}} \ (\mathrm{min}^{-1}) \tag{5.6}$$

For passenger cars with steel springs the body vibration rate should be

Front: n_f = 60–80 min^{-1}
Rear: n_r = 70–90 min^{-1}

The natural frequency (vibration frequency) of the body over the rear axle is chosen to be 10–20% higher than that of the body over the front axle. Thus the vibrating motion results from vibration of the front axle caused by unevenness of the road surface is ‘overtaken’ by the more quickly vibrating rear axle. Thus causes the bouncing motion, which is desirable for comfort reasons, instead of the pitching motion which is uncomfortable for occupants of the car. Particular importance is attached to this design in vehicles with a short wheelbase and a high seat position.

For reasons of comfort, $n_{f\ or\ r}$ should be approximately 60 min^{-1}, which is rarely achieved on the front axles of small to medium-sized passenger cars and can only be achieved at the back if the vehicle is fitted with an automatic level control. The load difference between the loading conditions ‘one person’ and 'full load' (Figs 5.14 and 5.15) makes it difficult to design the springing on the rear axle to be soft, as would be required for comfort.

There are further limitations on the front axle. The fact that the engine bonnet/hood is low, both for aesthetic reasons and because of the requirement for low air drag values, limits the space available for the springs, particularly in the case of McPherson struts. So as not to exceed the material stresses, soft springs are longer and their block length is therefore larger; harder springing does not have this disadvantage (Fig. 5.13). However, this reduces the comfort; on the other hand, strut dampers do allow longer spring travel (Figs 1.41 and 5.12).

The spring rate $c_{\mathrm{f\ or\ r}}$ can be calculated on the basis of a specified vibration level $n_{\mathrm{f\ or\ r}}$ using the transformed equation 5.4a:

$$c_{\mathrm{f\ or\ r}} = 0.011\, n^2_{\mathrm{f\ or\ r}}\, m_{1,\mathrm{Bo,\ f\ or\ r}} \left(\mathrm{N\ m^{-1}}\right) \tag{5.7}$$

The frequency figure in min^{-1} and the mass in kg must be inserted.

The front wheel springing of a front-wheel drive vehicle can be used as an example; the specified loads correspond to the lower limit, i.e. with one person in the vehicle:

Front axle load	m_f	= 455 kg
Axle mass	$m_{\mathrm{U,f}}$	= 55 kg
Specified vibration level	n_f	= 60 min^{-1}

In accordance with Equations 5.2 and 5.7:

$$m_{1,\mathrm{Bo,f}} = (475 - 55)/2 = 210\ \mathrm{kg}$$
$$c_\mathrm{f} = 0.011 \times 60^2 \times 210 = 8316\ \mathrm{N\ m^{-1}}$$
$$= 8.3\ \mathrm{N\ mm^{-1}}$$

Figure 5.9 shows the springing curve with the calculated rate (and associated long paths). The design or zero position (i.e. when there are three people each weighing 68 kg in the vehicle, see Section 5.3.4), is entered as a further point of reference and the weighed wheel load as a function of the wheel travel is shown. This load is observed in the centre of tyre contact.

In the reverse situation, the springing rate can be calculated from an existing springing curve as a function of the various loading conditions. If a curve is linear in the middle range (as shown in Fig. 5.9), it only needs to be extended over the whole spring travel in order to read the load difference at the end points (here 3.32 kN and 1.61 kN). This, divided by the total travel (s_t = 207 mm), gives the spring rate.

In the case of a progressive curve, a tangent must be drawn to the curve for the loading condition to be observed and for it to be possible to read the difference values of loads and paths from it. Figure 5.12 shows an example relating to the design position.

The vibration rate can then be calculated from spring rate, axle load and estimated axle weight. This is usually more precise than settling because most vehicles have McPherson struts, strut dampers or spring dampers and the inherent friction in these parts means a correct result is unlikely.

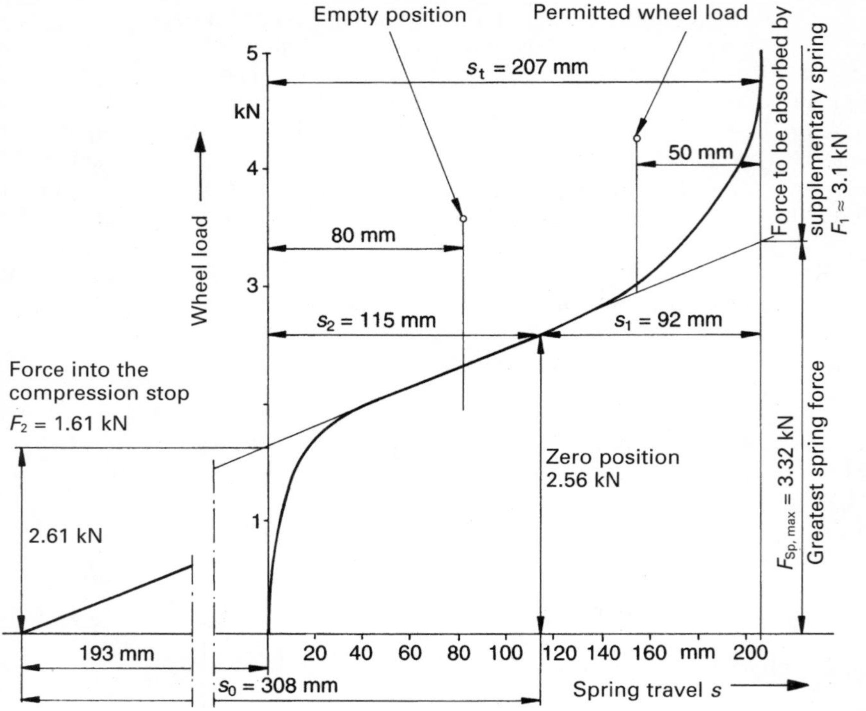

Fig. 5.9 Curve of the front wheel springing of a Renault model, the wheel load (in kg) is entered as a function of the wheel travel (in mm). The soft springing shown requires stops; if the bump stops were missing (Fig. 5.48), the front wheel could jounce from the zero position (the vehicle occupied with three people each weighing 68 kg) by s_0 = 308 mm. Where there is no supplementary spring (Figs 5.21 and 5.50), at $F_{Sp,max}$ = 3.32 kN the axle would make a hard contact. The residual forces to be absorbed by the spring travel limiters are entered in kN. The progressivity achieved by the supplementary spring can be seen clearly. If the stops are in the shock absorber (Fig. 5.29), the compliance of the suspension parts also appears in the curve. The rate of the body springing is:

$$c_{f,pl} = \frac{\Delta F}{\Delta s} = \frac{3.32 - 1.61}{0.0207}$$

$$c_{f,pl} = 8.26 \text{ kN m}^{-1} = 8.3 \text{ N mm}^{-1}$$

5.3 Weights and axle loads

Without knowledge of the weight in the empty or loaded condition and distribution of the load to the two axles, springing on a passenger car can neither be configured nor evaluated. The variables of weight and load laid down in German standard DIN 70 020, page 2, relate to the mass (in kilograms or tons) of the vehicle occupants, the transportable items or goods and the vehicle itself. For details, see Section 1.1 in Ref. 3 and Ref. 8.

The following information and details relate only to vehicles of class M_1 in accordance with the directive 71/320/EEC of the European Union. These vehicles must be used for carrying passengers and may not, apart from the driver's seat, have more than eight seats. They must have at least four (or three) wheels and a total mass of $m_{V,\,t,\,max}$ which does not exceed 1 t (ton force) when fully loaded.

5.3.1 Curb weight and vehicle mass

The actual weighed curb weight $m_{V,ul}$ of the vehicle is essentially determined by:

- the weight of the body with interior trim and the fuel tank;
- the engine and gearbox weight with all necessary accessories, such as starter motor, generator, exhaust system, etc.;
- the weight of the chassis;
- the optional equipment such as automatic gearbox, air-conditioning system, sun roof, etc. (see Equation 5.8a).

According to German standard DIN 70 020 the curb weight also includes:

- the charged battery
- lubricant, coolant and brake fluid
- the standard tool set
- a fuel tank at least 90% full.

However, Section 42 of the German Straßen verkehrs-Zulassungsordnung (StVZO, the regulations for vehicle approval) requires a full tank.

There are also various loose pieces of equipment, such as jack, spare wheel, etc. which must be carried in the vehicle and, in most countries, the triangular safety reflector and first aid kit. The international recommendation ISO/R 1170 contains further details.

5.3.1.1 Curb weight according to manufacturer's data

As the curb weight information required by law allows a tolerance of $\pm 5\%$ – which for a vehicle means a weight range of $\Delta m_V = 110$ kg $m_{V,\,m} = 1100$ – vehicle manufacturers try to set the curb weight $m_{V,\,ul,\,o}$ shown in the vehicle identification card such that it is as low as possible (this governs the balance weight class, which itself is important for the vehicle's fuel consumption and emission rating) and yet still covers as many model versions as possible to keep the technical expenditure low (for example, for type approval). This leads to the optional and supplementary equipment sometimes being ignored. In such a case it is not easy for the vehicle's registered keeper to calculate the actual permissible luggage, roof and trailer load and he or she will be held responsible if the maximum permissible gross vehicle weight is exceeded.

5.3.1.2 Mass of driveable vehicle

Since 1/1/1996, all new models of class M_1, and from 1/1/1998, all newly registered vehicles must be tested in accordance with EU Directive 92/21/EEC and 95/48/EC.

These specify that all vehicle manufacturers must quote the mass $m_{V,\,dr}$ in driveable condition, i.e. the weight of the vehicle driver at $m_p = 68$ kg and the baggage mass at $m_b = 7$ kg must be included. Until now, in Germany, this approval condition was only specified for vans and commercial vehicles (class N in EU Directive 71/320/EEC; see Section 5.3.6.3).

5.3.1.3 Mass of the driveable vehicle when towing a trailer

If the vehicle is intended for towing a trailer, the weight m_{Th} of the towing device and the permissible tongue load ΔM_{Tr} under static conditions must be added to the mass $m_{V,\,dr}$ (see Section 5.3.3.4). The permissible rear axle load must in this case usually be increased.

5.3.2 Permissible gross vehicle weight and mass

This is specified by the vehicle manufacturer taking into consideration the minimum load – which corresponds to the nominal payload m_t (see Equation 5.7c) in accordance with ISO 2416 – required by law, based on the number of seats provided.

5.3.3 Permissible payload

The permissible payload $m_{t,max}$ of a passenger car is the load that the driveable vehicle can carry without exceeding the permissible gross vehicle weight. It therefore results from the difference between the permissible gross vehicle weight $m_{V,\,t,\,max}$ and the actual curb weight $m_{V,\,ul}$:

$$m_{t,max} = m_{V,\,t,\,max} - m_{V,\,ul} \tag{5.7a}$$

Vehicle manufacturers generally specify the payload higher than the regulations demand. This is reflected in a larger permissible gross vehicle weight. The calculation takes into account the component and material stress to be guaranteed, the tyre and wheel bearing load capacity and the loss of braking capacity and handling usually associated with a higher load. The distribution of the goods being transported and the spring travel limitation also play a part in this loss of handling (see Sections 5.3.6 and 5.5.3 and Ref. 9).

There is also the risk of the permissible rear axle load $m_{V,\,r,\,max}$ being exceeded with a full boot and it is possible that the front axle then might lift. This is bound to lead to reduced steerability. On a front-axle drive vehicle, traction and climbing capacity are reduced (see Sections 1.1.7 and 6.4). The EU Directive 92/21 EEC therefore specifies that the front axle load $m_{V,\,f}$ may not be less than 30% of the vehicle total weight $m_{V,f}$, i.e.

$$m_{V,f} \geqslant 0.3\; m_{V,t} \tag{5.7b}$$

5.3.3.1 According to ISO 2416

This standard specifies the minimum payload for passenger cars, i.e. the nominal payload m_t. This depends on the number n of seats provided by the vehicle manufacturer and the passengers' luggage or on the number n_0 of the actually occupied seats and the luggage mass m_{tr} of the goods then transportable.

To determine the number n, a weight of $m_p = 68$ kg for each person – including clothing – must be assumed, plus a luggage mass of $m_b = 7$ kg per person. The nominal payload m_t must then be

$$m_t \geqslant (m_p + m_b)\, n \tag{5.7c}$$

The greatest value – i.e. the luggage mass actually transportable m_{tr} – is then

$$m_{tr} = m_t - m_p \times n_0 \quad \text{or} \tag{5.7d}$$

$$m_{tr} = m_{V,t,max} - m_{V,ul} - m_p \times n_0 \tag{5.8}$$

Experience has shown that the actual or weighed curb weight $m_{V,ul}$ exceeds the manufacturer's stated curb weight m_0 by the weight of the optional equipment Δm_V found in the vehicle

$$m_{V,ul} = m_{x0} + \Delta m_V \tag{5.8a}$$

A five-seater passenger car with a permissible payload of $m_{t,max} = 400$ kg and 20 kg optional equipment can be used as an example:

$$m_{tr} = m_{t,max} - \Delta m_V - m_p \times n \tag{5.8b}$$
$$m_{tr} = 400 - 20 - 68 \times 5 = 40 \text{ kg}$$

The transportable luggage mass m_{tr} is therefore above the minimum value:

$$m_b = 7 \times 5 = 35 \text{ kg}$$

5.3.3.2 Nominal payload

It is the manufacturer who specifies the payload – and therefore also the permissible gross vehicle weight – taking into consideration the expected use of the vehicle (saloon, estate car, sports coupé, etc.) while complying with the legally required nominal payload m_t, i.e. based on the number n of seats provided. In accordance with Equation 5.7c, m_t will be

2 people: 136 kg + 14 kg luggage = 150 kg
3 people: 204 kg + 21 kg luggage = 225 kg
4 people: 272 kg + 28 kg luggage = 300 kg
5 people: 340 kg + 35 kg luggage = 375 kg
etc.

This means that for a nominal payload of $m_t = 375$ kg a saloon will still be legally approved as a five-seater. The precondition is that the other requirements are met, e.g. in respect of seat belt anchoring.

If five people, each weighing 75 kg, occupy a five-seater passenger car, the permissible payload of which, at 375 kg, is at the lower limit, this already gives a figure of 375 kg. If the vehicle has retrofitted accessories not included in the weight calculation or optional equipment Δm_V beyond the normal amount (see Equation 5.8a), the vehicle is already overloaded and it would not be possible to carry any luggage. If, without being aware of the situation, the driver nevertheless puts items of luggage into the boot, the vehicle will exceed the permissible total weight and probably also the permissible rear axle load. If the resulting deterioration in handling or the now insufficient tyre pressure leads to an accident, the driver would be regarded under law in Germany as responsible for the overload. Legal decisions back this up.

5.3.3.3 According to EU directives 92/21/EEC and 95/48/EC

Contrary to Section 5.3.1.2 the curb (empty) weight (i.e. without occupants) is assumed rather than the ready-to-drive weight.

5.3.3.4 When towing a trailer

When the vehicle is towing a trailer, EU Directives 92/21/EEC and 95/48/EC specify that the weight m_{Th} of the towing device and the maximum drawbar-imposed load Δm_{Tr} allowed by the manufacturer must be included in the calculation of the necessary nominal payload (Section 5.3.1.3). A five-seater passenger car would then be permissible for the following nominal payload (Section 1.1.6 in Ref. [3]):

Minimum value for five people	m_p	=	375 kg
Weight of optional equipment including towing device (assumed)	$\Delta m_V + m_{Th}$	=	70 kg
Drawbar-imposed load when towing a trailer	Δm_{Tr}	=	75 kg
Required nominal payload	m_t	=	520 kg

If the payload is 420 kg, the relationships are different:

Nominal payload	420 kg
Optional equipment	−30 kg
Towing device	−15 kg
Drawbar-imposed load	−75 kg
Minimum value	300 kg

According to Equation 5.7c, the vehicle would just about count as a four-seater and the number of permissible seats would therefore have to be altered in the vehicle identification papers.

The maximum static torque load is generally Δm_{Tr} = 50–75 kg; however, according to Directive 92/21/EEC the maximum permissible load must not be less than 25 kg.

5.3.4 Design weight

The design weight $m_{V,t,pl}$ determines the axle weights $m_{V,f,pl}$ and $m_{V,r,pl}$, as well as the design position of the vehicle, also known as the normal or zero position. Under the specified payload, starting from the empty condition, the body compresses front and rear and the result is a particular position *vis-à-vis* the ground. ISO/IS 2958 'Road vehicles: Exterior protection for passenger cars' therefore internationally specifies the design position in relation to the number of seats (specified/allowable number of passengers) as follows:

Number of seats	*Distribution*
2 and 3	two people each weighing 68 kg on the front seats
4 and 5	two people on the front seats and one person on the rear seat
6 and 7	two people on the front and rear seats

Luggage is ignored. The vehicle should be shown with this number of passengers on the drawing board.

When vehicle manufacturers are exchanging vehicle dimensions, the design weight is always specified for determining the design position. The German Directive VDA 239-01 (Verband der Automobilindustrie – Automobile Industry Federation) and Ref. [11] cover all aspects relating to this field.

5.3.5 Permissible axle loads

5.3.5.1 According to Section 34 of the German Straßenverkehrs-Zulassungsordnung (StVZO)

The permissible axle loads front and rear are specified by the vehicle manufacturer. Several points on which the axle loads have a direct effect must be considered:

- component strength of the body and wheel suspension or axles;
- load capacity and therefore minimum size of the tyres;
- configuration of the brake and brake force distribution (Ref. 6);
- springing and damping.

The permissible axle loads are included in the ABE (Allgemeine Betriebserlaubnis or General Operating Approval) in type-testing in Germany or, in the case of the approval of an individual vehicle in accordance with Section 21 of the StVZO, are included in the report of an officially approved expert. The values are indicated on the type plate.

To date, for passenger cars, this specification has not been governed by any particular legal regulations, with the result that only the nominal payload m_t (Equation 5.7c) in accordance with the number of seats approved had to be considered and that the sum of permissible axle loads front $m_{V,f,max}$ and rear

$m_{V, r, max}$ has to be greater than, or at least equal to, the permissible gross vehicle weight (see also Equation 5.1):

$$m_{V,f,max} + m_{V,r,max} \geqslant m_{V,t,max} \tag{5.9}$$

To be able to match the payload to the load compartment in the vehicle better, the gross vehicle weight is usually kept larger than the permissible total value $m_{V,t,max}$ (see Fig. 5.11).

On drive tests and in vehicle behaviour simulations (see Sections 6.3 and 6.4), the least favourable loading condition, i.e. the permissible rear axle load $m_{V,r,max}$ must be assumed. The front axle load $m_{V,f,lo}$ which arises, is then usually below the permissible $m_{V,f,max}$ (Equation 5.1).

The vehicle manufacturer is given the option of the residual hub paths, i.e. there are no regulations on how far a fully laden axle may compress the springs. If this is less than s_{Re} = 50 mm, the desired springing effect would be compromised. Furthermore, the body can barely go any further down on the outside of the bend when cornering, so its centre of gravity rises and the cornering behaviour changes and tends to oversteer, as a result of which situations can arise which are beyond the competence of the driver (Figs 2.42, 5.15 and 5.16).

5.3.5.2 According to EU Directive 92/21/EEC

Directive 92/91/EEC (see Section 5.3.1.2) made the loading of the vehicle and therefore the axle loads subject to stricter regulations. The permissible payload $m_{t,max}$ (see Equations 5.7a and 5.8a) to be calculated from the difference between the permissible gross vehicle weight $m_{V,t,max}$ and the actual curb weight $m_{V,ul}$ must be divided up as a percentage into flat rate mass:

> 91% (90.7%, to be precise) were then allocated to the seats and 9% (or 9.3%) evenly distributed throughout the boot (Section 5.3.6).

The manufacturer had to certify the resulting axle loads as permissible values. Directive 65/48/EEC, which was subsequently issued, withdrew this measure and again requires the values according to ISO 2316 (see Section 5.3.3.1).

5.3.5.3 When towing a trailer

If the vehicle has a towing device, a reduced loading by its component weight must be assumed and, furthermore, the maximum static drawbar-imposed load Δm_T of the trailer must also be included (see Section 5.3.1.3 and Section 1.1.7 in Ref. [3]). The remaining payload is then set at 100% and distributed to the seats and boot.

The permissible rear axle load is then greater. Two options can be derived:

- The manufacturer specifies the higher axle load for all vehicles. This means that the vehicle components listed above must be designed with this in mind, with the disadvantage that stiffer springs reduce the comfort and, under certain circumstances, tyres, axle parts and wheel bearings with a higher load capacity may become necessary.
- The manufacturer specifies two separate axle loads with and without a trailer towing device; the manufacturer must then ensure that the requirements listed under Section 5.3.5.1 are met.

Shock absorbers with variable damping (see Section 5.9) or an automatic level control system (see Chapter 9 in Ref. 5) or supplementary springs (Figs 5.20 and 5.49) can balance the springing.

5.3.6 Load distribution according to ISO 2416

The springing of a vehicle, irrespective of whether it is a passenger car, commercial vehicle or trailer, can only be designed if the axle load distribution has previously been calculated or determined by weighing. The important thing is how many kilograms of payload (and not what percentage) will be on the respective axle, and whether the permissible axle load is fully utilized or exceeded.

The permissible roof load is between 50 kg and 100 kg (see Ref. [3]); it can be taken from the service manual of the respective vehicle.

5.3.6.1 On passenger cars with a non-variable boot volume

Figure 1.36 shows the axle load distribution as a percentage. Where the axle load weight is known, once the weight of the people has been added, the axle loads in the various loading conditions can be calculated.

Section 5.3.3 describes calculation of the permissible axle load which gives the axle load distribution. In industry, and at the TÜV, this is determined with weights placed on the seats at the hip centre H, i.e. the centre of gravity of a person. The position of this point is laid down internationally in the standards SAE-J 826a, ISO 6549 and in DIN 33408. See Sections 1.1.3 in Ref. [3] and 7.2 in Ref. [20] for details.

The adjustable front (and, where applicable, also rear) seats must be moved into the end position for calculating the load distribution and, in accordance with ISO 2416, the weight of the occupants arranged in such a manner that their H-points act 100 mm in front of the respective H-point of the seats. Where the rear seats are not adjustable, the distance is only 50 mm. However, EU Directive 92/21/EEC specifies exclusively the furthest back steering or sitting position and no shifting forwards of the H-points (see Section 5.3.5.2). Both cases are therefore a purely theoretical determination of the load distribution, which ignores whether the vehicle can be steered and operated at all with the sitting position set.

The permissible payload $m_{t,max}$ calculated using Equations 5.7a and 5.8a has to be distributed in accordance with Section 5.3.3.1, and the luggage mass must be put into the centre of the boot. The standard design passenger car shown in Fig. 5.10 would, at $m_{t,max}$ at = 427 kg, m_p = 68 kg and m_b = 87 kg, would have the following loads and axle loads.

For practical reasons, and because it would calculate the difference values afterwards, less tiring than lifting many individual weights into the boot and the passenger compartment, it would be easier to do the weighting with people of any weight. In order to work as precisely as possible, the driver (who should weigh around 68 kg and be approximately 1.70 m tall) should adjust the seat into a comfortable position. Because of the centre of gravity of the passengers, the weight of all the people should not deviate too greatly from this standard mass m_p. (See Sections 1.1.3 and 1.1.4 in Ref. [3] for details.)

Fig. 5.10 Axle load distribution determined on a standard medium-size passenger car by means of weighing. The vehicle was fitted with an electric sun roof. This and further special features meant it weighed 1173 kg empty (instead of 1100, as specified by the manufacturer).

Manufacturer's details	*Number of seats* 5 *Curb weight* 1100 kg *Payload* 500 kg *Permissible gross weight* 1600 kg				*Permissible axle load* *Front* 750 kg *Rear* 850 kg *Total* 1600 kg	
State of loading	*Load*	*Weight of vehicle*	*Axle load*		*Axle load distribution*	
			Front	*Rear*	*Front*	*Rear*
	(kg)	(kg)	(kg)	(kg)	(%)	(%)
Empty	0	1173	623	550	53.1	46.9
2 passengers	136	1309	692	617	52.8	47.2
2 passengers in front and 1 in rear	204	1377	705	672	51.2	48.8
4 passengers	272	1445	718	727	49.6	50.4
5 passengers	340	1513	731	782	48.4	51.6
Maximum load	427	1600	721	879	45.1	54.9

The table (Fig. 5.10) shows the load distribution of a mid-range passenger car which, because it is carrying additional equipment, has an unladen (empty) weight 73 kg heavier than in the specified 'as-delivered' condition. In consequence the permitted luggage capacity is reduced from 500 kg to 427 kg. Although the weight of luggage that can be carried now is still 87 kg, with 5 passengers on board, each with an average weight of 68 kg (= 5 × 68 kg, the allowable rear axle load is exceeded by 29 kg. However, the 185/65 R 15 88 H tyre fitted carries 490 kg at $v \leqslant 190$ km h^{-1} with the specified air pressure for full load of $p_T = 2.5$ bar (Fig. 2.15 and Equation 2.14), so the overload would affect neither the tyres nor, as shown in Fig. 5.14, the springs.

The axle load distribution at 45%/55% (front to rear) in the fully laden condition is likely to cause a slight deterioration in the driving properties of this standard vehicle, while significantly improving the traction.

The situation on a front-wheel drive vehicle also studied in the Laboratory for Chassis Engineering of the University of Cologne shows a different picture (Fig. 5.11). The axle load distribution of 46%/54% calculated under full passenger load, indicates such a severe load alleviation on the driven front wheels that difficulties will be encountered in wet weather conditions, during uphill driving and when the vehicle is towing a trailer (Fig. 6.22). Passenger weights of 70 kg were used to compensate somewhat for the manufacturer's specified excessively high additional load of 500 kg. When empty, the vehicle weighs 6 kg more than shown on the logbook; nevertheless, 144 kg of luggage weight had to be taken into consideration. If this luggage is in the boot, the handling, braking and cornering properties deteriorate (see Figs 5.13, 5.15, 5.16 and 6.15). The ideal load distribution in accordance with EU Directive 92/21/EEC would certainly give significantly better results.

Fig. 5.11 Axle load distribution determined on a front-wheel drive compact family car by means of weighing. Empty, the vehicle weighed only 6 kg more than quoted. The manufacturer's approved high payload of 500 kg (or here 494 kg) would be extremely difficult to achieve. If it is fully utilized, serious effects on the driving safety cannot be ruled out (Fig. 5.16). The rear axle load can be up to 780 kg which, at the total weight of maximum 1400 kg, would mean a load of 620 kg on the driven front wheels and an unreasonable axle load distribution of 44.2%/55.8% on a front-wheel drive vehicle (Fig. 1.36 and Equation 5.7b).

Manufacturer's details	*Number of seats*	*5*	*Permissible axle load*	
	Curb weight	*893 kg*	*Front*	*770 kg*
	Payload	*500 kg*	*Rear*	*750 kg*
	Permissible gross weight	*1393 kg*	*Total*	*1550 kg*

State of loading	*Load*	*Weight of vehicle*	*Axle load*		*Distribution of axle load*	
			Front	*Rear*	*Front*	*Rear*
	(kg)	(kg)	(kg)	(kg)	(%)	(%)
Empty	0	899	548	351	60.9	39.1
2 passengers	140	1039	623	416	60.0	40.0
2 passengers in front and 1 in rear	210	1109	635	474	57.2	42.8
4 passengers	280	1179	647	532	54.8	45.2
5 passengers	350	1249	659	590	52.7	47.3
Maximum load	494	1393	643	750	46.1	53.9

The 155 R 13 78 S tyres fitted have a load capacity of 410 kg at speeds of up to 160 km h^{-1} with a tyre pressure $p_T = 2.1$ bar. The total of the two wheels (820 kg) is above the permissible rear axle load of 780 kg.

5.3.6.2 On passenger cars with a variable boot volume

On all estate cars, hatchback and fastback saloons (and some notchbacks) the boot volume can be increased by folding the rear seats forwards. In this type of passenger car design, the load distribution must be calculated in accordance with ISO 2416, both for when the vehicle is carrying passengers only and when it has been converted to carry goods. As specified by the vehicle manufacturer, to do this the rear seat cushion must be folded forwards and the seat backs folded down (or seat backs alone folded forwards) or the entire row of seats taken out. One disadvantage can be that, on some vehicles the front seats cannot then be pushed back far enough; the driver seat travel is limited by the seat cushion which has been folded forwards.

The axle loads must be calculated with two people, each weighing 68 kg, on the front seats and the mass of luggage (or goods) determined in accordance with Equation 5.7d. The numerical values of Equation 5.8b (and n_0 for the number of seats occupied) with two people in the vehicle give:

$$m_{tr} = m_{t,max} - \Delta m_V - \Delta m_p \times n_0$$
$$m_{tr} = 400 - 20 - 68 \times 2 = 244 \text{ kg}$$

This large luggage mass can lead to the rear axle load $m_{V,r,max}$ being exceeded. To avoid this, ISO 2416 allows the weight to be distributed in accordance with the manufacturer's instructions.

Folding the rear seats forward can result in slight axle load changes of the empty and driveable condition (including the driver), or if the rear row of seats is removed, to a lower curb weight and a higher payload.

5.3.6.3 On vans and lorries

Where they have three or more wheels and a total weight exceeding 1 ton, these types of commercial vehicle meet the conditions of class N in the EU Directive 71/320/EEC; the weight of 75 kg of the driver here, is therefore included in the curb weight (see Section 5.3.1.2). Only the load distribution with any mass in the centre of gravity of the cargo area and in the fully laden state needs to be determined, to calculate from this the axle loads at the design weight – calculated on these vehicle types at 85% of the payload and in the fully laden condition.

5.4 Springing curves

5.4.1 Front axle

The springing on the front axle of a passenger or estate car should be soft, to give a high level of comfort to the occupants, making it possible to transport goods without them being shaken around and to give good wheel grip (see Section 5.1.1). At extremely low vibration frequency ($n \approx 30$ min^{-1}) people notice the vibration paths and speeds 80% less than they do on hard springing with frequencies around 100 min^{-1}. However, the softness of the springing is limited by the overall spring travel available:

$$s_{t,f} = s_{1,f} + s_{2,f} \tag{5.9a}$$

which comprises the compression and rebound travel of the wheels and should be at least:

$$s_{t,f} \geqslant 160 \text{ mm}$$

It is almost as important that, on the front and rear axles, a residual spring travel of $s_{Re} \geqslant 50$ mm is specified to keep the body centre of gravity from rising too much when the vehicle is cornering (see Fig. 5.11). Measurements on a variety of passenger car models have shown that on comfortable vehicles (fitted with steel springs), frequencies on the front axle are between n_f = 60 min^{-1} and 70 min^{-1}, with a total travel path (from stop to stop) of 200 mm; Fig. 5.9 shows a springing curve of this type.

In automotive engineering, presentation of the paths on the x-axis and the wheel loads on the y-axis has become the norm. To make it possible to read path differences and the associated load changes on each wheel easily, it is necessary for them to be entered in a sufficiently large scale, at least 1:1 for the x-axis and 100 kg $\approx$ 40 mm for the y-axis.

In Fig. 5.9, the spring rate in the linear range is $c_f = 8.3$ N mm^{-1}, and the wheel would travel a path of $s_0 = 308$ mm as it rebounds – starting from the zero position ($F_{z,W,pl} = 2.56$ kN). The travel can be calculated easily using the units of N and mm.

$$s_0 = \frac{F_{Z,\,W,\,pl}}{c_f} = \frac{2560}{8.3} = 308 \text{ mm} \qquad (5.10)$$

From a ride and handling point of view, such a long travel is unnecessary and cannot be designed in. For this reason, a rebound stop limits the rebound travel s_2 on all vehicles; on passenger cars and light lorries, this component is inside the shock absorber (Figs 5.31, 5.51 and 5.54) or in the McPherson strut or strut damper. In Fig. 5.9 s_2 is relatively large at 115 mm. The kink in the curve at around $s = 30$ mm indicates the point where the stop comes into operation. Soft springing also demands that the compression travel be limited. If there were no buffers the axle would make a hard contact. The buffer force (or load) in Fig. 5.9 is

$$F_{Sp,max} = F_{Z,W,pl} + c_f \times s_1 = 2560 + 8.3 \times 92 = 3324 \text{ N} \qquad (5.10a)$$
$$F_{Sp,max} = 3.32 \text{ kN (or 338.5 kg)}$$

On roads with potholes, an impact factor of 2.5 is easily possible, i.e. based on the normal force $F_{Z,W,pl}$ in the zero position, the maximum value $F_{Z,W,max}$ could be:

$$F_{Z,W,max} = 2.5 \times F_{Z,W,pl} = 2.5 \times 2.56 = 6.4 \text{ kN} \qquad (5.10b)$$

The main spring, designed with a spring rate of 8.3 N mm^{-1} absorbs $F_{Sp,max} = 3.32$ kN, whilst the additional rubber or polyurethane spring absorbs the residual force $F_1 \approx 3.1$ kN. Figures 5.21 and 5.50 show various configurations and characteristic curves; Fig. 5.9 shows where it comes into play after 140 mm spring travel. If the vehicle compresses over a path of 67 mm from the zero position, the spring begins to act in a way that is not noticed by the occupants and then becomes highly progressive.

Figure 5.12 shows the curve of a soft-sprung standard passenger car (and Fig. 5.10 the associated load distribution); the frequency $n_{f,pl} = 63$ min^{-1} is in the soft range desired and, at $s_t = 196$ mm, there is a large total spring travel. In contrast, the front-wheel drive vehicle shown in Fig. 5.13 has a high frequency (i.e. stiff springing) at $n_{f,pl} = 84$ min^{-1} and, at $s_t = 156$ mm, still reasonable total spring travel. The residual spring travel (54 mm) when there are five people in the vehicle is sufficient but, if the very high, permissible front axle load of 770 kg is fully utilized (Fig. 5.11), s_{Re} returns to the too low a value of 36 mm.

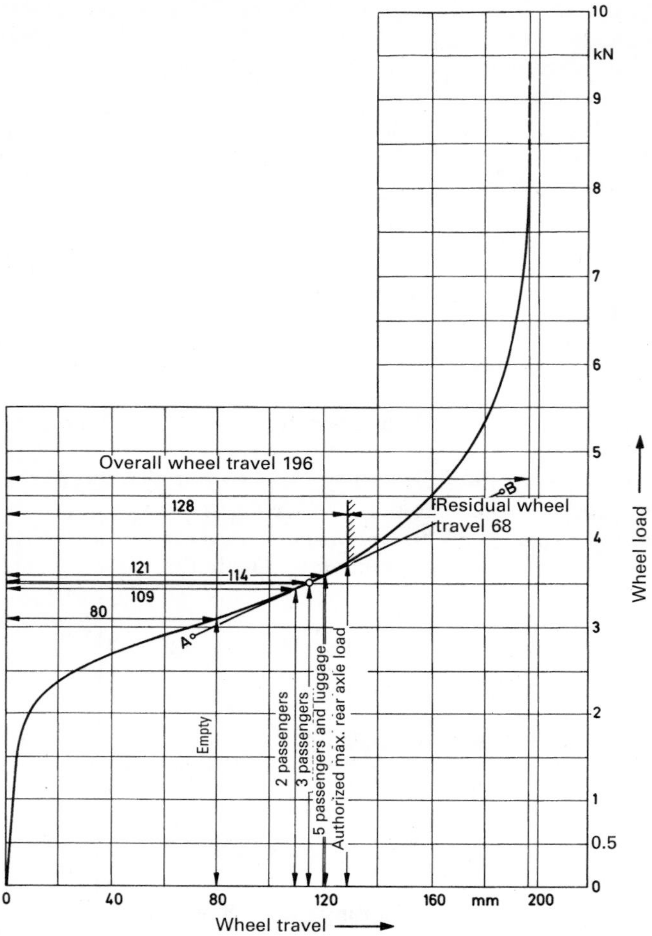

Fig. 5.12 Soft front wheel springing with long travel and linear coil springs, measured on a medium-size standard passenger car. The progressive characteristic curve is achieved with supplementary spring (see Fig. 5.21); Fig. 5.10 contains the wheel loads. To be able to determine the spring rate on the design weight (three people each weighing 68 kg), a tangent must be drawn to the progressive curve (path AB) which is then used to read off two points:

wheel load 4.5 kN, wheel travel 183 mm
wheel load 3.0 kN, wheel travel 78 mm

The spring rate in the partly laden condition (index pl) is then:

$$c_{f,pl} = \frac{150 \times 9.81}{105} = 14.0 \text{ N mm}^{-1}$$

The axle weight needed to calculate the frequency figure is 59 kg and, in accordance with Equation 5.4, it becomes $n_{f,\,pl} = 63 \text{ min}^{-1}$.

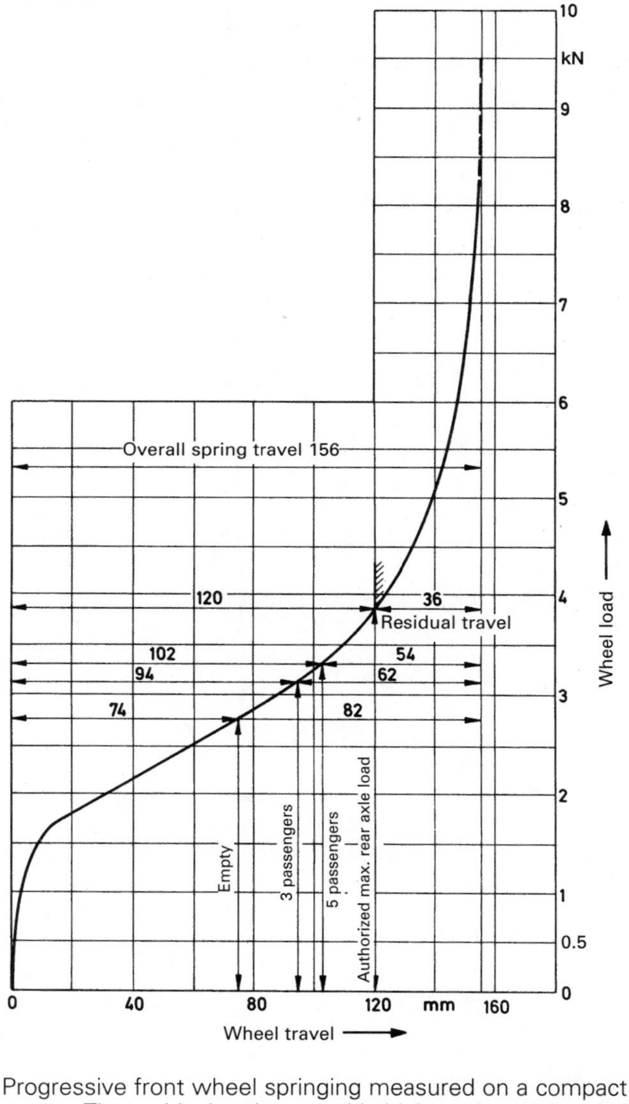

Fig. 5.13 Progressive front wheel springing measured on a compact front-wheel drive passenger car. The residual spring travel is high, and at 156 mm the total path is sufficient, which also applies to the residual bump travel of 54 mm when there are five people in the vehicle. Luggage load in the boot would result in the front end rebounding, i.e. it would increase the bump travel. As can be seen in Fig. 5.11, the manufacturer allows a front axle load of 770 kg, which will be impossible to utilize fully. On the wheel load of 385 kg then possible, the residual bump travel, at 36 mm, is clearly too low. Frequency and rate indicate relatively hard springing; on the design weight it is:

$$c_{f,pl} = 21.8 \text{ N mm}^{-1} \text{ and } n_{f,pl} = 84 \text{ min}^{-1}$$

5.4.2 Rear axle

The springing configuration on the rear axle is more difficult because of the greater loading difference. Furthermore, the residual rebound travel $s_{2,\mathrm{Re}}$ is also included in the observation. The fuel tank is located in front of, behind or over, the axle. If it is only part-full and there is only one person in the vehicle, the axle load corresponds to the empty condition. The road-holding can be compromised if the wheel cannot rebound far enough; ideally,

$$s_{2,\mathrm{Re}} \geqslant 50 \text{ mm}$$

At the front, the permissible axle load can be taken up, at most, by the roof luggage. The amount of the difference between occupancy with one and five people actually utilized is only

$$\Delta m_{\mathrm{V,f}} = 73 \text{ kg}$$

in accordance with Figs 5.10 and 5.11.

The weight of the people sitting on the front seats is distributed approximately equally between the front and rear axle. However, if passengers sit on the rear seat, on average 75% of their weight is carried on the rear axle springing.

Both standard design and front-wheel drive vehicles have the boot at the back. When they are loaded, around 100% of the luggage weight is carried on the rear axle. This is the reason for the significantly higher load difference between the empty and permissible axle weight of

$$\Delta m_{\mathrm{V,r}} = 300 \text{ daN or almost } 400 \text{ daN}$$

on the rear axles of the two vehicles studied. The result would be the value $\Delta m_{\mathrm{V,r}}$ = 400 daN for each axle side. This would correspond to a wheel force difference of $\Delta F_{\mathrm{Z,W,r}}$ = 2000 N. If we assume linear springing at a rate c_{r} = 20 N mm^{-1}, due to $\Delta F_{\mathrm{Z,W,r}}$ a path of

$$\Delta s = \Delta F_{\mathrm{Z,W,r}}/c_{\mathrm{r}} = 1962/20 = 98.1 \text{ mm}$$

would be needed. There is also a residual jounce and bump-travel path of 50 mm in each case so that total travel can barely be less than $s_{\mathrm{r,t}}$ = 200 mm.

Figure 5.14 shows the linear rear-wheel springing of a standard passenger car. In spite of the soft springing at a rate of $c_{\mathrm{r,pl}}$ = 18.9 N mm^{-1}, there is residual travel of 86 mm or 50 mm. The frequency on a partly laden vehicle (with three people) is $n_{\mathrm{r,pl}}$ = 77 min^{-1} and, with additional loading, it reduces, increasing the comfort (the spring rate remains constant but the mass increases, see Equation 5.4). This type of favourable configuration is achieved by:

- a large total spring travel ($s_{\mathrm{r,t}}$ = 220 mm)
- a payload level which only corresponds to 45% of the curb weight
- a long wheelbase (l = 2665 mm)
- a boot that does not protrude too far at the back.

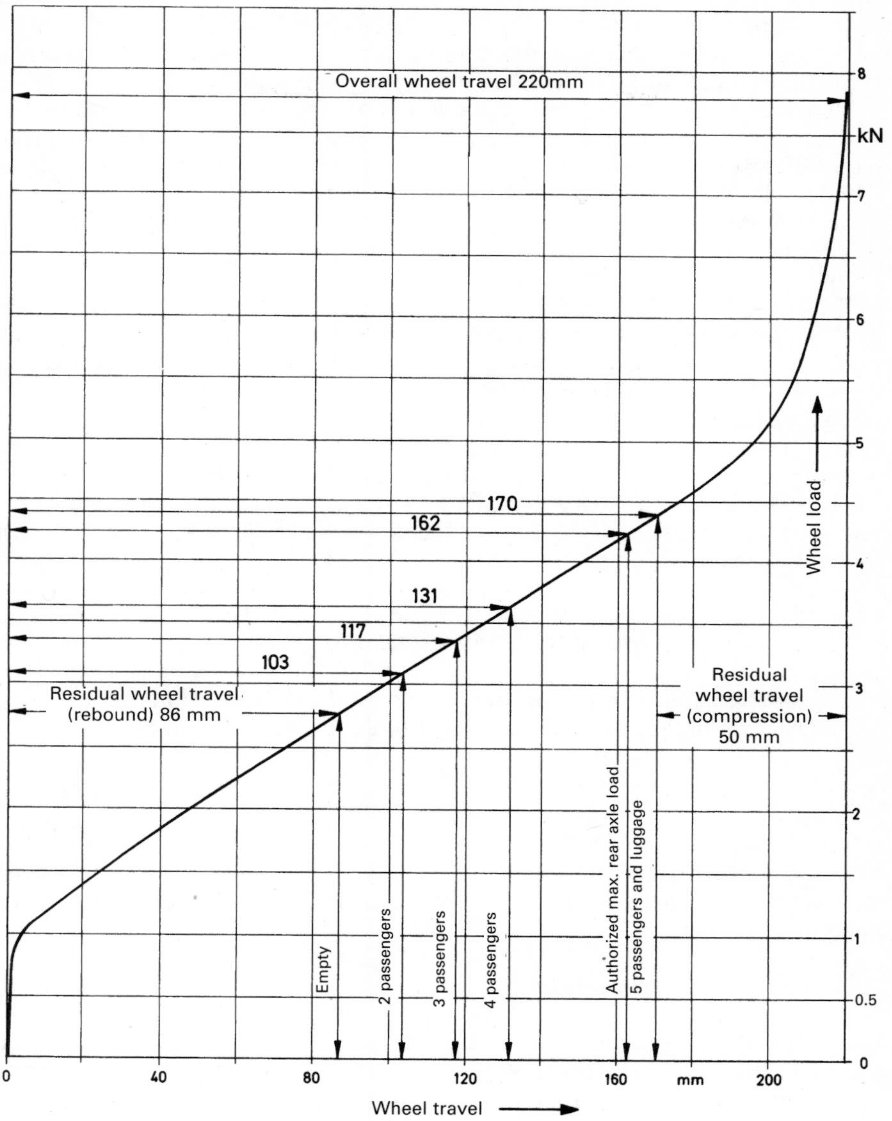

Fig. 5.14 Almost linear and soft rear wheel springing with large travel, measured on a standard passenger car; rebound stop and supplementary springs are located in the shock absorber. Figure 5.10 contains the associated wheel loads. With a loading of five people plus luggage (427 kg), the rear wheels still have a residual compression travel of 50 mm. The springing rate (with a design weight, $m_{V,r,pl}$ = 672 kg) is $c_{r,pl}$ = 18.9 N mm^{-1} and the frequency $n_{r,pl}$ = 77 min^{-1}. The manufacturer specifies $m_{U,r}$ = 91 kg as the weight of the unsprung mass.

The disadvantages are the dropping of the tail when the vehicle is laden and the associated pitch angle θ (Fig. 3.137), although the danger of dazzling other road-users, which would otherwise be a problem, can be overcome by the headlight height adjustment, which is fitted as standard.

Shortening of the spring travel and a less pronounced squat on the tail can be overcome with progressive springing for bump travel beyond normal ride height. Figure 5.15 shows this type of curve, measured on a front-wheel drive vehicle. The frequency $n_{r,pl}$ = 93 min^{-1} (where there are three people in the vehicle) points to harder springing. In spite of the very high load difference of 399 kg, which can be seen in Fig. 5.11, the axle bump travel of only Δs_r = 76 mm.

The possible loading of 500 kg represents 56% of the manufacturer's stated curb weight (893 kg). This unfavourable ratio leads to the severe load alleviation on the driven front wheels, described in Fig. 5.11 and to the highly loaded rear axle with a residual spring travel of only $s_{1,Re}$ = 28 mm, whereas at $s_{2,Re}$ = 89 mm, the rebound travel is large.

The springing curves (of front and rear axles seen together) lead to the assumption that the vehicle initially stood higher by $\Delta s \approx$ 20 mm.

The vehicle having been lowered by the owner (this is assumed due to the unusually large jounce travel) and the high load are likely to be the reasons for the too low residual compression paths front and back.

5.4.3 Springing and cornering behaviour

5.4.3.1 Wheel load change on independent wheel suspensions

As can be seen in Fig. 1.6, the centrifugal force related to the front axle

$$F_{c,V,f} = m_{m,f}\, a_y = \mu_{Y,W}\, F_{Z,V,f} \tag{5.11}$$

acts at the level of the vehicle centre of gravity. The wheel force change that arises during cornering (outside of the bend $+\Delta F_{Z,W}$ and on the inside of the bend $-\Delta F_{Z,W}$) can be approximated separately for the two axles. For the rear axle the equation is

$$\pm\Delta F_{Z,W,r} = \mu_{Y,W}\, F_{Z,V,r}\, h_V/b_r \tag{5.12}$$

The values of the front-wheel drive vehicle inserted with a permissible axle load and with the centre of gravity height $h_V \approx$ 530 mm, the tread width b_r = 1425 mm and the lateral coefficient of friction $\mu_{Y,W}$ = 0.7, give a force of

$$\pm\Delta F_{Z,W,r} = 0.7 \times 780 \times 9.81 \times 530/1425 = 1993 \text{ N}$$

The wider the tread width and the lower the centre of gravity, the smaller $\pm\Delta F_{Z,W}$ will be.

The equation for the front axle is:

$$\pm\Delta F_{Z,W,f} = \mu_{Y,W}\, F_{Z,V,f}\, h_V/b_f \tag{5.12a}$$

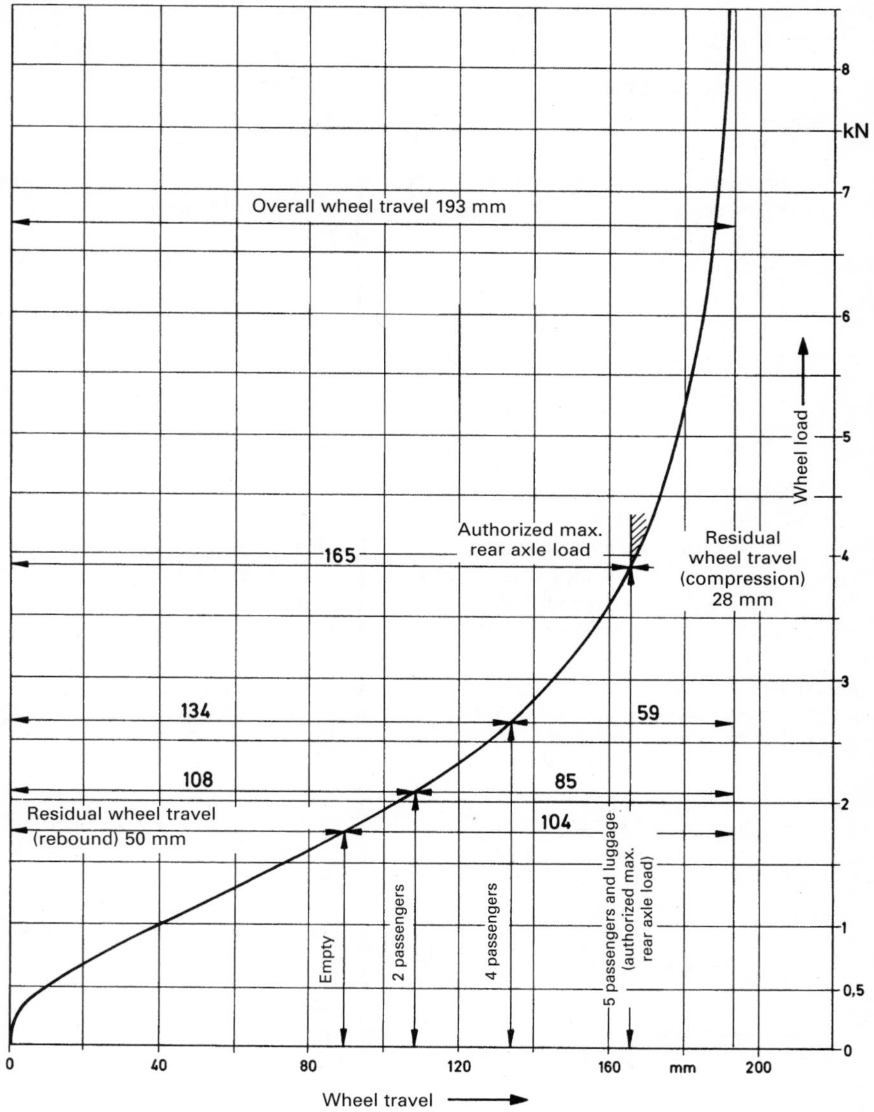

Fig. 5.15 Progressive rear wheel springing, measured on a front-wheel drive vehicle; a poor example in respect of springing design with permissible axle load. Only 28 mm residual spring travel, in association with the very high load of 494 kg, jeopardizes the driving safety (see Fig. 5.16). The associated wheel loads are shown in Fig. 5.11. With a design weight of $m_{V,r,pl}$ = 474 kg, the spring rate $c_{r,pl}$ is 20.2 N mm^{-1} and the frequency $n_{r,pl}$ = 93 min^{-1}.

5.4.3.2 Spring travel on independent wheel suspensions

The calculated value of 1993 N corresponds to a load change of 203 daN and therefore a wheel load

on the outside of the bend	593 daN
on the inside of the bend	187 daN

Assuming a permissible value of 390 daN in Fig. 5.15, this leads to a compression travel of $\Delta s_{1,r} = 20$ mm and a jounce travel of $\Delta s_{2,r} = 69$ mm.

5.4.3.3 Change in the height of the centre of gravity of the body

The values inserted into the formula that is valid for the rear axle,

$$\Delta h_{Bo,r} = (\Delta s_{2,r} - \Delta s_{1,r})/2 \tag{5.13}$$

give the amount by which the body is pushed upwards above the rear axle (Fig. 6.15):

$$\Delta h_{Bo,r} = (69 - 20)/2 = 24.5 \text{ mm}$$

On the outside of the bend, the body only tilts downwards a little, but on the inside it moves upwards. For this reason, Δs_1 must be deducted from Δs_2 (Fig. 5.16). The higher the centre of gravity R rises, the greater the wheel force change (see Equation 5.12), particularly at the axle with the greater travel difference $\Delta h_{Bo,f \text{ or } r}$. This is usually the rear axle. A tendency to oversteer and the torque steer effect, increase, particularly when the tyre on the outside of the bend is highly compressed, i.e. when its stress goes far beyond the possible load capacity (Figs 2.42 and 2.52).

The difference travel

$$\Delta h_{Bo,f} = (\Delta s_{2,f} - \Delta s_{1,f})/2 \tag{5.13a}$$

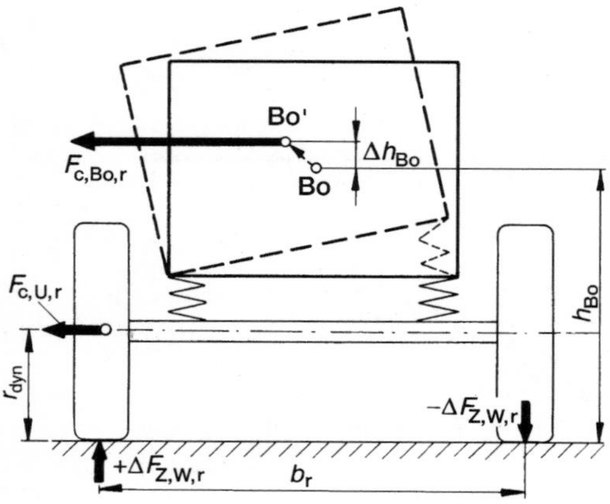

Fig. 5.16 When the vehicle corners, the centrifugal forces $F_{c,Bo,r}$ relating to the body act at its centre of gravity. If the vehicle has too low a residual bump travel, it is not able to move in bump as much on the outside of the bend as it jounces on the inside. This means that the body centre of gravity moves up from Bo to Bo′ by the path Δh_{Bo}, and critical oversteering, which is difficult to control, can be the result.

Sections 3.4.5 and 5.4.3.5 and Fig. 1.25 contain further details.

by which the body moves upwards in the centre of the front axle, when (as shown in Fig. 5.13) the spring curve is also progressive, can be calculated in the same way.

The distances $l_{Bo,f}$ and $l_{Bo,r}$ from the centre of the front or rear axles should also be taken into consideration when calculating the path Δh_{Bo}, around which the body centre of gravity R changes position (Fig. 6.1, see also Equation 6.2.4):

$$\Delta h_{Bo} = \frac{\Delta h_{Bo,f} l_{Bo,r} + \Delta h_{Bo,r} l_{Bo,f}}{l} \tag{5.14}$$

If the axle loads or the weights are known, they can be inserted into the equation:

$$l_{Bo,r}/l = m_{V,f}/m_{V,t} = F_{Z,V,f}/F_{Z,V,t} \text{ and} \tag{5.14a}$$

$$l_{Bo,f}/l = m_{V,r}/m_{V,t} = F_{Z,V,r}/F_{Z,V,t} \tag{5.14b}$$

5.4.3.4 Body roll angle on independent wheel suspensions

The body roll angle of a torsionally stiff body is the same over the front and rear axles. It can therefore only be determined for the vehicle as a whole, taking into consideration any anti-roll bars fitted and the body roll centres front and rear (see Section 3.4). A two-wheel trailer, which has the springing shown in Fig. 5.15, can therefore be used as an example.

The body roll angle φ (Fig. 1.6) can easily be calculated in such cases:

$$\varphi = \frac{\Delta s_{1,r} + \Delta s_{2,r}}{b_r} \text{ (radian) and, for } 360/2\pi,$$

$$\varphi = 57.3 \frac{\Delta s_{1,r} + \Delta s_{2,r}}{b_r} \text{ (degrees)} \tag{5.15}$$

If the values of the examples are inserted on the existing progressive springing (see also equation 6.23), the result is

$$\varphi = 57.3 \frac{20 + 69}{1425} = 3.58° = 3°35'$$

In the case of linear springing over the whole range, compression and rebound travel are equal and the level of the centre of gravity does not change. The larger travel can be easily calculated using Equation 5.10:

$$\Delta s_{1,r} = \Delta s_{2,r} = \Delta F_{Z,W,r}/c_{r,pl}$$

If the values of Fig. 5.15 are inserted the result is:

$$\Delta s_r = 1993/20.2 = 99 \text{ mm}$$

This corresponds to a body roll angle of $\varphi = 8°$. The example in the calculation should indicate the advantage of the progressive springing.

5.4.3.5 Body roll angle on rigid axles

The springs sit on the axle housing (Fig. 1.23) and the basis of support of the body is the now smaller effective distance b_{Sp}. Furthermore, unlike on all independent wheel suspensions, the rigid axle does not support the tendency of the body to roll. Therefore the shortened body roll lever arm $(h_{Bo} - h_{Ro,r})$ is included in the equation, and this comprises the level h_{Bo} of the body centre of gravity and the body roll centre height $h_{Ro,r}$ (rear, Fig. 1.25) (see Section 3.4.5 and Fig. 5.16) plus the proportion of weight $F_{Z,Bo,r}$ of the body ($F_{Z,Bo,r} = m_{Bo,r}g$, see Equation 6.6b) and the weight $F_{Z,U,r}$ of the axle (see Section 6.1.3). In accordance with the laws of static, parts that are flexibly connected must be separated. Almost all previous equations are altered by this. The wheel force change (Equation 5.12) becomes smaller,

$$\pm\Delta F_{Z,W,r} = \mu_{Y,W} F_{Z,Bo,r} \frac{h_{Bo} - h_{Ro,r}}{b_{Sp}} + F_{Z,U,r} \frac{r_{dyn}}{b_r} \tag{5.16}$$

and the travels $\Delta s_{1,Sp}$ and $\Delta s_{2,Sp}$ calculated from it relate to the springs that are further to the inside. The ratio i_φ is needed to obtain the values related to the centre of tyre contact

$$i_\varphi = b_r / b_{Sp} \tag{5.17}$$

$$\Delta s_{1,r} = \Delta s_{1,Sp}\, i_\varphi \text{ and } \Delta s_{2,r} = \Delta s_{2,Sp}\, i_\varphi \tag{5.18}$$

In line with Equation 5.15, the equation for a single-axle vehicle with a rigid axle (for the increased angle φ' in degrees) would then be:

$$\varphi' = 57.3 \frac{(\Delta s_{1,r} + \Delta s_{2,r})}{b_r} i^2_\varphi \tag{5.18a}$$

The further out the springs can be positioned, the smaller (i.e. more favourable) i_φ becomes; this applies in particular to the drawbar axle (see Fig. 1.60 and Section 3.4 in Ref. [2]).

5.4.3.6 Rates on reciprocal springing

Apart from slight deviations, the spring rates with parallel and reciprocal springing are equal on all independent wheel suspensions, if we ignore the (symmetric and asymmetric bumps) influence of the anti-roll bar:

$$c_{\text{f or r}} = c_{\varphi,\text{f or r}}$$

The picture is different for rear (or front) rigid axles:

- If the springing is parallel, the rate on the centre of tyre contact c_r is equal to that on the spring mounting $c_{Sp,r}$.
- However, if the springing is reciprocal (first wheel rises as second falls), the rigid axle takes on an inclined position (Fig. 1.21).

As Equation 5.18 shows, the differences in travel $\Delta s_{1,r}$ and $\Delta s_{2,r}$ are greater than those at the springs (Δs_{Sp}); however, the changes in force $\Delta F_{Z,W,r}$ are smaller:

$$\Delta F_{Z,W,r} = \Delta F_{Sp,r}\, b_{Sp}/b_r = \Delta F_{Z,W,r}/i_{\varphi,r}$$

With a spring rate $c'_{\varphi,r}$ related to the centre of tyre contact, this yields

$$c'_{\varphi,r} = \frac{\Delta F_{Z,W,r}}{\Delta S_{1,2}} = \frac{\Delta F_{Sp,r}}{i_{\varphi,r} s_{Sp,r} i_{\varphi,r}}$$

$$c'_{\varphi,r} = c_{Sp,r}/i^2_{\varphi,r} \tag{5.19}$$

In the case of reciprocal springing, the elastic parts in the guide joints and struts are tensioned; the actual reciprocal springing rate $c_{\varphi,r}$ is around 7% higher than the values

$$c_{\varphi,r} = 1.07\, c'_{\varphi,r} \tag{5.19a}$$

calculated with Equation 5.19. The equations for a rigid front axle are similar; except the index f appears.

5.4.4 Diagonal springing

Front and rear axle trailing links and longitudinal pairs of links mostly have a vehicle pitch pole $O_{f\text{ or }r}$ (per side, see Section 3.11). The wheels then no longer move vertical to the ground when they rebound and compress, but on arcs $\pm f$ around the existing pole (Fig. 3.158). The driving safety is not impeded by the wheels moving out, to the back or the front. When the wheels compress, they move Δl in the direction of the radius-arm axes (if these are at the height of the wheel centre or below it) and away from them as long as they are above the wheel centres. When the body moves downwards, the axes go down with it.

The diagonal springing angle χ on trailing and semi-trailing links is entered in Figs 3.158 and 3.160. This angle also applies to compound crank suspensions and multi-link suspensions and to appropriately sprung rigid suspension (Figs 1.1, 1.2, 1.43 and 1.61). However, on double wishbones disposed at an angle (Fig. 3.155) this becomes:

$$\chi = (\alpha + \beta)/2 \tag{5.19b}$$

On McPherson struts and strut dampers, the change in the caster angle, i.e. $\Delta\tau_k$ is a consideration (Fig. 3.137).

5.5 Spring types

Two springs, four stops, two shock absorbers and one anti-roll bar usually control the springing of a pair of road wheels, the limitation of spring travel and the reduction of body roll inclination for each wheel-station on passenger cars and light commercial vehicles.

An exact description of the individual components and calculations is given in Refs [2] and [5]. Here, the intention is merely to show the arrangement of components and what they do.

The springs can be distinguished by media and materials as follows:

- steel springs
- air and gas springs
- composite (leaf) springs
- rubber springs
- springs of polyurethane elastomer.

These last two types are mainly used on passenger car two-wheel trailers or as additional springs parallel to the steel spring. The polyurethane is stressed in compression and the rubber in tension.

5.5.1 Air- and gas-filled spring devices

For reasons of comfort, the natural frequency of the body above the rear axle should be between 10% and 20% higher than on the front axle and should not be dependent on the load. Unlike steel springs, air-spring systems allow the natural frequencies of the body to remain unchanged even when there are changes in vehicle load, and this is associated with other advantages:

- vibration and suspension properties not dependent on load;
- simple level adjustment;
- a guarantee of complete wheel travel even with a load;
- compact design with large spring and shock-absorber track widths.

Depending on the design of the system, it is also possible to ensure the following:

- the balance of unsymmetrical loading conditions on the left/right;
- a specific effect on vehicle roll movement, and thus dynamic wheel loads, by support of the rolling moment;
- damping, which depends on the driving conditions (speed range, longitudinal, vertical and lateral acceleration).

Because of these advantages, air-spring systems have long found almost universal application in buses and commercial vehicles used for long-distance travel (in which the adjustment of the height of the loading surfaces is important as

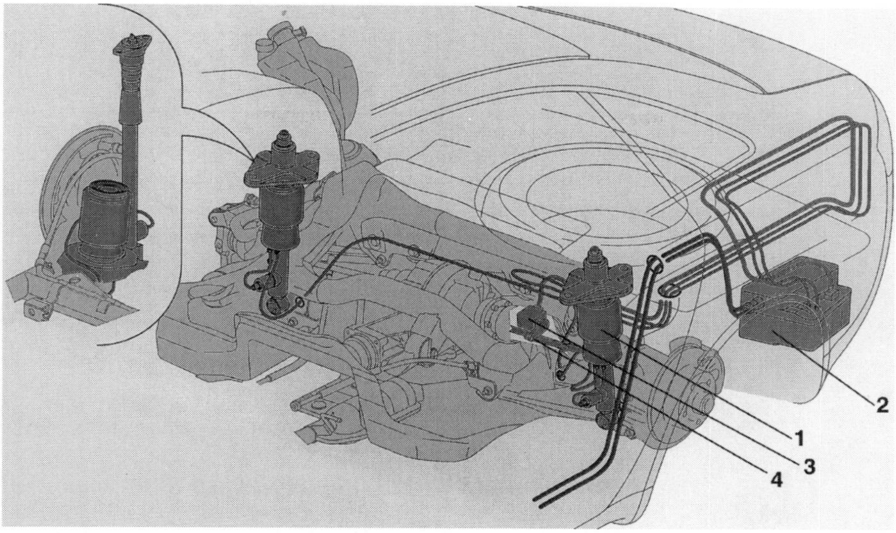

Fig. 5.17 Air-sprung double-wishbone axle of the Audi A6 Quattro (1997). With axle components that are as similar in design as possible (wheel carriers made of nodular graphite cast iron, upper transverse link and auxiliary frame made of hydroformed tubes, lower transverse link and rod-shaped link in a sheet-steel shell design), Audi supply the driven rear axle of the A6 Quattro with an air-spring and shock absorber 1 instead of the single-tube gas-pressure absorber with coil spring.

The air is supplied by a vibration- and noise-damped air supply unit in the rear end 2 consisting of a 280 watt compressor, air dryer and control valves. A contactless Hall rotational angle sensor 3 is attached to the levelling-value 4 in the middle of the vehicle for the purpose of establishing the height of the vehicle. The time to adjust from empty running to maximum load is about 60 s and the average current consumption with a 1% on-period is 5 W.

The system was developed by Continental AG and is supplied as a ready-assembled complete system.

well as protection of the load and road). Air-springs are also increasingly being used in the van sector (Mercedes-Benz Vito). In the case of passenger cars, air-springs have up until now only occasionally been used in comfort-orientated vehicles (e.g. Mercedes 600) or off-road vehicles (Range Rover) for reasons of cost; air suspension has recently been offered as an optional extra in vehicles of the upper middle class (Audi A6 with front-wheel or all-wheel drive, BMW 5 series tourer). The traditional springs are replaced with air-springs and sometimes air-spring-and-shock absorbers (Figs 5.17 and 5.18). The front and rear axles of the Mercedes Benz S class W220 are for the first time being air-sprung as standard (Figs 1.39 and 5.19).

Citroën has installed gas springing – where the forces are transmitted hydraulically via oil pressure, in its larger models since 1953 – as its so-called hydro-pneumatic springing.

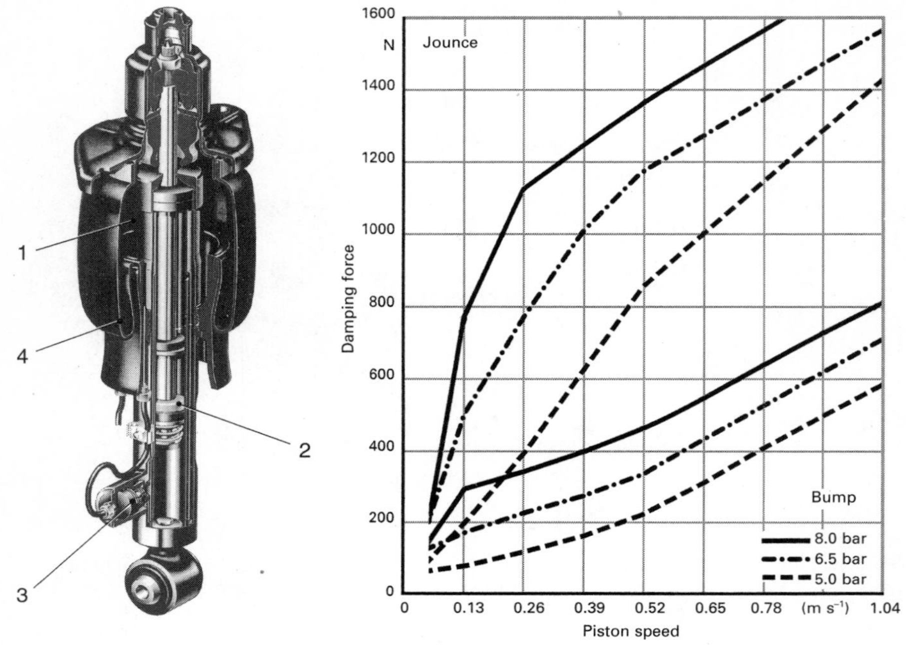

Fig. 5.18 Air-spring-and-shock-absorber assembly and damping characteristic for Audi A6 Quattro (manufacturing diagram of Continental AG). Air-spring 1 and shock absorber 2 are coaxially arranged and form a spring-and-shock-absorber strut. By means of a valve 3 connected to the air-springs, the air-spring pressure is used as a load-proportional correcting variable for the continuously variable, load-dependent adjustment of the damping force in rebound and compression, shown in the diagram for the regions of 5.0, 6.5 and 8.0 bar.

With small excitation amplitudes (less than 3–5 mm), the dynamic rigidity of the rubber bellows 4 reduces comfort. This hardness of response depends on the strength of the material – durability and comfort have conflicting aims here – and the nature of the material. The response was able to be improved by replacing polyamide with aramide.

Fig. 5.19 Air-spring-and-shock-absorber assembly of the front axle of the Mercedes Benz W220 series (1998). In order to reduce the inner friction which is disadvantageous in air-sprung vehicles at small amplitudes and the dynamic hardening of the hose reel bellows with two X-shaped layers of textile fibres, Daimler Chrysler uses single-layer rubber bellows with not very thick walls (1.6 mm) in the Mercedes Benz W220 series. As a result, the hardening related to the total spring rate of the system is reduced from 80% to 25% and the quality of the response considerably improved at small excitation amplitudes. The limited capability for expansion between the fibres of the single-layer rubber bellows requires a small gap between the rolling piston and outer guide and protection against fouling because of the lack of self-cleaning of the original bellows; the latter is ensured by the polyurethane folding sleeve with labyrinth ventilation. The radial deflection of the lower strut point which occurs during wheel travel and particularly steering movements must be taken into consideration when establishing the position of the rolling piston and outer guide. On the front axle, this occurs by means of the deepset guided connection of the rolling piston to the shock absorber by way of a cardan soft sliding bearing. The pressure for the system is supplied by a compressor with an output of 400 W. In order to ensure short filling times, a 4 ℓ central pressure tank whose tank pressure of 16 bar clearly lies above the system pressure of 10 bar is used. The weight-dependent spring rates are 15 N mm^{-1} on the front axle and 17 N mm^{-1} on the rear axle; this results in natural body frequencies of 0.8 Hz and 0.9 Hz. The air-spring system controls the height of the vehicle regardless of the load, while taking into account the driving speed (reduction of 15 mm above 140 km h^{-1}), lifting of 25 mm caused by a bad road surface and maintenance and wheel change positions. The adjustable damping system operates automatically, taking into account the driving conditions which are established by means of the driving speed, ABS signal, body acceleration, steering angle signal and braking pedal signal. The following shock absorber characteristics can be produced by switching on the valves which are flange-mounted on the single-tube gas-pressure shock absorbers:

Step 1: rebound and compression soft – maximum driving comfort.
Step 2: rebound soft, compression hard – increased damping.
Step 3: rebound hard, compression soft – increased damping.
Step 4: rebound and compression hard – maximum damping for minimization of wheel load fluctuations.

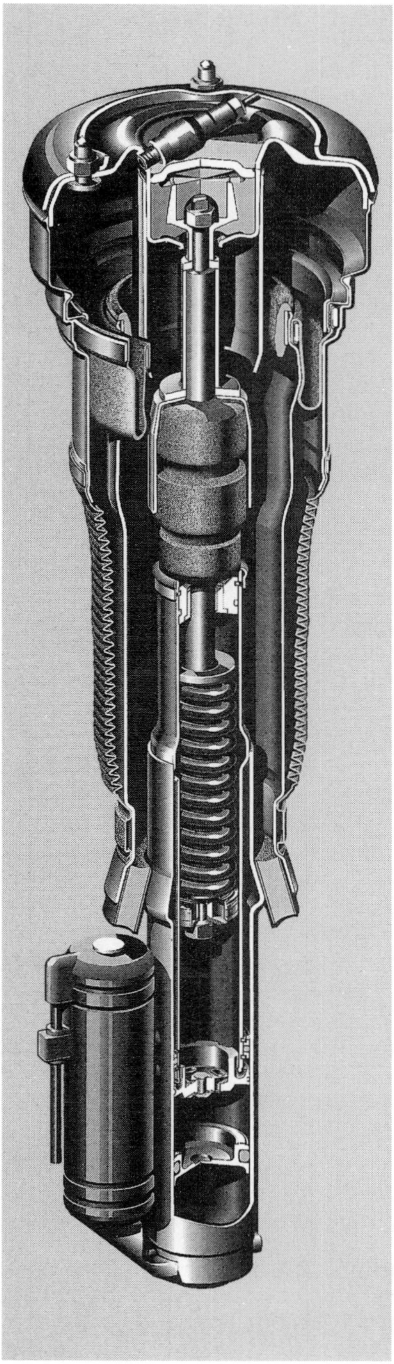

5.5.2 Steel springs

The following are manufactured in steel:

- leaf springs
- coil springs
- torsion bars
- anti-roll bars.

5.5.2.1 Leaf springs

Leaf springs are subdivided into longitudinal and transverse leaf springs. Longitudinal leaf springs are used only on rigid axles, more commonly on commercial vehicles and trailers. Figure 5.20 contains a weight comparison between the previously exclusively used multi-layer leaf springs and modern parabolic springs; Figs 1.20, 1.26 and 1.37 show various designs and also the advantages. For reasons of cost and weight, springs with only a single layer, so-called single-leaf springs, are fitted to an increasing number of passenger cars and light commercial vehicles; Fig. 1.24 shows these on the non-driven rear axle of a van.

Transverse leaf springs, by contrast, can provide the springing on both sides of the axle; they were previously used in independent wheel suspensions of

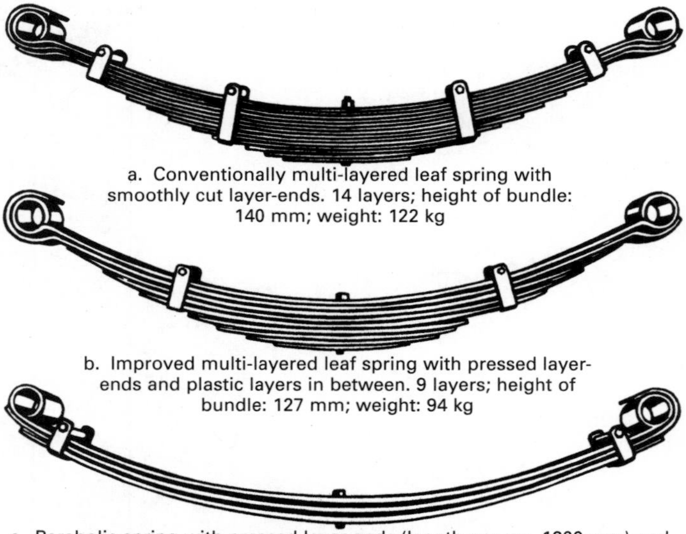

a. Conventionally multi-layered leaf spring with smoothly cut layer-ends. 14 layers; height of bundle: 140 mm; weight: 122 kg

b. Improved multi-layered leaf spring with pressed layer-ends and plastic layers in between. 9 layers; height of bundle: 127 mm; weight: 94 kg

c. Parabolic spring with pressed layer-ends (length approx. 1200 mm) and plastic layers in between. 3 layers; height of bundle: 64 mm; weight: 61 kg

Fig. 5.20 Weight comparison between three different commercial vehicle rear springs with the same data, carried out by Krupp-Brüninghaus; eye distance $L = 1650$ mm, spring rate $c_r = 200\ \mathrm{N\ mm^{-1}}$ and loadability $F_{Sp} = 33$ kN; however, the designs are different.

passenger cars (see Section 5.2.3 in Ref. [2] and are also built into (the 1955) Daimler-Benz-Transporter Sprinter.

5.5.2.2 Coil springs

Coil springs with a linear curve over the entire wheel travel are used on the front and rear axles of passenger cars (Figs 5.9 and 5.14). If necessary, a certain progression can even be achieved by using various formings of conical spring wire. Figures 1.7, 1.15, 1.39, 1.41, 1.60 and 1.81 show springs in their fitted condition, (see also Section 2.1.4 in Ref. [2]).

5.5.2.3 Torsion bars

Cylindrical torsion bars made of round steel are used to spring the body and as anti-roll bars (see Section 5.5.4). To transfer the springing movement, both ends have warm-upset heads, which carry a toothed profile or a square. U-shaped brackets can also be butt-welded and can be very easily fitted to the suspension links. Figures 1.2 and 1.63 show torsion bars in the installed condition.

5.5.3 Stops and supplementary springs

The following are differentiated:

- jounce stops
- bump stops
- supplementary springs.

As shown in Fig. 5.9, the jounce stops limit the jounce travel of the wheels on soft and medium-hard springing. Apart from a few exceptions, jounce stops are found in shock absorbers or McPherson struts and strut dampers (Figs 5.26, 5.47, 5.51 and 5.55). In this case, under tensile forces, the elastic attachment parts of the damper and the elastomer, polyurethane or hydraulic jounce stop, all flex (see Fig. 5.31 and Section 5.6.8.1; also Ref. [5]).

Bump stops limit the bump travel; they absorb high forces over a short path. The elastic stops can also be accommodated in the shock absorber (Fig. 5.47). They can sit within the coil springs (Figs 1.7 and 1.13) or be fixed on the axle housing (Fig. 1.20) or come into contact with it when the springs reach full bump.

In comparison with the relatively flat, hard bump stop, the supplementary or additional springs are much longer, but act softly and, as shown in Figs 5.21, 5.50 and 5.51, have a favourable springing curve and absorb high forces when fully compressed. The parts are made of rubber or polyurethane elastomer. The air bubbles in the elastomer enable the bumpers to be compressed by 77% where the diameter is increased by only 35%. Like compression stops, they then absorb a force of $F_1 = 7$ kN (Fig. 5.49). Figures 1.24, 1.40, 1.41, 1.60 and 5.29 show supplementary springs in the installed condition.

Almost any springing curve can be achieved by combining a linear steel spring with a highly progressive supplementary spring (Figs 5.9 and 5.14).

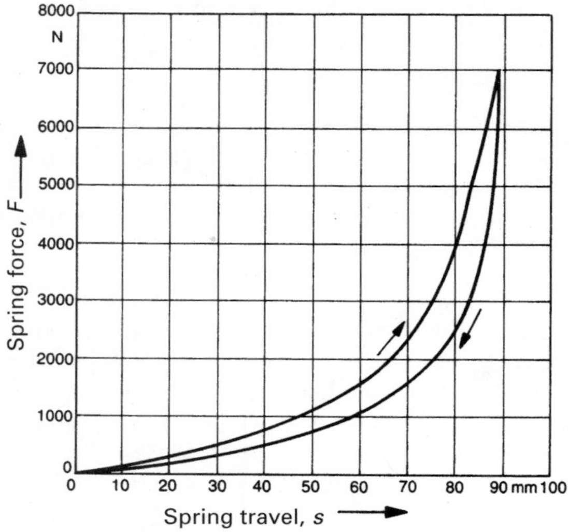

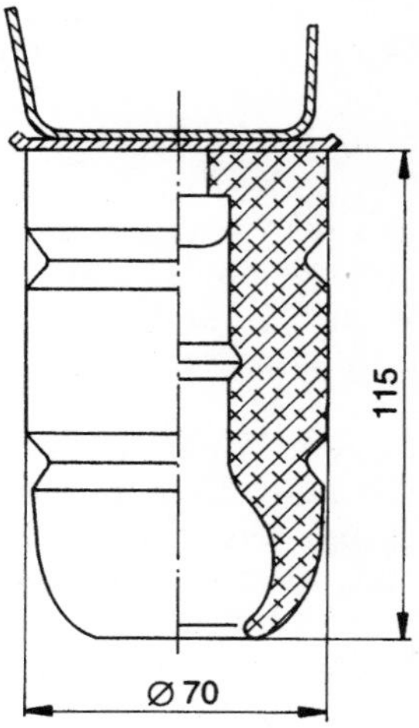

Fig. 5.21 Supplementary spring manufactured by the company Elastogran and fitted by Ford. The part is made of a polyurethane elastomer and remains flexible when cold down to an ambient temperature of –40°C. The 'buckling lip', which can be seen at the lower end, ensures a soft contact and a low initial spring rate. The upper end is kept tight against the body within the coil spring.

5.5.4 Anti-roll bars

The function of the anti-roll bars is to reduce the body roll inclination during cornering (Figs 1.6 and 5.16) and to influence the cornering behaviour in terms of under- or oversteering (Fig. 5.2), i.e. increasing the driving safety. In the case of parallel springing, the back 1 turns (Fig. 5.22) in the bearings L; the anti-roll bar remains inactive. The anti-roll bar rate $c_{S,\varphi}$ on reciprocal springing, which is important for reducing roll inclination, related to both wheels of an axle, depends, for independent wheel suspensions, on the ratio of the wheel joint G to the attachment point T_2 on the suspension link or on rigid axles of distances b_r

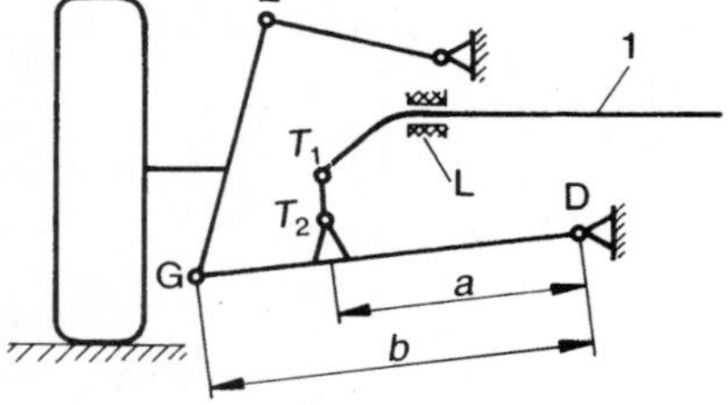

Fig. 5.22 The anti-roll bar is mounted to pivot with its centre-part 1 in the points L. The connection between leg ends T_1 and wishbones (points T_2) is made by an intermediate rod. The ratio $i_S = b/a$ is much greater than one and therefore increases the forces in the suspension links and their mountings.

and b_S (Fig. 1.23). With the rate c_S on the leg ends T_1 of the anti-roll bar, the rate, related to the centre of tyre contact, becomes

$$c_{S,\varphi} = c_S/i_\varphi^2 \tag{5.20}$$

and $$i_S = b/a \text{ or } b_r/b_S \tag{5.21}$$

The closer to the wheel the anti-roll bar operates, the lighter and less expensive it becomes and the lower the forces that occur in all the components.

An underslung-type anti-roll bar, shown in Fig. 1.8 and used only on McPherson struts to date, provides a solution in this direction. The connecting rod 5, whose path is around the same size as that of the wheels, is fixed to the outer tube 1. The disadvantage of this arrangement is the effect of the anti-roll bar on the McPherson struts during steering.

Figures 1.12, 1.42, 1.43, 1.54, 1.56, 1.57 and 1.63 show the configurations of normal anti-roll bars and the various ways in which they are mounted. Apart from the body roll inclination, the cornering behaviour can also be influenced by anti-roll bars. The following rules will apply (see Fig. 5.2 and Section 5.2.1 in Ref. [9]):

- A front-axle-mounted harder anti-roll bar promotes the tendency to understeer and improves the behaviour when changing lanes.
- Higher rear axle stabilization means the front-wheel drive vehicle can become more neutral, whereas the rear-wheel drive vehicle oversteers more.

However, the anti-roll bar also has disadvantages. The more the rate $c_{S,\varphi}$ related to the wheels increases, and the more highly the elastic parts are pre-tensioned in the various mountings (positions L, T_1 and T_2 in Fig. 5.22 and positions 17 and 19 in Fig. 1.12), the less the total springing responds when the vehicle is moving over a bumpy road; the vehicle 'copies' the road. Furthermore, the engine begins to vibrate on its mountings (especially on front-wheel drive vehicles) and front-end shake occurs. The ride comfort also deteriorates (see Sections 5.1.3 and 5.1.2). There is also an unfavourably harder reciprocal springing when the vehicle is moving along a pot-holed road (Fig. 1.21).

5.6 Shock absorbers (suspension dampers)

Running vehicles are exposed to almost constant vibration excitation; shock (i.e. vibration) absorbers are consequently required for reasons of driving safety and riding comfort. These aims partly conflict, because a taut suspension prevents wheel hopping and thus a loss of road contact, whereas a soft suspension is supposed to reduce body vibration and thus the annoying effects of acceleration on the occupants. The establishment of the damping force is also made more difficult by its dependence on the driving and load conditions, so that vehicle manufacturers usually work on the assumption of an average load (two people and 75 kg of luggage) as well as road surface excitation which is typical for the use of the vehicle. Electronic components such as antilock, slip and stability control systems must have operative shock absorbers, as wheel hopping with a transmission of longitudinal forces as a result of a brief lack of normal power, results in wheel lock (or spin) and thus gives a false input signal to the control system.

Together with tyres and disc wheels, shock absorbers are one of the parts of the chassis that are most frequently exchanged for models of the owner's choice. The owner believes that the handling characteristics of the vehicle can be improved. This can apply, although associated with the risk of premature wear to the stops, if the dampers also have to limit spring travel (see Section 5.6.8). If exchanging this part causes a change to the driving, steering or braking characteristics of the vehicle, and therefore represents a danger to other road-users, in Germany the vehicle type approval and therefore also the insurance protection would automatically lapse.

The correct tyre can be recognized from the size marking and the ECE index (Fig. 2.18), just as a worn profile, the depth of which is no longer permissible, is clearly visible. The shock absorber, in contrast, is located inside the chassis, the type marking is embossed on to it, but mostly covered by dirt and barely legible. Furthermore, with the variety of dampers available on the market, it is likely only to be possible to find out whether the type fitted has been approved by the manufacturer or is serviceable for the vehicle by referring to manuals.

The fact that a visual inspection only indicates failure where dampers are leaky, and that inspections are rarely carried out when they are in the installed condition is likely to be one of the reasons why there are more cars on our roads with defective shock absorbers than ones with inadequate tyres.

For more details on the various systems and their practical applications, see Ref. [5].

5.6.1 Types of fitting

The top of the shock absorber is fixed to the body or the frame and the bottom to a suspension link or the axle itself. Both fixing points should be as rigid as possible, so that the shock absorber also functions at more sensitive levels. When the wheels rebound and compress, the rebound stage and the compression stage usually come into play; in both cases vibration is dampened (see Section 5.2).

The shock absorber should be arranged vertically; if it is at the angle ξ_D to the rigid axle (Fig. 5.23), the ratio i_D is included in the calculation of the damping related to the wheel on parallel springing:

$$i_D = 1/\cos \xi_D \tag{5.22}$$

The larger ξ_D becomes, the smaller the force at the wheel and the lower the path in the damper; the ratio i_D is therefore squared in the damping calculation. In the case of reciprocal springing, the distance b_D also plays a role on rigid axles:

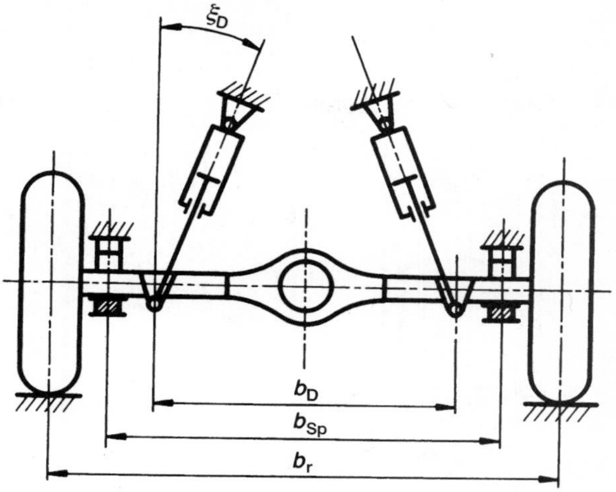

Fig. 5.23 If the dampers are fixed to a rigid axle at an angle, the angle ξi_D increases with compression with the advantage of a more unfavourable damping in the loaded condition. Moreover, the further in the dampers are positioned, the less they prevent the body roll movement.

$$i_{D,\varphi} = \frac{b_r}{b_D \cos \xi_D} \tag{5.23}$$

The further inside are the dampers, the less their effective distance b_D in comparison with the tread width b_r. The ratio $i_{D,\varphi}$ for reciprocal springing increases, leading to reduced body roll damping, the effect of which is unfavourable, particularly on high bodies.

The deviation of the damping position from the vertical is a disadvantage – in the rear and side view – even on individual wheel suspension and compound crank axles (Fig. 1.2), except that here Equation 5.22 is valid both for parallel and for reciprocal springing. All i_D and $i_{D,\varphi}$ data can be found in Section 5.3 of Ref. [3].

When establishing the damping forces, changes in the position of the shock absorber with wheel travel are to be taken into consideration (Fig. 1.13). Changes in the angle of the shock absorber can result in an unwanted decrease in the damping force with an increase in jounce. The shock absorber connection points (eye and pin bearings) must also be designed for such changes in angle.

5.6.2 Twin-tube shock absorbers, non-pressurized

5.6.2.1 Design of the damper

Figure 5.24 shows the design principle. The damper consists of the working chamber A, the piston 1 fixed to the inner end of the piston rod 6, the bottom valve 4 and the rod guide 8 (Figs 5.25 to 5.28); this also takes the seal 5 and, together with the piston 1, transmits any bending moments that occur through lateral forces to the eye-type joint of the damper. The reservoir C, also known as

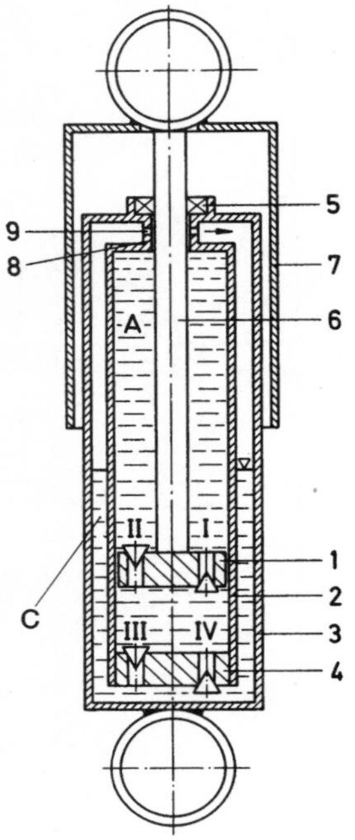

Fig. 5.24 Diagram of the twin-tube principle to explain the function.

1 piston
2 cylinder tube
3 outer tube
4 bottom valve
5 piston rod seal
6 piston rod
7 protective sleeve
8 piston rod guide
9 return holes
A working chamber
C equalization chamber

the equalization chamber, which is around half filled with oil, is located between cylinder 2 and outer tube 3. The remaining volume is used for taking both the oil volume, which expands when it warms (temperatures up to +120°C are possible and briefly up to +200°C where viton seals are used), and the oil volume which is evacuated by the entry of the piston rod.

The level of the oil column in the equalization chamber must be at half full to avoid air being sucked into the working chamber through the bottom valve in the

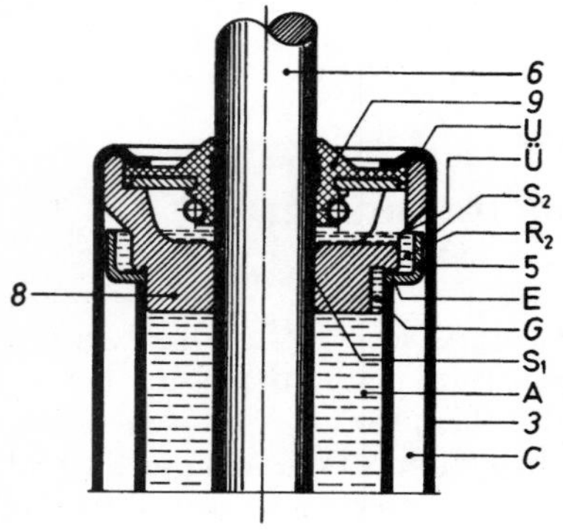

Fig. 5.25 Guide and seal set used by Sachs Boge in series production of twin tube dampers. The finished damper is closed by rolling the outer tube 3 around the edge U of the piston rod guide 8.

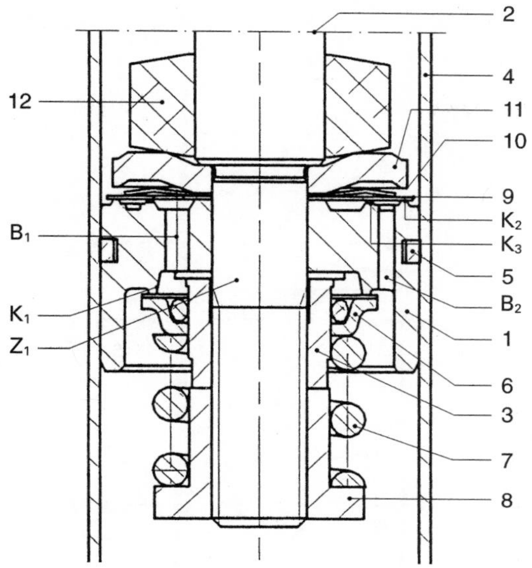

Fig. 5.26 Valve combination used by Sachs Boge on twin-tube dampers (item 1 in Fig. 5.24).

1 piston
2 piston rod
3 nut
4 cylinder tube
5 piston ring
6 valve disc
7 coil spring
8 nut
9 } rebound valve { (cover plate)
10 } rebound valve { Y-spring
11 } rebound valve { washer
12 top out (stop)
K_1 sealing edge 1
K_2 sealing edge 2
K_3 sealing edge 3
B_1 drill hole
B_2 channel
Z_1 spigot

case of extreme driving conditions. This could occur if the piston rod extends fully at extremely cold temperatures (–40°C).

The inclined position of the shock absorber in the vehicle, which leads to the oil level in the equalization chamber C falling on one side, must also be considered. There is therefore a limit to the amount by which the angle ξ_D can deviate

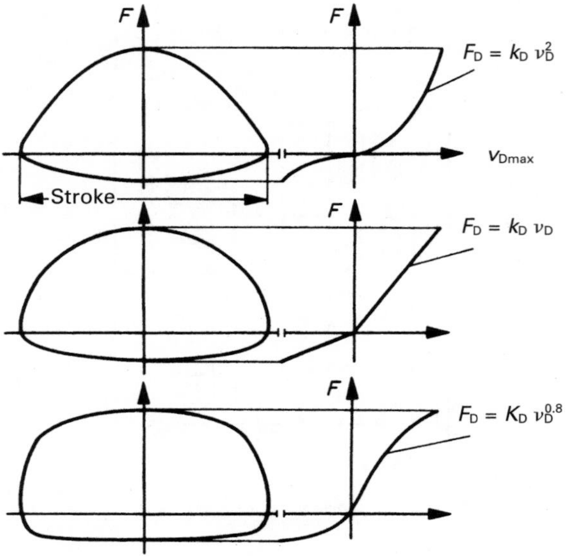

Fig. 5.27 The damping curve can be progressive (top), linear (centre) or degressive (bottom). The curve shape and diagram shape are directly related. The smallest area and therefore the lowest mean damping is in the diagram of the progressive curve, while the largest area is that of the degressive damping. The shape of the damping curve can be expressed in an equation by the exponent n:

$$F_D = k_D v_D^n$$

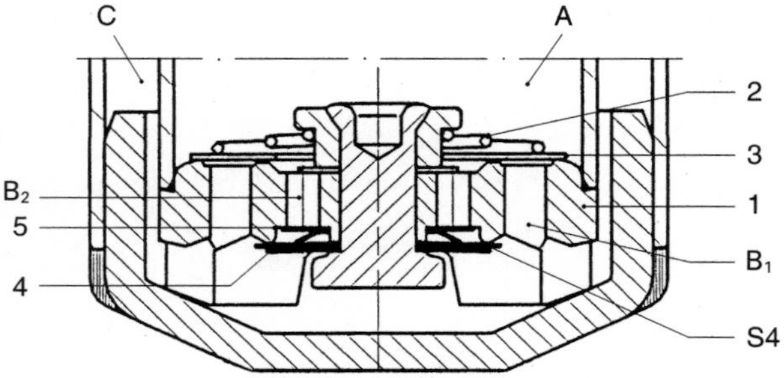

Fig. 5.28 Bottom valve of the Sachs S27, S30 and S32 twin-tube dampers.

from the vertical (Fig. 5.23); a maximum of 45° may be reached in the fully compressed condition.

5.6.2.2 Function

When the wheels are subject to bump motion, the damper shortens, piston 1 moves down and part of the oil flows out of the lower working chamber through the valve II into the upper one A (Fig. 5.24). The volume corresponding to the immersed piston rod volume is thereby pushed into the equalization chamber C through the valve IV in the valve body 4. This produces the main forces necessary for the compression damping and only if this does not suffice can the valve II on the piston valve become effective.

As Fig. 5.26 shows, the valve II consists only of the Y-spring 10 loaded covering plate 9.

When the axle rebounds, there is overpressure between the piston 1 and piston guide 8. As this happens, the main oil volume is pushed to the adjustable valve I, which causes the jounce damping. The minor fluid volume is squeezed through the gap between the guide and piston rod, indicated as S_1 in Fig. 5.25, and the corner channels E and G. If the rod extends, this leads to a lack of oil in the working chamber A. The missing volume is sucked from the chamber C (Fig. 5.24) and flows through the valve III, which is also only a simple return valve. The oil pulsing back and forth between the working and equalization chamber is cooled on the outer tube 3.

5.6.2.3 Air vent and volume equalization

Twin-tube shock absorbers have to be air vented, because air bubbles can form in the working chamber – unavoidable in this type of damper. This happens when:

- the damper is stored or transported horizontal prior to installation;
- the oil column in the working chamber falls when the vehicle has been standing for a long time;
- the shock absorber cools at the end of a journey, the oil in the working chamber contracts and air is sucked through the piston rod and rod guide.

Without special measures, an air pocket would arise and, particularly during cold weather, unpleasant knocking, known as ‘morning sickness’, could occur. Designers must ensure that the oil reaching to the top of the working chamber cannot flow back into the equalization chamber when the vehicle is standing and, furthermore, that fluid fills the space that has been freed as the oil has contracted. Sachs Boge solves this problem with the angular ring 5, shown in Fig. 5.25, and several channels E and G, disposed at a right angles and pressed to the outside of the rod guide. Ring 5 creates the reservoir R_2 from which the oil can flow back via the two channels as it cools.

Another advantage is that the air that has been captured inside the working chamber can escape better. Channels E and G are used for evacuation in such cases; the air cushion quickly dissipates through the channels as a result of wheel movements. The angular ring also prevents the oil jets, which shoot from the channel E as the piston rises, from hitting the outer tube 3 directly and foaming up.

As the piston lifts, over-pressure arises above the piston, which also pushes some oil out upwards through the gap S_1 (between piston rod and guide) and the corner channels E and G. This small amount lubricates, amongst other things, the rod, collects in the reservoir R_2 and flows through the ring gap S_2 (formed by the angle ring 5 and the outer tube 3) back into the equalization chamber C. It is then cooled in the tube 3 by the wind blast of the moving vehicle. Ring gap S_1 as well as the size and number of the transverse channels G nevertheless represent a constant by-pass and their cross-sections must be considered when designing the orifices in the piston.

When subjected to compressive forces, the piston rod moves in, displacing a certain volume and thereby creating over-pressure in the working chamber A, i.e. in the compression stage oil is also pushed through the gap S_1 and the channels E and G and cools down on the outer tube 3 as it flows back.

5.6.2.4 Rebound valve

The rebound valve in twin-tube shock absorbers is generally a combination of constant orifices and bores closed by spring-loaded valve discs (Fig. 5.26). Piston 1 is fixed by the nut 3 to the lower end of the piston rod 2. Sealing to the side of the cylinder tube 4 is provided by the piston ring 5, the mid-centring of the piston by the spigot Z_1. The valve consists of the valve disc 6, which is pushed by the coil spring 7 against the sealing edge K_1. The valve spring force is regulated with the nut 8. Bypass or advance opening sections whose areas of passage ensure a constant flow are impressed between the sealing edge K_3 and the top cover disc. As the piston rises, oil flows through hole B_1 in order to bypass the constant flow as well as the actual valve after the rising of the valve head.

The height of the jounce damping is determined by several factors:

- the number and cross-sections of the impressed advance openings and (see Fig. 5.25), the gap S_1 between piston rod 6 and the hole in the guide A, as well as the ventilation ducts E and G at a low piston speed;
- at medium speeds by the aperture of the valve disc 6, i.e. by the stiffness and initial tension of the spring 7;
- at high piston speeds and with the valve wide open, by the number and cross-section of the holes B_1.

By combining these options, any valve curve, from degressive via linear through to progressive curves (Fig. 5.27) can be set.

The jounce movement of the shock absorber and hence the jounce travel of the wheel suspension are limited by the jounce stop 12 which sits on the supporting plate 11; see Section 5.6.8.1.

The oil first flows through outer duct B_2 in the direction of the pressure and then lifts valve plate 9. This thin plate which only serves as a non-return valve is movable in an axial direction and normally provides a seal along edges K_2 and K_3. The pressure force is applied by the soft star-shaped spring which is attached between valve plate 9 and supporting plate 11; the latter also serves as a stop and prevents too wide opening of the valve at high piston speeds.

5.6.2.5 Compression stage valve

Parts 9 to 11 shown in Fig. 5.26, which sit on top of the piston, are simply a check valve, as described at the start of Section 5.6.2.2; the bump damping forces are primarily produced by the compression valve in the bottom of the damper (part 4 in Fig. 5.24). Figure 5.28 shows a section through the configuration fitted by Sachs in shock absorber types S27, S30 and S32. The actual valve body 1 has the holes B_1, through which oil is sucked when the piston moves upwards as the wheel jounces and the volume of the extending piston rod must be replaced. The covering disc 3 loaded by the conical spring 2 lifts.

The piston rod has a diameter of 11 mm on passenger car and light van dampers; the small cross-section area of only 95 mm^2 must provide the fluid displacement which then produces the bump damping (in comparison 478 mm^2 is available for the jounce stage, corresponding to a 27 mm piston diameter minus the rod cross-section surface).

When the piston rod enters, the bump stage valve is charged by the displacing oil. This valve consists of the set of spring washers 4, the upper washer of which has the grooves S_4 as a constant orifice. The required setting can be achieved by means of the diameter of the hole B_2, the number and thickness of the spring washers and the size of the by-pass grooves S_4.

However, the constant by-pass has the disadvantage that when the vehicle is standing, the oil in the working chamber A, which is at a higher level, can flow back into the equalization chamber C. If the vehicle moves off again, after it has travelled a certain distance this equalizes out, although it may be linked to a certain unpleasant knocking noise, known as 'morning sickness'. Until the air bubble at the top of the working chamber escapes, when the wheels rebound, the oil is drawn suddenly against the piston guide. To avoid the noises this causes, Sachs has added the anti-communication valve 5. This is upstream of the spring washers 4, covers the holes B_2 and therefore prevents the oil from flowing back.

The compression damping curve arises through the interplay of the bottom valve with the opening at the front of the piston S_4 shown in Fig. 5.26 and the check valve 9 on the piston. There are also the air vent channels E and G shown in Fig. 5.25 and the nozzle clearance S_1 between piston rod and guide.

In order to prevent the quantity of oil being pushed out through the bottom valve during compression and hence the introduction of the damping force, the bottom valve must oppose the oil to be displaced with a fairly high level of resistance as the non-return valve 9 located on the piston opposes the quantity of oil which flows through the piston.

5.6.3 Twin-tube shock absorbers, pressurized

The most economical form of damper design is the one that operates on the non-pressurized twin-tube system. Where certain vehicles or chassis conditions make it appear sensible or necessary to use a gas-pressure damper, the low-pressure twin-tube shock absorber is a good solution. The additional costs remain reasonable. Because compression damping continues to be provided via the bottom valve, gas pressure of around 4 bar is sufficient. This means that the piston rod extension force F_{Pi}, described in Section 5.6.4.1, remains low. This makes it

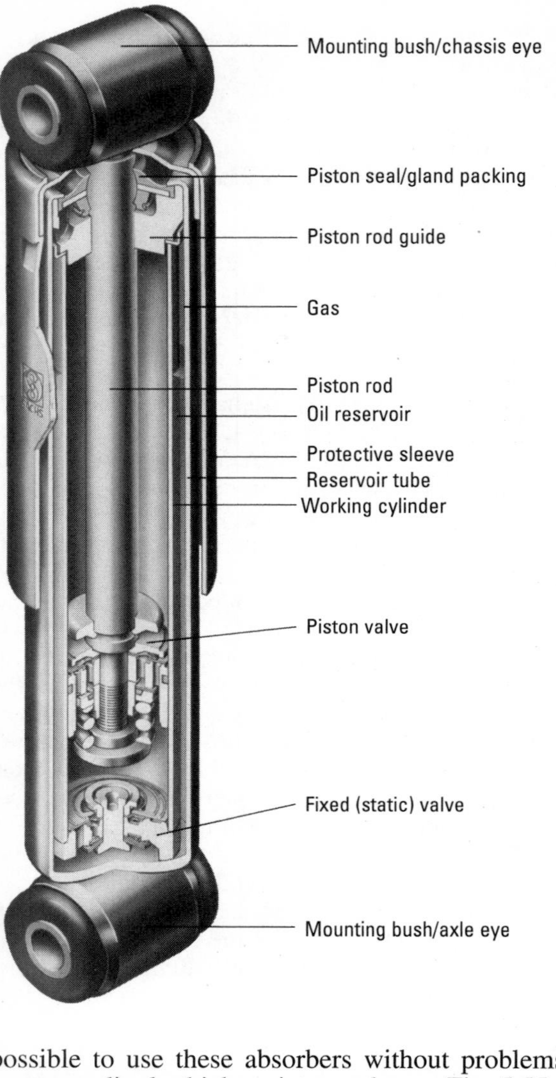

Fig. 5.29 Low-pressure twin-tube shock absorber of Sachs. In the shock absorber preloaded with a gas pressure of 6–8 bar, particular importance is attached to the function of the piston rod seal, because this must provide a secure seal in all operating conditions. Guided by the piston rod, the jounce stop which comes into play on the piston-rod guide 8 during rebound sits above the piston valve and thus limits the jounce travel. The rigidity properties of the jounce stop are particularly important for reasons of comfort, as the jounce movement may not suddenly be stopped. Figure 5.56 shows a sectional view of the piston-rod guide.

possible to use these absorbers without problems on McPherson struts, with correspondingly thicker piston rods; see Fig. 5.55.

The basic design, the length and dimensions of the non-pressurized and pressurized designs are the same, so it does not matter which variety is used on a vehicle (e.g. for sports models), as no change to the vehicle is necessary.

The advantages of the low-pressure twin-tube design are:

- more sensitive valve response at small amplitudes;
- ride comfort increases;
- damping properties under extreme conditions (e.g. on pot-holed roads) are better;

- hydraulic hissing noises are reduced;
- shorter lengths and less friction than monotube gas pressure dampers – as the required gas reservoir is not accommodated in the cylinder tube, but between cylinder tube and tank reservoir tube;
- the shock absorbers remain functional even after loss of gas.

Unlike the unpressurized design described in the previous section, the oil reserve (or equalization) chamber is also preloaded to 1/3 with gas of 6–8 bar with a pressure-loaded twin-tube shock absorber (Fig. 5.29). As the gas chamber is not located in the cylinder tube, as is the case with the single-tube gas-pressure shock absorber, twin-tube shock absorbers have the advantage of a particularly short length.

5.6.4 Monotube dampers, pressurized

5.6.4.1 Design and function

The design, used almost exclusively today, with a separator piston (position 1) can easily be explained on the basis of the schematic diagram in Fig. 5.30. At the top is the evacuation chamber 3, which (as in the twin-tube system) must absorb the volume equalization by the oil warming and the volume displaced by the piston rod. Gas and oil are separated by the piston 1, which seals off the actual working chamber 2.

The damper piston 5 usually has a diameter of 30, 36, 45 or 46 mm and is fixed to the piston rod 8. It carries the valves 6 and 7. As shown, the piston rod can extend upwards or downwards (Fig. 5.31); the separator piston 1 makes it possible to install the shock absorber in any position. If the damper cylinder is fixed to the body or frame, the cylinder weight forms part of the sprung mass and

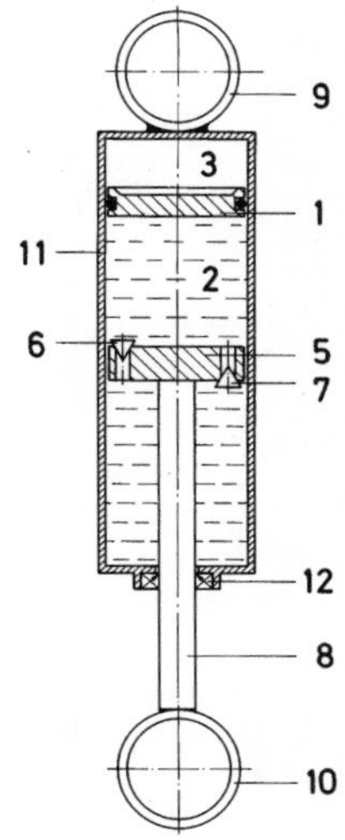

Fig. 5.30 Diagram of the pressurized monotube principle with separator piston (position 1).

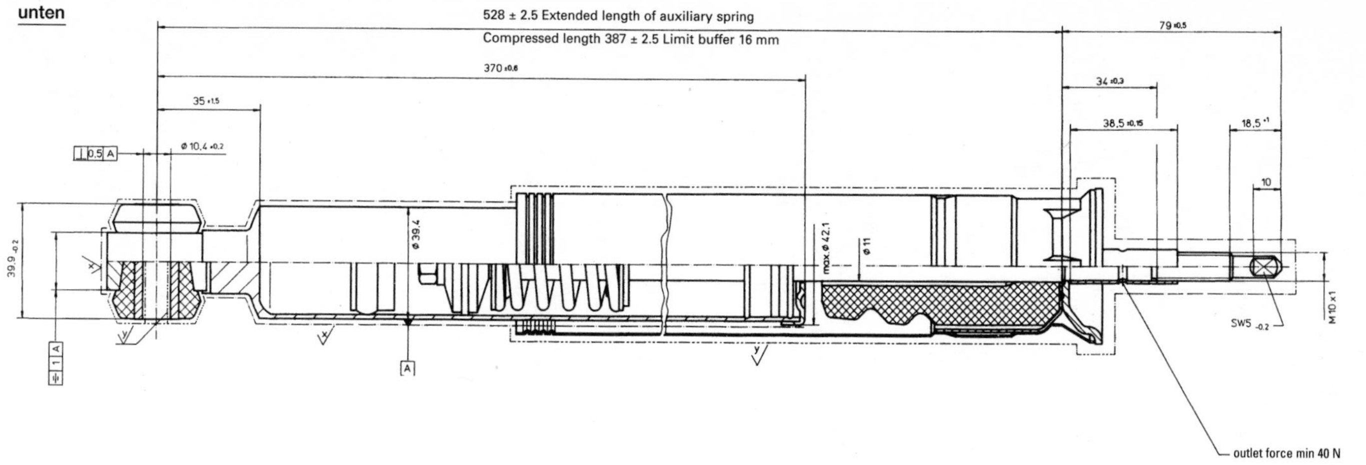

Friction 100N max.	Abnahmewerte VDA Maschine Bilstein				
	$v_{max.}$ m/s	0,065	0,13	0,26	0,52
	Zug N	215±35	470±80	910±110	1730±140
	Druck N	130±25	210±35	345±50	660±65

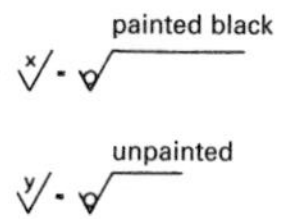

Fig. 5.31 After an original drawing by Bilstein of the front axle damper of the Mercedes Benz C (1997) class with a stroke of s_D = 141 mm, the fixed length L_{fix} = 246 mm, pin-type joint at the top (with a crimped spacer) and eye-type joint at the bottom; the piston rod comes out at the top. The supplementary spring shown on the right is surrounded by a short, stable tube and comes into contact with this and the support disc located above the piston rod guide when the spring compresses. The tube also carries the actual plastic sleeve, which reaches up to the damper centre.

The mechanical compression stop sited above the piston is only fitted in this form on the performance models. The coil spring, which forms part of this, helps to reduce pitch and roll movement on the body and its top is carried by the piston rod by means of a washer. When the wheel rebounds, the washer comes into contact with the piston rod guide (Fig. 5.32). Further details are given in Ref. [5], Section 8.3. Settings and tolerance values are shown at the bottom on the left of the figure. These are between 7.5% and 18%.

only the light piston rod contributes to the unsprung mass. This is a reason for preferring the installation position shown in Fig. 5.30.

When the wheel jounces, the oil flows through the jounce stage valve 6, shown in Fig. 5.33 from the bottom to the top part of the operating chamber. The gas pressure in the gas chamber 3 forces the separator piston to follow, equalizing out the reduction in volume (caused by the piston rod extending). If the wheel goes into bump travel, the compression valve 7 is charged (Fig. 5.34) as the dividing piston 1 moves upwards through the oncoming piston rod volume. The entire piston surface is available for bump damping; this is then significantly more effective than on the twin-tube system, and the valve 7 produces high forces at lower fluid pressure – without loss of comfort – an advantage on vehicles with heavy rigid axles. The road-holding can be improved here by means of responsive and correspondingly high compression damping.

The gas pressure at ambient temperature (20°C) is at least 25 bar. This value is required to counteract the compression damping forces. If these exceed the opposed force exercised by the gas pressure on the separator piston, the oil column will rip off at the compression valve. Therefore, for a 36 mm piston diameter, 2.8 kN are needed, and for a 46 mm diameter piston, 4.6 kN.

A disadvantage of the high gas pressure is the piston rod extensive force, which amounts to

$$F_{Pi} = 190 \text{ N to } 250 \text{ N}$$

If a vehicle has soft springing (e.g. $c_f = 15$ N mm^{-1}), where gas pressure dampers are retrofitted, this can raise the body by

$$s_2 = F_{Pi}/c_f = 250/15 = 17 \text{ mm}$$

if the dampers are positioned close to the wheels. When the vehicle is running, they warm up and, at an oil temperature of 100°C, extension force and body lift increase by up to

$$F_{Pi} \approx 450 \text{ N} \quad \text{and} \quad s_2 \approx 30 \text{ mm}$$

If gas pressure dampers are fitted as standard, this influence has already been taken into consideration by the vehicle manufacturer. If the owner subsequently changes over from twin-tube to pressurized monotube dampers it is recommended that appropriately shortened springs be fitted.

5.6.4.2 Piston rod and rod guide

Figure 5.32 shows a section through the seal package with a piston rod guide above the seal and therefore only slightly lubricated.

Unlike twin-tube dampers, a detachable piston rod guide (position 1), held by the wire snap ring 2, is used to plug the damper. The guide can be pushed down to the second snap ring 3 and the ring 2 can then be laid into the free groove in tube 4. When the load is removed, the oil pushes the guide back against ring 2.

The O-ring 5 seals the rod guide to the outside and the mono-lip seal 6 to the piston rod. The flange of this seal sits inside the guide 1 with its neck in the 'perbunane' disc 7. Internal pressure and clamping load of the closure disc 8,

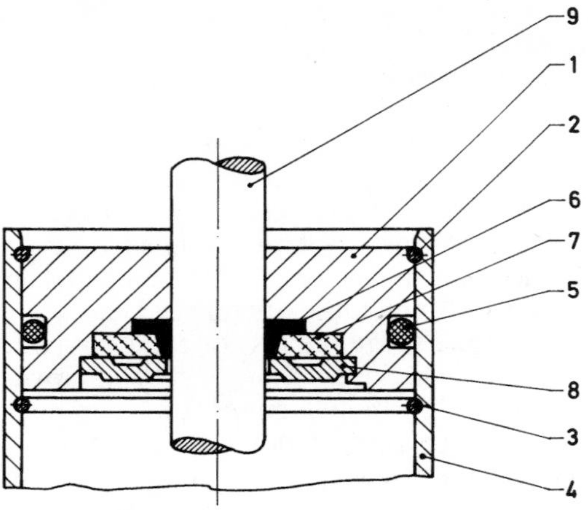

Fig. 5.32 Seal package developed by Bilstein, which keeps the temperature range of –40°C to +200°C demanded by the automobile industry. The outer piston rod guide 1 has a hard-coated hole and is made of an aluminium wrought alloy (e.g. AlMgSi 1 F 28). The piston rod 9 has the diameter $d = 11^{-0.02}$ and the hole has the tolerance range

$$d = 11\,^{+0.07}_{+0.05}$$

which corresponds approximately to the ISO fit D7/h7 with a play between 0.05 mm and 0.09 mm.

which is secured to the guide, ensure that the sealing neck is also pressed against the piston rod 9. The more the oil warms up when the vehicle is moving, the more the inner pressure increases and the more tightly the seal is pressed on. If a compression stop is fitted into the damper, it comes into contact with disc 8 when the wheel jounces.

The fluid seal on the pressurized monotube damper is more dependent on the surface condition of the piston rod than on the gasket 6. The rod is therefore manufactured with particular precision. In passenger cars and light commercial vehicles monotube dampers made by Bilstein Ltd, the rods have 11 mm diameter and are made of the heat-treatable Ck 45 QT steel. The strength properties are:

$$R_m = 750 \text{ N mm}^{-2} \text{ to } 900 \text{ N mm}^{-2},\ R_e \geqslant 530 \text{ N mm}^{-2} \text{ and } A_5 \geqslant 6\%$$

The surface is raised by induction hardening to a Rockwell hardness of 58^{+2} HRC and is then ground to achieve a roughness depth of $R_t = 0.8$ μm to 1 μm. A hard chrome layer over 20 μm thick, subsequently applied, raises the surface hardness to 70 ± 2 HRC and the subsequent super-finish treatment reduces the roughness depth to the value $R_t = 0.2$ μm needed for the seal.

5.6.4.3 Pistons and valves

Due to the equalization chamber being above the working chamber, the monotube damper is longer than the one operating in the twin-tube system. To minimize this

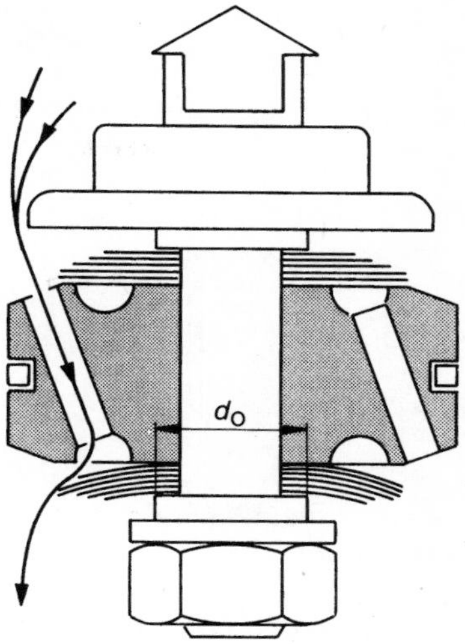

Fig. 5.33 Space-saving compression stage valve with spring plates and a supporting washer found on almost all monotube dampers. If, as shown, the piston rod moves upwards, the lower valve is achieved. The piston ring shown in the illustration is used to prevent any unwanted flow in the gap between this and the cylinder walls.

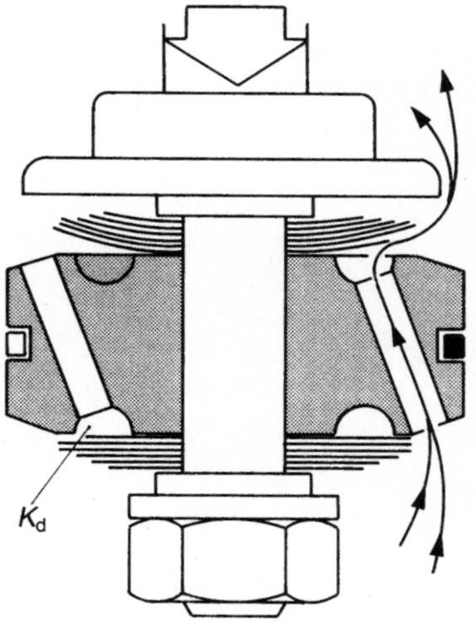

Fig. 5.34 If the piston rod moves upwards, the spring plate valve for the compression stage under the piston moves.

disadvantage, the separator piston 1 (Fig. 5.30) is hollowed out in the centre and a flatter working piston fitted (flatter than the one in the twin-tube system). Flat plate valves are also used.

When the piston rod extends, the oil flows past the compression valve at the top through diagonal holes to the rebound stage valve (Fig. 5.33). Both thickness and number of valve plates, as well as the support disc diameter d_0 and the amount of the constant orifices K_d, are critical for the level of the damping forces. The constant by-pass is created by a bottom valve plate on the compression valve (Fig. 5.34) which is smaller in diameter and does not completely cover the inclined holes. Unlike in the twin-tube system, when the piston enters,

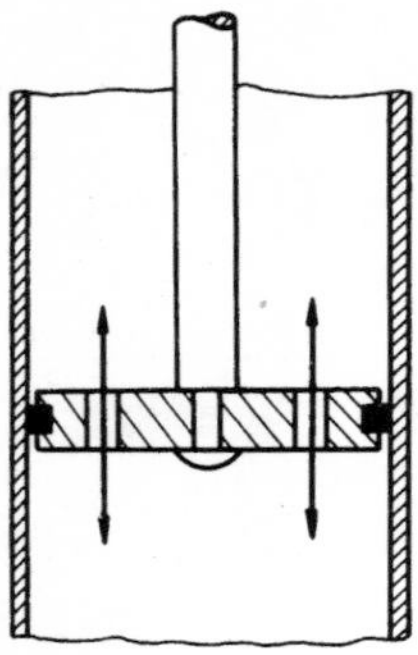

Fig. 5.35 Unshielded holes in the piston correspond to a constant flow, also known as the advanced opening cross-section or by-pass. On the monotube system they give the highly progressive damping curve shown in Fig. 5.36. The compression and rebound forces are the same size and have very high terminal values.

its larger diameter valve plates are charged by the entire oil column; this causes much more intensive damping and prevents the wheels from oscillating – without reducing the ride comfort.

In all monotube dampers, the characteristic of the damping curve is determined exclusively by the valves on the piston and the holes.

If these just have constant orifices (Fig. 5.35) there is a highly progressive curve shape with high forces (Figs 5.36 and 5.27 top) on both the compression and the extension side; this also applies when there is a by-pass between the piston and cylinder tube, i.e. if the piston ring were missing, or as is the case in several variable dampers, if a by-pass nut is fitted to the cylinder tube (see Fig. 5.57).

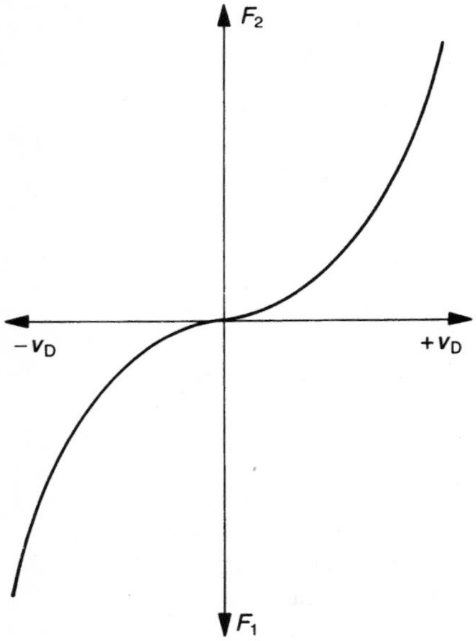

Fig. 5.36 Highly progressive damping curve achieved by holes in the piston or a gap between the piston and cylinder wall.

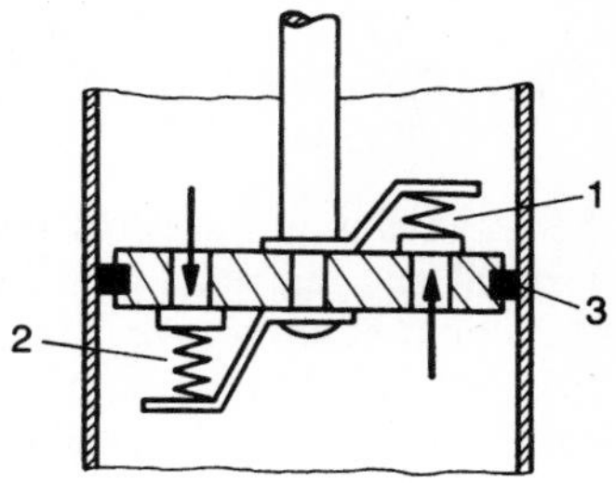

Fig. 5.37 Spring-loaded valves over large holes give a degressive damping curve. The forces in the compression and rebound side can be set to different levels. The piston ring 3 prevents an additional by-pass.

Pre-tensioned valve plates over large holes (Fig. 5.37) cause the curve to take on a degressive shape with the additional advantage of being able to set different forces on extension and compression sides (Fig. 5.38). At higher piston speeds these only increase a little. The linear curve shown in Fig. 5.27 is achieved either through low pre-tensioned valve plates or by using a combination of constant orifices and spring-loaded valve discs (Fig. 5.26).

5.6.4.4 Advantages and disadvantages

The pressurized monotube damper has a series of advantages over the non-pressurized twin-tube damper:

- good cooling due to the cylinder tube 11 (Fig. 5.30) with direct driving air contact;
- a larger piston diameter is possible with the same tube diameter (e.g. 36 mm instead of 27 mm), reducing the operating pressures;
- the compression stage valve 7 sits on the piston 5 and is charged by the entire oil column;
- the oil level in the oil column does not fall as it cools, so no 'morning sickness' occurs (see Section 5.6.2.3);

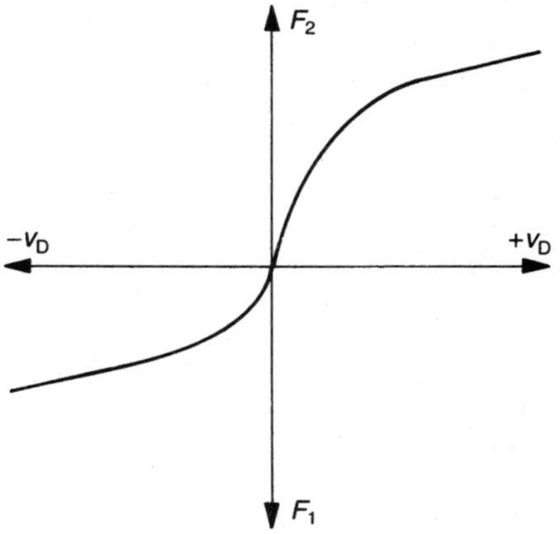

Fig. 5.38 Degressive curve with different force levels on the compression and rebound side, achieved by spring-loaded valves (see also Fig. 5.27).

- due to the pressurized oil column, the oil cannot foam, resulting in good damping of even small high-frequency vibrations;
- where there is a separator piston, the installation position is not restricted.

The disadvantages are that the high degree of manufacturing precision and the essential gas seal lead to higher costs. Furthermore, the greater space requirement can amount to over 100 mm in the stroke length.

As a result of the sometimes considerable pressure preloading (25–30 bar), the forces acting on the seals are greater; this results in unwanted friction which reduces the response properties of the shock absorbers.

5.6.5 Monotube dampers, non-pressurized

Non-pressurized monotube dampers generally have a piston of only 20 or 22 mm diameter, an 8 to 9 mm thick piston rod and therefore absorb correspondingly lower forces. They are used as:

- engine vibration dampers (see Chapter 10 in Ref. [5])
- driver seat dampers
- steering dampers (see Section 4.5).

The first two designs are installed vertically and it is only necessary to fit a compression valve (Fig. 5.28) instead of the separator piston (Fig. 5.30). As in the twin-tube system, this ensures the necessary back-pressure when the piston rod enters. The equalization chamber is above the working chamber and is around half-filled with oil and air; the two media could mix if there were no separator part, which is common on engine dampers.

Steering dampers must not have any extension force at the piston rod, otherwise the steering would be assisted in the compression direction and pulled to one side. The dampers are fitted in a lying position, so only non-pressurized monotube dampers (where the oil and air are separated) can be used.

Figure 5.39 shows a standard design, on which the flexible hose 1 performs this function; it is fixed by rolling the outer tube 3. Part 3 is bevelled off on both sides and presses the hose into pointed grooves of the cylinder tube to provide a good seal. At the same time this measure prevents displacement when the vehicle is moving. When the piston rod 17 moves in, the oil flows through the two apertures 4 of the valve in the valve body 5 and lifts the valve plate, which is loaded by the spring 7; this produces part of the compression damping.

The area between the protective tube 3 and the hose 1 acts as an equalization chamber. The hose 1 flexes when oil flows through the hole 9. As in the case of all monotube dampers, the damping valve unit (consisting of the rebound stage and the actual compression valve) is situated on the piston 10 (Figs 5.33 and 5.34). The piston ring 11 seals this off to the cylinder tube 2. The piston rod guide 12, seal 13 and support disc 14 sit between the two rolled-in grooves; the longitudinal hole in the guide acts as a pressure equalization. The eye-type joints 15 and 16 provide the installation.

The advantage of this design is the short length; increasing the stroke only

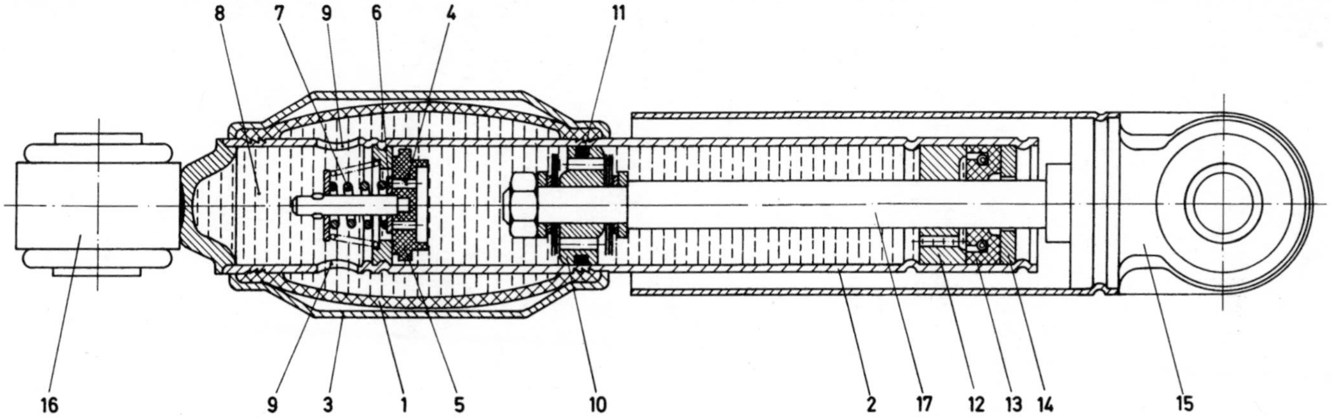

Fig. 5.39 Section through the Stabilus steering damper used on passenger cars and light vans, with its equalization chamber consisting of the elastomer tube 1 and the upper part 8 above the working chamber. The piston 10 is 20 mm and the rod is 8 mm in diameter.

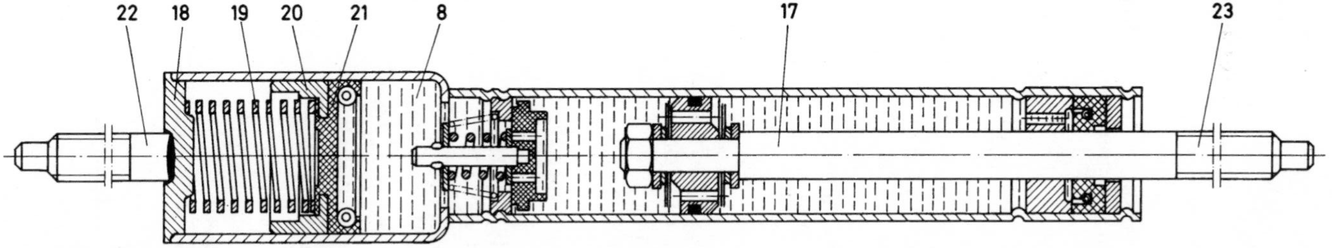

Fig. 5.40 Stabilus compact steering damper with pin-type joints on both sides (position 22 and 23), butt-welded equalization chamber 8 and spring-loaded cup seal 20.

makes it necessary to extend the tube 2 and the equalization hose 1 with the protective tube 3. The longer tube 3 can then be a disadvantage. If it should not prove possible to house this, an alternative design with a separator sleeve could be used (Fig. 5.40), which has the same functional parts but also has an in-line, welded-on, equalization chamber 8 with its inside diameter increased to 26 mm. The coil spring 19 in flat rolled steel supported on the top 18 flexes under the pressure of the oil displaced when the piston rod 17 enters. The opposed force of the spring 19 is measured such that a light pressure is applied to the oil column, but no extension force occurs. The seal between air and oil is provided by the cup seal 21, which is inserted into the guiding part 20.

5.6.6 Damping diagrams and characteristics

The spring force is a function of the wheel travel, whereas the damping force depends on the speed at which the two fixing points are pulled apart or pushed together. A damper, which is subject to a constant force F_D, flexes at a constant speed over the whole stroke, whereas a spring flexes immediately, but only up to a certain travel s_1, the length of which depends on the quotients of force and spring rate $c_{\text{f or r}}$ (see Fig. 5.27):

$$F_{Sp} = c_{\text{f or r}}\, s_1 \text{ and } F_D = k_D\, v_D^n$$

The spring therefore stores work and usually releases it at a moment that is not conducive to driving safety, whereas the damper annuls mechanical energy by converting it into heat. The more energy that the damper absorbs, the hotter it gets. In diagrams, the damping force F_D appears as a function of the piston speed v_D in m s^{-1}.

Figure 5.41 shows diagrams recorded on a standard test rig. At a constant rev speed ($n_D = 100$ min^{-1}), the stroke is changed step by step, but it is also possible to keep the stroke fixed and to vary the engine and therefore the test rig speed (Fig. 5.42). To record the damping curve, in both cases the maximum forces of each stroke are taken and, as shown in Fig. 5.42, entered upwards and downwards on the y-axis as a function of a maximum piston speed. The equation for calculating the individual values is

$$v_{D,max} = \frac{\pi\, s_D\, n_D}{60} \text{ (m s}^{-1}\text{)} \tag{5.24}$$

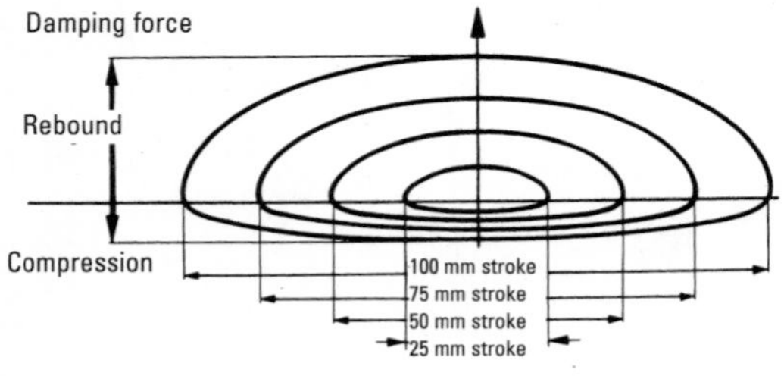

Fig. 5.41 The damping forces on the production test stand can be measured at $n = 100$ min^{-1} with increasing strokes to determine the curve.

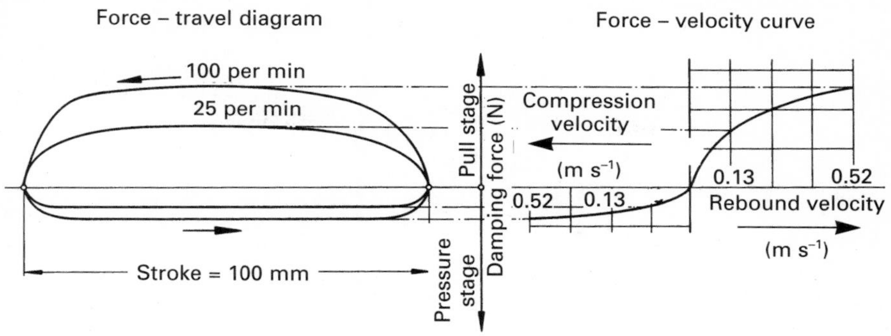

Fig. 5.42 The maximum compression and rebound forces are taken from the individual diagrams to create the damping curve formerly known as the force speed curve.

The value $n_D = 100\ \text{min}^{-1}$ and $s_D = 100$ mm gives the following speed:

$$v_{D,max} = \frac{\pi \times 0.1 \times 100}{60} = 0.524\ \text{m s}^{-1}$$

Figure 5.43 shows the curve of the rear axle damping of a front-wheel drive vehicle. Damping curve and diagram shape are closely related. A progressive curve (Fig. 5.27, top, and 5.36) has a cornered diagram with a relatively small surface, i.e. the actual mean damping, which is important for the springing behaviour, is low. The degressive curves shown in Figs 5.27 (bottom), 5.38 and 5.44 have a rounded shape and so a high mean damping.

It would be correct, but too time-consuming with conventional methods, to determine the size of the diagram's area in order to plot the resulting mean damping over the corresponding mean piston velocity, or to oppose the mean damping force to the mean piston speed $v_{D,med}$ by calculations:

$$v_{D,med} = v_{D,max}/1.62 \quad (5.25)$$

5.6.7 Damper attachments

5.6.7.1 Requirements

The damper attachments are used for fixing the damper to the frame, suspension subframe or body at the top, and to the axle housing itself or a suspension control arm at the bottom. Certain requirements must be fulfilled:

- maintenance-free and inexpensive to manufacture;
- angular flexibility (to absorb the movements in fixing points) with only low reaction torque, so as not to subject the piston rod to bending stress;
- noise insulation (to prevent the transfer of road noise);

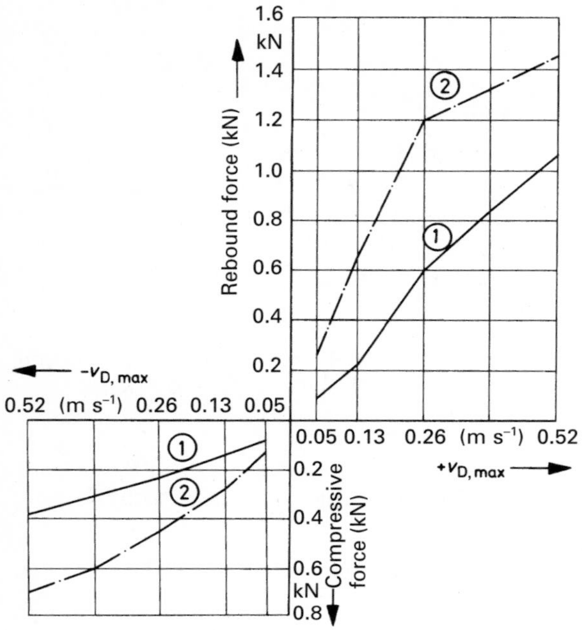

Fig. 5.43 Rear axle damping curve; 1 is the standard setting and 2 that for the heavy-duty version.

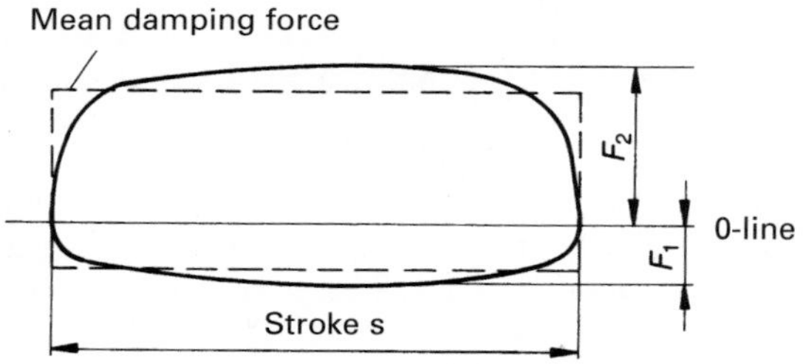

Fig. 5.44 The maximum piston speed $v_{D,max}$ and the greatest force F_2 in the rebound and F_1 in the compression direction are included in simplified form in the determination of the wheel and body damping; both are easily measurable. The actual form of the diagram, in this instance that of degressive damping (Fig. 5.27, bottom) is ignored.

- precisely defined flexibility towards the damping forces – any unwanted loss of travel in the rubber components reduces damping precision and road harshness.

On the vehicle side it must be ensured that the upper and lower fixing points align with one another in the design (normal ride height) position (i.e. when there are three people each weighing 68 kg in the vehicle); only in this way can

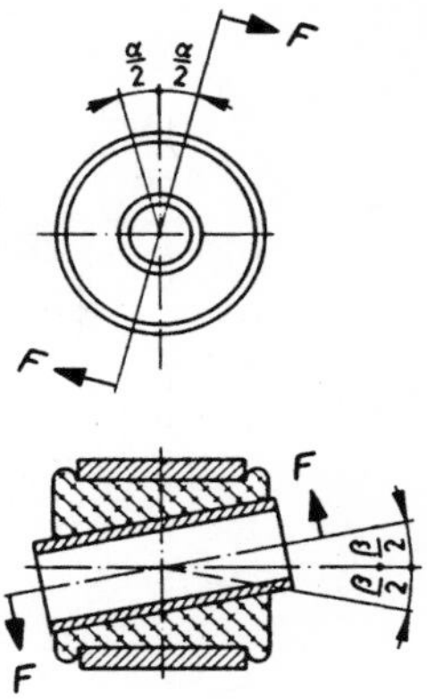

Fig. 5.45 The eye-type joint has 35 mm to 36 mm outside diameter, a hole of $10^{+0.15}$ mm or $12^{+0.15}$ mm and is 32 mm wide. The maximum approved distortion angles are $\alpha/2 = \pm 15°$ and the cardan (conical) angles $\beta/2 = \pm 4°$.

distortion, when the vehicle is running, and premature shock absorber wear be avoided.

5.6.7.2 Eye-type joints

The requirements are best met by rubber joints. Figure 5.47 shows, on the top and bottom of the damper, the type of suspension most used: the eye-type joint, sometimes also known as a ring joint. The most common size in passenger cars is 32 mm wide, 35 mm to 36 mm diameter and has a 10 mm or 12 mm fixing hole with a +0.15 mm tolerance (Fig. 5.45). If compression stops are housed in the shock absorber or if spring forces are also concentrated in the mountings, 40–60 mm wide joints may be necessary (Fig. 5.29).

The joint itself consists of a rubber bush that is in high radial pre-tension between the outermost ring and the pressed-in inner tube. The rubber part has beads at both sides as a measure to stop it sliding out when the vehicle is moving. The size mostly used and shown in the illustration allows twisting angles up to $\alpha/2 = \pm 15°$ and cardan (conical) deviations of up to $\beta/2 = \pm 4°$. Greater twist angles would increase the bending moment in the piston rod and therefore need different configurations (Fig. 5.31 and Section 5.2 in Ref. [5]).

5.6.7.3 Pin-type joints

If the same angle movement occurs in all planes at the upper or lower suspension when the vehicle moves, the design solution is to use a pin-type joint (Figs 5.46 and 5.40). This allows deviations up to $\pm 6°$ in all directions and consists of two rubber snubbers, one above and one below the fixing point; the snubbers can be separated or manufactured in one piece as a 'knob snubber'. The guide pin usually has a cold-formed 10 mm diameter and an M 10×1 thread at the end. The rubber parts are pre-tensioned via a dished washer and (as shown in the figures) using a self-locking nut or two lock nuts. The distance between the lower edge of washer and the damper, which is important for the function, can be achieved using a loose spacer tube (usually of 2 mm wall thickness, i.e. 14 mm outside diameter) or by means of a rolled-in tube, as shown in Fig. 5.31.

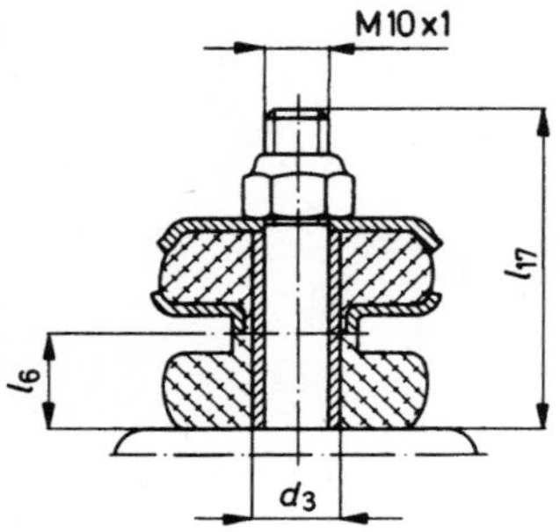

Fig. 5.46 On a pin-type joint, the preload on the rubber parts should be ensured by a spacer tube. Usually this has a wall thickness of 2 mm and 14 mm outside diameter. To avoid contact in the location hole, the upper snubber can be centred by a washer. A self-locking nut is frequently used for clamping the parts together (illustration: Sachs).

From a design perspective, it must be ensured that even at its greatest compression and twist, the side of the pin or the spacer does not come into contact with the bodywork or axle; this would lead to unpleasant noises and increased bending stress. As shown in Fig. 5.46 on the upper snubber, contact can be avoided by the use of a washer, the outer collar of which surrounds the rubber part and grips into the hole with an edge that is turned downwards. In the case of the lower snubber, the same effect is achieved by a vulcanized collar. The fixing point itself can also be designed as a 'shim'.

5.6.8 Stops and supplementary springs

Installation of any end-stops means both the damper and the suspension strut increase in length and there must be enough space in the vehicle to allow this.

5.6.8.1 Jounce stop

Figure 5.43 shows the maximum jounce force 1.45 kN at $v_{D,\,max} = 0.52$ m s^{-1}. However, piston speeds of 3 m s^{-1} can occur, which lead to higher forces. If these forces can no longer be absorbed hydraulically in the shock absorber valves, jounce stops come into action (Fig. 5.9). On passenger cars and light commercial vehicles, the most economic solution is to locate the elastic limitation of the jounce travel or the 'hydraulic stop' in the damper (see also Sections 5.3 and 8.3.1 in Ref. [5]).

The other advantage is that the slight springing effect of the top and bottom damper mountings can be additionally used to damp the jouncing wheel, and so a relatively flat, more easily manufactured bumper 5 made of rubber, polyurethane or Viton, polyamide or a similar plastic is completely adequate (Figs 5.47 and 5.26). All that is needed to fit this is a groove turned into the piston rod in which the collar on the stop disc 4 is rolled or a lock washer inserted.

In the twin-tube system, when the piston rod is extended, the snubber 5 comes into contact with the piston rod guide 6 which is smooth at the bottom (Fig. 5.47), or into contact with a disc 8 protecting the set of gaskets on monotube dampers (Fig. 5.32). Figure 5.48 shows the shapes and the progressive springing curve of the 4–12 mm high snubbers.

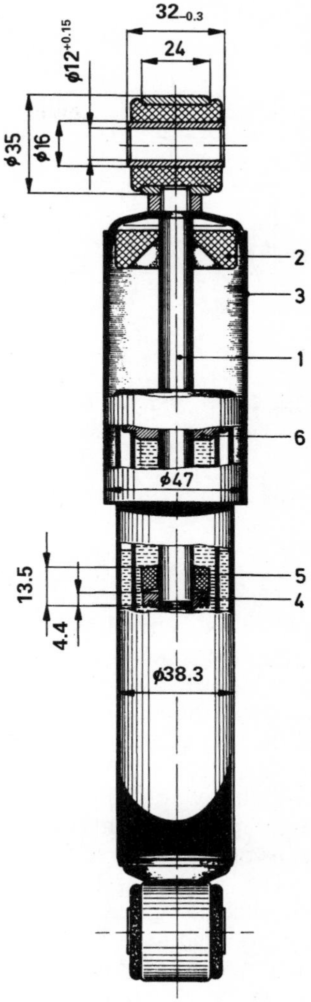

Fig. 5.47 Sachs S27 twin-tube damper with a bump stop 2 carried by the piston rod 1. The rebound stop 5 is supported on the disc 4 rolled into a groove. The upper eye-type joint and the outer and protective tube are also dimensioned and toleranced.

The durability of the elastic compression stop is determined by the shape and material used. It must be able to withstand oil temperatures between −40°C and +140°C without detrimental changes of elasticity and, in the case of sudden loads, neither scuffing nor fissures may occur. Particle abraded off would get into the valves and cause the damping to fail or lock.

Endurance tests carried out jointly by the respective vehicle and shock absorber manufacturers, ensure that this type of damage does not occur. For this reason, and to ensure wheel rebound travel is maintained, where dampers with snubbers are used, only those authorized by the manufacturer should be fitted.

The same applies to spring dampers which, as an assembly unit, contain the compression stop and the supplementary spring as shown in Fig. 5.51.

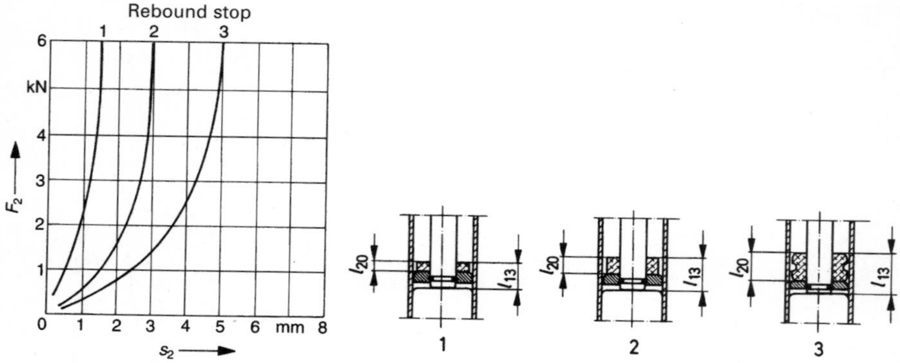

Fig. 5.48 Sachs rebound stop in a twin-tube damper with 27 mm and 30 mm piston diameter (types S26 and S30); shown here are body shapes and bump travel s_2 as a function of the tensile force F_2 up to 6 kN. The heights l_{20} are at: position 1, 4 mm; position 2, 9 mm; and position 3, 12 mm. Snubbers up to 18 mm high are used.

5.6.8.2 Bump stops

Bump stops act close to the end of the wheel travel and are designed to limit bump travel without generating noise. The stop parts are housed in the top of the protective tube (Fig. 5.47), which represents a low-cost solution and today creates no difficulties, either from a technical point of view or in respect of the service life. As explained in Section 5.6.8.1, the damper mountings are designed in such a way that they can transfer relatively large forces and usually only a slight reinforcement is necessary if additional forces occur through compression stops or supplementary springs.

The bump stop 2 shown in Fig. 5.47 is carried by the piston rod 1; when the wheels bottom out, it comes into contact with a cap surrounding the outer tube and is supported – at full bump – on the steel protective tube 3. In the case of an incorrect shape or non-wearproof rubber or plastic mixture, dust can get into the piston rod seal and render it ineffective (Fig. 5.24). The consequence would be escaping oil, a reduction in the damping effect and destruction of the (not always oil-proof) bumper.

Figure 5.49 shows the progressive springing curves of three compression stops of different length and the shape of those shown in part 2 of Fig. 5.47.

5.6.8.3 Supplementary springs

Flat compression stops barely allow any reasonably shaped springing curve. Reduced impacts or the desire for a soft cushioning necessitate installation of a supplementary spring made of polyurethane elastomer or a hollow bumper (Figs 5.9 and 5.14). Figure 5.49 contains at position 4 a springing curve of a 44 mm high supplementary spring suitable for twin-tube dampers and Fig. 5.50 shows a design used for strut dampers. As shown in Fig. 5.51, this is

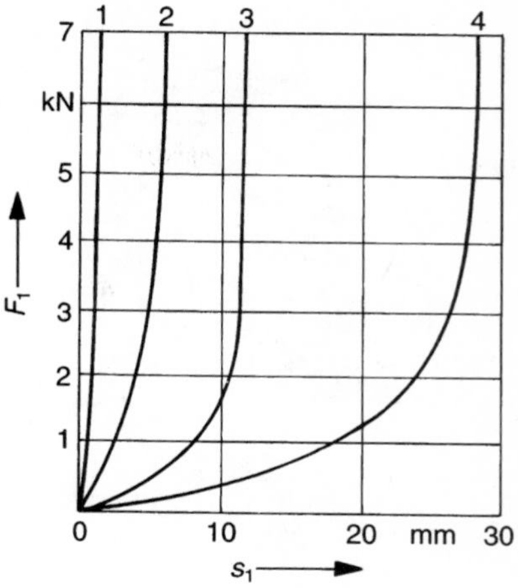

Fig. 5.49 Bump travel s_1 on the Sachs bump stops for the S27, S30 and S32 twin-tube dampers at forces up to F_1 = 7 kN. Configurations 1, 2 and 3 are l_{12} = 8 mm, 15 mm or 23 mm high in their unladen condition and are the same shape as part 2 in Fig. 5.47. The supplementary spring (position 4) is 44 mm high.

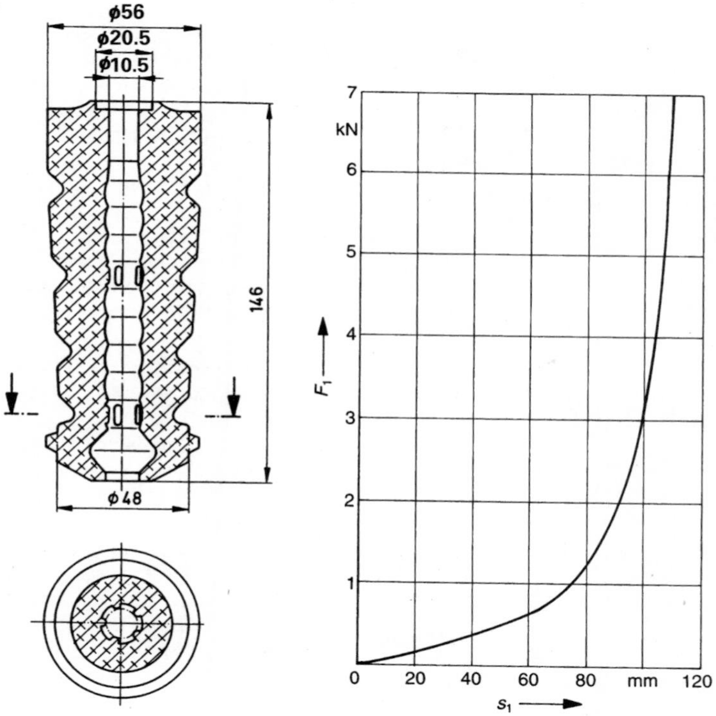

Fig. 5.50 Supplementary spring manufactured by Elastogram in Cellasto polyurethane elastomer on the rear spring dampers of the VW Golf (III, 1996). Material properties and shape make the highly progressive springing curve possible. At 146 mm overall height, it can be compressed by 110 mm and accept an impact load of over 700 kg or a force of $F_1 \geqslant 7$ kN.

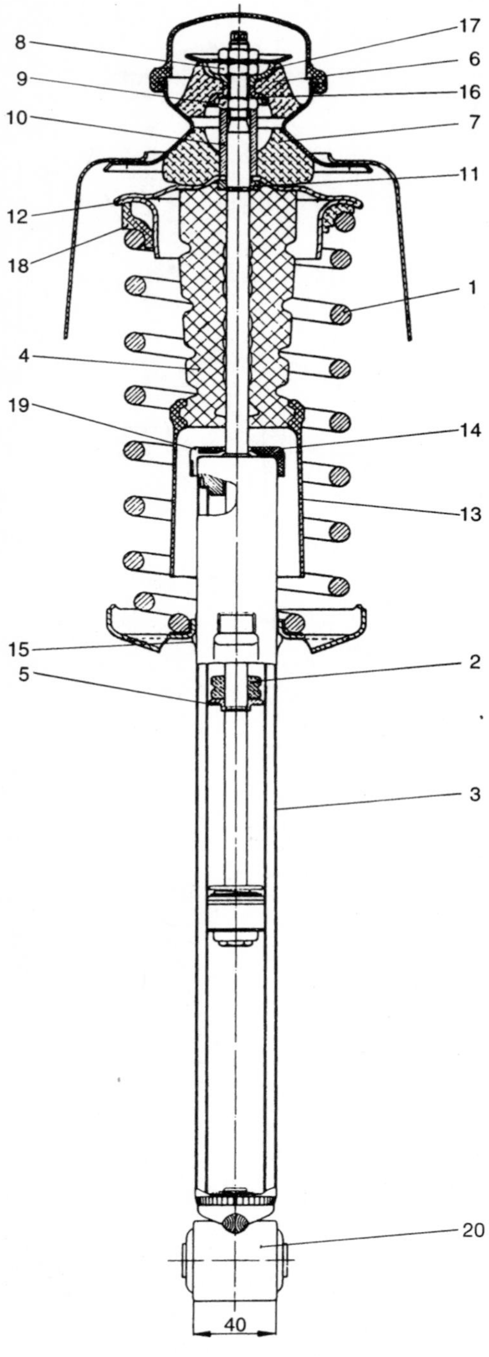

Fig. 5.51 Sachs rear spring damper on the VW Golf (III, 1996) and Vento with coil spring 1 and jounce stop 2 visible in the cross-section. This is carried by the 11 mm thick piston rod and is located 107 mm above the 27 mm diameter piston so that it has an adequate minimum bearing span in the fully extended condition; the stop ring 5 is rolled into a groove of the piston rod.

The upper fixing is a pin-type joint that transfers the springing and impact forces to the body via the large noise-insulating rubber snubbers 6 and 7. The two parts are drawn together by the hexagonal nuts 8 and 9; the tube 10 and the bushes 16 and 17 ensure that a precise preload is achieved. The lower washer 11 comes into contact with a wire snap ring (which sits in a half-round groove) and both the spacer tube 10 and the upper spring seat 12 come into contact with the washer. The spring seat supports the coil spring 1 via the elastic ring 18 and also the polyurethane supplementary spring 4, which has a circular bead at the bottom to take the plastic protective tube 13.

If the suspension is in bump travel, part 4 comes into contact with the cap 14. This ensures the piston rod seal is not damaged. The cap has a groove (position 19) through which the air in the supplementary spring can escape when it is compressed. The lower spring seat is supported at three points (position 15), which protrude from the outer tube and the outside diameters of which must have a tolerance of ±0.5 mm.

To ensure the rubber part only flexes a little under vertical forces, the eye-type joint 16 was made 40 mm wide.

carried by the piston rod and comes into contact with a cap or disc when it compresses.

5.7 Spring/damper units

The spring/damper unit, which is described in detail in Ref. [5] Section 6.2, is a device carried over from the motor cycle. It is used by more and more passenger car manufacturers, not only on independent wheel suspensions, but also on rigid and compound crank axles. This force centre, formerly described as a suspension strut, does not carry the wheel-like McPherson struts, but comprises all parts of a wheel suspension that are necessary for springing and damping. These are the coil spring 1, jounce stop 2, supplementary spring 4 (Fig. 5.51) and, as the supporting element, the shock absorber.

The coil spring can be retrofitted and supported with rubber insulators on the body or pre-assembled into the unit, in which case two bolts are used to fix the entire assembly.

Installed spring/dampers can be seen in Figs 1.54, 1.55, 1.61, 1.62 and 1.77.

5.8 McPherson struts and strut dampers

5.8.1 McPherson strut designs

The McPherson strut also carries and controls the wheel. The piston rod, which is strengthened from 11 mm to 18 mm up to 25 mm diameter on passenger cars (and up to 28 mm on light commercial vehicles), can absorb longitudinal and lateral forces and replaces the upper suspension link, including its three mountings. The designs, which are known today as McPherson struts, are divided into two groups:

- those with the steering knuckle solidly fixed to the outer tube (Fig. 5.52):
- those with a bolted-on steering knuckle (Figs 1.8, 1.56, 5.54 and 5.55).

And, in terms of the damper part, into:

- those with wet suspension struts on which the damper part is directly mounted into the carrier tube (Figs 5.54 and 5.55);
- cartridge designs in which the damper part is inserted into the carrier tube and screwed together (Fig. 5.53).

A decision in favour of one of the solutions is mainly a question of the manufacturer's preferences, although whether the outer tube needs to be included for transferring steering forces, i.e. whether the steering arms sit on it, is also a consideration (Figs 1.57, 3.102, 4.1, 4.47 and 5.52).

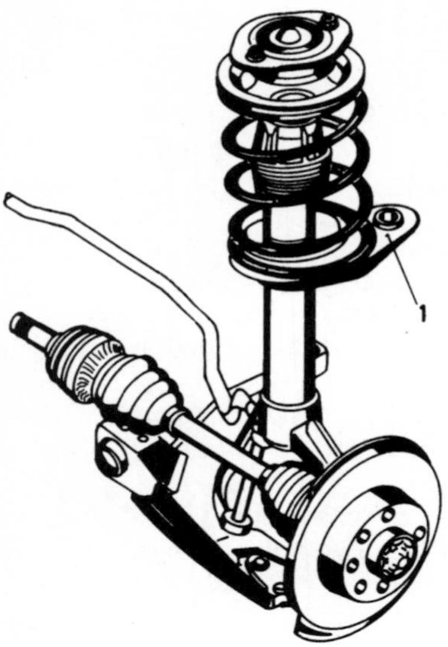

Fig. 5.52 McPherson front drive axle and suspension of an Opel/Vauxhall model. The outer tube is press-fitted to the steering knuckle, with the steering arm 1 relatively high up.

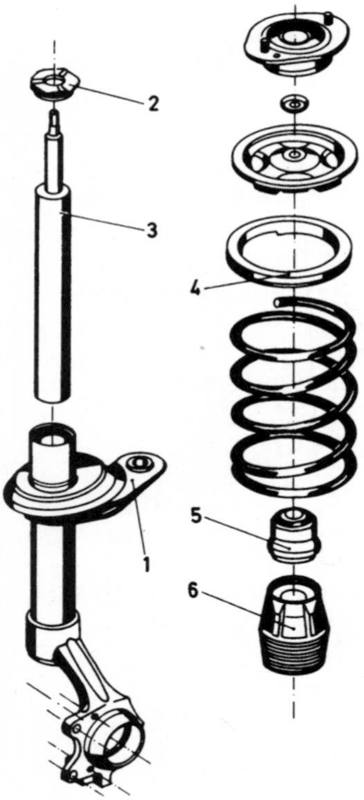

Fig. 5.53 If the damping on the Opel/Vauxhall suspension strut fails, the bolted closure cap 2 must be undone and the shock absorber cartridge 3 changed. The elastic ring 4, located above the coil spring, the supplementary spring 5 and the dust bellow 6 can be seen clearly.

Wet suspension struts are better at conducting heat away from the damper and, where they are detachably linked to the steering knuckle, offer the advantage that they do not need to be able to be dismantled and that, if the damping fails, the actual damping part can be easily exchanged. This design also makes it possible to close the strut by means of indentations in the outer tube (against cover 5, Fig. 5.56), rolling it (edge 6 in Fig. 5.54) or welding it to the sealing cap.

If, as shown in Fig. 5.53, the steering knuckle is press-fitted to the suspension strut, a screwed closure cap is necessary for exchanging the damper cartridge.

5.8.2 Twin-tube McPherson struts, non-pressurized

The suspension strut shown in Fig. 5.54 operates on the twin-tube principle; it operates in the same way as the non-pressurized twin-tube damper (see Section 5.6.2). To have a sufficient minimum bearing span *l–o* (Fig. 1.11) in the fully jounced condition, the jounce stop 13 has been set high. This measure, together with the PTFE-coated guide bush 11, reduces friction.

5.8.3 Twin-tube McPherson struts, pressurized

The development of the pressurized McPherson strut has met with significant difficulties for many years. Direct transfer of the monotube principle, as used in the shock absorber, is not possible because of the high extension force. There are solutions that keep the rod small and transfer the wheel control to the cylinder tube, but these are expensive and involve high levels of damper friction (see Section 6.4.6 in Ref. [5]).

The pressurized twin-tube system is a good compromise. Here, the oil is only under a pressure of 6–10 bar (depending on the manufacturer) and the extension force of the 18–28 mm thick piston rod is therefore limited.

Figure 5.55 shows a section through a McPherson strut. The spring seat 22 and the lower bracket for fixing to the steering knuckle are welded to the outer tube 2. The piston rod 1 is solid but can be hollow to reduce the weight; the piston has valve plates on both sides, depending on the desired damping curve, or a twin-tube damper valve operating only in the extension direction (see Sections 5.6.2.4 and 5.6.4.3). This can be an advantage where degressive valve curves are requested.

The studs of the hollow piston rod are made in a special cold-forming process; essentially, the upper one is given a hexagonal socket (or two flat surfaces) for holding during assembly and the lower tappet must be oil- and gas-tight. The rebound stop is made of plastic, is tight to the rod and transmits the vertical forces via the tube 14 to a zone of the rod that is not subject to bending. To keep friction low, the seal between the piston and cylinder wall is the broad PTFE ring 15. The extension stage valve 16 is similar to that shown in Fig. 5.26. The forces in the pressure stage are applied jointly by the valves 18 and 20 (see Section 5.6.2.5).

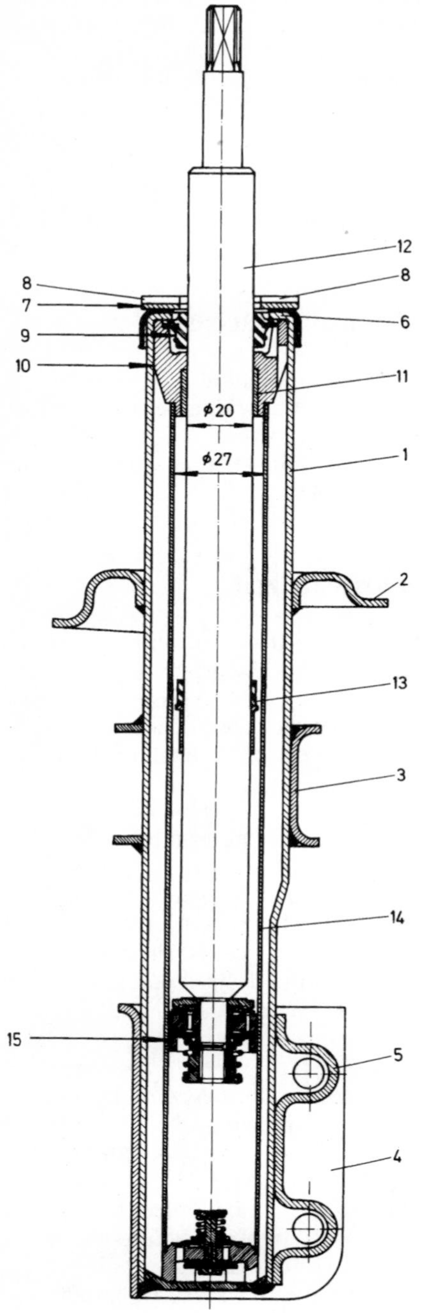

Fig. 5.54 McPherson strut of the Fiat Panda (1995) manufactured by Monroe. The spring seat 2 for taking the coil spring, the tab 3 (for fixing the steering arm) and the bracket parts 4 and 5 to which the steering knuckles are bolted to the outer tube 1. The stop disc 7 is supported on the rolled edge 6 of the outer tube, and its two transverse grooves 8 ensure that the supplementary spring cannot create overpressure in the interior; this would press dirt and deposits into the seal 9. The bush 11 is pressed into the sintering iron rod guide 10 from the bottom and its surface conditioned to reduce friction (to the piston rod 12). The rod is 20 mm diameter and, in the mid-range, carries the jounce stop 13; when the wheel is fully extended, the minimum bearing span (centre bush 11 to centre piston) is 120 mm.

The rod 12 is drawn in at the bottom to provide space for the rebound stage and check valve (see Fig. 5.26). The low-friction ring 15 provides the seal between the piston, which is 27 mm diameter, and the cylinder tube 14.

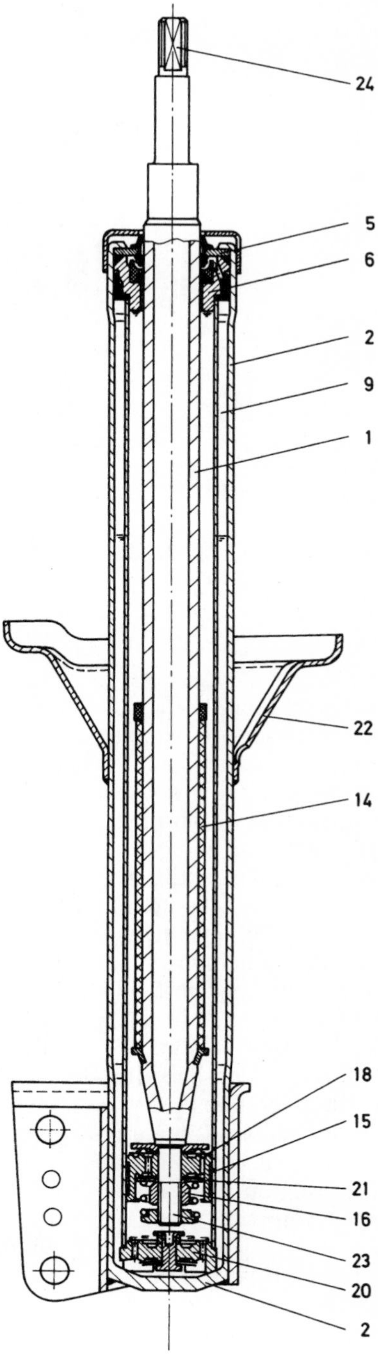

Fig. 5.55 Low-pressure twin-tube McPherson strut by Sachs, drawn with the piston rod 1 fully in. The lower end 23 is drawn in and threaded to mount the compression stage valve; the upper tappet 24 gripping into the upper strut mount on the wheel house has two surfaces for retention.

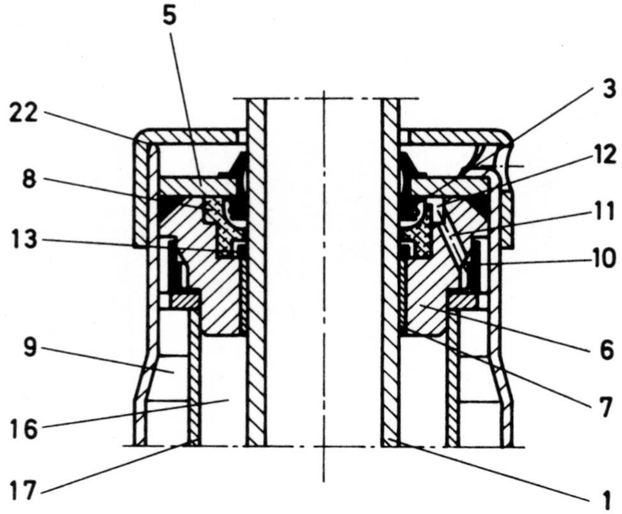

Fig. 5.56 Rod guide and seal unit of the Sachs low-pressure twin-tube McPherson strut.

The constant orifice on the piston, also known as a by-pass or advanced opening cross-section, is created by punched holes in the lower valve plate 21 and a similar by-pass plate for the compressive stroke is used on the compression valve 20. In order not to influence the efficiency of this constant opening on the damping curve too much, the clearance area between guide bush 7 and piston rod 1 is sealed in a controlled manner using the PTFE ring 13 (Fig. 5.56). In the non-operative condition (as shown) it is at the bottom, but during operation, i.e. when there is pressure in the working chamber 16, it comes into contact with the spacer 8. This has transverse grooves of a precisely fixed cross-section which provide the necessary ventilation.

As described in Section 5.6.2.3, when the oil cools after a journey, an air bubble can form in the top of the pressurized twin-tube damper. On the strut damper, the pressure in the oil column in the equalization chamber 9, together with the inner tube valve 10, should significantly delay this. However, if at very low temperatures pressure is reduced and the oil concentrated, the ventilation facility becomes important again.

The internal pressure in the upper part 16 of the working chamber increases on both jounce and bump damping. The residual oil volume flowing through the clearance between rod 1 and bush 7 collects in the high ring channel 12 and is passed through the inclined holes 11 into the lower channel, which is formed by an angle ring and the tube valve 10. This latter part lifts and allows the oil to flow back into the equalization chamber 9. The chamber is around half full of oil and is pressurized by gas. The tube 10 acts as a lock and prevents ingress of gas in the reverse direction on the rod seal 3.

Foaming of the oil and forming of air bubbles in the valve, known as cavitation, is prevented by the inner pressure of $p = 6–10$ bar. If, for some reason, the gas should escape, the damping function remains largely intact due to the existent bottom valve 20 (a designed-in safety feature). The suspension strut can be closed by welding, or as shown in Fig. 5.56, by several beads, which press the closure plate against the guide unit 6, and press this in turn against the cylinder tube 17, which then presses the valve body 20 shown in Fig. 5.55 against the bottom of the cold-sunk outer tube 2. The gasket 3, with the dust lip that protrudes upwards, forms a unit with the closure plate 5, which is covered from the top by the cap 22. The supplementary spring comes into contact with this cap at full jounce.

5.8.4 Damper struts

Damper struts only carry the wheel without transferring vertical springing forces; there is no spring seat. However, rebound stops and supplementary springs are arranged as in suspension struts (Fig. 1.41).

5.9 Variable damping

The dampers and McPherson struts described in the previous sections have a fixed curve over the entire operating range that depends only on piston velocity. It is determined by the vehicle manufacturer for a given vehicle type and the loading condition, usually two people and 75 kg of luggage in the case of passenger vehicles. This characteristic represents a compromise between driving comfort and riding safety, i.e. soft and hard shock absorption.

Different loading and driving situations ideally require a damping characteristic specifically geared to these.

Figure 5.57 shows a shock absorber with impressed longitudinal grooves in the cylinder tube. These grooves produce a by-pass flow around the shock-absorber piston in the normal state of the vehicle which corresponds to driving conditions with a low load and a small roll angle and thus result in reduced damping forces, that is to say, greater comfort. Outside the normal situation, with a strongly jouncing or rebounding wheel, this bypass opening is not available; the damping force increases. Apart from the by-pass cross-section and the length and position of the grooves, the shock absorber can be made to suit the individual vehicle with regard to comfort and riding safety requirements (Fig. 5.58).

Almost any adjustment of the damping characteristic is possible with the continuously adjustable twin-tube shock absorber shown in Fig. 5.59 depending on driving and loading conditions. As an input quantity for the control of the

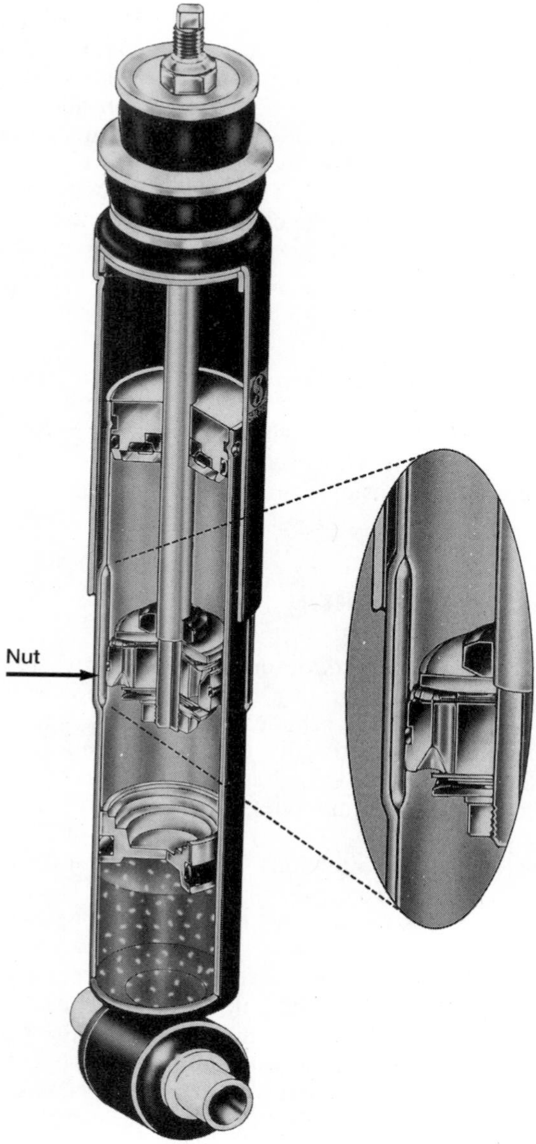

Fig. 5.57 Pressure-loaded single-tube shock absorber with bypass technology (Sachs).

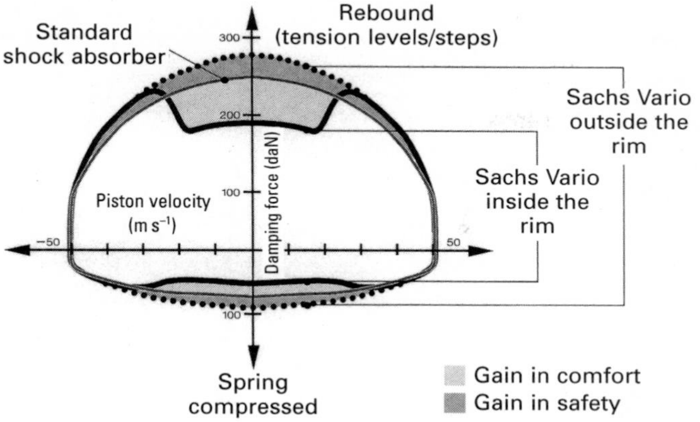

Fig. 5.58 Damping characteristic of a shock absorber with by-pass technology. Compared with a traditional shock absorber, the damping force can be reduced for the purposes of increasing comfort within the normal range of the vehicle, whereas a higher damping force is made available for the purposes of riding safety outside the normal range.

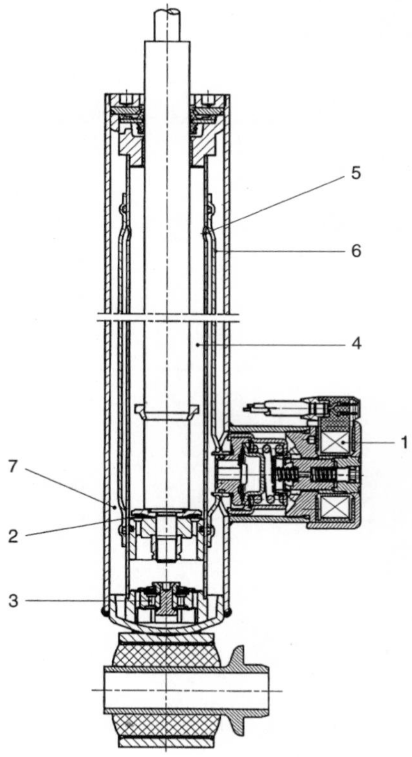

Fig. 5.59 Continuously adjustable shock absorber of Sachs. The piston valve 2 acts as a non-return valve during rebound, so that the oil in the ring chamber 4 is displaced and directed through the openings 5 and the intermediate tube 6 by way of the proportional solenoid valve 1 into the gas-filled equalization chamber 7. Since the floor valve 3 closes during compression, the oil displaced by the volume of the piston rod must also flow over the solenoid valve during compression. Compression and rebound damping is largely ensured by this solenoid valve.

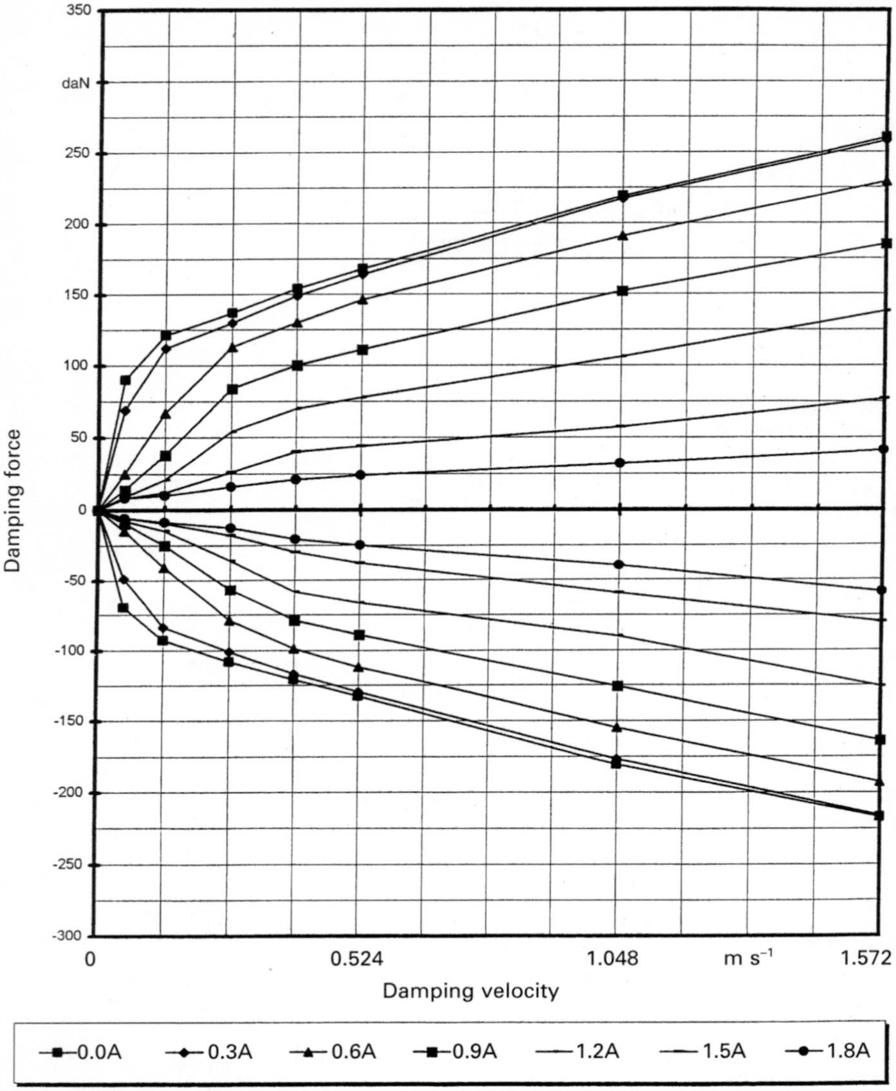

Fig. 5.60 Damping characteristics of a continuously adjustable shock absorber of Sachs. In the region of low shock absorber speeds, a slightly rising characteristic can be set if low damping is required for reasons of comfort. If a high damping force is required for reasons of riding safety, a very much steeper characteristic can be chosen. The characteristics and height can be varied within a wide range for the purposes of comfort or riding safety.

electrically operated proportional valve, driving speed, lateral acceleration, acceleration of the body at the front/rear, deceleration, accelerator and brake actuation as well as steering angle are used in the same way as the bump movements on the wheels themselves, so that characteristic-controlled, adaptive adjustment of the damping force takes place (Fig. 5.60).

6

Chassis and vehicle overall

6.1 Vehicle and body centre of gravity

6.1.1 Centre of gravity and handling properties

Depending on the problem posed and the topic, the following are important variables in vehicle engineering:

- vehicle centre of gravity V
- body (sprung-mass) centre of gravity Bo
- axle (unsprung mass) centres of gravity U_f or U_r.

The distance of the centres of gravity V and Bo from the front or rear axle and their height above ground are crucial for

- braking and acceleration capacity
- calculating the climbing ability
- designing brake systems and four-wheel drives
- designing body centre of gravity and aspects of vibration stability
- driving stability investigations
- determining mass moment of inertia.

Low centres of gravity are always desirable, as they are associated with fewer driving dynamic problems and increased vehicle performance during cornering and braking, but in practice the design options are relatively restricted.

The position of a vehicle centre of gravity V and the body centre of gravity Bo is highly dependent on the load; when people get into the vehicle or luggage is loaded in the boot or onto the roof, the centre of gravity changes *vis-à-vis* the empty condition, both in the longitudinal direction (x-axis) and upwards (z-direction). The body lowers when it is loaded, i.e. its centre of gravity Bo drops. The centre of gravity of the people and, in particular, that of the luggage carried on the roof, is higher than that of the body so the end result is usually a higher overall centre of gravity V (distance h_V, Fig. 6.1).

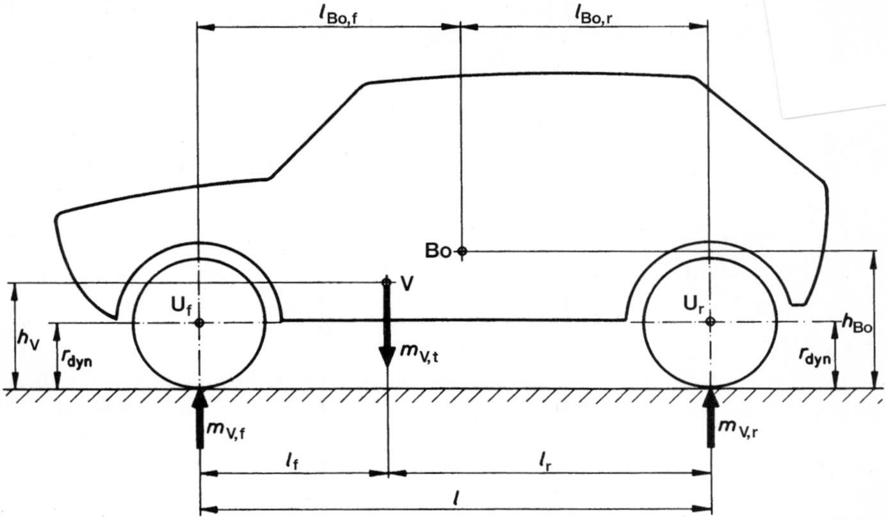

Fig. 6.1 Designation of the paths for determining the centres of gravity V of the overall vehicle and Bo of the body. The centres of gravity U_f and U_r of the front and rear axles can be regarded as being in the centres of the wheels.

Details on these problems are dealt with exhaustively and further simple ways of ascertaining centres of gravity are given in Chapter 1 of Ref. [3].

6.1.2 Calculating the vehicle centre of gravity

Calculating the position of the centre of gravity is likely to be possible only with great difficulty and considerable effort. If the vehicle and all its individual components are shown on a computer in the form of a digital model including body surfaces and properties (digital surfaced model), modern CAE tools make it possible to calculate the position of the centres of gravity of the components and the whole vehicle.

It is much simpler to determine the position experimentally by weighing. For this, both the empty vehicle should be observed and when it is occupied by two or four people, approximately 170 cm tall and weighing around 68 kg.

6.1.2.1 Centre of gravity distance to front and rear axle

Figure 6.1 contains the paths and angles necessary for calculating the centres of gravity and Fig. 3.3 the position of the coordinate system. When the vehicle is weighed, it must be standing on a completely horizontal plane and with each axle on a weighbridge. So as not to distort the weighbridge, it must be possible to turn the wheels freely. The weighed front axle load $m_{V,f}$ and the rear axle load $m_{V,r}$ give the total weight $m_{V,t}$ of the vehicle:

$$m_{V,t} = m_{V,f} + m_{V,r} \text{ (kg)} \tag{6.1}$$

The balance of moments around $m_{V,f}$ or $m_{V,r}$, in conjunction with the wheelbase l in the longitudinal direction, gives the centre of gravity distances l_f to the front and l_r to the rear axle:

$$l_f = \frac{m_{V,r}}{m_{V,t}} l; \; l_r = \frac{m_{V,f}}{m_{V,t}} l = l - l_r \tag{6.2}$$

If the lateral distance of the centre of gravity (y-direction) from the vehicle centre-line is required the wheel loads must be weighed to be able to calculate first of all the lateral offset of the centres of the front and rear axles from the centre-line via similar equations made up from the rear view, and then similarly for the vehicle centre of gravity from the top view (see Equations 5.14 and 6.24).

6.1.2.2 Centre of gravity height

To calculate h_V, first the front and then the rear axle must be lifted as high as possible (by the amount h) with an elevating mechanism (autohoist, jack, crane), with the other axle standing in the centre of a weighbridge (Fig. 6.2). The following would need to be ensured:

- The vehicle must be prevented from falling off by inserting wedges from the outside on the axle to be raised. The brake must be released and the gearbox must be in neutral. It must be possible to turn the wheels on the platform easily; the platform would otherwise distort and the result be imprecise.
- The wheels are held still on the centre of the platform, the vehicle forward movement must be even when the vehicle is raised, in order to prevent wrong measured values as a result of different force application positions on the horizontal surface.
- If the change in axle load during lifting is measured by means of a crane over a load cell, it is possible to ensure that the direction of lifting is completely vertical.
- The vehicle should be in the on-road condition, i.e. full tank, tools, spare wheel, etc. (as per the curb weight, see Section 5.3.1).
- Both axles must be prevented from compressing or rebounding before the vehicle is raised. The locking device must be of an adjustable variety so that the amount by which the body sinks when there are two or four people and luggage in the vehicle can be taken into consideration.
- To eliminate tyre springing during the measurement, it is recommended that the tyre pressure on both axles be increased to 3.0 to 3.5 bar.

Mathematical observation of the measurement is as follows (Fig. 6.2):

$$h/l = \sin\alpha$$

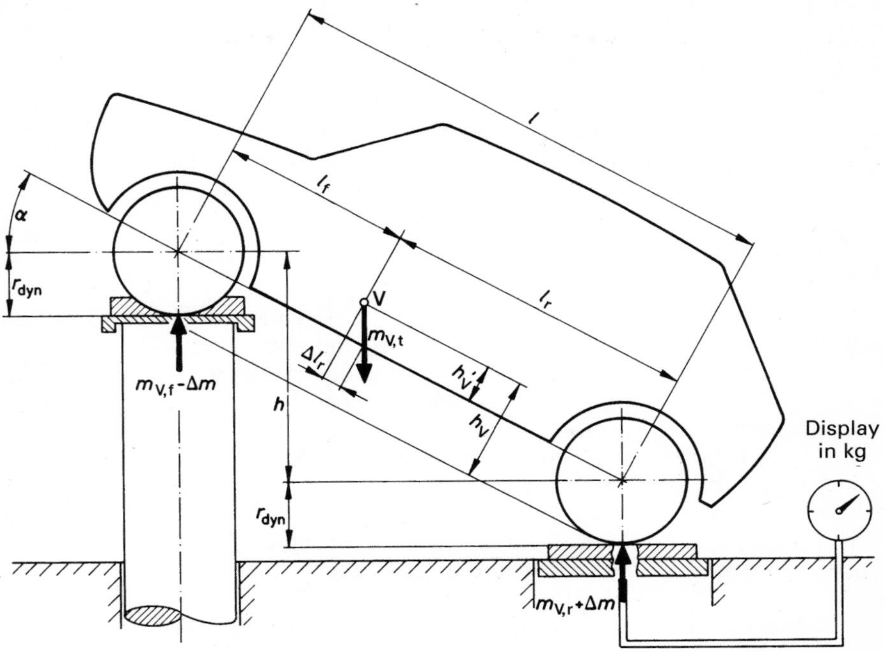

Fig. 6.2 Vehicle on a weighbridge with forces and paths for deriving the equation for vehicle centre of gravity height h_V included.

The angle α is known; but $h_V = h_V' + r_{dyn}$ is sought, whereby

$$h_V' = \Delta l_r / \tan \alpha$$

To be able to determine Δl_r, the equation of moments produced around the centre of the front axle is set up:

$$m_{V,t}\,(l_f + \Delta l_r)\cos\alpha = (m_{V,r} + \Delta m)\,l\cos\alpha$$

Eliminating cos α

$$\Delta l_r = \frac{(m_{V,r} + \Delta m)\,l}{m_{V,t}} - l_f, \text{ whereas } l_f = \frac{m_{V,r}}{m_{V,t}}\,l$$

therefore,

$$\Delta l_r = \frac{\Delta m}{m_{V,t}}\,l, \text{ hence } h_V' = \frac{l\,\Delta m}{m_{V,t}\tan\alpha} \text{ and}$$

$$h_V = \frac{l}{m_{V,t}} \frac{\Delta m}{\tan \alpha} + r_{dyn} \tag{6.3}$$

In Equation 6.3 the angle α can be expressed through the easily measurable vehicle stroke height h and so the equation can be simplified:

$$h_V = \frac{l}{m_{V,t}} \frac{\Delta m}{h} (l^2 - h^2)^{1/2} + r_{dyn} \tag{6.4}$$

With $\Delta m/h$ or $\Delta m/\tan \alpha$ there is a constant in the equation. When it is weighed, in each instance, only the changes caused by the vehicle lifting on one side, namely Δm and the raised dimension h, need to be determined. The other values such as wheelbase l, vehicle weight $m_{V,t}$ and the dynamic rolling radius r_{dyn} remain the same. The centre of gravity height is required for calculating various vehicle conditions, i.e. for the travelling vehicle, so the dynamic rolling radius r_{dyn} of the tyre must be added to h'_V and not the somewhat lower static rolling radius that only applies to the standing vehicle. In accordance with Equation 2.2, r_{dyn} must be calculated from the rolling circumference C_R (or $C_{R,dyn}$). The values of C_R can be found in Fig. 2.15 and in Ref. [4].

As Equation 6.3 shows, the sensitivity of error is very large for small lifting heights (small values of α). Extensive tests have shown that exact, reproducible results can only be obtained with large lifting heights. The vehicle should consequently be raised to the maximum possible lifting height several times. Measurements of intermediate values at smaller heights can be dispensed with.

The representation of axle load differences as a function of the lifting height shown in Fig. 6.3 makes it possible to identify possible outliers which are not taken into consideration in the evaluation. The assessment in the form of a linear regression is computer-assisted, so that information about the accuracy of the established result can also be provided.

6.1.2.3 Ratio i_{ul} for the empty condition

The known centre of gravity height $h_{V,0}$ of the empty vehicle can be compared with the empty height h_{ul} of the unladen vehicle. For the passenger car, this would be $h_{ul} = 1380$ mm, so that the ratio would be

$$i_{ul} = h_{V,0}/h_{ul} = 0.377$$

If the centre of gravity height $h_{V,0}$ of a four- or five-seater passenger car is not known, it can be judged using i_{ul}:

$$h_{V,0} = (0.38 \pm 0.02)\, h_{ul} \tag{6.4a}$$

6.1.2.4 Influence of loading

The value $h_{V,0}$ applies to the curb weight; when the vehicle is laden, the centre of gravity generally moves upwards, i.e. the path h_V increases, unlike the vehicle

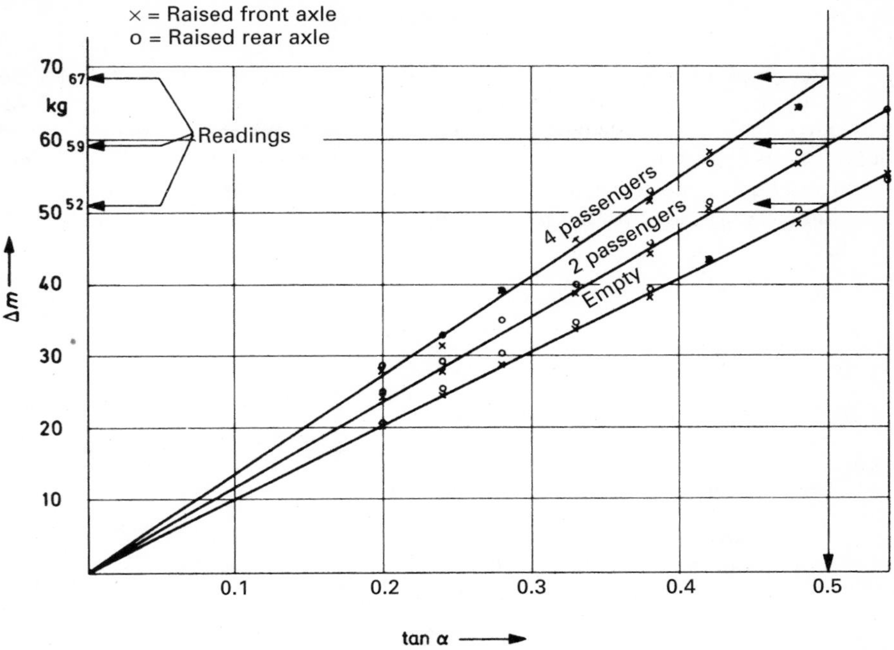

Fig. 6.3 The measured axle load differences Δm are entered separately for the front and rear axles as a function of tan α, depending on which axle was on the weighbridge. A straight line, determined by linear regression and which must go through the origin, can be used for the most precise possible determination of the quotient $\Delta m/\tan\alpha$.

height, which reduces. The amount by which the centre of gravity of the vehicle as a whole rises when there are two, four or five people in it, is a question of the spring rate on the front and rear axles, the seat heights and the weights and sizes of the occupants (Figs 5.12 to 5.15). The following can be an approximate figure for the centre of gravity height $h_{V,pl}$ (index pl = partial loaded or partly laden):

$$h_{V,pl} = h_{V,0} + \Delta h_V \tag{6.4b}$$

two people $\Delta h_{V,2} \approx +12$ mm
four people $\Delta h_{V,4} = -8$ mm to $+29$ mm

A fifth person on the rear seat or load in the boot causes the body to go down, so the overall centre of gravity sinks (Fig. 6.4).

6.1.2.5 Roof load

Roof load will raise the vehicle and body centre of gravity. Section 1.3 in Ref. [3] contains details.

6.1.3 Axle weights and axle centres of gravity

If, instead of the height of the centre of gravity h_V of the vehicle as a whole, the height h_{Bo} of the body centre of gravity is required, it can be determined by assuming that the centre of gravity of the unsprung mass $m_{U,f}$ (front) $m_{U,r}$ (rear) is approximately at the centre of the wheel, i.e. at the distance of the dynamic rolling radius r_{dyn} to the ground (Figs 6.1 and 6.5). Furthermore, their weight should be known, determined by weighing or calculated by approximation.

$$m_{U,f} = \frac{i_{m,f}\, m_{V,t}}{1 + i_{m,f}} \tag{6.4c}$$

$$m_{U,r} = \frac{i_{m,r}\, m_{V,r}}{1 + i_{m,r}} \tag{6.4d}$$

The following approximate values can be included in the equation:

front axle	$i_{m,f} \approx 0.12$
non-driven rear axle	$i_{m,r} \approx 0.13$
driven rear independent wheel suspension	$i_{m,r} \approx 0.14$
driven rear rigid axle	$i_{m,r} \approx 0.22$

A passenger car which has a front axle load $m_{V,f} = 609$ kg in the unladen condition can be used as an example:

$$m_{U,f} = \frac{i_{m,f}\, m_{V,f}}{1 + i_{m,f}} = \frac{0.12 \times 609}{1 + 0.12} = 65.3 \text{ kg}$$

Section 5.2 and Ref. 3 (also Section 5.2) contain further details.

6.1.4 Body weight and body centre of gravity

Taking into consideration both axles, the body weight is:

$$m_{Bo} = m_{V,t} - (m_{U,f} + m_{U,r}) \tag{6.5}$$

and the distances of the centres of gravity to the axle centres shown in Fig. 6.1 become:

$$l_{Bo,f} = \frac{m_{Bo,r}}{m_{Bo}}\, l; \; l_{Bo,r} = \frac{m_{Bo,f}}{m_{Bo}}\, l = l - l_{Bo,f} \tag{6.6}$$

where $m_{Bo,f}$ and $m_{Bo,r}$ are the proportions of the body weight over the front or rear axle:

Fig. 6.4 Measuring sheet with values of a passenger car entered with additional information on size and weight of the people in the vehicle during the measurement. Source: Technical laboratory, Polytechnic of Cologne.

Vehicle	Passenger car				
Year of manufacture	1994				
l	2570 (mm)				
r_{dyn}	296 (mm)				
Tyres	195/65R14				
State when measured	M	*Passengers in front*		*Passengers in rear*	
		Weight (kg)	Size (cm)	Weight (kg)	Size (cm)
Empty	0				
2 passengers	2	72.5	183	60	170
4 passengers	4	68.2	180	71	178
		Σ140.7		Σ131	

M	h (mm)	$\sin\alpha = \frac{h}{l}$	α (°)	$\tan\alpha$	$m_{V,f}$ (kg)	$\Delta m_{V,f}$ (kg)	$m_{V,r}$ (kg)	$\Delta m_{V,r}$ (kg)
0	1200	0.47	27.92	0.53	683.2	54.5	577.4	54.2
	1100	0.43	25.42	0.48	678.0	49.3	573.6	50.4
	1000	0.39	22.96	0.42	672.6	43.9	567.4	44.2
	900	0.35	20.56	0.38	667.0	38.3	562.8	39.6
	800	0.31	18.06	0.33	662.3	33.6	558.8	35.6
	700	0.27	15.66	0.28	657.6	28.9	553.5	30.3
	600	0.23	13.30	0.24	653.2	24.5	549.3	26.1
	500	0.20	11.54	0.20	648.9	20.2	544.1	20.9
	0	0	0	0	628.7	0	523.2	0
2	1200	0.47	27.92	0.53	—	—	654.9	64.3
	1100	0.43	25.42	0.48	760.9	57.3	648.5	57.9
	1000	0.39	22.96	0.42	755.1	51.5	642.4	51.8
	900	0.35	20.56	0.38	748.1	44.5	636.4	45.8
	800	0.31	18.06	0.33	742.7	39.1	630.8	40.2
	700	0.27	15.66	0.28	742.9	39.3	625.8	35.2
	600	0.23	13.30	0.24	731.6	28.0	620.0	29.4
	500	0.20	11.54	0.20	727.7	24.1	615.3	24.7
	0	0	0	0	703.6	0	590.6	0
4	1100	0.43	25.42	0.48	—	—	759.6	63.8
	1000	0.39	22.96	0.42	785.3	58.2	752.2	56.4
	900	0.35	20.56	0.38	772.9	45.8	739.3	43.5
	800	0.31	18.06	0.33	772.9	45.8	739.3	43.5
	700	0.27	15.66	0.28	765.7	38.6	734.6	38.8
	600	0.23	13.30	0.24	759.3	32.3	728.8	33.0
	500	0.20	11.54	0.20	754.6	27.5	724.0	28.2
	400	0.16	8.98	0.16	749.0	21.9	—	—
	0	0	0	0	727.1	0	695.8	0

$m_{V,t}$ (kg)	0	1151.9	$l_f = \frac{m_{V,r}}{m_{V,t}}\, l$	0	1164.1	$l_r = l - l_f$	0	1398.9	$\frac{l}{m_{V,t}}$	0	2.23
	2	1294.2		2	1169.6		2	1393.4		2	1.98
	4	1422.9	(mm)	4	1253.3	(mm)	4	1309.7	(mm kg^{-1})	4	1.80

Δm by $\tan\alpha = 0.5$	0	52	$h_V = \frac{l}{m_{V,t}} \times \frac{\Delta m}{\tan\alpha} + r_{dyn}$	0	$2.23 \times \frac{52}{0.5} + 296$	$h_{V,0} = 528$ mm
	2	59	$(\text{mm} = \frac{\text{mm}}{\text{kg}} \times \text{kg} + \text{mm})$	2	$1.98 \times \frac{59}{0.5} + 296$	$h_{V,2} = 530$ mm
	4	67		4	$1.8 \times \frac{67}{0.5} + 296$	$h_{V,4} = 537$ mm

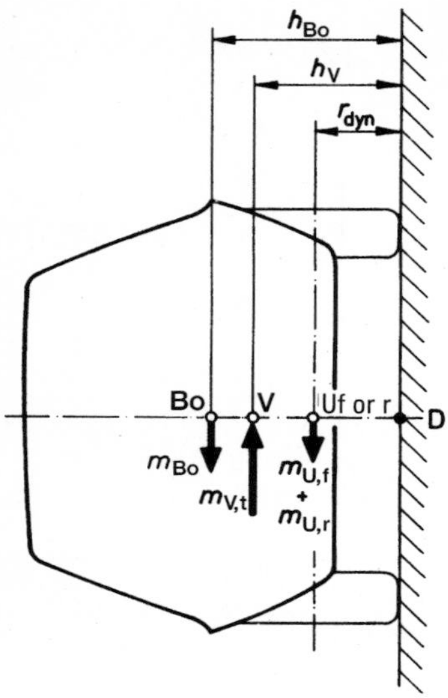

Fig. 6.5 Vehicle shown tipped to derive the equations of moments for the height h_{Bo} of the body centre of gravity.

$$m_{Bo,f} = m_{V,f} - m_{U,f} \tag{6.6a}$$

$$m_{Bo,r} = m_{V,r} - m_{U,r} \tag{6.6b}$$

The height h_{Bo} of the body centre of gravity B is easy to calculate by observing the vehicle when it is tipped forwards (Fig. 6.5) using an equation of moments, assuming that the individual weights act as forces at their respective distance on the ground:

$$h_{Bo} = \frac{m_{V,t} h_V - (m_{U,f} + m_{U,r})\, r_{dyn}}{m_{Bo}} \tag{6.7}$$

Depending on the loading condition and the weight of the unsprung mass, the body centre of gravity h_{Bo} is 20–40 mm higher than that of the vehicle as a whole h_V.

6.2 Mass moments of inertia

From the theory of mechanics it is known that when a body is accelerated in a straight line the inertia F_c is given by

$$F_c = m\, a_x = \text{mass} \times \text{acceleration (N)}$$

In comparison to this, in the case of accelerated rotational movement, the acceleration moment is influenced by the rotation mass J.

The rotation mass – equivalent to the mass moment of inertia J (kg m^2) and also known as second degree mass moment – is a measure of inertia on rotating bodies. In vehicles, three important rotational movements occur in the various vehicle conditions, to which the variables of the mass moments of inertia J are related.

- The vehicle moment of inertia $J_{Z,V}$ around the vertical axis (z-axis, Fig. 3.3) is required for driving stability studies or even for reconstructing road traffic accidents.
- The body moment of inertia $J_{X,Bo}$ around the vehicle's longitudinal axis (x-axis) is essential for generally studying body movement (roll behaviour) during fast lane changes in the driving direction.
- The body moment of inertia $J_{Y,Bo}$ around the transverse axis (y-axis) is the determining variable for calculating pitch vibration behaviour.

In addition to this, in general, the inertia moments of power units (engine–gearbox unit) and individual rotationally symmetrical elements, such as steering wheels, tyred wheels, etc. are of importance. (See also Section 1.5 in Ref. [3].)

The position of its centre of gravity and the variables of the moment of inertia are usually determined with the basic design of a vehicle (drive, wheelbase, dimensions and weight).

In addition to the type of drive, the vehicle's moment of inertia $J_{Z,V}$ around the vertical axis is the determining factor for its cornering performance. Manoeuvrability increases as the inertia moment decreases, whereas driving stability when the vehicle is moving in a straight line and on S bends decreases by the same amount.

$J_{Z,V}$ comprises the mass $m_{V,T}$ of the vehicle as a whole and the radius of gyration $i_{Z,V}$ squared:

$$J_{Z,V} = m_{V,t}\, i^2_{Z,V} \text{ (kg m}^2\text{)} \tag{6.8}$$

The magnitude of the radius of gyration $i_{Z,V}$ depends on the length, width and weight distribution of the body, the length and weight of aggregate units (engine, gear box, differential) and the position and weight of the occupants and the luggage. Series tests with saloons have shown that the radius of gyration is mainly a function of the load status and only varies within narrow limits from vehicle to vehicle. Figure 6.6 shows the average values. Only the vehicle weight $m_{V,t}$ in the occupancy or load condition to be investigated is necessary for determining the approximate moment of inertia $J_{Z,V}$ (see Section 5.3.6). The values shown in Fig. 6.6 relate to medium-sized saloons. If the vehicle has a five-, six- or eight-cylinder engine, a difference value must be added:

$$\Delta i \approx 0.05 \text{ m to } 0.1 \text{ m}$$

If vehicle length L_t and wheelbase l are included in the following equation, an accuracy of at least 98% can be achieved; only a correction factor needs to be added:

Fig. 6.6 The approximate radius of gyration $i_{X,Bo}$ or $i_{Y,Bo}$ (valid for medium-sized saloon cars) for the inertia moment $J_{X,Bo}$ or $J_{Y,Bo}$ of the body or $J_{Z,V}$ of the vehicle as a whole, shown as a function of the loading condition and the pivot axis (Fig. 3.3).

Load	*Inertia radius in metres*		
	Car body only		*Whole vehicle*
	x-axis	*y-axis*	*z-axis*
Empty	0.65	1.21	1.20
2 passengers in front	0.64	1.13	1.15
4 passengers	0.60	1.10	1.14
4 passengers and luggage	0.56	1.13	1.18
Formula sign	$i_{X,Bo}$	$i_{Y,Bo}$	$i_{Z,V}$

$$J_{Z,V} = 0.1269\ m_{V,t}\ L_t\ l\ (\mathrm{kg\ m^2}) \tag{6.9}$$

Nevertheless, this equation only applies to the usual vehicle loading. Higher loads in the boot (or a roof load) must be considered separately:

$$J^*_{Z,V} = 0.1269\ m_{V,t}\ l\ L_t + \Delta m\ l_X^2\ (\mathrm{kg\ m^2}) \tag{6.10}$$

where l_X is the distance of the loading mass Δm to the vehicle centre of gravity.

The moment of inertia $J_{X,Bo}$ of the body is not so easy to calculate. In this instance, the weights $m_{U,f}$ and $m_{U,r}$ of the unsprung masses must be known and their distances to the respective coordinate axis drawn through the vehicle centre of gravity (see Equations 6.4c, 6.4d and 6.6, and Fig. 3.3); it is easier, in this case, to use Fig. 6.6:

$$J_{Y,Bo} = m_{Bo}\ i^2_{Y,Bo}\ (\mathrm{kg\ m^2}) \tag{6.11}$$

A front-wheel drive passenger car with two occupants can be used as an example for the pitch vibration calculation (around the y-axis):

axle load front, partly laden (index pl) $m_{V,f,pl} = 609$ kg
axle load rear, partly laden (index pl) $m_{V,r,pl} = 393$ kg

The weight of the axle mass is:

front $m_{U,f} = 67$ kg and rear $m_{U,r} = 59$ kg

The radius of gyration is $i_{Y,Bo} = 1.13$ m. Equation 6.5 gives:

$$m_{Bo,pl} = m_{V,f,pl} + m_{V,r,pl} - (m_{U,f} + m_{U,r}) = 609 + 393 - (67 + 59)$$
$$m_{Bo,pl} = 876\ \mathrm{kg}$$

In accordance with Equation 6.11 the mass moment of inertia of the body is then:

$$J_{Y,Bo} = m_{Bo,pl}\, i^2_{Y,Bo} = 876 \times 1.13^2,\ J_{Y,Bo} = 1118.6 \text{ kg m}^2$$

The same applies to body roll movements around the x-axis. The values in the table should also be used here:

$$J_{X,Bo} = m_{Bo}\, i^2_{X,Bo} \text{ (kg m}^2\text{)} \tag{6.12}$$

For further details, see Ref. [3], Section 1.5.

6.3 Braking behaviour

Braking path s_B in metres, starting speed v in metres per second (see Equation 2.1c) and delay a_X are related as follows:

$$s_B = v^2/2 - a_X \tag{6.12a}$$

6.3.1 Braking

When the driver brakes, the equivalent braking force acts as a reaction force at the centre of gravity V of the vehicle as a whole (Fig. 6.1):

$$F_{X,V,B} = \mu_{X,W}\, F_{Z,V,t} \tag{6.13}$$

i.e. the coefficient of friction $\mu_{X,W}$ times the weight force $F_{Z,V,t}$ of the vehicle as a whole, whereas $\mu_{X,W}$ can be equated to the deceleration a_X in m s^{-2}, divided by gravity:

$$\mu_{X,W} = a_X/g \tag{6.13a}$$

At an international level (DIN 74 250), the formula z is used for the braking function:

$$z = \mu_{X,W} \text{ and as a percentage: } z = \mu_{X,W}\, 100(\%) \tag{6.13b}$$

i.e. during braking $z = 80\%$ (corresponding to $a_X = 7.85$ m s^{-2}) the coefficient of friction $\mu_{X,W} = 0.8$ is necessary (Fig. 2.33 and Section 1.3 of Ref. [6]).

The braking force $F_{X,V,B}$ acting at the vehicle's centre of gravity causes longitudinal forces $F_{X,W,B,f}$ and $F_{X,W,B,r}$ at the centres of wheel contact of the front and rear axles, and an increase in axle load $+\Delta F_{Z,V,0}$ at the front and a reduction $-\Delta F_{Z,V,0}$ at the back when the vehicle is observed as a rigid body. In accordance with Fig. 6.7 the equations would then be

$$\chi = h_V/l \tag{6.13c}$$

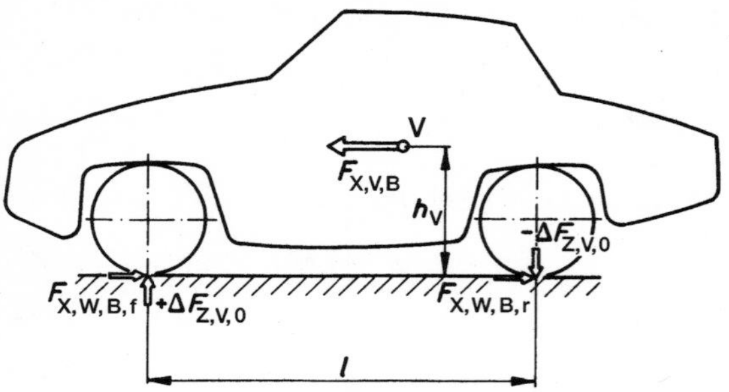

Fig. 6.7 A braking force $F_{X,V,B}$ acting at the centre of gravity V of the vehicle causes the axle load transfer $\pm\Delta F_{Z,V,0}$ and the braking forces $F_{X,W,B,r}$ on front and $F_{X,W,B,r}$ on rear axle. If the aerodynamic and rolling resistances are ignored, the forces can be easily calculated.

$$\Delta F_{Z,V,0} = \mu_{X,W}\, F_{Z,V,t}\, \chi \text{ (kN)} \tag{6.14}$$

$$F_{Z,V,f,dyn} = F_{Z,V,f} + \Delta F_{Z,V,0} \text{ and } F_{Z,V,r,dyn} = F_{Z,V,r} - \Delta F_{Z,V,0} \tag{6.15}$$

The lower the centre of gravity and the longer the wheelbase, the less is the (undesirable) load transfer $\Delta F_{Z,V,0}$. The braking force related to one axle is then

$$\text{front } F_{X,W,B,f} = \mu_{X,W}\, F_{Z,V,f,dyn} \text{ and} \tag{6.16}$$

$$\text{rear } F_{X,W,B,r} = F_{X,V,B} - F_{X,W,B,r} = \mu_{X,W}\, F_{X,V,r,dyn} \tag{6.17}$$

Half the braking forces per axle multiplied by the dynamic rolling radius r_{dyn}, gives the braking moments M_b at the wheels (see Equation 6.25a), which are:

$$\text{front } M_{b,f} = 0.5\, F_{X,W,B,f}\, r_{dyn} \tag{6.18}$$

$$\text{rear } M_{b,r} = 0.5\, F_{X,W,B,r}\, r_{dyn} \tag{6.19}$$

The larger is r_{dyn}, the higher is the moment to be generated by the brake. This is one reason for using tyres with an $r_{dyn} \leqslant 300$ mm on a medium size passenger car ($r_{dyn} = C_R/2\pi$, see Equation 2.2).

The sizes of $F_{X,W,B,f}$ and $F_{X,W,B,r}$ depend both on the vehicle and its loading condition and on the road, i.e. the coefficient of friction $\mu_{X,W}$ possible on it (see Section 2.7). A front-wheel drive vehicle and the calculation of the braking forces for two possible cases with an unfavourable loading can be used as an example to indicate the range of the braking force distribution:

$$\Phi_f = F_{X,W,B,f}/F_{X,V,B} \text{ (times 100 as a percentage)} \tag{6.20}$$

$$\Phi_r = (1 - \Phi_f) = F_{X,W,B}/F_{X,V,B} \tag{6.20a}$$

with an unchanged centre of gravity height h_V. As described in Sections 6.1.2.4 and 6.3.3.5, however, h_V alters based on the load and the pitch angle.

6.3.1.1 Braking on dry concrete with only two people in the vehicle

$F_{Z,V,f} = 6.9$ kN; $F_{Z,V,r} = 4.2$ kN; $l = 2.49$ m
$\mu_{X,W} = 0.9$; $h_V = 0.58$ m

$$\Delta F_{Z,V,0} = 0.9 \times 11.15 \times \frac{0.58}{2.49} = 2.34 \text{ kN}$$

$$F_{X,W,B,f} = 0.9\ (6.95 + 2.34) = 8.36 \text{ kN}$$
$$F_{X,W,B,r} = 0.9\ (4.2 - 2.34) = 1.67 \text{ kN}$$

6.3.1.2 Braking on ice with a fully laden vehicle

$F_{Z,V,f} = 7.1$ kN; $F_{Z,V,r} = 7.0$ kN; $\mu_{X,W} = 0.15$; l and h_V as previously
$\Delta F_{Z,V,0} = 0.49$ kN, $F_{X,W,B,f} = 1.14$ kN and $F_{X,W,B,r} = 0.98$ kN

In the first case, the front axle must accept as a percentage share:

$$\Phi_f \times 100 = \frac{8.36}{11.15} \times 100 = 75\%$$

and accordingly the rear axle 25%. In the second example, the braking force distribution is 54% and 46%. In the usual distribution of 75–80% on the front and 20–25% on the rear in the case of non-ABS fitted cars, the axle could lock in the first case (because its brake applies too high a torque); on ice it would be the front axle. For details, see Chapter 3 of Ref. [6].

6.3.2 Braking stability

If both wheels of an axle lock (if ABS is not fitted), i.e. if they slide on the road, there is not just reduced friction in the longitudinal direction (Fig. 2.33), but also lower friction in the lateral direction. If the rear axle locks, as shown in Fig. 6.8, lateral forces $F_{Y,W,f}$ will occur at the rolling wheels of the front axle, which will intensify the problem, even in the case of a minor yawing effect, i.e. the condition is unstable.

Lateral forces or irregularities in the road acting on the body can cause the vehicle which, to this point, has been travelling in a straight line to leave its direction of travel. A reinforcing yawing moment M_ϕ occurs (Fig. 6.9), which seeks to turn the vehicle sideways to its previous direction. There is a danger of lateral roll over. However, if the front axle locks, the rear wheels, which will still be rolling, will produce stabilizing lateral force $F_{Y,W,r}$. The condition is stable (Fig. 6.10).

The position is different if the braking moments on the wheels of one axle are of different sizes. The brakes pull to one side due to different lining coefficients of friction or unequal coefficients of friction on the left and right wheels (μ split, see Section 1.7.1 and Ref. [6] Section 2.4.4).

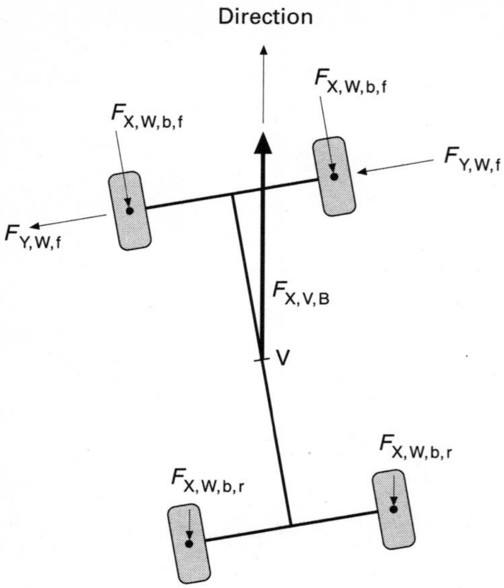

Fig. 6.8 Locking rear wheels lead to an unstable driving condition.

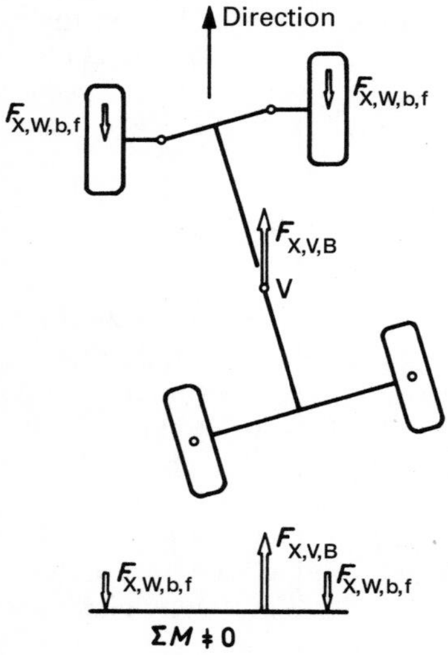

Fig. 6.9 When the rear wheels lock, a reinforcing yawing moment occurs even when the vehicle only slightly leaves the direction of travel.

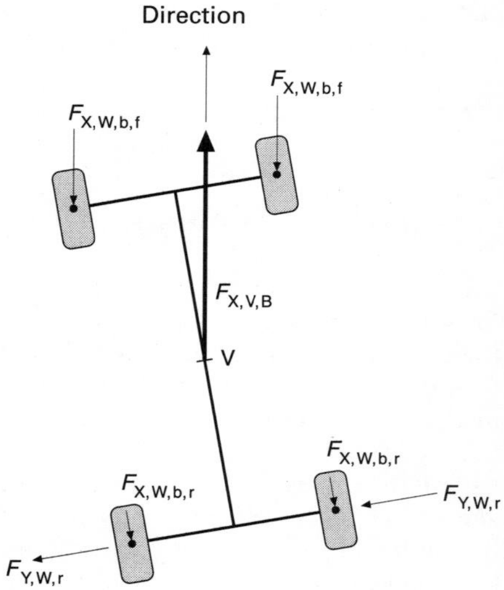

Fig. 6.10 When the front wheels lock, the vehicle condition remains stable although the vehicle can no longer be steered.

Fig. 6.11 As the static calculation below indicates, unequal braking forces $F_{X,W,b,f,l}$ and $F_{X,W,b,f,rs}$ at the centres of tyre contact of the front wheels cause the vehicle to rotate around the vertical axis. In the case of a positive wheel offset at ground (positive scrub radius), there is also a steering input in the same direction of rotation.

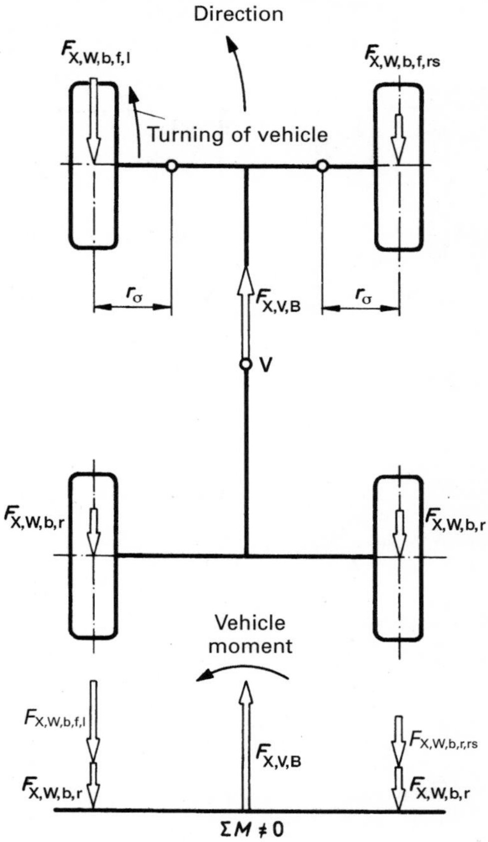

Figure 6.11 shows a higher braking force $F_{X,W,b,f,l}$ on the left front wheel (than on the right one). The difference force of the two wheels $\Delta F_{X,W,b,f} = F_{X,W,b,f,l} - F_{X,W,b,f,rs}$, with the lever of half the tread width, gives the yawing moment $M_\psi = \Delta F_{X,W,b,f}\ 0.5\ b_f$ which introduces rotation to the left into the vehicle. In addition, there is also the steering moment $M_{Z,W,b}$, which causes the steering to turn in the same direction.

Where the brake is on the outside (at the wheel), the size of this moment depends on the length of the wheel offset $+r_\sigma$, and is:

$$+M_{Z,W,b} = \Delta F_{X,W,b,f}\ r_\sigma \cos\sigma \tag{6.21}$$

In the case of negative $-r_\sigma$, there is counter steering (Fig. 6.12), and if $r_\sigma = 0$, only the yawing moment ($-M_\psi$, Fig. 6.13) occurs. This also applies to centre axle steering (Fig. 3.114).

A differential braking torque is less noticeable on the rear axle. First, the braking forces $F_{X,W,b,r}$ are smaller and second there is a stable condition. The

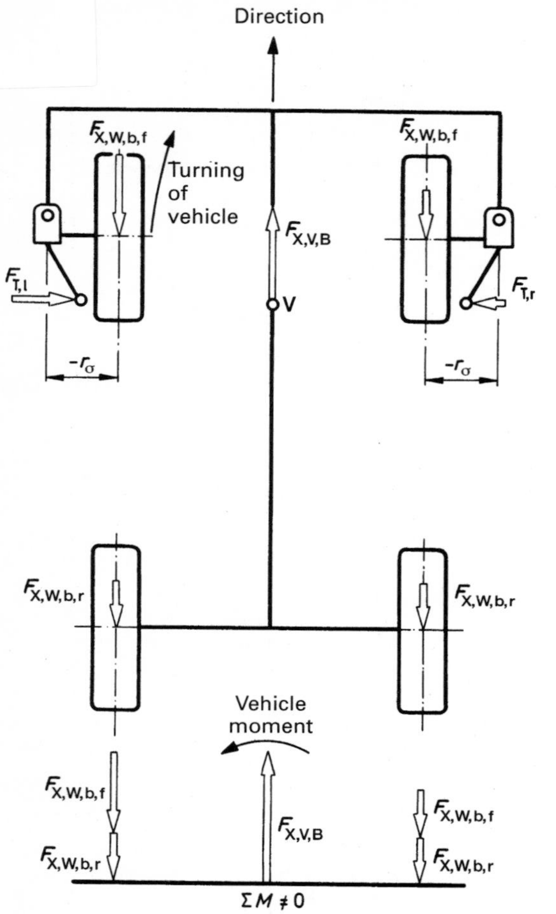

Fig. 6.12 In the case of a negative wheel offset (or elastokinematic toe-in alteration (Figs 3.82 and 3.102) the steering is turned by the front wheel, which must transfer the greater braking force $F_{X,W,b,f}$ (the left wheel in the illustration) opposite to the direction in which the vehicle is turned by the outer yawing moment. The static calculation shown indicates this. This leads to an equalization which, even in the case of different braking forces at the front, largely prevents deviation from the direction of travel.

different sized forces $F_{X,W,b,r}$ and $F_{X,W,b,f}$ are behind the centre of gravity V (Fig. 6.14).

For further details, see Ref. [6] Section 2.4 and Ref. [9] Chapter 3.

6.3.3 Calculating the pitch angle

The pitch angle, i.e. the angle Θ_B by which the body turns around the lateral axis when the brakes are applied, can be calculated as a function of the braking force $F_{X,V,B}$ (Fig. 6.15, see also Section 5.4.3).

6.3.3.1 Data for the example calculation

A passenger car with the following data can be used to clarify the relationships:

Fig. 6.13 If, where the brake is on the inside, the longitudinal force lever $r_a = 0$ or, where the brake is on the outside, the wheel offset at ground (positive scrub radius) is 0, then unequal braking forces $F_{X,W,b,f}$ on the front wheels have practically no effect on the steering. The steering rod forces F_T are almost zero.

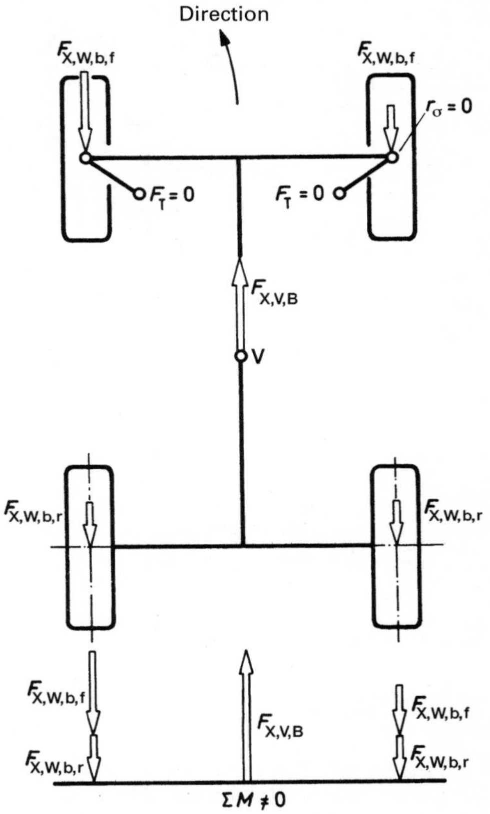

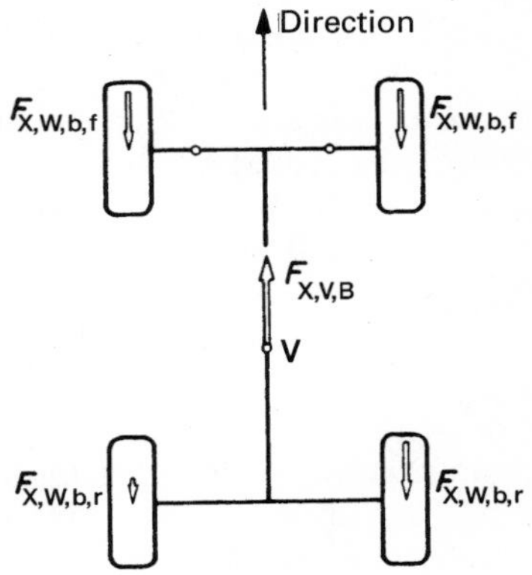

Fig. 6.14 A rear wheel brake, which is unevenly pulled, hardly has any effect on the steerability of a vehicle.

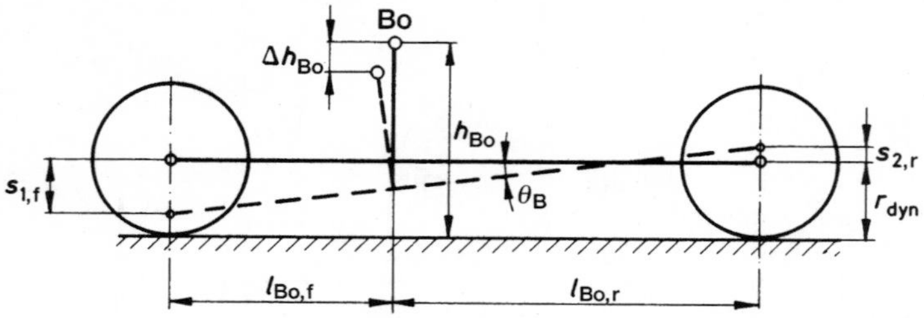

Fig. 6.15 If the body goes down more at the front than it rebounds at the rear, the body centre of gravity Bo moves down by Δh_{Bo}. The braking force $F_{X,Bo,B}$ would then act at the height $(h_{Bo} - \Delta h_{Bo})$ at point Bo. The pitch angle θ_B is also shown (see also Fig. 3.137).

Overall weight force (vehicle)	$F_{Z,V,t} = 11.15$ kN
axle load front	$F_{Z,V,f} = 6.95$ kN
axle load rear	$F_{Z,V,r} = 4.20$ kN
axle weight force front	$F_{Z,U,f} = 0.80$ kN
axle weight force rear	$F_{Z,U,r} = 0.70$ kN
spring rate based on only one axle	front $c_f = 11.5$ N m^{-1}, rear $c_r = 14$ N m^{-1}
dynamic rolling radius of tyre	$r_{dyn} = 0.288$ m
braking	$z = 0.8$, i.e. $\mu_{X,W} = 0.8$ (see Equation 6.13b)
wheelbase	$l = 2.50$ m
centre of gravity height	$h_V = 0.58$ m

Details relating to the numerical values can be found in Sections 2.2.5.4, 5.3.6.1, 6.1.3 and 6.12.

6.3.3.2 Opposed springing forces

When the body is observed as a rigid mass, the spring opposed forces (related to one axle's wheel track considered as one, front and rear, Fig. 6.7) correspond to half the axle load transfer $\pm\Delta F_{Z,V,0}$ and, irrespective of whether the brakes are outside of the wheel or inside on the differential, the forces can be calculated easily on the basis of Equation 6.14.

$$\pm F_{Z,V,0} = 0.8 \times 11.15 \times (0.58/2.50) = 2.07 \text{ kN}$$

The vehicle goes down at the front and rebounds at the back. The spring rates are quoted in newtons per millimetre and also relate to one wheel. That is, 1 N mm^{-1} = 1 kN m^{-1}, so we can assume $\Delta F_{Z,V,0}$ is multiplied by two. In accordance with Equation 5.10, with linear springing the following theoretical values would result:

Bump travel front $s_{1,f} = \Delta F_{Z,V,0}/(2\ c_f) = 2.07/23 = 0.09$ m
Jounce travel rear $s_{2,r} = \Delta F_{Z,V,0}/(2\ c_r) = 2.07/28 = 0.074$ m

6.3.3.3 Pitch angle with linear springing

The pitch angle θ_B is (see also Equation 5.15):

$$\theta_B = \frac{s_{1,f} + s_{2,r}}{l} \tag{6.22}$$

and (times 360°/2π)

$$\theta_B = 57.3 \times \frac{s_{1,f} + s_{2,r}}{l} \tag{6.23}$$

In this example the result is then:

$$\theta_B = 57.3 \times \frac{0.09 + 0.074}{2.50} = 3.76° = 3°46'$$

6.3.3.4 Pitch angle with progressive springing

In order to determine the travel on the front and rear axles, the spring characteristics must be known. Travel is entered here in mm and wheel load in kg. The required values should therefore be calculated from the axle load:

wheel load, front, normal $$m_{1,V,f} = \frac{F_{Z,V,f}}{2g} = \frac{6950}{2 \times 9.81} = 354 \text{ kg}$$

wheel load, front, maximum $$m_{1,V,f,max} = \frac{F_{Z,V,f} + \Delta F_{Z,V,0}}{2g} = \frac{6950 + 2070}{2 \times 9.81} = 460 \text{ kg}$$

wheel load, rear, normal $$m_{1,V,f} = \frac{F_{Z,V,f}}{2g} = \frac{4200}{2 \times 9.81} = 214 \text{ kg}$$

wheel load, rear, minimum $$m_{1,V,min} = \frac{F_{Z,V,f} - \Delta F_{Z,V,0}}{2g} = \frac{4200 - 2070}{2 \times 9.81} = 108 \text{ kg}$$

In spite of the harder springing, the highly progressive curve shown in Figs 5.13 and 5.15 can be used as an example. The spring travel is

front at 354 kg = 112 mm, and at 460 kg = 134 mm

rear at 214 kg = 110 mm, and at 108 kg = 44 mm

The vehicle response would therefore be

front, goes down by $s_{1,f} = 134 - 112 = 22$ mm and

rear, rebounds by $s_{2,r} = 110 - 44 = 66$ mm

This would then give a pitch angle of only

$\theta_B = 2.02° \approx 2°1'$

6.3.3.5 Change of the centre of gravity height

Point B rises or falls when the brakes are applied based on how far the vehicle rebounds front and rear and how far the body centre of gravity is away from the axle centres. With the path entered in Fig. 6.15 and the weight forces, plus Equations 5.14 to 5.14b, the result is then the change of height

$$\Delta h_{Bo} = -s_{1,f} \frac{F_{Z,Bo,f}}{F_{Z,Bo}} + s_{2,r} \frac{F_{Z,Bo,r}}{F_{Z,Bo}} \tag{6.24}$$

When the springs compress, the body goes down and so $s_{1,f}$ becomes negative. The values in accordance with Equations 6.5 to 6.6b are:

$$F_{Z,Bo,f} = F_{Z,V,f} - F_{Z,U,f} \text{ and} \tag{6.24a}$$

$$F_{Z,Bo,r} = F_{Z,V,r} - F_{Z,U,r} \tag{6.24b}$$

$$F_{Z,Bo} = F_{Z,Bo,f} + F_{Z,Bo,r} \tag{6.24c}$$

With the numerical values of the calculation example (see Section 6.3.3.1) and with linear springing, the result is then:

$F_{Z,Bo,f} = 6.95 - 0.80 = 6.15$ kN; $F_{Z,Bo,r} = 4.20 - 0.70 = 3.5$ kN

$F_{Z,Bo} = 6.15 + 3.5 = 9.65$ kN, $s_{1,f} = -0.09$ m and $s_{2,r} = +0.074$ m

$$\Delta h_{Bo} = -0.09 \times \frac{6.15}{9.65} + 0.074 \times \frac{3.5}{9.65} = -0.03 \text{ m}; \ \Delta h_{Bo} = -0.03\text{m}$$

The static centre of gravity height of the body, calculated using Equations 6.7 and 6.24a to c, is $h_{Bo} = 0.625$ m and that which occurs when the brakes are applied is:

$$h_{Bo} - \Delta h_{Bo} = h'_{Bo} = 0.595 \text{ m} \tag{6.24d}$$

The centre of gravity therefore goes down 4.8%. The resulting height h'_V of the vehicle centre of gravity can be calculated from the value $h'_{Bo} = 0.595$ m using the transformed Equation 6.7. This can be more easily done if the axle weight forces $F_{Z,U,f}$ and $F_{Z,U,r}$ are ignored. The error involved is less than 0.5%:

$$\Delta h_V = -s_{1,f}\frac{F_{Z,V,f}}{F_{Z,V,t}} + s_{2,r}\frac{F_{Z,V,r}}{F_{Z,V,t}} \tag{6.25}$$

The axle weight forces (unsprung masses, see Section 6.1.3) must be known if the pitch poles are to be included in the equation.

6.3.4 Influence of radius-arm axes

6.3.4.1 Pre-conditions for calculations

Radius-arm axes poles are only effective when the brakes are outside the wheel. The entire calculation has to be done differently because not only do the braking force portions of the body $F_{X,Bo,B,f}$ and $F_{X,Bo,B,r}$ act at these axes O_f (front) and O_r (rear), but so do the vertical forces $\Delta F_{Z,Bo,B,f}$ and $\Delta F_{Z,Bo,B,r}$ acting against brake dive (B = axis-related).

Figures 3.108 and 3.113 show the static situation and, with Equations 6.13 to 6.16, the braking force related to one wheel can be calculated:

$$F_{X,W,b,f} = F_{X,W,B,f}/2 \tag{6.25a}$$

$$F_{X,W,b,r} = F_{X,W,B,r}/2 \tag{6.25b}$$

6.3.4.2 Forces on the radius-arm axes of both axles

Figure 3.155 shows how the forces are calculated on one wheel station with double wishbones and Fig. 6.16 shows the calculation based on the entire axle-suspension and using the pitch poles required in this instance.

To be able to calculate the forces $F_{Z,Bo,B,f}$ supporting the body vertically when the brakes are applied, the equation of moments must be formed with the pivot at the centre of tyre contact. Paths e and c define the point O_f (present on both sides) in the illustration:

$$\Delta F_{Z,Bo,f} = \frac{F_{X,B,Bo,f}\, e + F_{X,B,U,f} r_{dyn}}{c} = \Delta F_{Z,V,f,2} \tag{6.26}$$

The axle load difference $\Delta F_{Z,V,f,2}$ is the same size as $\Delta F_{Z,Bo,B,f}$ and opposes compression when the brakes are applied. The forces that also appear in Equations 6.26 and 6.29 can be determined using Equations 6.6a and 6.6b:

$$F_{X,Bo,B,f} = \mu_{X,W}\, F_{Z,Bo,f} = \mu_{X,W}\, m_{Bo,f}\, g$$
$$F_{X,Bo,B,r} = \mu_{X,W}\, F_{Z,Bo,r} = \mu_{X,W}\, m_{Bo,r}\, g \qquad \text{(6.27 a and b)}$$

$$F_{X,U,B,f} = \mu_{X,W}\, F_{Z,U,f} = \mu_{X,W}\, m_{U,f}\, g$$
$$F_{X,U,B,r} = \mu_{X,W}\, F_{Z,U,r} = \mu_{X,W}\, m_{U,r}\, g \qquad \text{(6.28 a and b)}$$

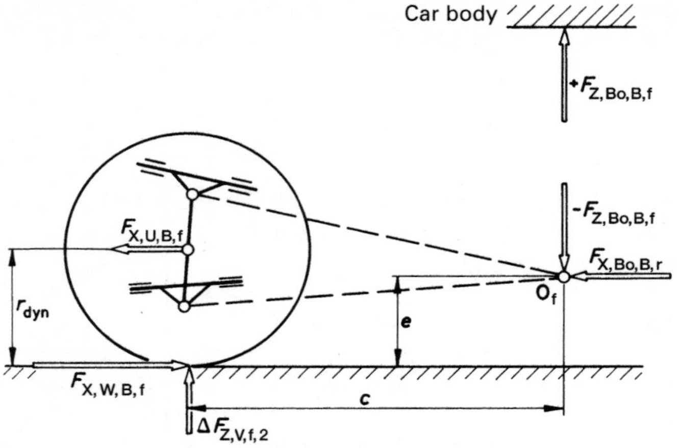

Fig. 6.16 Paths and forces when the body is supported around the front virtual pitch axis. The higher O_f (path *e*) and the closer to the wheels (path *c*), the larger the difference in force $\Delta F_{Z,Bo,B,f}$ supporting the body and the smaller the pitch angle θ_B.

To calculate $F_{Z,Bo,B,f}$ (and also $F_{Z,Bo,B,r}$) only the braking forces $F_{X,W,B,f}$ and $F_{X,W,B,r}$, which occur in the centres of tyre contact and relate to the axle as a whole, need to be divided up into the proportion affecting the wheel suspension $F_{X,U,B,f}$ and $F_{X,Bo,B,f}$, which is critical to the body. The same applies to the rear axle. The index r appears in this instance.

6.3.4.3 Numerical values

With the values of Section 6.3.3.1 the result for the front axle is then:

$$F_{X,W,B,f} = \mu_{X,W} F_{Z,V,f} = 0.8 \times 6.95 = 5.56 \text{ kN}$$
$$F_{X,U,B,f} = \mu_{X,W} F_{Z,U,f} = 0.8 \times 0.8 = 0.64 \text{ kN}$$
$$F_{X,Bo,B,f} = F_{X,W,B,f} - F_{X,U,B,f} = 4.92 \text{ kN}$$

and for the rear axle:

$$F_{X,W,B,r} = \mu_{X,W} F_{Z,V,r} = 0.8 \times 4.20 = 3.36 \text{ kN}$$
$$F_{X,U,B,r} = \mu_{X,W} F_{Z,U,r} = 0.8 \times 0.70 = 0.56 \text{ kN}$$
$$F_{X,Bo,B,r} = F_{X,W,B,r} - F_{X,U,B,r} = 2.8 \text{ kN}$$

The following dimensions should apply to the pitch poles (Figs 6.16 and 6.17):

front $c = 1.0$ m, $e = 0.15$ m
rear $d = 0.5$ m, $g = 0.25$ m which gives the result

$$\text{front} \quad F_{Z,Bo,B,f} = \frac{F_{X,Bo,B,f}\, e + F_{X,U,B,f}\, r_{dyn}}{c} \tag{6.26}$$

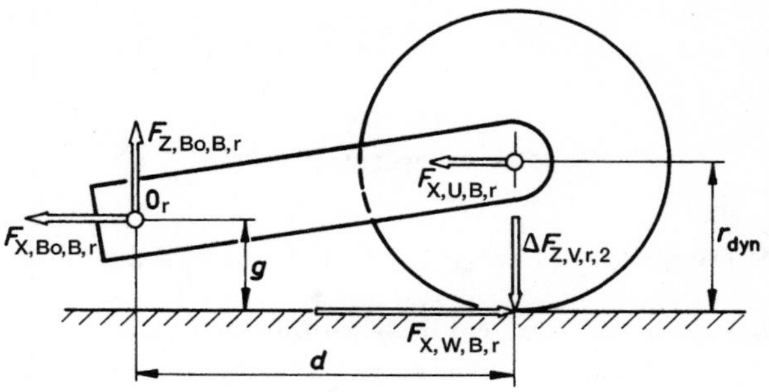

Fig. 6.17 Paths and forces when the body is supported at the rear pitch axis O_r which are relatively close to the wheels. Figure 3.154 shows the forces, based on only one axle side, and Fig. 3.159 shows further kinematic aspects.

$$\Delta F_{Z,Bo,B,f} = \frac{(4.92 \times 0.15) + (0.644 \times 0.288)}{1.0} = 0.922 \text{ kN} = \Delta F_{Z,V,f,2}$$

rear $$\Delta F_{Z,Bo,B,r} = \frac{F_{X,Bo,B,r}\, g + F_{X,U,B,r}\, r_{dyn}}{d} \tag{6.29}$$

$$\Delta F_{Z,Bo,B,r} = \frac{(2.8 \times 0.25) + (0.56 \times 0.288)}{0.5} = 1.723 \text{ kN} = \Delta F_{Z,V,r,2}$$

The pitch poles should be as close as possible to the wheel and as high as possible. Because of the more favourable position of the rear poles, $\Delta F_{Z,Bo,B,r} > \Delta F_{Z,Bo,B,f}$, i.e. also $\Delta F_{Z,V,r,2} > \Delta F_{Z,V,f,2}$.

6.3.4.4 Spring-opposing forces

The spring-opposing forces $\Delta F_{Z,V,1}$ that result from the axle load transfer $\Delta F_{Z,V,0}$ (see Equation 6.14) and the weight force differences $\Delta F_{Z,V,2}$ with existing radius-arm axes are critical to the pitch angle desired θ_B:

front $$\Delta F_{Z,V,f,1} = \Delta F_{Z,V,0} - \Delta F_{Z,V,f,2} \tag{6.30}$$

rear $$\Delta F_{Z,V,r,1} = \Delta F_{Z,V,0} - \Delta F_{Z,V,r,2} \tag{6.31}$$

Where $\Delta F_{Z,V,0} = 2.07$ kN (see Section 6.3.3.2) related to the entire axle, the values are

$\Delta F_{Z,V,f,1} = 1.148$ kN
$\Delta F_{Z,V,r,1} = 0.347$ kN

6.3.4.5 Pitch angles

At the spring rates $2 \times c_f = 23$ N mm^{-1} or $2 \times c_r = 28$ N mm^{-1} based on one axle side and in accordance with Equation 5.10, the paths needed for determining θ_B (see also Section 5.4.3) are:

front $s_{1,f} = 0.05$ m and rear $s_{2,r} = 0.012$ m

With the linear springing assumed, the angle decreases. Without considering the axis it was 3°46′ and now in accordance with Equation 6.23 it is

$$\theta_B = 57.3 \times \frac{0.05 + 0.012}{2.50} = 1.42° = 1°25'$$

Section 6.3.3.4 contains the calculation of θ_B with progressive springing.

6.3.4.6 Radius-arm axes on one axle only

If, for example, only the rear axle suspension has radius-arm axes $s_{2,r}$ must be determined using $\Delta F_{Z,V,r,1}$, whereas $\Delta_{Z,V,f,0}$ alone is critical on the front axle. At the value $\Delta F_{Z,V,0} = 2.07$ kN in Section 6.3.3.2 the bump travel was $s_{1,f} = 0.09$ m.

6.3.5 Anti-dive control and brake reaction support angle

Automobile manufacturers frequently quote the anti-dive control k_ε as a percentage in publications. It can easily be calculated on the basis of Figs 6.15 and 6.16.

front $k_{\varepsilon,f} = \Delta F_{Z,V,f,2}/\Delta F_{Z,V,0}$
$k_{\varepsilon,f} = 0.922/2.07 = 0.45$ (6.32)
$k_{\varepsilon,f} = 45\%$

rear $k_{\varepsilon,r} = \Delta F_{Z,V,r,2}/\Delta F_{Z,V,0}$
$k_{\varepsilon,r} = 1.723/2.07 = 0.83$ (6.32a)
$k_{\varepsilon,r} = 83\%$

The brake reaction support angle ε entered in Fig. 3.160 in the examples (Figs 6.16 and 6.17) is

front $\tan \varepsilon_f = e/c$ (6.33)
$\tan \varepsilon_f = 0.15/1.0 = 0.15;\ \varepsilon_f = 8°30'$

rear $\tan \varepsilon_r = g/d$ (6.33a)
$\tan \varepsilon_r = 0.25/0.5 = 0.5;\ \varepsilon_r = 25°32'$

On production passenger cars, ε_f is usually below 10° and ε_r between 30° and 40°.

6.4 Traction behaviour

6.4.1 Drive-off from rest

The relationships when the vehicle moves off and accelerates are somewhat different to those when the brakes are applied. As shown in Fig. 3.113, the tractive force $F_{X,W,A}$ must be shifted to the centre of the rolling wheel if the differential is fixed to the body or the engine (i.e. separately from the wheel suspension) and the drive-off moment is concentrated in its suspension (Fig. 3.110). This applies on all front independent wheel suspensions and is equivalent to one virtual radius-arm axis O_f coming into effect. As shown in Fig. 3.154, in such cases, the squat can be reduced by angling the two double wishbones in the same direction. The same applies to the rear axle in terms of the take-off dive (Fig. 3.160). The diagonal springing angle χ is then positive.

The picture is different when the differential is in the axle housing on a driven rigid axle (Fig. 1.43). The drive pinion connected to the prop shaft is vertical to the axle shaft connected to the wheels (Fig. 1.22), i.e. torque in and output form a 90° angle. The result is that the tractive force $F_{X,W,A}$, which occurs at the centres of wheel contact, is supported exclusively in the suspension system of the axle. Where there are pitch poles, the body is pushed upwards into these points O_r and the tail only dives a little, as shown in Fig. 3.159 using the example of the opposed braking forces. The same effect is achieved with trailing link pairs at an angle to one another (Fig. 3.161); there are pitch poles here too.

Figure 6.18 shows the forces generated during acceleration. Those acting on the body are:

- the aerodynamic force F_L, which can be ignored at speeds below 25 km h^{-1}, and the
- excess force $F_{X,ex}$, which is equal to the inertia in the x-direction.

The rolling resistance forces are

$$F_{R,t} = F_{R,f} + F_{R,r} = k_R\,(F_{Z,V,f} + F_{Z,V,r}) = k_R\,F_{Z,V,t} \tag{6.34}$$

and the opposed tractive forces $F_{X,W,A}$ act at the wheels. As described in Ref. [9], Section 2.1.4, the additional forces necessary to accelerate the turning masses, can be determined using the rotating mass factor.

The equations for calculating the drive-off are:

$$F_{X,ex} = F_{X,W,A} - (F_L + F_{R,t})\ \text{(N)} \tag{6.35}$$

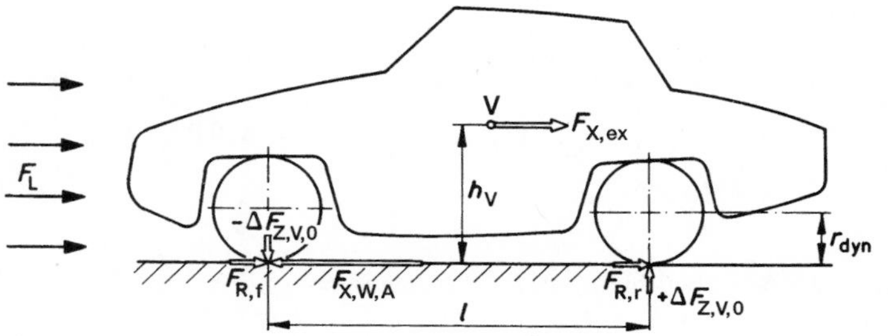

Fig. 6.18 Forces occurring in the vehicle centre of gravity V and at the centres of tyre contact when a front-wheel drive vehicle accelerates.

$$F_{X,W,A} = \frac{M_{M,max}\ \eta\ i_D i_G}{r}\ (N) \tag{6.36}$$

The following terms are used in the equation:

$M_{M,max}$	the maximum engine torque in N m
η	the total efficiency (Figs 6.19 and 6.20)
i_D	the ratio of the final drive (differentials)
i_G	the ratio of the gear engaged
r	the static rolling radius r_{stat} must be inserted in m at speeds below 25 km h^{-1}, and above 60 km h^{-1}; the dynamic rolling radius $r_{dyn} = C_{R,dyn}/2\pi$ ($C_{R,dyn}$ = rolling circumference in m, see Equations 2.2 and 2.2e).

Chapters 2 and 3 in Ref. [3] give details on resistances.

A compact front-wheel drive passenger car with a 1.3 l transverse engine can be used as an example. When the acceleration in first gear from around 5 km h^{-1} is observed, the necessary data are

$M_{M,max} = 94$ Nm, $i_D = 3.94$, $i_{G,1} = 3.55$, $\eta = 0.90$ (Fig. 1.50)

tyres 155 R 13 78 S, $r_{stat} = 0.263$ m

Fig. 6.19 When standard vehicles are in the direct (mostly fifth) gear, no pair of gears of the manual gearbox is engaged. However, the lower gears require two pairs of gears to transfer the engine moment.

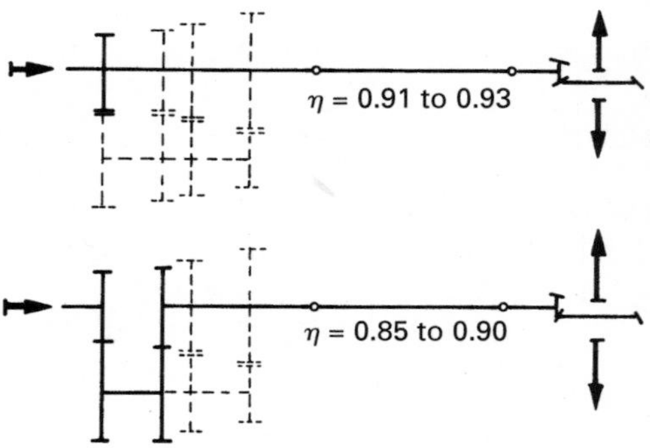

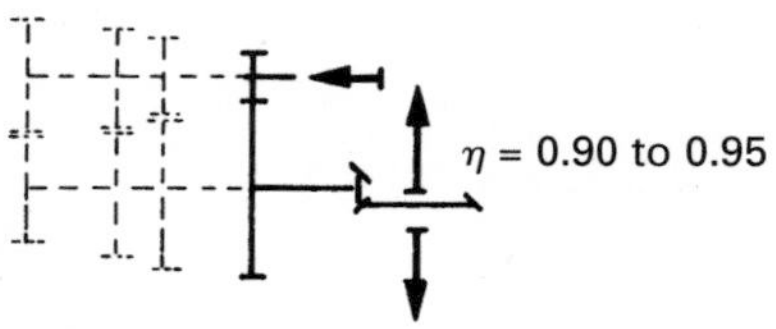

Fig. 6.20 If, on a front-wheel drive or rear-engine vehicle, the engine is longitudinal, on a manual gearbox one pair of gears is always engaged to transfer the drive moment, regardless of what gear has been selected and whether the vehicle has a four-, five- or six-speed box. On transverse engines (Fig. 1.50), the degree of efficiency can be better than $\eta = 0.9$.

$$F_{X,W,A} = \frac{94 \times 0.90 \times 3.94 \times 3.55}{0.263} = 4499 \text{ N} = 4.5 \text{ kN}$$

Because the weight is taken off the front axle as the vehicle moves off (see Equation 6.37), r becomes 10–15 mm greater than r_{stat} and the driving force around 5% smaller.

The vehicle has a curb weight of $m_{V,ul} = 875$ kg. With two people each weighing 68 kg in the vehicle the actual weight would be

$$m_{V,t} = 1011 \text{ kg} \quad \text{and} \quad F_{Z,V,t} = m_{V,t} \times g = 9918 \text{ N} = 9.92 \text{ kN}$$

The forces in the longitudinal direction at k_R from Fig. 2.31 are

$$F_{R,t} = k_R \, F_{Z,V,t} = 0.012 \times 9.92 = 0.12 \text{ kN}, \quad F_L = 0$$

and $F_{X,ex} = F_{X,W,A} - F_{R,t} = 4.5 - 0.12 = 4.38$ kN

The axle load transfer $\Delta F_{Z,V,0}$ is determined using Equation 6.14; the vehicle data are

$$l = 2.52 \text{ m}, \quad h_{ul} = 1.4 \text{ m} \quad \text{and} \quad \mu_{X,W} = 1.05 \text{ (see Fig. 2.33)}$$

The height of the centre of gravity $h_{V,0}$ can be obtained, using Equation 6.5, from the unladen height h_{ul} of the vehicle:

$$h_{V,0} \approx 0.38 \times 1.4 \approx 0.532 \text{ m}$$

When there are two people in the vehicle the centre of gravity rises by 10 to 15 mm (see Section 6.1.2.4); therefore $h_{V,2} = 0.546$ m is assumed:

$$\Delta F_{Z,V,0} = \mu_{X,W} \, F_{Z,V,t} \frac{h_{V,2}}{l} = 1.05 \times 9.92 \times \frac{0.546}{2.52}$$

$$\Delta F_{Z,V,0} = 2.26 \text{ kN}$$

As shown in Fig. 1.36, when there are two people in the vehicle, approximately 60% of the weight is carried on the front axle:

$$F_{Z,V,f} = 0.6 \, F_{Z,V,t} = 5.95 \text{ kN}$$

Unlike when the brakes are applied, when the vehicle accelerates the weight is taken off the front axle by $\Delta F_{Z,V,0}$:

$$F_{Z,V,f,dyn} = F_{Z,V,f} - \Delta F_{Z,V,0} = 5.95 - 2.26 = 3.69 \text{ kN} \tag{6.37}$$

The coefficient of friction required is then

$$\mu_{X,W} = F_{X,\,ex}/F_{Z,V,f,dyn} \tag{6.37a}$$

$$\mu_{X,W} = 4.38/3.69 = 1.19$$

When the vehicle accelerates fast from slow speeds, the driven front wheels would spin due to the load alleviation. This disadvantage is particularly evident in the range of maximum engine torque. The coefficient of friction needed $\mu_{X,W} = 1.19$ is too high. Taking into consideration the load alleviated and therefore larger tyre radius r, μ_H would drop to around 1.13 but not solve the problem. With both values, $\Delta F_{Z,V,0} > 2.26$ kN, which would be an even greater load alleviation of the front axle than the assumed $\mu_{X,W} = 1.05$.

On rear-wheel drive vehicles, Equation 6.35 is exactly the same. It is simply a matter of shifting force $F_{X,W,A}$ shown in Fig. 6.18 to this axle and adding the axle load shift to $F_{Z,V,r}$. The result would be:

$$F_{Z,V,r,dyn} = F_{Z,V,r} + \Delta F_{Z,V,0} \tag{6.38}$$

$$\Delta F_{Z,V,r,1} = \Delta F_{Z,V,0} - \Delta F_{Z,V,r,2} \tag{6.31}$$

If the driven rigid axle of the vehicle under investigation has pitch poles (Figs 1.43 and 3.161), $\Delta F_{Z,V,0}$ and $\Delta F_{Z,V,r,2}$ must first be calculated to obtain $\Delta F_{Z,V,r,1}$ (Equations 6.14 and 6.31).

Equation 6.23 is again used for calculating the pitch angle θ_A and Equations 6.32a and 6.33a can be used for determining the take-off drive control $k_{\varepsilon,r}$ and the drive-off reaction support angle χ_r as these values here are of the same size as $k_{\varepsilon,r}$ and ε, i.e. are produced when the brakes are applied (Fig. 3.160). Only in the case of independent wheel suspensions and rigid axles with a separate differential (De Dion axles) does the actual angle χ need to be taken into consideration (Fig. 3.154 and see also Ref. [2] Section 3.6).

6.4.2 Climbing ability

The climbing ability q is quoted as a percentage and relates to the vertical height h_z reached at the end of a path s_x measured on the horizontal:

$$q = h_z/s_x \, 100(\%) \tag{6.39}$$

$$\tan \alpha = h_z/s_x \tag{6.40}$$

The inclination the vehicle can theoretically climb in first gear (e.g. in the range of the greatest engine torque) can be calculated using the excess force $F_{X,ex}$ and the total weight $F_{X,V,t}$ of the vehicle. In the previous example, in Equation 6.35 only the rolling resistance would be somewhat smaller. The force $F_{R,t}$ must be multiplied by $\cos \alpha \approx 0.9$. $F_{X,ex}$ would increase from 4.38 kN to 4.39 kN, a negligibly small difference of only 0.2%. The climbing ability of the example vehicle is:

$$\sin \alpha = F_{X,ex}/F_{Z,V,t} = 4.38/9.92 = 0.44$$
$$\sin \alpha = 26.1°, \ \tan \alpha = 0.49 \ \text{ and } \ q = 49\%$$

On inclines, an axle load transfer of $\pm F_{Z,V,3}$ occurs, i.e. a reduction of $F_{Z,V,f}$ on the driven front axle. In accordance with Fig. 6.21 and Equation 6.35, it is

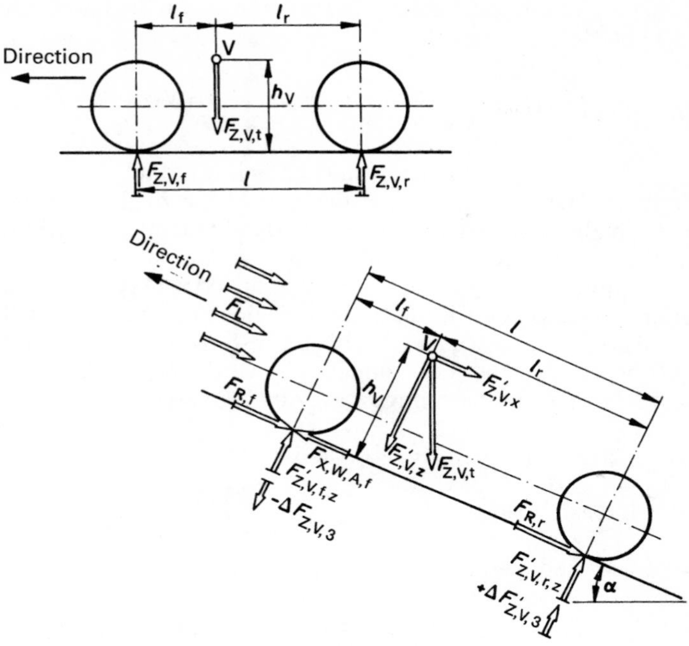

Fig. 6.21 Paths and forces necessary for calculating the skid point, shown on a front-wheel drive vehicle that is travelling up an incline at a constant speed.

$$\Delta F_{Z,V,3} = F'_{Z,V,x}\, h_V/l = F_{Z,V,t} \sin\alpha\, (h_V/l)\ (\mathrm{kN}) \tag{6.41}$$

and

$$\mu_{X,W} = F_{X,ex}/(F_{Z,V,f} - \Delta F_{Z,V,3}) \tag{6.42}$$

Insertion of the example values give

$$\Delta F_{Z,V,3} = 0.92 \times 0.44 \times \frac{0.546}{2.52},\ \Delta F_{Z,V,3} = 0.946\ \mathrm{kN}$$

and

$$\mu_{X,W} = \frac{4.38}{5.952 - 0.946} = 0.87$$

To travel a 49% incline evenly the example vehicle only needs a coefficient of friction of $\mu_{X,W} = 0.87$ in the range of the greatest engine torque, a value to be found on dry concrete.

For further details on climbing ability and resistance, see Section 2.3 and 3.3, and Section 3.11 in Ref. [3].

6.4.3 Skid points

Theoretically, more powerful engines would be able to climb steeper inclines with either front- or rear-wheel drive, if the grip of the road surface were to permit it. To have realistic values the skid points should therefore be determined, i.e. the inclination (as a percentage) on the road surface of which the driven wheels do not yet quite slip; $\mu_{X,W} = 0.8$ would be the correct coefficient of friction as an initial value. Using Fig. 6.21 the equations can be derived that are necessary for calculating $\alpha = f(\mu_{X,W})$. In this the x'-direction is in the climbing plane (slope of the incline) and the z'-direction is vertical to it. Breaking down the total weight force $F_{Z,V,t}$ at the centre of gravity V gives

$$F'_{Z,V,z} = F_{Z,V,t} \cos\alpha \quad \text{and} \quad F'_{Z,V,x} = F_{Z,V,t} \sin\alpha$$

and $F'_{Z,V,z}$ causes the axle loads

$$F'_{Z,V,f,z} = F_{Z,V,f}\, \alpha \quad \text{and} \quad F'_{Z,V,r,z} = F_{Z,V,r} \cos\alpha$$

As can be seen in Fig. 6.21, $F_{Z,V,f}$ or $F_{Z,V,r}$ are the axle loads applied to the vehicle standing on the flat. The component $F'_{Z,V,x}$, known as the vehicle load downhill, is the same as the excess force $F_{X,ex}$ previously calculated. This causes a load reduction on the front axle by $-\Delta F_{Z,V,3}$ (Equation 6.41) and an increase in axle load on the rear axle of $+\Delta F_{Z,V,3}$. The value h_V/l, which appears in the equation, shows that the longer the wheelbase l and the lower the centre of gravity V, the smaller is the axle load transfer (which is unfavourable on front-wheel drive).

The condition that the sum of all forces in the x'-direction equals 0, would be met if:

$$F_{X,W,A} = F'_{Z,V,x} + F_{R,f} + F_{R,r} + F_L$$

$F_{R,f}$ and $F_{R,r}$ together give

$$F_{R,t} = k_R\, F'_{Z,V,z}$$

The solution, based on a driven front axle, in accordance with Equation 6.42 is:

$$\mu_{X,W} = \frac{F'_{Z,V,x} + k_R\, F'_{Z,V,z} + F_L}{F'_{Z,V,f,z} - \Delta F_{Z,V,3}}$$

$$\mu_{X,W} = \frac{F_{Z,V,t} \sin\alpha + k_R\, F_{Z,V,L} \cos\alpha + F_L}{F_{Z,V,f} \cos\alpha - F_{Z,V,t} \sin\alpha\, (h_V/l)}$$

Numerators and denominators divided by $F_{Z,V,t}$ give

$$\mu_{X,W} = \frac{\sin\alpha + k_R\cos\alpha + F_L/F_{Z,V,t}}{(F_{Z,V,f}/F_{Z,V,t})\cos\alpha - \sin\alpha\,(h_V/l)}$$

The speeds achievable on steep inclines do not exceed 25 km h^{-1} so F_L can be ignored. However, on flatter inclines this counter force must be included in the equation. To simplify matters, numerators and denominators are divided by cos α:

$$\text{front wheel drive } \mu_{X,W} = \frac{\tan\alpha + k_R + F_L(F_{Z,V,t}\cos\alpha)}{F_{Z,V,f}/F_{Z,V,t} - \tan\alpha\,(h_V/l)} \qquad (6.43)$$

The coefficient of friction needed to travel a given incline can be determined using this equation. To obtain the gradient-ability, i.e. tan α, as the result, it is necessary to transform the equation, as follows:

front wheel drive

$$\tan\alpha = \frac{\mu_{X,W}(F_{Z,V,f}/F_{Z,V,t}) - k_R - F_L(F_{Z,V,t}\cos\alpha)}{1 + \mu_{X,W}(h_V/l)} \qquad (6.44)$$

The value tan $\alpha \times 100$ = gradient-ability q as a percentage (see Equations 6.39 and 6.40).

If F_L needs to be considered, α needs to be estimated provisionally for it to be possible to insert cos α. It may be necessary to correct this later in such cases. However, the numerical value is relatively small.

The formula clearly indicates that the higher is the front axle load $F_{Z,V,f}$ and the smaller the value h_V/l, the greater the angle α becomes. The picture is completely reversed on a rear-wheel drive vehicle (Fig. 1.36); in this instance the equation is:

$$\text{rear wheel drive } \tan\alpha = \frac{\mu_{X,W}(F_{Z,V,r}/F_{Z,V,t}) - k_R - F_L(F_{Z,V,t}\cos\alpha)}{1 - \mu_{X,W}(h_V/l)} \qquad (6.45)$$

On this type of drive h_V/l and rear axle load should be large. If the coefficient of friction necessary for a given incline is required, the following formula applies:

$$\text{rear wheel drive } \mu_{X,W} = \frac{\tan\alpha + k_R + F_L(F_{Z,V,t}\cos\alpha)}{F_{Z,V,r}/F_{Z,V,t} + \tan\alpha\,(h_V/l)} \qquad (6.46)$$

To produce the diagram of the driving and climbing performance, half the payload in the total weight force should be considered, whereas to determine the skid point the different loading conditions must be assumed. These do not just lead to a change in $F_{Z,V,t}$ but also in the axle load distribution, which is included in the equation as $F_{Z,V,f}/F_{Z,V,t}$ or $F_{Z,V,r}/F_{Z,V,t}$. The three most important loading conditions are (see Section 5.3.6 and Fig. 1.36):

- two people each weighing 68 kg in the front
- four people each weighing 68 kg
- full payload.

The payload $F_{Z,t,max}$ (see Section 5.3.3) must be distributed so that, in accordance with Equation 5.1, the permissible rear axle load $F_{Z,V,r,max}$ is achieved. Therefore, the front axle is usually not fully loaded. The wheelbase l and the changing centre of gravity heights:

$$h_{V,2},\ h_{V,4} \text{ and } h_{V,max}$$

must also be known (see Section 6.1.2.4). Figure 6.22 shows a diagram of the (calculated) skid points on three different coefficients of friction:

$$\mu_{X,W} = 0.8 \text{ (dry)},\quad \mu_{X,W} = 0.5 \text{ (wet)} \quad \text{and} \quad \mu_{X,W} = 0.15 \text{ (ice)}$$

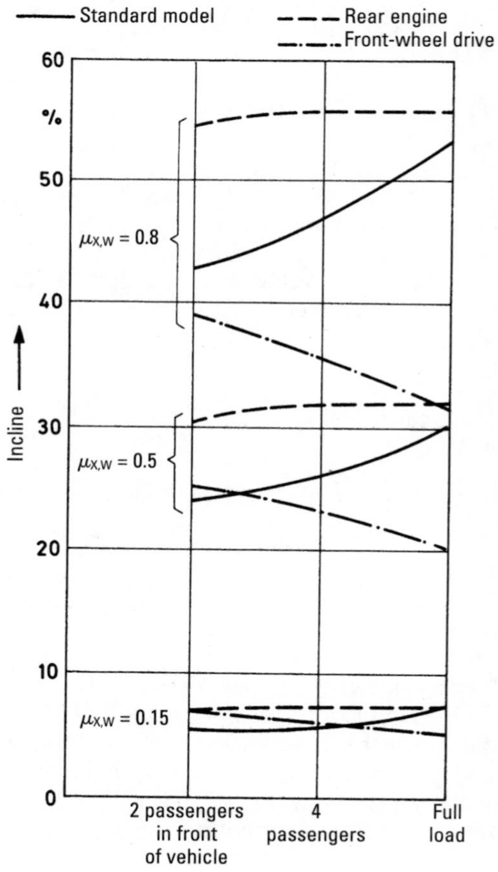

Fig. 6.22 Skid points as a function of three different coefficients of friction $\mu_{X,W}$ = 0.8, 0.5 and 0.15 and the loading condition and type of drive.

Fully loaded rear-wheel drive vehicles can negotiate the largest inclines, whereas front-wheel drive vehicles have the best climbing capacity at low loads, i.e. with only two passengers in the front and a relatively empty fuel tank, particularly on ice.

as a function of loading condition and configuration drive. The mean percentage axle load distribution from Fig. 1.36 and $h_V/l = 0.23$ is considered in this calculation, whereas F_L is ignored. As a series of investigations showed, on standard passenger cars the gradient-ability at $\mu_{X,W} = 0.8$ with two people in the vehicle averages 45%, increasing to around 52% when the vehicle is fully laden. Gradient-abilities over 60% quoted by manufacturers are not realistic. The wheels would spin because of the lack of friction (see the calculation in Section 6.4.1).

Publications should therefore not base their values on the (purely theoretical) engine performance, but rather on the climbing capacity as a function of the coefficient of friction $\mu_{X,W} = 0.8$ produced by the road. This applies even more so to a front-wheel drive vehicle.

The picture for four-wheel drive vehicles is rather different. Here, the engine torque and the ratio in the manual gearbox and differential are the deciding factors along with the higher rolling resistance on uneven roads k_R (see Fig. 2.31 and Equation 6.36). These are:

$$\tan \alpha = \mu_{X,W} - k_R \tag{6.47}$$

In the case of $\mu_{X,W} = 0.8$ and $k_R = 1.5 \times 0.012$ an incline of 38% could be climbed.

With added roof load, the change in axle load distribution should first be calculated, then the $h_{V,t}$ of the common centre of gravity, so that these values can be used in Equations 6.41 to 6.46 (instead of h_V). Section 1.3.5 of Ref. [3] gives details. Further information, including far trailers, is also contained in Ref. [3], Sections 3.13 and 3.14.

6.5 Platform, unit assembly and common part systems

The high cost pressure placed on vehicles makes it necessary to have systems able to provide the individuality of products required by clients at a low cost.

As in other areas of mechanical engineering, unit assembly and common part systems are more and more frequently being used in vehicle technology within vehicle series or even by different manufacturers within a concern for the purpose of meeting these requirements.

These approaches offer the following benefits:

- acquisition of experience of the system and component characteristics of complex functions;
- creation of a basis for structural and crash calculations which can be used repeatedly;
- shorter development periods (e.g. from 30 to 19 months according to vehicle manufacturers Nissan);
- reduced developments costs;
- lower and more easily calculable development risks;

- enormous cost savings as a result of the considerable decrease in the numbers of units used for different series.

Apart from the cost of the drive train, the highest development and machine costs relate to the platform, i.e. the basic structure of the vehicle which consists of the floor of the vehicle and the support structure for the assemblies. The realization of the platform concept results in the same basic platform being adapted, for instance, for the suspensions or the drive train in different models of vehicle with different wheelbases and different track widths by extension of the support panels with unchanged connection conditions. Figure 6.23 shows this in the example of the B/C platform for the Audi A4, Audi A6 and VW Passat models. Using only four instead of the 17 original platforms, the Volkswagen group with its Audi, Seat, Skoda and VW makes is able to manufacture over 40 different types of vehicle. Even the Fiat group has been able to reduce its number of platforms from the original 20 to four. Nissan is seeking to achieve a reduction from 25 to five by the year 2005, with corresponding cost benefits.

The same part concept is not only confined to the actual platform with floor panel and side rail, but includes the chassis with the front and rear axle, the complete propulsion system including the engine and gearbox, the tank, the steering system, the seat frame and even the central electrics and hence a total of 60% of all the development costs (Fig. 6.24). The front axle developed for the

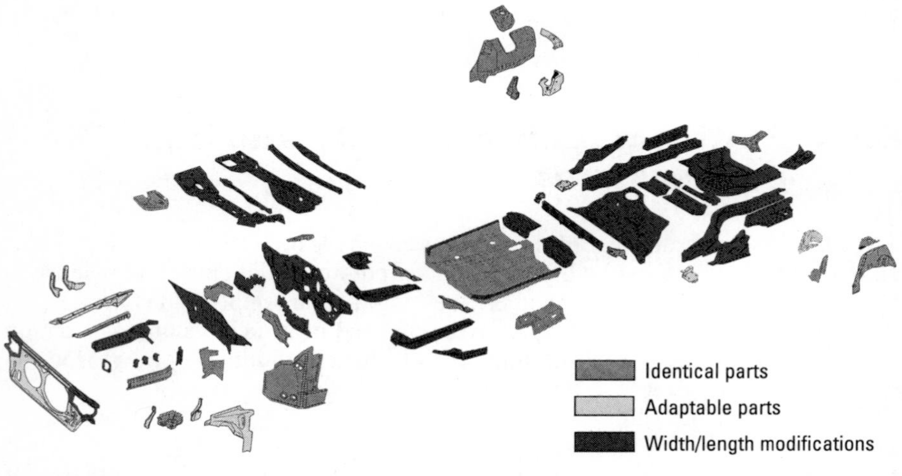

Fig. 6.23 Same and matching parts of the B/C platform of the Volkswagen AG. With unchanged connection conditions, the track width is increased by 42 mm and the wheel base by 143 mm from the Audi A4 to the Audi A6. The resultant increase in the overall resilience of the bodywork is completely offset by the increased width of the sills with the use of identical sheet-metal moulded parts. The lowest bending natural frequency is 46 Hz in both vehicles and the first torsional natural frequency is 48 Hz in the A4 and 46 Hz in the A6. These data define the rigidity of the bodywork and are thus essential for safety, comfort and driving accuracy.

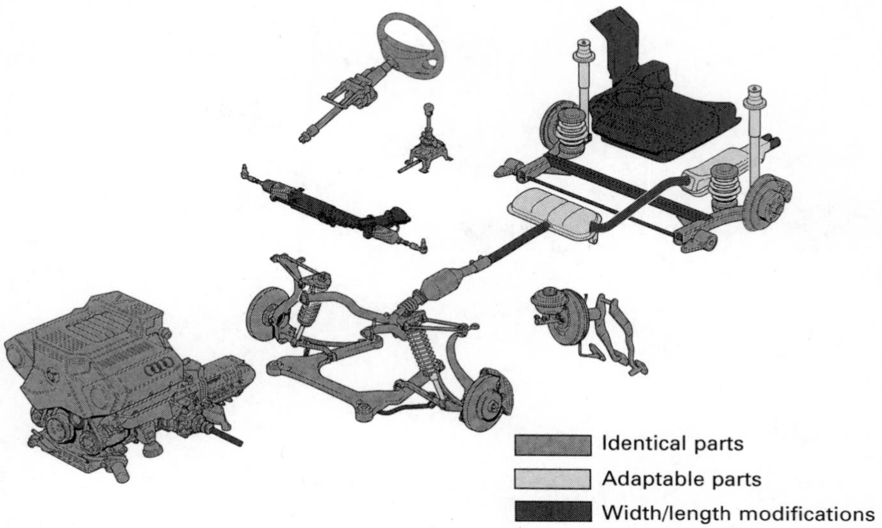

Fig. 6.24 Common and matching parts (chassis, drive train, steering system, tank) for the B/C platform of Volkswagen AG. The Audi A4 and A6 as well as the VW Passat are, for example, built on this platform.

Audi A8 is thus used with the necessary modifications in the Audi A4, Audi A6 and Volkswagen Passat models (Fig. 1.54); BMW uses the front axle of the 7 series (1994) in the 535i and 540i vehicles (1996); and Porsche uses similar wheel carriers and hubs as well as transverse links on the front and rear axles of the Boxster (Fig. 1.46) and the front axle of the Boxster is also used in the 911 model (type 996, from 1997). By standardizing the cylinder-centre distance and confining themselves to two sizes of hole, Fiat have succeeded in obtaining 67 variants from eight basic engines. With the consistent application of the same part philosophy, companies say that up to 30% of components can be used on different types.

Bibliography

Chassis Reference Books

[1] STOLL, HELMUT: *Lenkanlagen und Hilfskraftlenkungen*. Würzburg: Vogel Buchverlag, 1992.

[2] REIMPELL, JÖRNSEN: *Radaufhängungen*. Würzburg: Vogel Buchverlag, 2. Aufl. 1988.

[3] REIMPELL, JÖRNSEN/HOSEUS, KARLHEINZ: *Fahrzeugmechanik*. Würzburg: Vogel Buchverlag, 2. Aufl. 1992.

[4] REIMPELL, JÖRNSEN/SPONAGEL, PETER: *Reifen und Räder*. Würzburg: Vogel Buchverlag, 2. Aufl. 1988.

[5] REIMPELL, JÖRNSEN/STOLL, HELMUT: *Stoß- und Schwingungsdämpfer*. Würzburg: Vogel Buchverlag, 2. Aufl. 1989.

[6] BURCKHARDT, MANFRED: *Bremsdynamik und Pkw-Bremsanlagen*. Würzburg: Vogel Buchverlag, 1991.

[7] BURCKHARDT, MANFRED: *Radschlupf-Regelsysteme*. Würzburg: Vogel Buchverlag, 1992.

[8] PREUKSCHAT, ALFRED: *Antriebsarten*. Würzburg: Vogel Buchverlag, 2. Aufl. 1988.

[9] ZOMOTOR, ADAM: *Fahrverhalten*. Würzburg: Vogel Buchverlag, 2. Aufl. 1991.

Reference Books

[10] PIPPERT, HORST: *Karosserietechnik*. Würzburg: Vogel Buchverlag, 2. Aufl. 1992.

[11] LEHMANN, WOLFGANG: *Reparatur- und Einstelltabellen 1999/2000*. Würzburg: Vogel Buchverlag, 1999.

[12] BOSCH: *Kraftfahrtechnisches Taschenbuch*. Berlin, Heidelberg: Springer-Verlag, 22. Aufl. 1998.

[13] *FAKRA-Handbuch*. Band 1 bis 4. Berlin: Beuth Verlag, 10. Aufl. 1987.

[14] TÜV Bayern: *Änderungen an Auto und Motorrad*. München: 2. Aufl. 1978.

[15] WALLENTOWITZ, HENNING: *Aktive Fahrwerkstechnik*. Braunschweig: Vieweg-Verlag, 1991.

[16] REIMPELL, JÖRNSEN; STOLL, HELMUT: *The Automotive Chassis: Engineering Principles*. London: Arnold Group, 1st Edition 1996.

[17] REIMPELL, JÖRNSEN; STOLL, HELMUT; BETZLER, JÜRGEN: *The Automotive Chassis: Engineering Principles*. Oxford: Butterworth-Heinemann, 2nd Edition 2001.

[18] SCHULÉ, ROLAND: *Fahrwerktechnik*. Würzburg: Vogel Buchverlag, 2000. (Lernprogramm).

[19] MATSCHINSKY, WOLFGANG: *Radführungen der Straßenfahrzeuge*. Berlin; Heidelberg; New York; London; Paris; Tokyo; Hong Kong, Barcelona; Budapest: Springer, 2. Aufl. 1998.

General books

DIXON, J. C.: *Tires, Suspension and Handling*. London: Arnold.
MILLIKEN, W.: *Race Car Vehicle Dynamics*, Society of Automotive Engineers, Warrendale, PA, 1995.
GILLESPIE, T. D.: *Fundamentals of Vehicle Dynamics*, SAE R-114, Society of Automotive Engineers, Warrendale, PA, 1992.
MATSCHINSKY, W.: *Road Vehicle Suspensions*, Professional Engineering Publishing, 1999.
STOLL, H.: *Lenkanlagen und Hilfskraftlenkungen*. Würzburg: Vogel Buchverlag, 1992.
REIMPELL, J., HOSEUS, K.: *Fahrzeugmechanik*. Würzburg: Vogel Buchverlag, 1992.
BURCKHARDT, M.: *Bremsdynamik und Pkw-Bremsanlagen*. Würzburg: Vogel Buchverlag, 1991.
BURCKHARDT, M.: *Radschlupf-Regelsysteme*. Würzburg: Vogel Buchverlag, 1992.
ZOMOTOR, A.: *Fahrverhalten*. Würzburg: Vogel Buchverlag, 1991.
BOSCH: *Automotive Handbook*, Society of Automotive Engineers, Warrendale, PA, 1993.

Journals

1. *Auto-zeitung*. Köln: Heinrich Bauer Verlag.
2. *auto, motor und sport*. Stuttgart: Vereinigte Motorverlage.
3. *Automobil-Industrie*. Würzburg: Vogel-Verlag.
4. *Automobiltechnische Zeitschrift (ATZ)*. Stuttgart: Franckh – Kosmos Verlag.
5. *kfz-betrieb*. Würzburg: Vogel-Verlag.
6. *konstruieren und giessen*. Düsseldorf: VDI-Verlag.
7. *Kugellager-Zeitschrift*. Schweinfurt: SKF GmbH.
8. *mot*. Stuttgart: Vereinigte Motorverlage.
9. *Verkehrsunfall und Fahrzeugtechnik*: Kippenheim: Verlag Information Ambs.

Glossary of symbols

The index of the symbols and dimensions follows the international standards:

ISO 31	Quantities
ISO 2416	Passenger cars – Mass distribution
ISO 8855	Road vehicles – Vehicle dynamics and road holding ability – Vocabulary
ISO 1000	SI – Units and recommendations
SAE J670e	Vehicle dynamics terminology

and the German standards:

DIN 1301	Einheiten
DIN 1304	Formelzeichen
DIN 70 000	Straßenfahrzeuge, Begriffe der Fahrdynamik

In a few cases, the connection between the various standards could not be achieved. In these cases priority was given to the ISO standards or specific suffixes or symbols have been selected.

1 Reference points in figures

In den Bildern werden die einzelnen Bezugspunkte mit großen, *nicht kursiven* Buchstaben, die gleichzeitig als «Index» bei Formelzeichen dienen, bezeichnet:

Bo	body centre of gravity
C to G	reference points, in general
M	centre point
O	pitchpole
P	rollpole
Q	centre of driving joint
Ro	roll centre
T and U	tie rod or linkage point

$U_{f\ or\ r}$	wheel centre point, front or rear
V	vehicle centre of gravity
W	center of tyre contact

2 Suffixes

The majority of symbols require the usage of suffixes for clear identification. In cases where more than one is needed, a comma is set between them. A few cases, where small letters follow capital ones, deviate from this rule. The various suffixes have the following meanings.

a	driven, accelerating (one wheel only)
ax	axial
A	drive-off condition, accelerating in general
A	Ackermann steering angle
b	braking (one wheel only)
b	baggage
B	braking (overall vehicle)
Bo	body
c	inertia
co	cornering
dr	drivable, incl. driver
dyn	dynamic
D	damping
D or	axle drive
e	due to the elasticity, (compliances)
ex	excess
E	earth fixed
f	front
fix	fixed, idle
fr	friction
F	fault, flaw
G	gearbox
H	steering-wheel
i	inside of curve, inner wheel
k	kinematic
kb	kerb
l or L	left, left side
lo	slipping, sliding or lock
lo	loaded condition
L	aerodynamic
m	mass
m or med	middle, mean, medium
max	maximum permissible
min	minimum
M	motor
o	outside of curve, outer wheel
O	orifice closing plate
P	person, passenger
pl	partial loaded (or partly laden) or design position

Pi	piston
Pr	piston rod
r	rear
rad	radial
rs	right, right side
rsl	resulting
R	rolling (wheel)
Re	residual, remaining
Ro	body roll center
S	steering
S	anti-roll bar, stabilizer
Sp	spring
t	total or nominal value
tc	turning circle
tr	transportable
T	tyre
T	rod, tie rod or linkage
Th	trailer hitch
Tr	trailer, single-axle
ul	unloaded, empty condition
U	unsprung weight or axle weight
V	overall vehicle
X or x	longitudinal direction (see also suffixes a and b)
Y or y	lateral direction
Z or z	vertical direction
0	zero-point position or starting point
1	to the top, in jounce, in compression, in or one
2	to the bottom, in rebound, out or two
χ	acceleration reaction support, angle or diagonal springing angle
δ	steer angle
Δ	static toe-in angle
ε	camber angle
φ	body roll angle
σ	kingpin inclination angle
τ	caster

3 Lengths and distances in mm, cm or m

a and *b*	distances and length in general
b	distance between vertical force $F_{Z,W}$ and G
b_D	distance of shock absorber (damper) attachment points (at rigid axles)
$b_{f \text{ or } r}$	track width, front or rear
b_S	distance of anti-roll bar (stabilizer) attachment points at rigid axles
b_{Sp}	effective spring distance at rigid axles
Δb	track-change or track offset at rigid axle
C_R	dynamic rolling circumference at 60 kmh^{-1}
$C_{R,dyn}$	dynamic rolling circumference at top speed
d or *D*	diameter, in general
D_S	track circle diameter, front
$D_{S,r}$	track circle diameter, rear

D_{tc}	turning circle diameter, wall to wall
$D_{tc,kb}$	turning circle diameter, kerb to kerb
e	wheel offset
f	diagonal spring travel
h or H	height, in general
h_{Bo}	height of body centre of gravity
$h_{Ro,f\ or\ r}$	height of roll centre at front or rear axle
h_{ul}	height of the unloaded vehicle
h_V	height of the vehicle center of gravity
$i_{X,Bo\ or\ Y,Bo}$	radius of inertia of the body center of gravity in X or Y-direction
$i_{Z,V}$	radius of inertia of the vehicle center of gravity in Z-direction
j	distance between the two steering axis at the ground
l	wheelbase
$l_{Bo,f\ or\ r}$	distance of body centre of gravity of the middle of the front or rear axle
$l_{f\ or\ r}$	distance of vehicle centre of gravity to middle of front or rear axle
L_{fix}	idle (fixed) length of the shock absorber
L_t	total length of the vehicle
n_τ	caster offset at wheel centre
$n_{\tau,k}$	kinematic lateral force lever arm due to caster
$n_{\tau,t}$	lateral force arm, in total
OD_T	outer diameter of the tyre
q	force lever of vertical force
r	effective control arm length or force lever in general
r_a	force lever of longitudinal or tractive force
r_b	force lever of brake force
r_{dyn}	dynamic rolling radius of the tyre at 60 kmh^{-1}
r_{stat}	static loaded radius of the tyre
r_T	force offset in the centre of tyre contact (+) inside or (-) outside of curve
r_Δ	static toe-in (one wheel only)
$r_{\Delta,t}$	total static toe-in (both wheels of one axis)
r_σ	transverse offset at ground, static
$r_{\sigma,t}$	total transverse offset at ground
$r_{\tau,e}$	elastokinematic caster offset at ground
$r_{\tau,k}$	kinematic caster offset at ground
$r_{\tau,t}$	total caster offset at ground
$r_{\tau,T}$	caster offset tyre
R	path radius
s	travel or stroke, in genral
s_{Re}	residual wheel travel
s_t	total wheel travel
s_T	static tyre deflection
s_1	wheel travel in jounce
s_2	wheel travel in rebound

4 Masses, loads and weights in kg

m	mass, load or weight in general
m_b	mass of luggage (baggage) related to one passenger
m_{Bo}	vehicle body weight

$m_{Bo, f or r}$	part of body mass on front or rear
m_p	mass of one passenger
m_t	nominal design pay mass (minimum required)
$m_{t,max}$	permissible payload
m_{Th}	weight of the trailer hitch
m_{tr}	nominal mass of transportable goods
m_{Tr}	trailer load
Δm_{Tr}	tongue load, trailer
$m_{U,f or r}$	unsprung axle mass, front or rear
$m_{V,dr}$	weight of drivable vehicle (with driver)
$m_{V,f or r}$	axle load, front or rear
$m_{V,f, lo or r,lo}$	axle load under full loaded condition, front or rear
$m_{V,f,max or r,max}$	maximum permissible axle load, front or rear
$m_{V,f,pl or r,pl}$	partial axle load (design load), front or rear
$m_{V,t}$	gross vehicle weight (GVW)
$m_{V,t,max}$	maximum gross vehicle weight
$m_{V,ul}$	kerb weight (actual weight of unloaded vehicle without driver)
$m_{V,ul,0}$	kerb weight as published by the car manufacturer (with driver)
Δm_V	weight of vehicle options
m_W	weight of one wheel
$m_{1,Bo,f or r}$	part of body mass on one side of the front or rear axle
$m_{1,U,f or r}$	weight of one side of front or rear axle
$m_{1,V,f or r}$	axle load front or rear

5 Forces in N and kN

A lower-case subscript letter after the symbol F means that the force refers only to one side of the axle; an upper-case letter refers to the whole axle. An exception is F_R, the rolling resistance of the tyre. However, this can also refer to a wheel, an axle or the whole vehicle; the subsequent further subscript enables the difference to be recognized. The forces at the reference points, or at the links C to U of the wheel suspension, are denoted by the letter of that particular point and the direction.

dF or ΔF	change of force
F_D	damping force
$F_{c,Bo or V}$	centrifugal force at the body centre or vehicle
F_{fr}	friction force in general or related to one side of the axle
F_H	steering-wheel force
F_L	aerodynamic drag
F_O	force at pitch center
F_{Pi}	piston rod extensive or aid force
F_{rsl}	resulting friction
F_R	rolling resistance of the tyre
F_{Sp}	spring force, one side of the axle
F_T	tie rod or push rod force
$F_{X,Bo,B,f or r}$	brake reaction force to the body, front or rear
$F_{X,ex}$	excess force
$F_{X,U,B,f or r}$	brake reaction force to the front or rear axle
$F_{X,V,B}$	brake force at the centre of gravity of the vehicle
$F_{X,W,a or A}$	accelerating force in the centre of tyre contact of one wheel (a) or both wheels (A)

$F_{X,W,b}$	brake force in the centre of tyre contact of one wheel
$F_{X,W,B,f\ or\ r}$	brake reaction force to the front or rear axle
$F_{Y,T,\varepsilon}$	lateral force dur to camber
$F_{Y,V}$	lateral force at vehicle
$F_{Y,W}$	lateral force at wheel
$F_{Z,Bo}$	static body weight (force)
$F_{Z,Bo,B,f\ or\ r}$	body lift or dive differential force during braking, front or rear
$F_{Z,t,max}$	force of maximum payload
$F_{Z,U,f\ or\ r}$	weight (force) of front or rear axle
$\Delta F_{Z,V}$	axle load transfer
$F_{Z,V,f\ or\ r}$	axle load front or rear
$F_{Z,V,f,dyn\ or\ r,dyn}$	dynamic axle load, front or rear
$F_{Z,V,t}$	cross vehicle weight
$F'_{Z,W}$	verticle force at the centre of tyre contact
$F_{Z,W}$	verticle force without the axle weight of one axle side
$\Delta F_{Z,W}$	change of verticle force at one wheel
F_1	compressive force
F_2	rebound force

6 Moments in NM

$M_{ZTX\ or\ W}$ denote the wheel torque around the steering axle (z), followed by either X, Y or Z for the direction of the (wheel) aligning force or the causal force, and left (l) or right (rt) may also be indicated.

If a lower case t appears, this signals that both axle wheels are meant, whereby r relates to the rear axle only. All other moments are indicated in section 7 below.

$M_{Z,T,X}$	(tyre) self aligning torque due to longitudinal force
$M_{Z,T,Y}$	(tyre) self aligning torque due to sideforces at the front wheels
$M_{Z,T,Y,r,t}$	(tyre) self aligning torque due to sideforces at the rear wheels
$M_{Z,W,a\ or\ A}$	(wheel) aligning torque due to the accelerating force at one wheel (a) or both wheels (A)
$M_{Z,W,b\ or\ B}$	(wheel) aligning torque due to the brake force at one wheel (b) or both wheels (B)
$M_{Z,W,Y}$	(wheel) aligning torque due to the lateral force
$M_{Z,W,Z}$	(wheel) aligning torque due to the vertical force
$M_{Z,W,\tau}$	steering torque due to differences in caster

7 Other moments in NM

$M_{a\ or\ A}$	driving torque related to one wheel (a) or axle (A)
$M_{b\ or\ B}$	braking torque related to one wheel (b) or axle (B)
M_{fr}	friction torque
M_H	steering-wheel torque
M_M	engine (motor) torque
M_R	rolling resistance torque
$M_{X,Bo}$	rolling torque, body
$M_{X,T,a}$	overturning torque
$M_{Y,Bo}$	pitching moment, body
$M_{Z,V}$	yawing moment, vehicle
T	torsional moment, torque

8 Spring rates in Nmm^{-1} or kNm^{-1}

$c_{\mathrm{f\ or\ r}}$	rate of the body supporting spring at parallel springing, related to the centre of tyre contact of one axle side, front or rear
c_S	rate of the anti-roll bar (stabilizer) at receptrocal springing
$c_{S,\varphi}$	rate of the anti-roll bar related to the centre of tyre contact
c_{Sp}	static rate of the spring
c_T	spring rate of the tyre
$c_{\varphi,\mathrm{f\ or\ r}}$	front or rear rate of the body supporting spring at reciprocal springing related to the centre of tyre contact

9 Angles in degrees or radians

α	torsional angle of a joint or bushing
α	top view angle of the semi-trailing arm twist axis
α	angle of gradient of the road
α or α'	inclination (rear view) angle of upper control arm (double wishbone axle)
$\alpha_{\mathrm{f\ or\ r}}$	slip angle of the front or rear wheel
β	rear view angle of the semi-trailing arm twist axis
β	sideslip angle (dynamic)
β'	driving angle of the axis (static)
β or β'	inclination (rear view) angle of lower control arm (double wishbone or McPherson axles)
χ	acceleration reaction support angle or diagonal springing angle
δ	steer angle
$\delta_{A,o}$	Ackermann steer angle, nominal value to outside of curve
δ_H	steering-wheel angle
δ_m	mean steer angle
$\delta_{\mathrm{o\ or\ i}}$	actual steer angle, outside or inside of curve
δ_r	steering or toe-in angle at rear wheels
$\delta_{V,0}$	static toe-in angle of one wheel
$\delta_{V,0,t}$	total static toe angle
Δ	static toe-in angle
$\Delta\delta_\theta$	part of steer angle due to suspension pitch
$d\delta$	change of steer angle of both wheels
$\Delta\delta$	differential steer angle (actual value)
$\Delta\delta_A$	differential steer angle according to Ackermann (nominal value)
$\Delta\delta_e$	part of steer angle due to compliances
$\Delta\delta_F$	steering flaw
$\Delta\delta_H$	part of steering-wheel angle due to manual steer
$\Delta\delta_{H,e}$	part of steering-wheel angle due to compliances
$\Delta\delta_{H,Re}$	residual angle at the steering-wheel
$\Delta\delta_k$	change of toe-in or steer angle due to kinematics
ε	brake reaction support angle
ε or ε_W	camber angle
$\Delta\varepsilon_{W,k}$ or $d\varepsilon_{W,k}$	part of camber angle due to kinematics
$\Delta\varepsilon_{W,\delta}$	part of camber angle due to steer
$\Delta\varepsilon_{W,\varphi}$	part of camber angle due to suspension roll

φ	body roll angle
$\Delta\varphi$ or $d\varphi_k$	kinematic change of body roll angle
λ	steering arm angle
$\theta_{A \text{ or } B}$	body pitch angle under accelerating or braking
$\Delta\theta$	pitch angle change due to load changes
σ	kingpin inclination angle
τ or τ_f	caster angle of the (steered) front wheels
τ_r	caster angle at rear wheels (not steered)
$\Delta\tau_k$	part of caster angle due to kinematics
$\Delta\tau_\delta$	part of caster angle due to steer
ψ	yaw angle
ξ	top view angle between two control arms or roads
ξ_D	inclination of the shock absorber

10 Characteristics and data with no dimensions

i_{dyn}	dnamic steering ratio
i_D	axle differential ratio
i_D	ratio of shock absorber (damper) to the wheel
i_G	gearbox ratio
i_l	ratio of wheelbase to vehicle length
i_m	mass ratio
i_R	roll resistance ratio coefficient
i_S	overall kinematic steering ratio
i'_S	steering gear ratio
i_{Sp}	ratio of spring to the wheel
i_{ul}	ratio of vehicle centre of gravity to height of the unloaded vehicle
i_φ	ratio of the wheel to the spring, shock absorber or anti-roll bar at reciprocal springing of a rigid axle
k_b	ratio of tread to width (breadth)
k_D	damping coefficient
k_m	load factor
k_R	rolling resistance coefficient
$k_{R,co}$	rolling resistance coefficient when concerning
$k_{R,O}$	rolling resistance coefficient measured on a tyre test rig
k_T	factor of the increase in tyre spring rate
k_v	velocity factor
k_x	anti-dive coefficient, accelerating
$k_{\delta,V,\varphi}$	suspension roll steering coefficient
k_ε	anti-dive coefficient, braking
$k_{\varepsilon,W,\varphi}$	suspension roll camber coefficient
k_μ	friction coefficient correction factor, tyre
n	number of specified seats
n_0	number of seats engaged
$S_{X,W,a \text{ or } b}$	longitudinal slip under accelerating or braking lateral slip
$S_{Y,W}$	lateral slip
z	braking factor
$\Phi_{f \text{ or } r}$	brake force fraction front or rear

η	total efficiency
μ_{rsl}	resulting coefficient of friction
$\mu_{X,W}$	coefficient of longitudinal force
$\mu_{X,W,lo}$	coefficient of sliding braking force
$\mu_{Y,W}$	coefficient of lateral force
$\mu_{Y,W,lo}$	coefficient of sliding lateral force

11 Other symbols with dimensions

a_X	longitudinal acceleration or deceleration	$m\ s^{-2}$
a_Y	lateral acceleration	$m\ s^{-2}$
A	area, cross-section area	m^2
A_5	ductile yield, elongation at rupture ($L_0 = 5 \times d_0$)	%
C_S	stiffness of the steering system	$Nm\ rad^{-1}$
E	modulus of elasticity	$N\ mm^{-2}$
f	frequency	Hz
g	acceleration due to gravity	$m\ s^{-2}$
HRC	Rockwell hardness	–
I	area moment of inertia	cm^4
$J_{X,Bo}$	dynamic moment of inertia of body around the longitudinal axis	$kg\ m^2$
$J_{Y,Bo}$	dynamic moment of inertia of body around the transverse axis	$kg\ m^2$
$J_{Z,V}$	dynamic moment of inertia of vehicle around the vertical axis	$kg\ m^2$
k_D	damping value	$N\ s\ m^{-1}$
n	revolutions per minute or vibration frequency	min^{-1}
P_{hyd}	hydraulic pressure	$N\ cm^{-2}$
P_T	tyre pressure	bar
q	climbing capability factor	%
R_e	yield strength	$N\ mm^{-2}$
R_m	tensile strength	$N\ mm^{-2}$
$R_{p0.2}$	0.2% yield strength	$N\ mm^{-2}$
v or v_X	longitudinal velocity	$m\ s^{-1}$ or $km\ h^{-1}$
v_D	piston velocity in shock absorber	$m\ s^{-1}$
v_W	circumferential tyre velocity	$m\ s^{-1}$
ω	circular frequency	Hz

Index of manufacturers

(As mentioned in the text)

Index of suppliers

(Suppliers to the car industry as mentioned in the text)

Subject index

Please note that:-
* Figure and Table numbers are given in italics (e.g. *2.6)* and come before page numbers
* Equation numbers have the prefix e (e.g. e3.23b)